Renewable Energy

Renewable Energy

Edited by Godfrey Boyle

OXFORD
UNIVERSITY PRESS
Oxford University Press in association with The Open University

OXFORD
UNIVERSITY PRESS
Great Clarendon Street, Oxford OX2 6DP

Published by Oxford University Press, Oxford in association with
The Open University, Milton Keynes

The Open University

Oxford University Press is a department of the University of Oxford.

It furthers the University's objective of excellence in research, scholarship, and education by publishing worldwide in

Oxford New York Auckland Bangkok Buenos Aires Cape Town Chennai Dar es Salaam Delhi Hong Kong Istanbul Karachi Kolkata Kuala Lumpur Madrid Melbourne Mexico City Nairobi São Paulo Shanghai Taipei Tokyo Toronto

Oxford is a registered trade mark of Oxford University Press in the UK and in certain other countries

Published in the United States by Oxford University Press Inc., New York

The Open University, Walton Hall, Milton Keynes MK7 6AA

First Published in the United Kingdom 2004

British Library Cataloguing in Publication Data

Data available

Library of Congress Cataloguing in Publication Data

Data available

ISBN 0-19-926178-4

5 7 9 10 8 6

Edited, designed and typeset by The Open University

Printed in the United Kingdom by Halstan Printing Group, Amersham.

T206 Energy for a Sustainable Future

This book was produced as a major component of the Open University's second level undergraduate course *T206 Energy for a Sustainable Future* and, through co-publication with Oxford University Press, is also available to students and staff at other Universities worldwide, and to a more general readership.

A companion text, *Energy Systems and Sustainability: Power for a Sustainable Future*, is also co-published by Oxford University Press and The Open University.

To order copies of either book, contact the Oxford University Press distribution centre at:

Oxford University Press, Saxon Way West, Corby, Northamptonshire, NN18 9ES or visit their web site at www.oup.com

T206 Energy for a Sustainable Future is also accompanied by two other textbooks, a set of Study Guides and a variety of other supplementary printed materials. This additional course material also includes several hours of material on video, produced by the BBC in partnership with the Open University, and a number of software-based exercises and simulations on CD-ROM. A selection of these are available from the Open University Worldwide Learning Resources webshop at www.ouw.co.uk/ or by writing to Open University Worldwide, Walton Hall, Milton Keynes, MK7 6AA.

Further details of *T206 Energy for a Sustainable Future* and how to enrol on the course are available via our T206 Taster Web Site: http://www.open.ac.uk/T206/index.htm

Alternatively, write to: The Course Manager, T206 Energy for a Sustainable Future, Faculty of Technology, The Open University, Milton Keynes MK7 6AA, United Kingdom.

Preface to the Second Edition

By the end of the twenty-first century, if current trends continue, the world's population is likely to have almost doubled and its wealth increased by a factor of between eight and sixteen times. World energy demand will probably have doubled and possibly quadrupled, despite major improvements in energy efficiency. How can this enormous demand be supplied, cleanly, safely and sustainably?

Even at current consumption levels, existing energy systems have many deleterious effects on human health and the natural environment, as described in our companion text *Energy Systems and Sustainability: Power for a Sustainable Future.* In particular, carbon dioxide and other greenhouse gases released by fossil fuel burning threaten to cause unprecedented changes in the earth's climate, with mainly adverse consequences. There is now a broad consensus that in the long term the world will need to shift to low- or zero-carbon energy sources if the impacts of climate change are to be mitigated.

The renewable energy sources are essentially carbon-free and appear to be generally more sustainable than fossil or nuclear fuels, though many technologies are still under development and the costs of some are currently high. This new edition of *Renewable Energy* reflects the remarkable progress that has been made in the field since the publication of the first edition in 1996. With that progress has come greatly increased recognition among professionals, politicians and the public that renewable energy could provide a major proportion of the world's needs by the middle of the twenty-first century, given adequate investment in research, development and deployment.

But if that potential is to be realised, the world will need many more professional people with a thorough knowledge of renewable energy systems, their underlying physical and technological principles, their economics, their environmental impact and how they can be integrated into the world's energy systems.

This book and its companion text aim to address these needs. Written initially for undergraduates studying the course T206 *Energy for a Sustainable Future* at the Open University, they are also aimed at students and staff in other universities, and at professionals, policy-makers and members of the public interested in sustainable energy futures. We hope that both books will contribute to an improved understanding of the sustainability problems of our present energy systems and of potential solutions. We also hope they convey something of the enthusiasm we feel for this complex, fascinating and increasingly important subject.

Godfrey Boyle
Course Team Chair, T206 *Energy for a Sustainable Future*

Contents

Chapter 1

Introducing Renewable Energy

by Gary Alexander and Godfrey Boyle

1.1 Introduction

The renewable energy sources, derived principally from the enormous power of the sun's radiation, are at once the most ancient and the most modern forms of energy used by humanity.

Solar power, both in the form of direct solar radiation and in indirect forms such as bioenergy, water or wind power, was the energy source upon which early human societies were based. When our ancestors first used fire, they were harnessing the power of photosynthesis, the solar-driven process by which plants are created from water and atmospheric carbon dioxide. Societies went on to develop ways of harnessing the movements of water and wind, both caused by solar heating of the oceans and atmosphere, to grind corn, irrigate crops and propel ships. As civilizations became more sophisticated, architects began to design buildings to take advantage of the sun's energy by enhancing their natural use of its heat and light, so reducing the need for artificial sources of warmth and illumination.

Technologies for harnessing the powers of sun, firewood, water and wind continued to improve right up to the early years of the industrial revolution. But by then the advantages of coal, the first and most plentiful of the fossil fuels had become apparent. These highly-concentrated energy sources soon displaced wood, wind and water in the homes, industries and transport systems of the industrial nations. Today the fossil fuel trio of coal, oil and natural gas provides three quarters of the world's energy.

Concerns about the adverse environmental and social consequences of fossil fuel use, such as air pollution or mining accidents, and about the finite nature of supplies, have been voiced intermittently for several centuries. But it was not until the 1970s, with the steep price rises of the 'oil crisis' and the advent of the environmental movement, that humanity began to take seriously the possibility of fossil fuels 'running out', and that their continued use could be destabilizing the planet's natural ecosystems and the global climate (see Section 3 below).

The development of nuclear energy following World War II raised hopes of a cheap, plentiful and clean alternative to fossil fuels. But nuclear power development has stalled in recent years, due to increasing concern about cost, safety, waste disposal and weapons proliferation.

Continuing concerns about the 'sustainability' of both fossil and nuclear fuels use have been a major catalyst of renewed interest in the renewable energy sources in recent decades. Ideally, a **sustainable energy source** is one that is not substantially depleted by continued use, does not entail significant pollutant emissions or other environmental problems, and does not involve the perpetuation of substantial health hazards or social injustices. In practice, only a few energy sources come close to this ideal, but as this and subsequent chapters will show, the 'renewables' appear generally more sustainable than fossil or nuclear fuels: they are essentially inexhaustible and their use usually entails much lower emissions of greenhouse gases or other pollutants, and fewer health hazards.

Before going on to introduce the renewables in more detail, let us now pause to introduce some basic energy concepts that may be unfamiliar to readers who do not have a scientific background. For a more detailed discussion, see the companion text *Energy Systems and Sustainability* (Boyle *et al*, 2003).

Force, energy and power

The word energy is derived from the Greek *en* (in) and *ergon* (work). The scientific concept of **energy** serves to reveal the common features in processes as diverse as burning fuels, propelling machines or charging batteries. These and other processes can be described in terms of diverse **forms of energy**, such as *thermal* energy (heat), *chemical* energy (in fuels or batteries), *kinetic* energy (in moving substances), *electrical* energy, *gravitational* potential energy, and various others.

BOX 1.1 **The SI system of units**

In 1960, the scientific world agreed on a single set of units: the SI system (Système Internationale d'Unités). There are three basic units: the *metre* (m), the *kilogram* (kg) and the *second* (s); the units for many other quantities are derived from these. For some of the derived units, such as *metres per second* (m s^{-1}), the unit for speed, the base units are obvious. Others have been given specific names – such as the newton, or the joule. (See Appendix A2 for more details.)

To change the motion of any object, a **force** is needed, and the formal **SI unit** (see Box 1.1) for force, the **newton** (**N**), is defined as that force which will accelerate a mass of one kilogram (kg) at a rate of one metre per second per second (m s^{-2}). More generally,

Force (N) = mass (kg) × acceleration (m s^{-2}).

In the real world, force is often needed to move an object even at a steady speed, but this is because there are opposing forces such as friction to be overcome.

Whenever a force is moving something, it must be providing **energy**. The unit of energy, the **joule** (**J**), is defined as the energy supplied by a force of one newton in causing movement through a distance of 1 metre. In general:

Energy (in joules) = Force (in newtons) × Distance (in metres)

The terms *energy* and *power* are often used informally as though they were synonymous (e.g. wind energy/wind power), but in scientific discussion it is important to distinguish them. **Power** is the *rate* at which energy is converted from one form to another, or transferred from one place to another. Its unit is the **watt** (**W**), and one watt is defined as one joule per second. A 100 watt light bulb, for example, is converting one hundred joules of electrical energy into light (and 'waste' heat) each second.

In practice, it is often convenient to measure energy in terms of power used for a given time period. If the power of an electric heater is 1 kW, and it runs for an hour, we say that it has consumed one **kilowatt-hour** (**kWh**) of energy. (Appendix A2 explains prefixes such as kilo-, mega-, etc.). A kilowatt is 1000 watts, i.e. 1000 joules per second, and there are 3600 seconds in an hour, so 1 kWh = 3600 × 1000 = 3.6×10^6 Joules (i.e. 3.6 MJ).

Energy is also often measured simply in terms of quantities of fuel, such as tonnes of coal or oil. National energy statistics often use the unit 'million tonnes of oil equivalent' (1 Mtoe = 41.9 PJ). The most common units and their conversion factors are listed in Appendix A2.

Energy conservation: The First Law of Thermodynamics

In any transformation of energy from one form to another, the total quantity of energy remains unchanged. This principle, that energy is always conserved, is called the **First Law of Thermodynamics**. So if the quantity of energy in the output of a power station, for example, is less than the quantity of energy in the fuel input, then some of the energy must have been converted to some other form (usually waste heat).

If the total quantity of energy is always the same, how can we talk of *consuming* it? Strictly speaking, we don't: we just convert it from one form into other forms. We consume fuels, which are sources of readily-available energy. We burn fuel in an engine, converting its stored chemical energy into heat and then into the kinetic energy of the moving vehicle. A wind turbine extracts kinetic energy from moving air and converts it into electrical energy, which can in turn be used to heat the filament of a lamp causing it to radiate light energy.

Forms of energy

At the most basic level, the diversity of energy forms can be reduced to four: *kinetic, gravitational, electrical* and *nuclear*.

The first of these is exemplified by the **kinetic energy** possessed by any moving object. This is equal to half the mass of the object times the square of its speed, i.e.:

$$\text{kinetic energy} = 0.5 \times \text{mass} \times \text{speed}^2$$

where energy is in joules (J), mass in kilograms (kg) and speed in metres per second ($m\ s^{-1}$).

Less obviously, kinetic energy *within* a material determines its temperature. All matter consists of atoms, or combinations of atoms called molecules. In a gas such as the air that surrounds us, these move freely. In a solid or a liquid, they form a more or less loosely linked network in which every particle is constantly vibrating. **Thermal energy**, or heat, is the name given to the kinetic energy associated with this rapid random motion. The higher the **temperature** of a body, the faster its molecules are moving. In the temperature scale that is most natural to scientific theory, the Kelvin (K) scale, zero corresponds to zero molecular motion. In the more commonly used Celsius scale of temperature (written as °C), zero corresponds to the freezing point of water and 100 to the boiling point of water. The two scales are related by a simple formula:

$$\text{temperature (K)} = \text{temperature (°C)} + 273.$$

A second fundamental form of energy is **gravitational energy**. On earth, an input of energy is required to lift an object because the gravitational pull of the earth opposes the movement. If an object such as an apple is lifted above your head, the input energy is stored in a form called **gravitational potential energy** (often just 'potential energy' or 'gravitational energy'). That this stored energy exists is obvious if you release the apple and observe the subsequent conversion to kinetic energy. The gravitational force pulling an object towards the earth is called the *weight* of the object, and is equal to its *mass*, m, multiplied by the acceleration due to gravity, g (which is 9.81 m s^{-2}). (Note that although everyday language may treat mass and weight as the same, science does not). The potential energy (in joules) stored in raising an object of mass m (in kilograms) to a height h (in metres) is given by the following equation

$$\text{potential energy} = \text{force} \times \text{distance} = \text{weight} \times \text{height} = m \times g \times h.$$

Gravity is not the only force influencing the objects around us. On a scale far too small for the eye to see, electrical forces hold together the atoms and molecules of all materials; gravity is an insignificant force at the molecular level. The **electrical energy** associated with these forces is the third of the basic forms. Every atom can be considered to consist of a cloud of electrically charged particles, electrons, moving incessantly around a central nucleus. When atoms come together to form molecules or solid materials, the distribution of electrons is changed, often with dramatic effect. Thus **chemical energy**, viewed at the atomic level, can be considered a form of electrical energy. When a fuel is burned, the chemical energy it contains is converted into heat energy. Essentially, the electrical energy released as the atomic electrons are rearranged is converted to the kinetic energy of the molecules of the combustion products.

A more familiar form of electrical energy is that carried by **electric currents** – organized flows of electrons in a material, most often a metal. In metals, one or two electrons from each atom can become detached and move freely through the lattice structure of the material. These 'free electrons' allow metals to carry electrical currents. To maintain a steady current of electrons requires a constant input of energy because the electrons continually lose energy in collisions with the metal lattice (which is why wires get hot when they carry electric currents). Voltage (in volts) is a measure of the energy required to maintain a current. The power (in watts) delivered by an electrical supply, or used by an appliance, is given by multiplying the voltage (in volts) by the current (measured in amperes, or 'amps'): i.e.

$$\text{power} = \text{voltage} \times \text{current}$$

In a typical power station, the input fuel is burned and used to produce high-pressure steam, which drives a rotating turbine. This in turn drives an electrical generator, which operates on a principle discovered by Michael Faraday in 1832: that a voltage is induced in a coil of wire set spinning in a magnetic field. Connecting the coil to an electric circuit will then allow a current to flow. The electrical energy can in turn be transformed into heat, light, motion or whatever, depending upon what is connected to the circuit. Electricity is thus a convenient *intermediary* form of energy: it allows energy released from one source to be converted to another quite different form, usually at some distance from the source.

Another form of electrical energy is that carried by electromagnetic radiation. More properly called **electromagnetic energy**, this is the form in which, for example, solar energy reaches the earth. Electromagnetic energy is radiated in greater or lesser amounts by every object. It travels as a wave that can carry energy through empty space. The length of the wave (its wavelength) characterizes its form, which includes X-rays, ultraviolet and infrared radiation, visible light, radio waves and microwaves.

The fourth and final basic form of energy, bound up in the central nuclei of atoms, is called **nuclear energy**. The technology for releasing it was developed during the Second World War for military purposes, and subsequently in a more controlled version for the peaceful production of electricity. Nuclear power stations operate on much the same principles as fossil fuel plants, except that the furnace in which the fuel burns is replaced by a nuclear reactor in which atoms of uranium are split apart in a 'fission' process that generates large amounts of heat.

The energy source of the sun is also of nuclear origin. Here the process is not nuclear fission but nuclear fusion, in which enormous quantities of hydrogen atoms fuse to form helium atoms, generating massive amounts of solar radiation in the process. Attempts to imitate the sun by creating power-producing nuclear fusion reactors have been the subject of many decades of research and development effort but have yet to come to fruition.

Conversion and efficiency

When we convert energy from one form to another, the useful output is never as much as the input. The ratio of the useful output to the required input (usually expressed as a percentage) is called the **efficiency** of the process. It can be as high as 90% in a water turbine or well-run electric motor, around 35–40% in a coal-fired power station (if the 'waste' heat is not put to use), and as low as 10–20% in a typical internal combustion engine. Some inefficiencies can be avoided by good design, but others are inherent in the nature of the type of energy conversion.

In the systems mentioned above, the difference between the high and low conversion efficiencies is because the latter involve the conversion of *heat* into mechanical or electrical energy. Heat, as we have seen, is the kinetic energy of randomly-moving molecules, an essentially chaotic form of energy. No machine can convert this chaos completely into the ordered state associated with mechanical or electrical energy. This is the essential message of the **Second Law of Thermodynamics**: that there is necessarily a limit to the efficiency of any heat engine. Some energy must always be rejected as low-temperature heat (This topic is discussed further in Box 2.4 of Chapter 2).

1.2 Present-day energy use

World energy supplies

World total annual consumption of all forms of **primary energy** increased more than ten-fold during the twentieth century, and in the year 2002 reached an estimated 451 EJ (exajoules), or some 10 800 Mtoe. As Figure 1.1 in Box 1.2 reveals, fossil fuels provided three quarters of the total. The world population in 2002 was some 6.2 billion, so the annual average energy consumption per person was about 74 GJ (gigajoules), equivalent to nearly 6 litres of oil per day for every man, woman and child.

But these figures conceal major differences. The average North American consumes nearly five times the world average: about 350 GJ per year. People in Europe and the former Soviet Union use about half this amount, and those in the rest of the world only about one fifth (see the companion text, Table 2.3).

BOX 1.2 World energy supplies: how much do renewables contribute?

As Figure 1.1 shows, renewables contributed some 18% of world primary energy supplies in 2002. But estimates of the renewables' contribution differ between sources, for two reasons. Firstly, the contribution of traditional biomass fuels is uncertain: this will be discussed in Chapter 4. Secondly, there are several different ways of calculating the primary energy contribution of hydroelectricity and other electricity-producing renewable sources. Some authors simply enter the contribution as equal to the annual electricity output; others calculate the equivalent input of fuel that would be needed by a power station producing this output at some notional plant efficiency. The result is that, with *exactly the same* original data, the hydro contribution might appear as 10 EJ in one report and 26 EJ in another. This can have a major effect, not only on the total but also on the percentage contributions from all the other sources. In this book, as noted in Figure 1.1, we adopt the second method.

Figure 1.2 gives a more detailed breakdown of the estimated contributions of individual renewable sources.

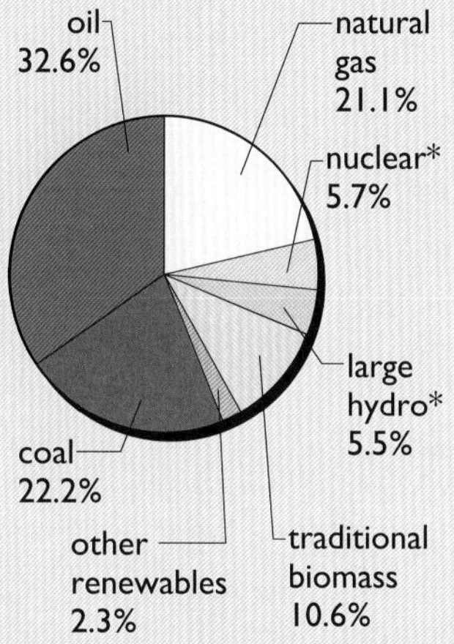

Figure 1.1 Percentage contributions of various energy sources to world primary energy consumption in the year 2002. Total consumption was 451 exajoules (EJ), equivalent to 10 800 Mtoe. The average rate of consumption was some 14.3 TW. Renewables (large hydro, traditional biomass and 'other') contributed 18.4% (sources: BP, 2003; United Nations, 2000)

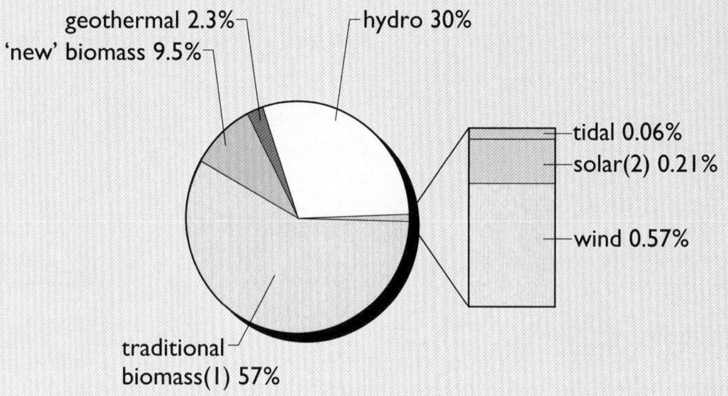

Figure 1.2 Estimated percentage contributions to world renewable energy supplies, 2001. Total: 83 EJ, or about 2000 Mtoe (main sources: IEA, 2003; BP, 2003). See Chapter 4 for discussion of the uncertainties in the traditional biomass contribution. Solar includes both solar thermal and solar photovoltaic energy

* **Note:** In Figures 1.1 and 1.2, the contributions from nuclear power, hydroelectricity and other electricity-producing renewables are the inputs that would be needed to produce the actual outputs at a notional 38% plant efficiency. Some other data sources (including the pre-2002 BP statistics used in the companion text) use a different convention, in which the primary energy contributions of hydroelectricity and other electricity-producing renewables are considered equivalent to their actual electricity output. This has the effect of reducing their contributions very substantially.

How long will the world's fossil fuel reserves last? At current consumption rates, proven world coal reserves should last for about 200 years, oil for approximately 40 years and natural gas for around 60 years. (BP, 2003) However, world production of liquid fuels, including non-conventional as well as conventional sources, seems likely to reach a *peak* between 2005 and 2015. Peak production of natural gas is likely around 2030. From then on, although large quantities of oil and gas will remain, the overall resource will be in decline (see Figure 1.3).

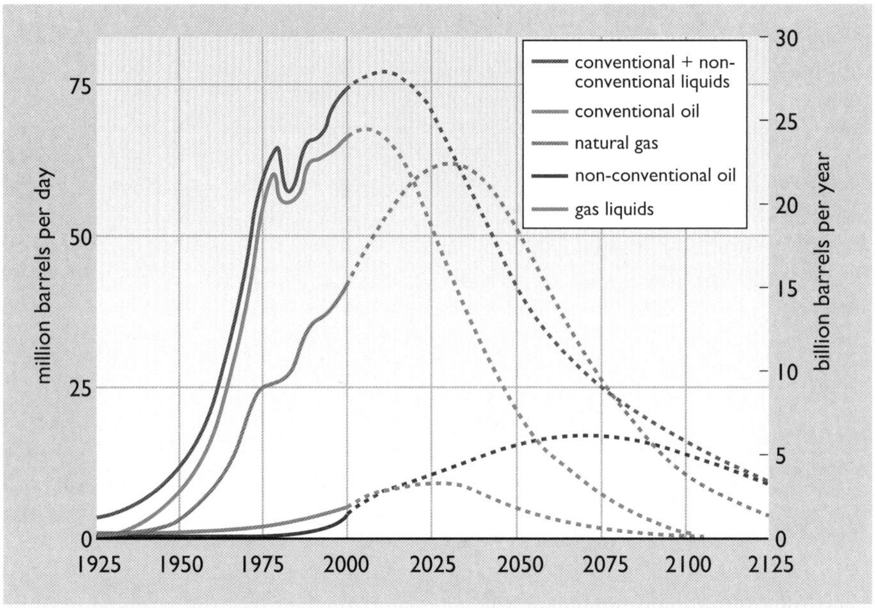

Figure 1.3 World production of oil and gas. The figures to the year 2000 are historical data; thereafter the dotted curves represent projections of future supply. These projections are based on ultimate recovery of 2000 billion barrels of conventional oil (including gas liquids), 750 billion barrels of non-conventional oil, and 2000 billion barrels oil equivalent of natural gas. The projected peak in conventional oil supply occurs around 2005; including non-conventional oil extends the peak to around 2015. The projected peak in natural gas production occurs around 2030 (source: based on Laherrere, 2001)

Energy use in the UK

In the UK, as in most countries, energy demand is categorized in official statistics into four main sectors: *domestic, commercial and institutional, industrial* and *transport.*

The energy used by the final consumers in these sectors is usually the result of a series of energy conversions. For example, energy from burning coal may be converted in a power station to electricity, which is then distributed to households and used in immersion heaters to heat water in domestic hot water tanks. The energy released when the coal is burned is called the **primary energy** required for that use. The amount of electricity reaching the consumer after transmission losses in the electricity grid is the **delivered energy**. After further losses in the tank and pipes, a final quantity, called the **useful energy**, comes out of the hot tap.

As Figure 1.4 shows, almost one third of UK primary energy is lost in the process of conversion and delivery – most of it in the form of 'waste' heat from power stations. These losses are greater than the country's total demand for space and water heating energy. And even when energy has been delivered to customers in the various sectors, it is often used very wastefully. Some of the ways in which this wastage could be reduced are described briefly in chapter 1 of the companion text. The UK energy system is described in more detail in Chapter 10 of this book.

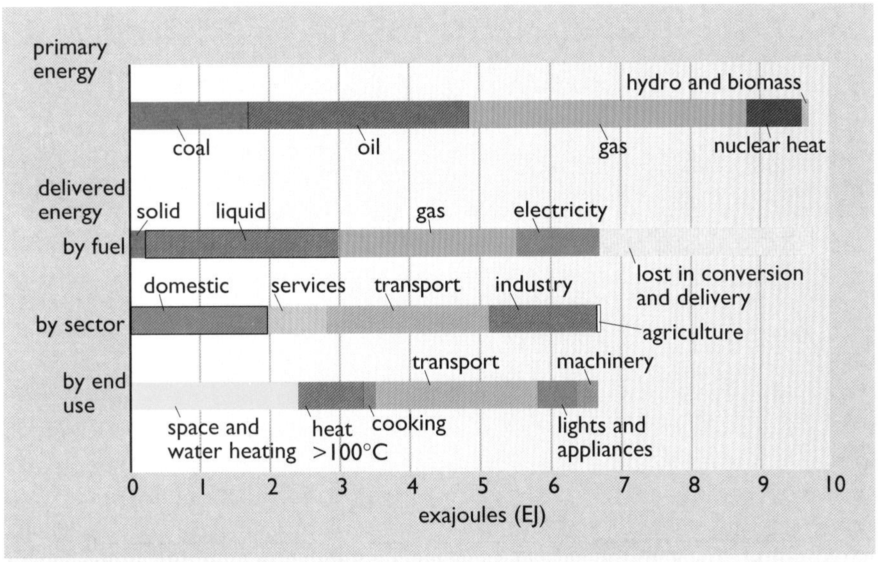

Figure 1.4 UK primary energy by fuel, and delivered energy by fuel, sector and end use, 2000 (sources: DTI, 2001a; DTI 2001b)

In the UK, the contribution of renewables to primary energy supply is quite small: just over 1% in 2002 (DTI, 2003a). The percentage contribution of renewables to *electricity* supplies was somewhat larger, however. Of the 375 TWh generated in 2002, 2.6% came from renewable sources, mainly in the form of hydropower, with smaller contributions from waste or landfill gas combustion and wind power (DTI, 2003). The UK Government aims to increase the proportion of electricity from renewables to 10% by 2010 and 20% by 2020. We return to this topic in Chapter 10.

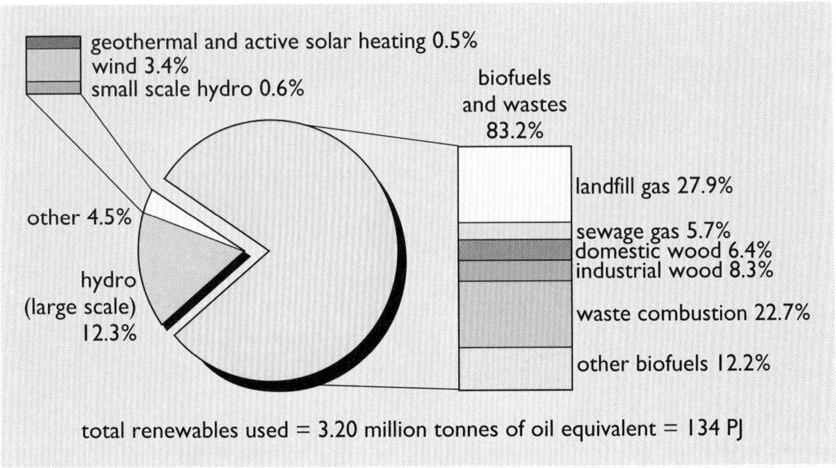

Figure 1.5 Primary energy contributions from renewable energy In the UK, 2002. The main contributors were biofuels in various forms, and hydro power (source: DTI, 2003a)

1.3 Fossil fuels and climate change

Society's current use of fossil and nuclear fuels has many adverse consequences. These include air pollution, acid rain, the depletion of natural resources and the dangers of nuclear radiation; these are described in detail in the companion text. In this brief introduction, we highlight only one problem: global climate change caused by emissions of greenhouse gases from fossil fuel combustion.

The surface temperature of the earth establishes itself at an equilibrium level where the incoming energy from the sun balances the outgoing infrared energy re-radiated from the surface back into space (see Chapter 2, Figure 2.5). If the earth had no atmosphere its average surface temperature would be −18 °C; but its atmosphere includes 'greenhouse gases', principally water vapour, carbon dioxide and methane. These act like the panes of a greenhouse, allowing solar radiation to enter but inhibiting the outflow of infrared radiation. The natural 'greenhouse effect' they cause is essential in maintaining the earth's surface temperature at a level suitable for life, around 15 °C.

Since the industrial revolution, however, human activities have been adding extra greenhouse gases to the atmosphere. The principal contributor to these increased emissions is carbon dioxide from the combustion of fossil fuels (Figure 1.6(a)). Scientists estimate (IPCC, 2001) that these 'anthropogenic' (human-induced) emissions caused a rise in the earth's global mean surface temperature of 0.6 °C during the twentieth century (Figure 1.6(b)). If emissions are not curbed, the surface temperature is predicted to rise by 1.4 to 5.8 °C (depending on the assumptions made) by the end of the twenty-first century. Such rises will probably cause an increased frequency of climatic extremes, such as floods or droughts, and serious disruption to agriculture and natural ecosystems. Mean sea levels are likely to rise by around 0.5 m by the end of the century, which could

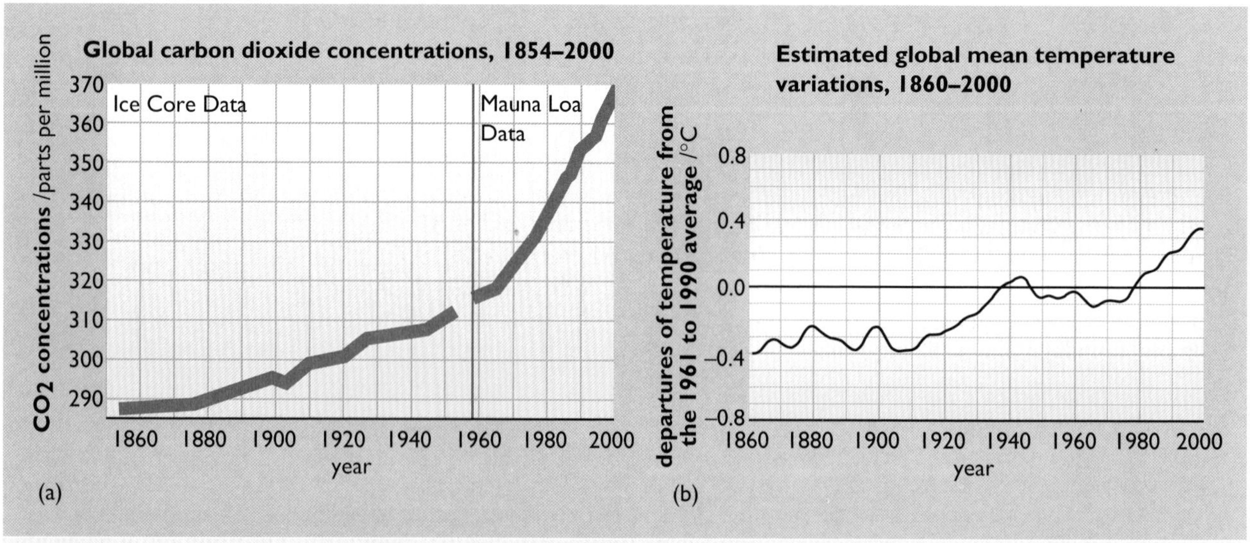

Figure 1.6 (a) Atmospheric concentrations of carbon dioxide (CO_2), 1854–2000. Carbon dioxide data from 1958 were measured at Mauna Loa, Hawaii; pre-1958 data are estimated from ice cores (b) estimated global mean temperature variations, 1860–2000 (source: Intergovernmental Panel on Climate Change, 2001)

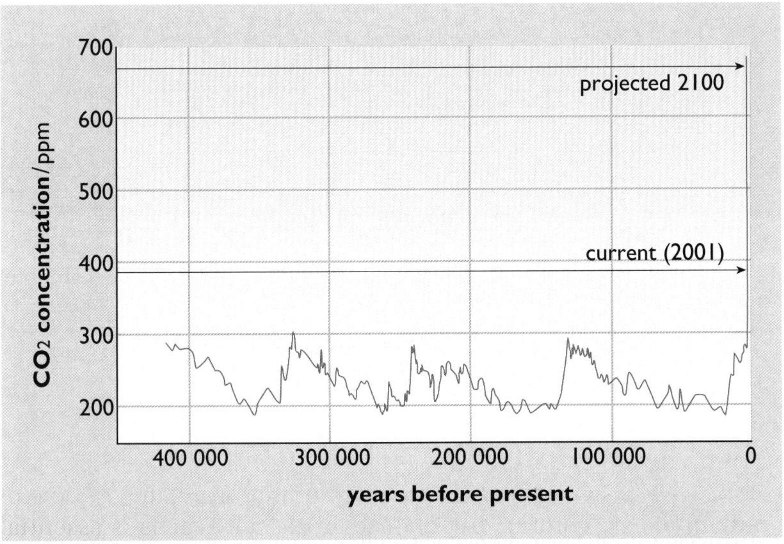

Figure 1.7 For half a million years before the twentieth-century, atmospheric concentrations of CO_2 have varied between about 200 and 300 parts per million by volume (ppmv). But during the twentieth century they rose beyond this range, increasing to 370 ppm by 2001. Projections by the Intergovernmental Panel on Climate Change (IPCC) suggest they could rise to around 700 ppmv by the end of the twenty-first century if no action is taken to limit emissions. The projection shown for 2100 is near the middle of the IPCC range (source: IPCC data)

inundate some low-lying areas. And beyond 2100, much greater sea level rises could occur if major Antarctic ice sheets were to melt.

The threat of global climate change caused by carbon dioxide emissions from fossil fuel combustion is one of the main reasons why there is a growing consensus on the need to reduce such emissions. Reductions in the range 60–80% may be needed by the end of the twenty-first century and, ultimately, a switch to low- or zero-carbon energy sources such as renewables.

1.4 Renewable energy sources

Renewable energy can be defined as 'energy obtained from the continuous or repetitive currents of energy recurring in the natural environment' (Twidell and Weir, 1986). Or as 'energy flows which are replenished at the same rate as they are "used"' (Sorensen, 2000). From Figure 1.8, which summarizes the origins and magnitudes of the earth's renewable energy sources, it is clear that their principal source is solar radiation.

Solar energy: Direct uses

Solar radiation can be converted into useful energy *directly*, using various technologies. Absorbed in solar 'collectors', it can provide hot water or space heating. Buildings can also be designed with 'passive solar' features that enhance the contribution of solar energy to their space heating and lighting requirements.

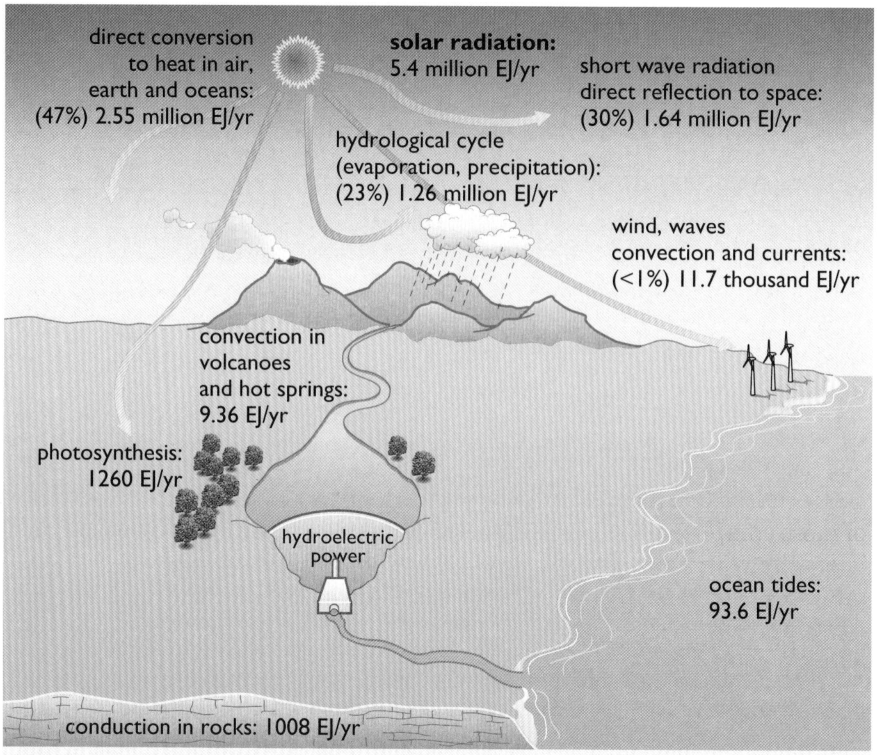

Figure 1.8 The various forms of renewable energy depend primarily on incoming solar radiation, which totals some 5.4 million exajoules (EJ) per year. Approximately 30% is reflected back into space. The remaining 70% is in principle available for use on Earth, and amounts to approximately 3.8 million EJ, more than 10 000 times the rate of consumption of fossil and nuclear fuels, some 370 EJ in 2002. Two non-solar renewable energy sources are shown: the motion of the ocean tides, caused principally by the moon's gravitational pull (with a small contribution from the sun's gravity); and geothermal heat from the earth's interior, which manifests itself in convection in volcanoes and hot springs, and in conduction in rocks

Solar energy can also be concentrated by mirrors to provide high-temperature heat for generating electricity. Such 'solar thermal-electric' power stations are in commercial operation in the USA. **Solar thermal energy** conversion is described in Chapter 2.

Solar radiation can also be converted directly into electricity using photovoltaic (PV) modules, normally mounted on the roofs or facades of buildings. Electricity from photovoltaics is currently expensive but prices are falling and the industry is expanding rapidly. **Solar photovoltaics** is described in Chapter 3.

Solar energy: Indirect uses

Solar radiation can be converted to useful energy *indirectly*, via other energy forms. A large fraction of the radiation reaching the earth's surface is absorbed by the oceans, warming them and adding water vapour to the air. The water vapour condenses as rain to feed rivers, into which we can put dams and turbines to extract some of the energy. **Hydropower**, described in Chapter 5, has steadily grown during the twentieth century, and now provides about a sixth of the world's electricity.

Sunlight falls in a more perpendicular direction in tropical regions and more obliquely at high latitudes, heating the tropics to a greater degree than polar regions. The result is a massive heat flow towards the poles, carried by currents in the oceans and the atmosphere. The energy in such currents can be harnessed, for example by wind turbines. **Wind power**, described in Chapter 7, has developed on a large scale only in the past few decades, but is now one of the fastest-growing of the 'new' renewable sources of electricity.

Where winds blow over long stretches of ocean, they create waves, and a variety of devices can be used to extract that energy. **Wave power**, described in Chapter 8, is attracting new funding for research, development and demonstration in several countries.

Bioenergy, discussed in Chapter 4, is another indirect manifestation of solar energy. Through photosynthesis in plants, solar radiation converts water and atmospheric carbon dioxide into carbohydrates, which form the basis of more complex molecules. Biomass, in the form of wood or other 'biofuels', is a major world energy source, especially in the developing world. Gaseous and liquid fuels derived from biological sources make significant contributions to the energy supplies of some countries. Biofuels can also be derived from wastes, many of which are biological in origin.

Biofuels are a renewable resource if the rate at which they are consumed is no greater than the rate at which new plants are re-grown – which, unfortunately, is often not the case. Although the combustion of biofuels generates atmospheric CO_2 emissions, these should be offset by the CO_2 absorbed when the plants were growing, but significant emissions of other greenhouse gases can result if the combustion is inefficient.

Non-solar renewables

Two other sources of renewable energy do not depend on solar radiation: *tidal* and *geothermal* energy.

Tidal energy, discussed in Chapter 6, is often confused with wave energy, but its origins (described in Figure 1.6) are quite different. The power of the tides can be harnessed by building a low dam or 'barrage' in which the rising waters are captured and then allowed to flow back through electricity-generating turbines.

It is also possible to harness the power of strong underwater currents, which are mainly tidal in origin. Various devices for exploiting this energy source, such as marine current turbines (rather like underwater wind turbines) are at the prototype stage.

Heat from within the earth is the source of **geothermal energy**, discussed in Chapter 9. The high temperature of the interior was originally caused by gravitational contraction of the planet as it was formed, but has since been enhanced by the heat from the decay of radioactive materials within the earth's core.

In some places where hot rocks are very near the surface, they can heat water in underground aquifers. These have been used for centuries to provide hot water or steam. In some countries, geothermal steam is used to produce electricity and, in others, hot water from geothermal wells is used for heating.

If steam or hot water is extracted at a greater rate than heat is replenished from surrounding rocks, a geothermal site will cool down and new holes will have to be drilled nearby. When operated in this way, geothermal energy is not strictly renewable. However, it is possible to operate in a renewable mode by keeping the rate of extraction below the rate of renewal.

1.5 Renewable energy in a sustainable future

We have seen that renewable energy sources are already providing a significant proportion of the world's primary energy. Chapter 10 *Integration* will describe a number of long-term energy studies suggesting that renewables are likely to be providing a much greater proportion of world energy in the second half of the twenty-first century. Meanwhile, we conclude this introductory chapter by looking briefly at the prospects for renewables in the UK in the next few decades.

The UK Government envisages a greatly-increased role for renewable energy, as can be seen from the extracts from its White Paper on Energy (DTI, 2003b) shown in Box 1.3. The technologies mentioned in the Box are described in detail in subsequent chapters, or in the companion text.

BOX 1.3 A possible scenario for the energy system in 2020

We envisage the [UK] energy system in 2020 being much more diverse than today (...)

- The backbone of the electricity system will still be a market-based grid, balancing the supply of large power stations. But some of those large power stations will be offshore marine plants, including wave, tidal and windfarms. Generally smaller onshore windfarms will also be generating. The market will need to be able to handle intermittent generation by using backup capacity when weather conditions reduce or cut off these sources.
- There will be much more local generation, in part from medium to small local/community power plant, fuelled by locally grown biomass, from locally generated waste, from local wind sources, or possibly from local wave and tidal generators. These will feed local distributed networks, which can sell excess capacity into the grid. Plant will also increasingly generate heat for local use.
- There will be much more micro-generation for example from CHP [Combined Heat and Power] plant, fuel cells in buildings, or photovoltaics. This will also generate excess capacity from time to time, which will be sold back into the local distributed network.(...)
- New homes will be designed to need very little energy and will perhaps even achieve zero carbon emissions. The existing building stock will increasingly adopt energy efficiency measures. Many buildings will have the capacity at least to reduce their demand on the grid, for example by using solar heating systems to provide some of their water heating needs, if not to generate electricity to sell back into the local network (...).

Source: Department of Trade and Industry, 2003b

Looking further ahead, to 2050, the UK Royal Commission on Environmental Pollution has produced a set of four energy 'scenarios', several of which envisage renewables playing an even bigger role in Britain's energy supply systems (RCEP, 2000). These scenarios are described in more detail in the companion text.

These and other expert studies suggest that the prospects for renewable energy, in the UK and elsewhere, look bright.

In the chapters that follow, we shall examine each of the principal renewable energy sources in turn, looking at their physical principles, the main technologies involved, their costs and environmental impact, the size of the potential resource and their future prospects. We start with the renewable source that, as we have seen, is the basis of most of the others: solar energy.

References and further information

Boyle, G., Everett, R. and Ramage, J. (eds) (2003) *Energy Systems and Sustainability: Power for a Sustainable Future*, Oxford University Press in association with the Open University, 620 pp.

BP (2003) BP Statistical Review of World Energy [online], BP. Available at www.bp.com/centres/energy2002/index.asp [accessed 27 October 2003].

Department of Trade and Industry (DTI) (2001a) *Digest of UK Energy Statistics (DUKES), 2000*, HMSO.

Department of Trade and Industry (DTI) (2001b) *UK Energy Sector Indicators 2001*, HMSO.

Department of Trade and Industry (DTI) (2003a) *UK Energy in Brief: July 2003*, HMSO.

Department of Trade and Industry (DTI) (2003b) *Our Energy Future – Creating a Low Carbon Economy* (the 2003 White Paper on Energy), HMSO. Available at: www.dti.gov.uk/energy/whitepaper/ourenergyfuture.pdf [accessed 28 October 2003].

Intergovernmental Panel on Climate Change (2001) *Climate Change 2001: the Scientific Basis*, Cambridge University Press. Summary for Policymakers and Technical summary at: www.ipcc.ch/pub/reports.htm [accessed 28 October 2003].

Laherrère, J. H. (2001) 'Forecasting future production from past discovery', *OPEC seminar* , 28 September 2001.

Twidell, J and Weir, A. (1986) *Renewable Energy Resources*, London, E & FN Spon.

RCEP (Royal Commission on Environmental Pollution) (2000) *Energy: the Changing Climate*, London, Stationery Office. Available at: www.rcep.org.uk/reports2.html [accessed 28 October 2003)].

Sorensen, B. (2000) *Renewable Energy* (Second Edition) Academic Press, p 3.

United Nations Development Programme (2000) *World Energy Assessment*, New York, United Nations. Available at: www.undp.org/seed/eap/activities/wea/ [accessed 10 November 2003].

Chapter 2

Solar Thermal Energy

by Bob Everett

2.1 Introduction

As we saw in Chapter 1, the sun is the ultimate source of most of our renewable energy supplies. Since there is a long history of the sun being regarded as a deity, the direct use of solar radiation has a deep appeal to engineer and architect alike.

In this section, we look at some of the methods employed to gather solar thermal or heat energy. Solar photovoltaic energy, the direct conversion of the sun's rays to electricity, is dealt with in the next Chapter. Solar thermal collection methods are many and varied, so we can only give the briefest introduction and supply points to further reading for those interested in studying the subject in greater depth.

What sorts of system can be used to collect solar thermal energy?

Most systems for low-temperature solar heating depend on the use of glazing, in particular its ability to transmit visible light but block infrared radiation. High-temperature solar collection is more likely to employ mirrors. In practice, solar systems of both types can take a wide range of forms.

Active Solar Heating. This always involves a discrete **solar collector**, usually mounted on the roof of a building, to gather solar radiation. Mostly, collectors are quite simple and the heat will be at low temperature (under 100 °C) and used for domestic hot water or swimming pool heating.

Solar Thermal Engines. These are an extension of active solar heating, usually using more complex collectors to produce temperatures high enough to drive steam turbines to produce electric power. They can come in a wide variety of types, but 90% of the world's solar thermally-generated electricity comes from a single plant in the Mohave desert in California.

Passive Solar Heating. This term has two slightly different meanings.

- In the 'narrow' sense, it means the absorption of solar energy directly into a building to reduce the energy required for heating the habitable spaces (or what is called **space heating**). Passive solar heating systems mostly use air to circulate the collected energy, usually without pumps or fans – indeed the 'collector' is often an integral part of the building.
- In the 'broad' sense, it means the whole process of integrated low-energy building design, effectively to reduce the heat demand to the point where small passive solar gains make a significant contribution in winter. A large solar contribution to a large heat load may look impressive, but what really counts is to minimize the total fossil fuel consumption and thus achieve the minimum cost.

In this chapter, for lack of space, we concentrate on the narrow view, although it is important to understand that implementing the broad view, with significant investment in insulation, can produce energy savings that are five or more times greater.

Daylighting. This means making the best use of natural daylight, though both careful building design and the use of controls to switch off artificial lighting when there is sufficient natural light available.

It must be stressed at the outset that making the best use of solar energy requires a careful understanding of the climate of any particular location. Indeed, many of our present energy problems stem from attempts to produce buildings inappropriate to the local climate. This can mean that the economics of solar technologies commonly used in southern Europe may be disappointing when transferred, for example, to northern Scotland.

However, most of the methods described in this chapter have been well tried and tested over the past century. Even the most spectacular of modern solar thermal electric power stations are just uprated versions of inventive systems built at the beginning of the twentieth century. The skill of using solar thermal energy, in all its forms, perhaps lies in producing systems that are cheap enough to compete with 'conventional' systems based on fossil fuels at current prices.

2.2 The rooftop solar water heater

Figure 2.1 Solar panels mounted on a roof (photo courtesy Arcon)

For most people, 'solar heating' means the rooftop solar water heater. At the end of 1999 there were a total of 8.5 million square metres of solar collectors installed in Europe, almost half of them in Germany (Weiss, 2001). Most use simple flat plate collectors. There are two basic forms of system: pumped or thermosyphon.

The pumped solar water heater

This is the form most common in Northern Europe. A typical flat plate pumped system consists of three elements, shown in Figure 2.2.

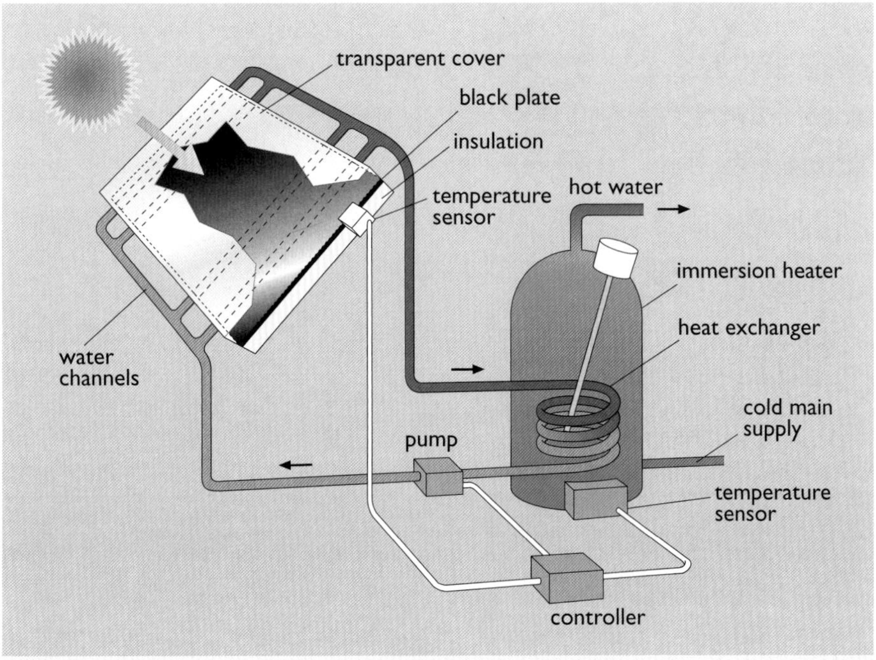

Figure 2.2 Pumped active solar water heater

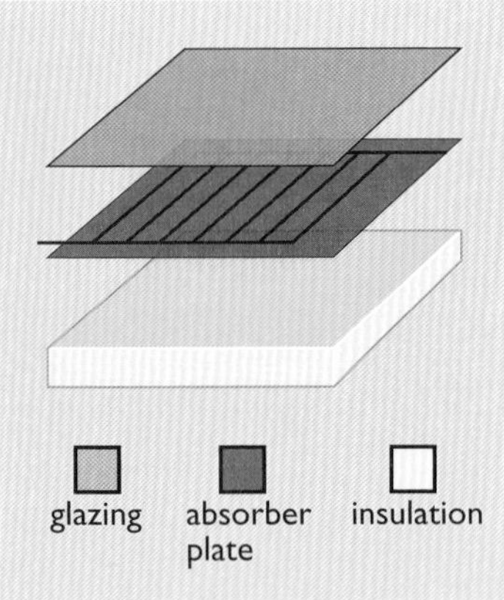

Figure 2.3 Components of a solar panel

1 A collector panel, typically of 3–5 square metres in area, tilted to face the sun and mounted on the normal pitched roof of a house, as in Figure 2.1. This panel itself normally consists of three components (see Figure 2.3). The main absorber might be a steel plate bonded to copper tubing through which water circulates. The plate is sprayed with a special black paint to maximize the solar absorption. It is normally covered with a single sheet of glass or plastic and the whole assembly is insulated on the back to cut heat losses.

2 A storage tank, typically of around 200 litres capacity, which often doubles as the normal domestic hot water cylinder. This usually contains an electric immersion heater for winter use. The tank is insulated all round typically with 50 mm of glass fibre or polyurethane foam. The hot water from the panel circulates through a heat exchanger at the bottom of the tank.

3 A pumped circulation system to transfer the heat from the panel to the store. Sensors detect when the collector is becoming hot and switch on an electric circulating pump. Since in northern Europe the collector has to be able to survive freezing temperatures, the circulating water contains an antifreeze. Non-toxic propylene glycol is often used (instead of the poisonous ethylene glycol commonly used in car engines).

Such a system in the UK can be expected to supply 40–50% of the hot water requirements of a typical house (Sadler, 1996).

The thermosyphon solar water heater

In frost-free climates where it is safe to mount the storage tank outdoors, a simpler **thermosyphon** arrangement can be used, as shown in Figure 2.4.

Figure 2.4 A typical Mediterranean thermosyphon solar water heater – the insulated storage tank is at the top

This design dispenses with the circulation pump. It relies on the natural convection of hot water rising from the collector panel to carry heat up to the storage tank, which must be installed above the collector. There is no need for a heat exchanger as the heated water circulates directly through the panel.

Normally the storage tank also contains an electric immersion heater for top-up and use on cloudy days. Mediterranean systems are usually designed to be free-standing for mounting on buildings with flat roofs. Given the higher levels of solar radiation in these countries, they are usually sold with only around 2 m^2 of collector area.

2.3 The nature and availability of solar radiation

The wavelengths of solar radiation

The sun is an enormous nuclear fusion reactor which converts hydrogen into helium at the rate of 4 million tonnes per second. It radiates energy by virtue of its high surface temperature, approximately 6000 °C. Of this radiation, approximately one-third of that incident on earth is simply reflected back. The rest is absorbed and eventually retransmitted to deep space as long-wave infrared radiation. The earth re-radiates just as much energy as it receives and sits in a stable energy balance at a temperature suitable for life.

We perceive solar radiation as white light. In fact it spreads over a wide spectrum of wavelengths, from the 'short-wave' infrared (longer than red light) to ultraviolet (shorter than violet). The pattern of wavelength distribution is critically determined by the temperature of the surface of the sun (see Figure 2.5).

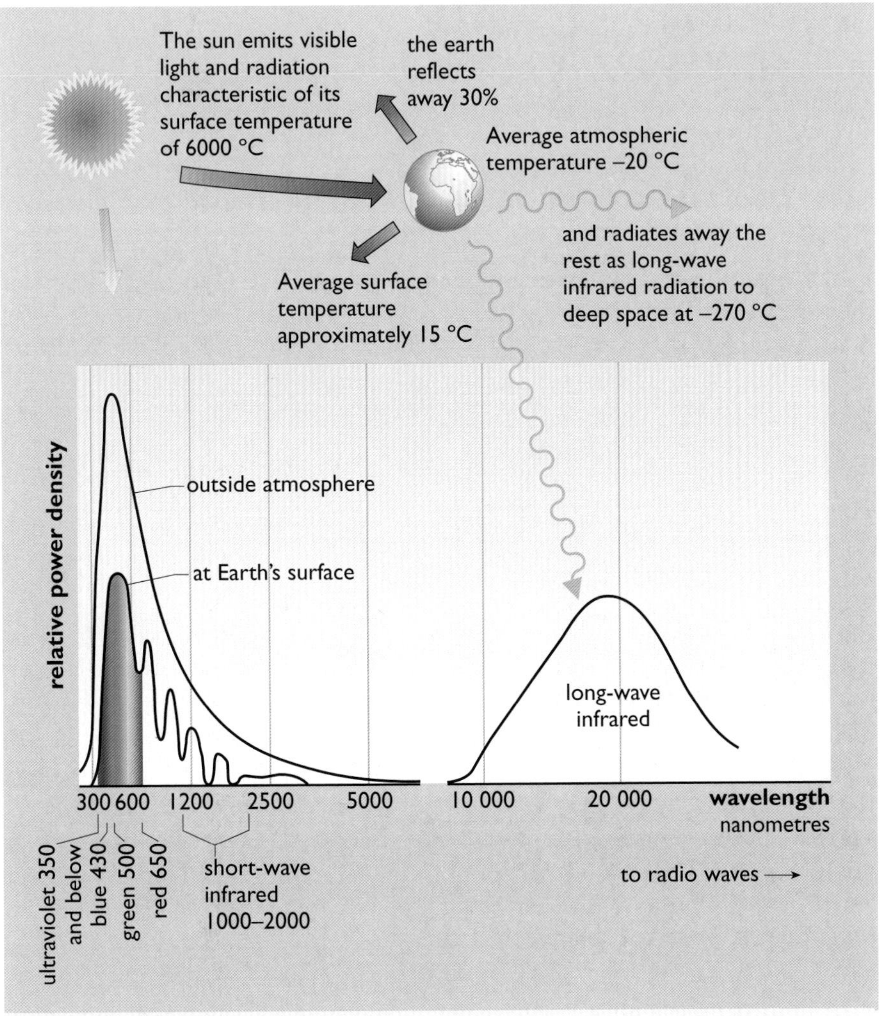

Figure 2.5 Radiation of energy to and from the earth

The earth, which has an average atmospheric temperature of –20 °C and a surface temperature of 15 °C, radiates energy as long-wave infrared to deep space, the temperature of which is only a few degrees above absolute zero, –273 °C. We tend to forget this outgoing radiation, but its effects can be observed on a clear night when a ground frost can occur as heat radiates first to the cold upper atmosphere and then out into space.

As we shall see, most of the art of low-temperature solar energy collection depends on our ability to use glass and surfaces with selective properties which allow solar radiation to pass through but block the re-radiation of long-wave infrared. The gathering of solar energy for high-temperature applications, such as driving steam engines, mainly involves concentrating solar energy using complex mirrors.

Direct and diffuse radiation

When the sun's rays hit the atmosphere, more or less of the light is scattered, depending on the cloud cover. A proportion of this scattered light comes to earth as diffuse radiation. On the ground this appears to come from all over the sky. Some of it we see as the blue colour of a clear sky, but most is the 'white' light scattered from clouds.

What we normally call 'sunshine', that portion of light that appears to come straight from the sun, is known as **direct radiation**. On a clear day, this can approach a power density of 1 kilowatt per square metre (1 kW m^{-2}), known as '1 sun' for solar collector testing purposes. Generally in northern Europe and in urban locations in southern Europe, practical peak power densities are around 900–1000 watts per square metre.

In northern Europe, on average over the year approximately 50% of the radiation is diffuse and 50% is direct. In southern Europe, where solar radiation levels are higher, most of the extra contribution is in direct radiation, especially in summer. Both diffuse and direct radiation are useful for most solar thermal applications, but only direct radiation can be focused to generate very high temperatures. On the other hand it is the diffuse radiation that provides most of the 'daylight' in buildings, particularly in north-facing rooms.

Availability of solar radiation

Figure 2.6 A solarimeter (also referred to as a pyranometer)

Interest in solar energy has prompted the accurate measurement and mapping of solar energy resources over the globe. This is normally done using **solarimeters** (see Figure 2.6). These contain carefully calibrated thermoelectric elements fitted under a glass cover, which is open to the whole vault of the sky. A voltage proportional to the total incident light energy is produced and then recorded electronically.

Most solarimeter measurements are recorded simply as **total energy incident on the horizontal surface**. More detailed measurements separate the direct and diffuse radiation. These can be mathematically recombined to calculate the radiation on tilted and vertical surfaces.

As we might expect, annual total solar radiation on a horizontal surface is highest near the equator, over 2000 kilowatt-hours per square metre per year (kWh m^{-2} per year), and especially high in sunny desert areas. These are more favoured than northern Europe, which typically only receives about 1000 kWh m^{-2} per year. Many experimental projects, such as solar thermal power stations, have been built in areas like southern France or Spain, where radiation levels are around 1500 kWh m^{-2} per year, or the southern US, where levels can reach 2500 kWh m^{-2} per year.

It is obvious that in Europe summers are sunnier than winters, but what does that mean in energy terms?

On average in July, the solar radiation on a horizontal surface in northern Europe (e.g. Ireland, UK, Denmark and northern Germany) is between 4.5 and 5 kWh m^{-2} per day (see Figure 2.7). Five kilowatt-hours is enough energy to heat the water for a (rather generous) hot bath. At 2003 UK domestic fuel prices, we would have to pay approximately 15 p for this amount of heat if we were using a normal gas boiler or off-peak electricity. In southern Europe (Spain, Italy and Greece), July solar radiation levels are higher, between 6 and 7.5 kWh m^{-2} per day.

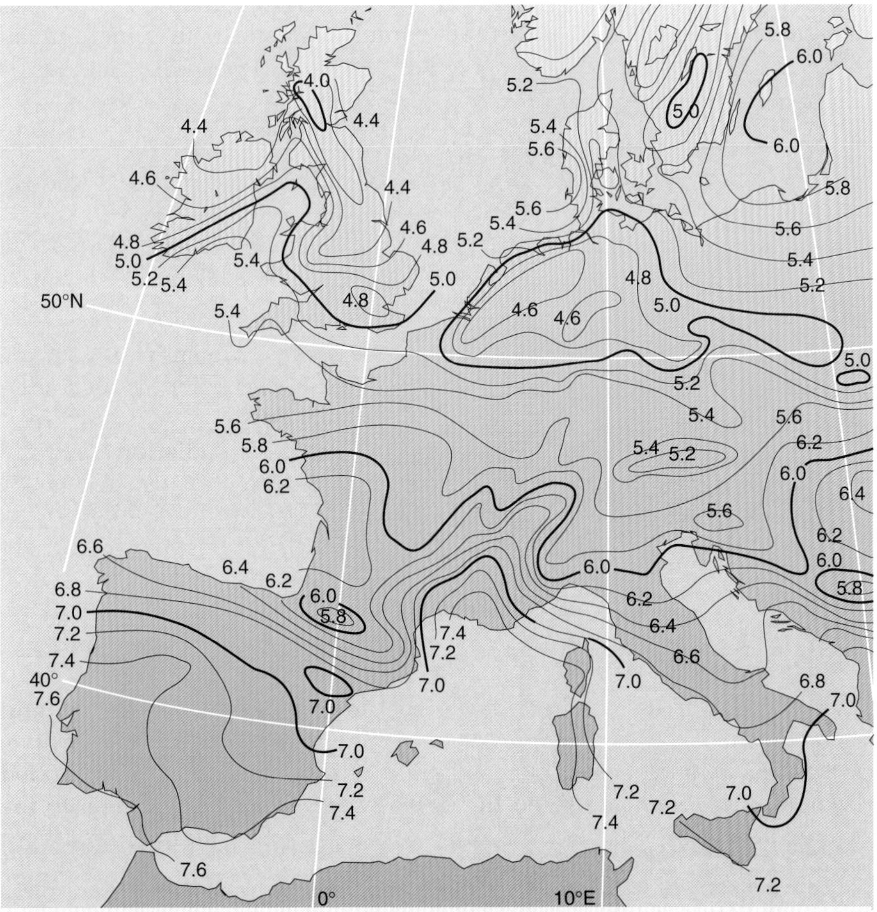

Figure 2.7 Solar radiation on horizontal surface (kWh per square metre per day), Europe, July

In winter, however, the amount of solar radiation is far lower. In January on average in northern Europe it can be only one tenth of its July value, around 0.5 kWh m^{-2} per day (see Figure 2.8), yet in southern Europe there may still be appreciable amounts, 1.5 to 2 kWh m^{-2} per day.

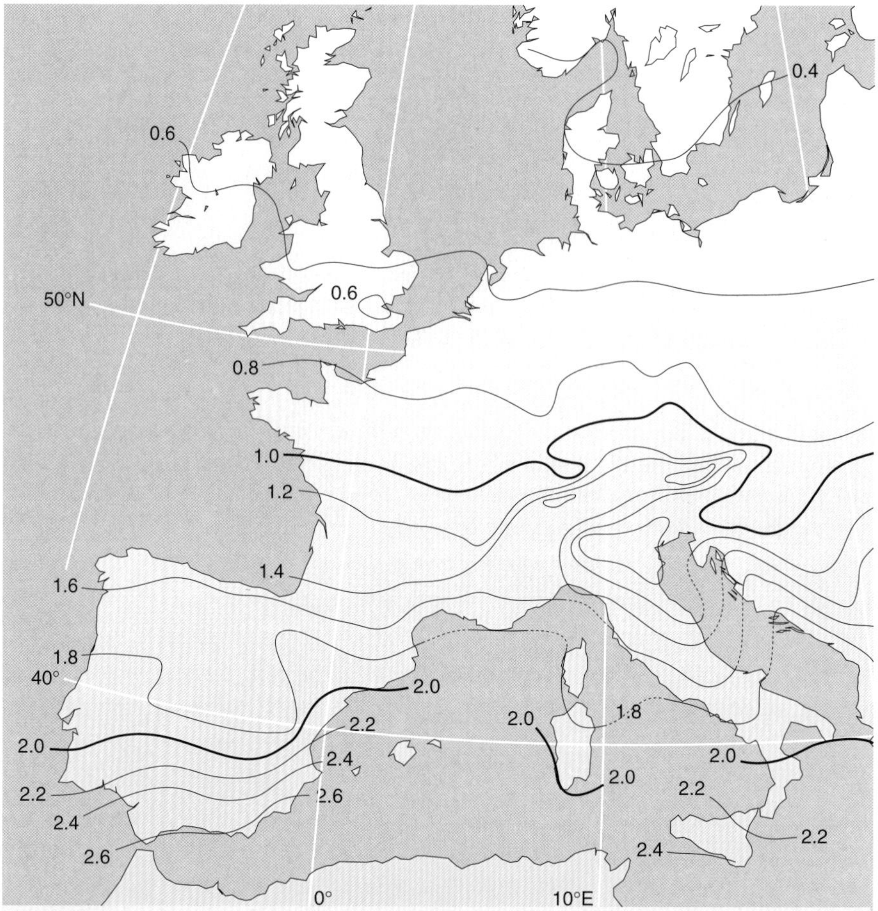

Figure 2.8 Solar radiation on horizontal surface (kWh per square metre per day), Europe, January

The implications of this are that in northern Europe we need to look for applications that require energy mainly in the summer. In southern Europe, there may be enough radiation in the winter to consider year-round applications.

Tilt and orientation

So far, we have talked about solar radiation on the horizontal surface. To collect as much radiation as possible, a surface should face south (assuming it is in the Northern hemisphere) and must be tilted towards the sun. How much it should be tilted is dependent on the latitude and at what time of year most solar collection is required.

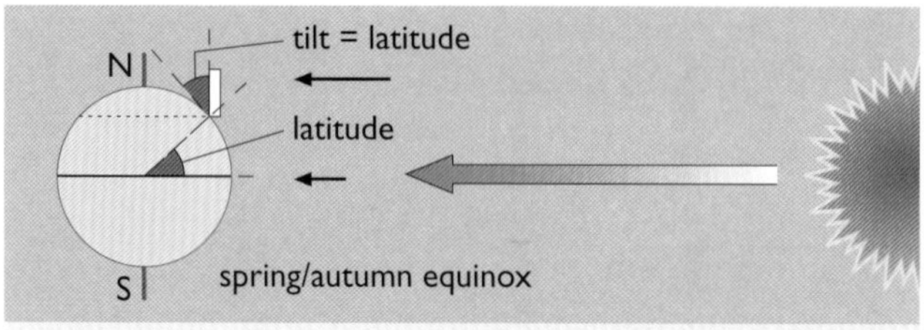

Figure 2.11 Surface tilted at latitude angle will be perpendicular to sun at spring or autumn

If the tilt angle between a surface and the horizontal is equal to the latitude, it will be perpendicular to the sun's rays at midday in March and September (see Figure 2.11). To maximize solar collection in summer (when there is most radiation to be had), the tilt angle should be less than the latitude.

BOX 2.1 Solar radiation and the seasons

If the energy output of the sun is constant, why do we receive more radiation in summer than in winter?

The earth circles the sun with its polar axis tilted towards the plane of rotation. In June, the North Pole is tilted towards the sun. The sun's rays thus strike the northern hemisphere more perpendicularly and the sun appears higher in the sky (Figure 2.10). In December the North Pole is tilted away from the sun and its rays strike more obliquely, giving a lower energy density on the ground (i.e. fewer kilowatt-hours reach each square metre of ground per day).

Another important factor is that the lower the sun in the sky, the further its rays have to pass through the atmosphere, giving them more opportunity to be scattered back into space. When the sun is at 60° to the vertical its peak energy density will have fallen to one-quarter of that when it is vertically overhead. This topic will be revisited in the next chapter.

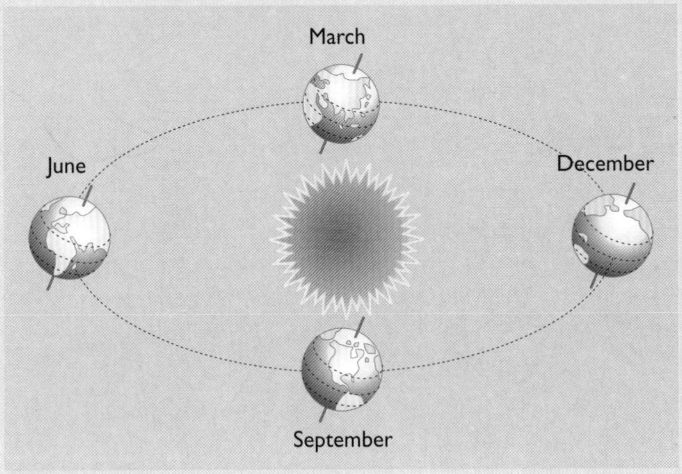

Figure 2.9 The earth revolves around the sun with its axis tilted at an angle of 23.5°

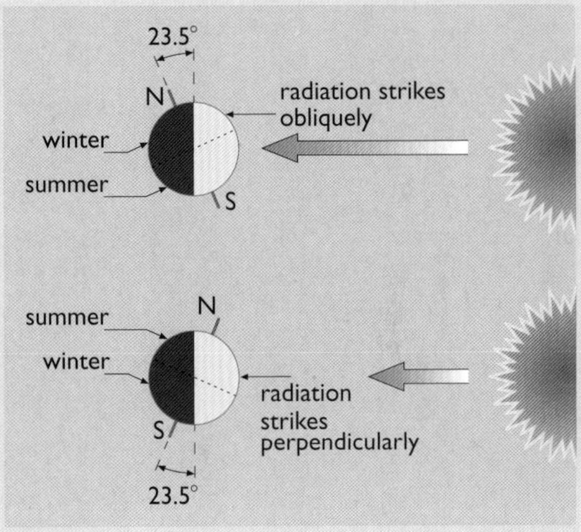

Figure 2.10 The tilt of the earth's axis creates summer and winter

To maximize in winter (when more solar radiation may be needed) the tilt angle should be greater (see Figure 2.12).

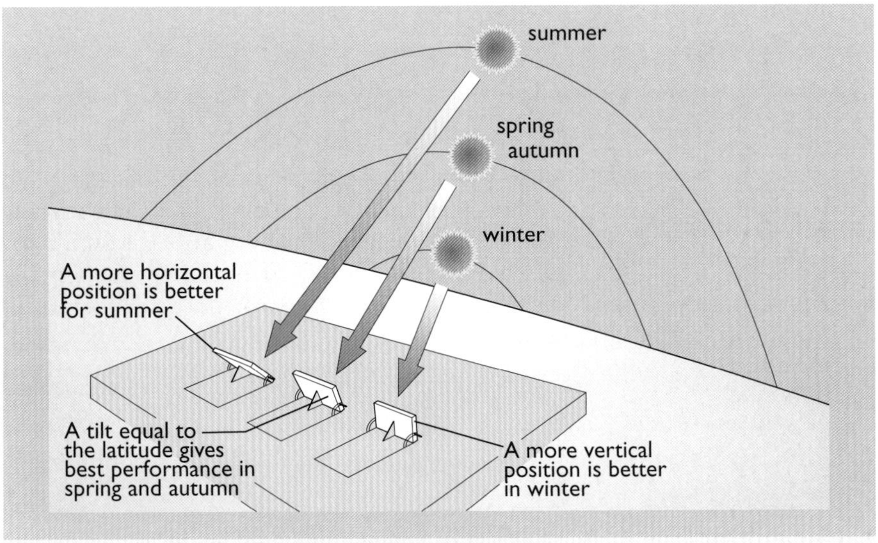

Figure 2.12 Optimizing the tilt for different seasons

Fortunately the effects of tilt and orientation are not particularly critical. Table 2.1 gives totals of energy incident on various tilted surfaces.

Table 2.1 Effect of tilting a south-facing collection surface (data for Kew near London, latitude 52°N)

Tilt (°)	Annual total Radiation (kWh m^{-2})	June total Radiation (kWh m^{-2})	December total Radiation (kWh m^{-2})
0 – Horizontal	944	153	16
30	1068	153	25
45	1053	143	29
60	990	126	30
90 – Vertical	745	82	29

(Source: Achard and Gicquel, 1986)

Similarly, the effects of orientation away from south are relatively small. For most solar heating applications, collectors can be faced anywhere from south-east to south-west. This relative flexibility means that a large proportion of existing buildings have roof orientations suitable for solar energy systems.

2.4 The magic of glass

Most low-temperature solar collection is dependent on the properties of one rather curious substance – glass. It is hard to imagine a world without glazed windows. They have been around since the time of the Romans, who invented a process for making plate glass:

> Certain inventions have come about within our own memory – the use of window panes which admit light through a transparent material.
>
> Seneca, AD 65, cited in Butti and Perlin, 1980

What he did not mention was that glass was also impervious to the wind, which would have blown through all the natural lighting openings in buildings of the period.

Transparency

The most important solar property of glass is that it is transparent to visible light and short-wave infrared radiation, but opaque to long-wave infrared re-radiated from a solar collector or building behind it (see Figure 2.13).

Over the past few decades, enormous effort has been put into improving the performance of glazing, both to increase its transparency to visible radiation, and to prevent heat escaping through it. A good summary of techniques under development will be found in Hutchins, 1997.

Manufacturers strive to make glass as transparent as possible, i.e. they try to maximize its **transmittance**, the fraction of incident light that passes

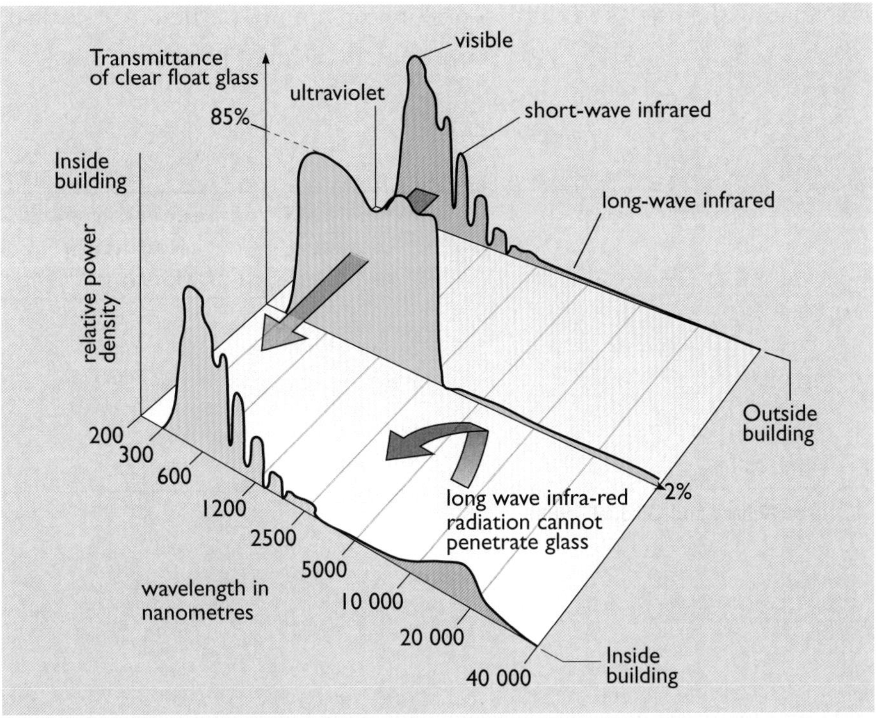

Figure 2.13 Spectral transmittance of glass

through it. They usually do this by minimizing the iron content of the glass. Certain plastics that have optical properties similar to glass can be used instead, although normally they must be protected from the damaging effects of ultraviolet light.

Table 2.2 below shows the optical properties of commonly used glazing materials. You will see that the solar transmittance is high (close to 1.0), but that the long-wave infra-red transmittance is very low by comparison.

Table 2.2 Optical properties of commonly used glazing materials

Material (mm)	Thickness	Solar Transmittance	Long-wave infrared transmittance
Float glass (normal window glass)	3.9	0.83	0.02
Low-iron glass	3.2	0.90	0.02
Perspex	3.1	0.82	0.02
Poly vinyl fluoride (tedlar)	0.1	0.92	0.22
Polyester (mylar)	0.1	0.87	0.18

Heat loss mechanisms

Much development work has also gone into reducing the heat loss through windows and solar collector glazing. Heat energy will flow through any substance where the temperature on the two sides is different.

The *rate* of this energy flow depends on:

- the temperature difference between the two sides
- the total area available for the flow
- the insulating qualities of the material

It is obvious that more heat is lost through a large window than a small one, and on a cold day than a warm one. In order to understand how this heat loss occurs, and how it can be minimized we need to look at the three mechanisms are involved in the transmission of heat: conduction, convection and radiation.

Conduction

In any material, heat energy will flow by **conduction** from hotter to colder regions. The rate of flow will depend on the first two factors above, and on the **thermal conductivity** of the material.

Generally, *metals* have very high thermal conductivities and can transmit large amounts of heat for small temperature differences. Where the frames of glazing systems are made of metal, they must be well insulated to minimize heat loss.

Insulators require a large temperature differential to conduct only a small amount of heat. Still air is a very good insulator. Most practical forms of insulation rely on very small pockets of air, trapped for example between the panes of glazing, as bubbles in a plastic medium, or between the fibres of mineral wool.

Convection

A warmed fluid, such as air, will expand as it warms, becoming less dense and rising as a result, creating a fluid flow known as **convection**. This is one of the principal modes of transfer heat through windows and out to the environment (see Figure 2.14). It occurs between the air and the glass on the inside and outside surfaces, and, in double glazing, in the air space between the panes. The convection effects can be reduced by filling double glazing with heavier, less mobile gas molecules, such as argon, krypton or carbon dioxide.

Convection can also be reduced by limiting the space available for gas movement. This is the principle used in the insulation materials mentioned above.

Various forms of **transparent insulation** are under development which use a transparent plastic medium containing bubbles of trapped insulating gas. These materials could eventually revolutionize the concept of windows (and walls), but at present the materials are not robust and need protection from the rigours of weather and ultraviolet light.

Alternatively the glazing can be evacuated. Convection currents cannot flow in a vacuum. However, a very high vacuum is required and it will have to last for the whole life of the window, 50 years or more. Also the window will need internal structural spacers to stop it collapsing inwards under the air pressure on the outside. These spacers conduct heat across the gap slightly reducing the overall performance.

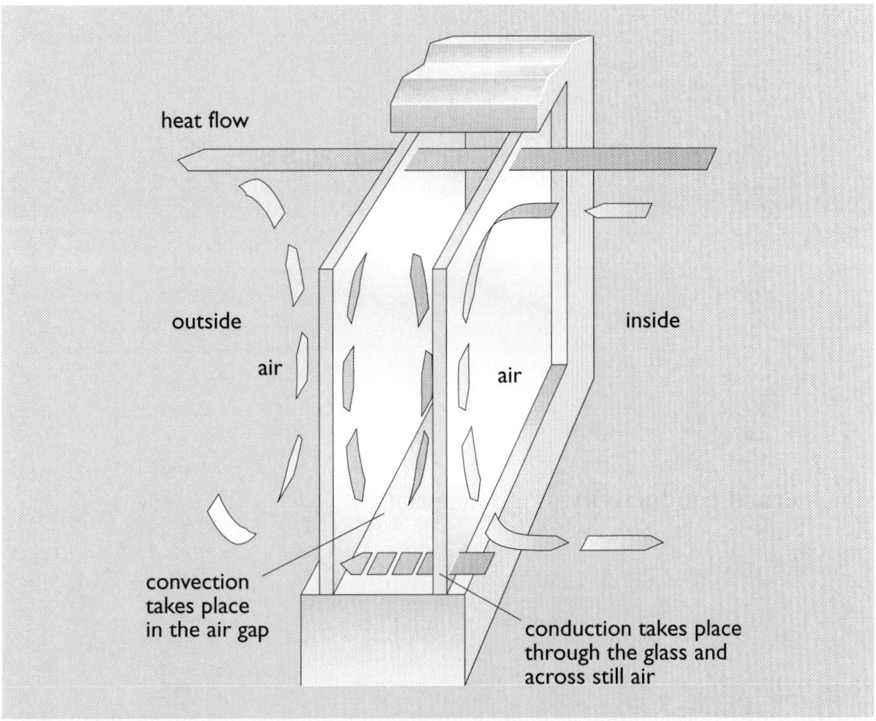

Figure 2.14 How heat escapes from a double-glazed window. The air space is normally 6–10 mm wide. If it is too narrow, convection will be difficult but conduction will be easy because there is only a small thickness of air to conduct across. If it is too wide, convection currents can easily circulate. In addition there is radiation across the air space which can be reduced by using low emissivity coatings

A simpler way to reduce the convection effects is to insert extra panes of glass or of transparent plastic film between the other two, turning double glazing into triple or quadruple glazing.

Radiation

Heat energy can be **radiated**, in the same manner as it is radiated from the sun to the earth. The quantity of radiation is dependent on the temperature of the radiating body. The roof of a building, for example, will radiate heat away to the atmosphere. It also depends on a quality of the surface known as **emissivity**. Most materials used in buildings have high emissivities of approximately 0.9, that is, they radiate 90% of the theoretical maximum for a given temperature.

Other surfaces can be produced that have low emissivities. This means that although they may be hot, they will radiate little heat outwards. **'Low-E' coatings** are now commonly used inside double glazing to cut radiated heat losses from the inner pane to the outer one across the air gap.

U-Value

Conduction, convection and radiation all contribute to the complex process of heat loss through a wall, window, roof, etc. In practice, the actual performance of any particular building element is usually specified by a

BOX 2.2 *U*-value and heat loss

What is the rate of heat loss through a large single-glazed window with an area of 2 square metres, on a day when the outdoor and indoor temperatures are 5°C and 20°C respectively?

Table 2.3 shows that the *U*-value for this window is 6 W m^{-2} $°C^{-1}$, so the loss rate is

$$2 \times 6 \times (20 - 5) = 180 \text{ W}$$

Note that, if the temperature difference remained the same throughout 24 hours, the total loss would be more than 4 kWh. For the best of the glazing types shown in Table 2.3, this would be reduced to about a quarter of a kilowatt-hour.

Table 2.3 Heat loss characteristics (*U*-values) of different types of glazing

Window type	*U*-Value/ W m^{-2} $°C^{-1}$
Single Glazed Window	6
Double Glazed Window	3
with 'Low-E' coatings	1.8
plus heavy gas filling	1.5
with 3 plastic films (low-E coated)	
plus heavy gas filling	0.35
evacuated with low-E coatings	1.0
For comparison: 10 cm opaque fibreglass insulation	0.4

Sources: Granqvist, 1989; Hutchins, 1997; and manufacturers' literature

U-value, defined so that:

heat flow through one square metre = *U*-value × temperature difference.

The units in which *U*-values are expressed are thus watts per square metre per degree Celsius (W m^{-2} $°C^{-1}$). The lower the *U*-value, the better the insulation performance. Table 2.3 gives typical *U*-values of various types of glazing system (the precise values will depend on construction details, particular the details of the frames).

2.5 Low-temperature solar energy applications

We have seen how solar radiation can produce low-temperature heat. Just how useful is this?

As we saw in Chapter 1, in the UK nearly one-half of all the end-use of fuel is for low-temperature heating. Over 80% of delivered energy use in the domestic sector is in this form (see Figure 2.15).

Although simple solar systems are in principle ideal for supplying this heat, there are other potential competitors. These include:

- district heating fed by waste heat from existing conventional power stations or from industrial processes;
- small-scale combined heat and power generation plant;
- heat pumps (see Box 2.3).

All of these merit further development, and unlike solar heating, most have the advantage of being able to run all year round. For various reasons, they are more likely to be available in commercial and industrial buildings than in the domestic sector.

Swimming pools are another potential application. They do not use a significant proportion of Europe's total energy consumption, since there are not very many of them, but individually they can be enormous energy users. A large, indoor leisure pool in northern Europe can use 1 kW for every square metre of pool area continuously throughout the year. This kind of establishment is a prime candidate for the technologies listed above.

Outdoor pools, usually unheated, are rather different. Here the aim is to make the water a little more attractive when people come to use them, which is usually on sunny, warm days. This is ideal solar heating territory.

BOX 2.3 Heat pumps

A heat pump is essentially very similar to a refrigerator, except that it is used for heating rather than cooling.

In a domestic refrigerator, heat is 'pumped' from an **evaporator** inside the refrigerated compartment, thus lowering its temperature, to a **condenser** on the back of the refrigerator, where the heat is released, warming one's kitchen in the process.

In a heat pump, the evaporator is located somewhere in the external environment. It can either be in the open air, or in the ground as in the case of ground source heat pumps, described in Chapter 9. Heat is then pumped from the environment to a condenser inside the building, which is warmed as a result. The outside environment surrounding the evaporator is cooled to an appreciable degree by this process, even to the extent of forming ice.

The heat pumping process is made possible by the use of a special **refrigerant** liquid that boils at low temperature. In order to convert a liquid to a vapour, it must be given energy – the so-called **latent heat of evaporation**. In the evaporator, this energy comes in the form of heat from the surrounding environment.

The vapour is then passed through pipes to a compressor (a form of pump, usually electrically powered), which raises its pressure, and then to the condenser. Here the vapour condenses (or changes back into a liquid), but this time at a higher temperature. It gives up the latent heat which it acquired during the evaporation process, but now at a high enough temperature to be used for space or water heating.

The heat pump, in effect, raises the temperature of the heat in the external environment to a level that is useful for heating purposes. Energy is, of course, required to operate the compressor, but typically the heat output is 2–3 times the electrical energy required by the compressor. Ultimately, the heat energy input to a heat pump comes from the sun, which has warmed the external environment to ambient temperature (typically, 0–15 °C).

Domestic water heating

Domestic water heating is perhaps the best overall potential application for active solar heating in Europe. It is a demand that continues all year round and still needs to be satisfied in the summer when there is plenty of sunshine. In the UK in 2000 it accounted for approximately 7% of the total national delivered energy use. A typical UK household uses approximately 15 kWh per day of delivered energy for this purpose. In practice, much of this can be simply lost as waste heat. Uninsulated hot water cylinders and unlagged pipework are common causes and even solar water heaters can suffer from this failing.

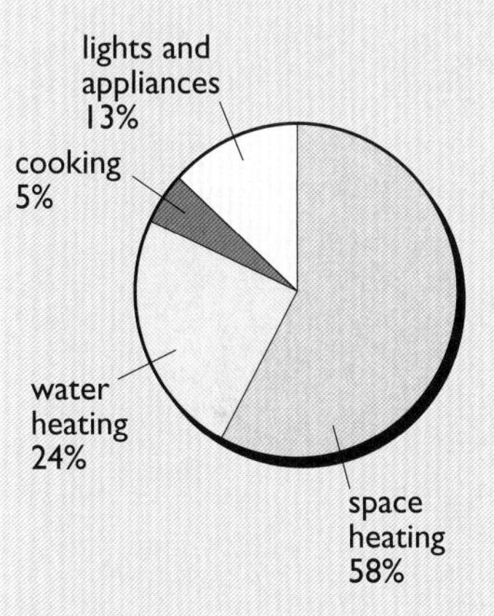

Figure 2.15 Breakdown of UK domestic sector energy use 2000 (DTI, 2003)

Incoming water is usually at a temperature close to that of the soil, approximately 12 °C in the UK, varying only slightly over the year, and it has to be heated up to 60 °C. In many books it is suggested that temperatures as low as 45 °C are adequate, but recent concerns over Legionnaires' Disease, caused by *Legionella pneumophila* bacteria multiplying in warm water, have highlighted the need for a higher temperature.

Domestic water heating is usually done in one of three ways:

- By electricity, with an immersion heater in a hot-water storage cylinder.
- Again using a storage cylinder, but with a heat exchanger coil inside connected to a central heating boiler (usually gas-fired) or possibly to a district heating supply system.
- By an 'instantaneous' heater, usually powered by gas or electricity.

In the UK, natural gas is the dominant fuel for domestic heating. As will be described in Chapter 4, obtaining heat by burning natural gas directly involves less CO_2 production than using electricity generated from fossil fuels. Given the current UK domestic heating fuel mix, 1 kWh of heat produced by a solar water heater in the UK is only likely to save on the emission of 0.19 kg CO_2 (see ESTIF, 2003).

However, in many sunnier countries the majority of homes use electric water heating throughout the year. In Greece, for example, a solar water heater is likely to be substituting for electricity generated in a coal or oil-fired power station and the emission savings may be closer to1 kg CO_2 per kWh of heat produced.

Also, in such countries, the national electricity demand peaks in the summer with ever-increasing demands for refrigeration and air-conditioning, rather than in the winter as in the UK. Thus every solar water heater installed saves not only on fuel, but, equally importantly, on building new power plants.

Put another way, where solar heat can be substituted for fossil-fuelled electricity used for low-temperature heating purposes, this is probably as beneficial as building a photovoltaic or solar thermal power system to generate an equivalent amount of extra electricity.

Domestic space heating

Space heating involves warming the interior spaces of buildings to internal temperatures of approximately 20 °C. In the UK, it consumes almost 20% of the country's delivered energy, yet with an appropriate heating system it can in principle be carried out with water at only 45 °C. It is an activity that only occurs over the **heating season**. For normal UK buildings this extends from about mid-September to April, although, as we shall see later in the section on passive solar heating, this can vary considerably with location and level of insulation.

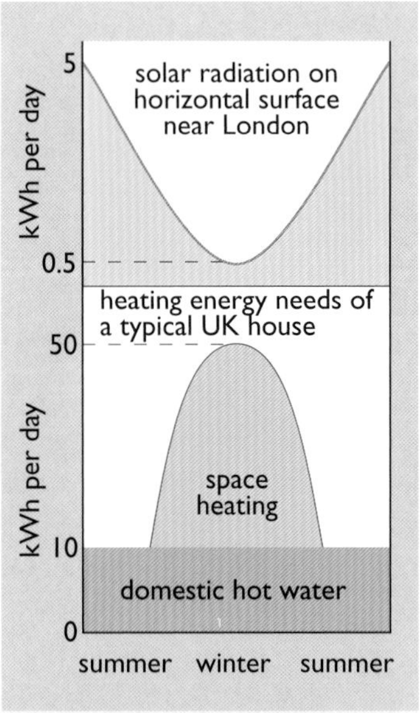

Figure 2.16 The availability of solar radiation is out of phase with space heating demand in the UK

However, there is a fundamental problem that for this application in the UK, as in much of northern Europe, the availability of solar radiation is completely out of phase with the overall demand for heat (see Figure 2.16). Although the total amount of solar radiation over a whole year on a particular site may far exceed the total building heating needs, that falling during the heating season may be quite small.

Even with south-facing vertical surfaces, the amount of radiation intercepted in the UK over the winter is relatively small. In London, for example, over a typical six-month winter period of October to March, 1 m^2 of south-facing vertical surface will only intercept 250 kWh of solar radiation.

It is important to emphasize that the suitability of solar energy for space heating is dependent on the local climate. Textbooks may show quite grandiose solar buildings, but these may only be appropriate in particular locations, often places that are both cold and sunny in winter.

In summer the UK has similar temperatures, and receives a similar amount of solar radiation, to other European countries on the same latitude.

In winter the picture is different. The UK has relatively mild winters. However, the winter solar radiation remains largely dependent on latitude alone. As can be seen in Figure 2.17, average January temperatures in London are virtually identical to those in the south of France (follow the 5 °C contour).

Why then do Northern Europeans go south for the winter? The answer is because it is sunnier. As we saw from Figure 2.8, the south of France receives three times as much solar radiation on the horizontal surface in mid-winter as does London.

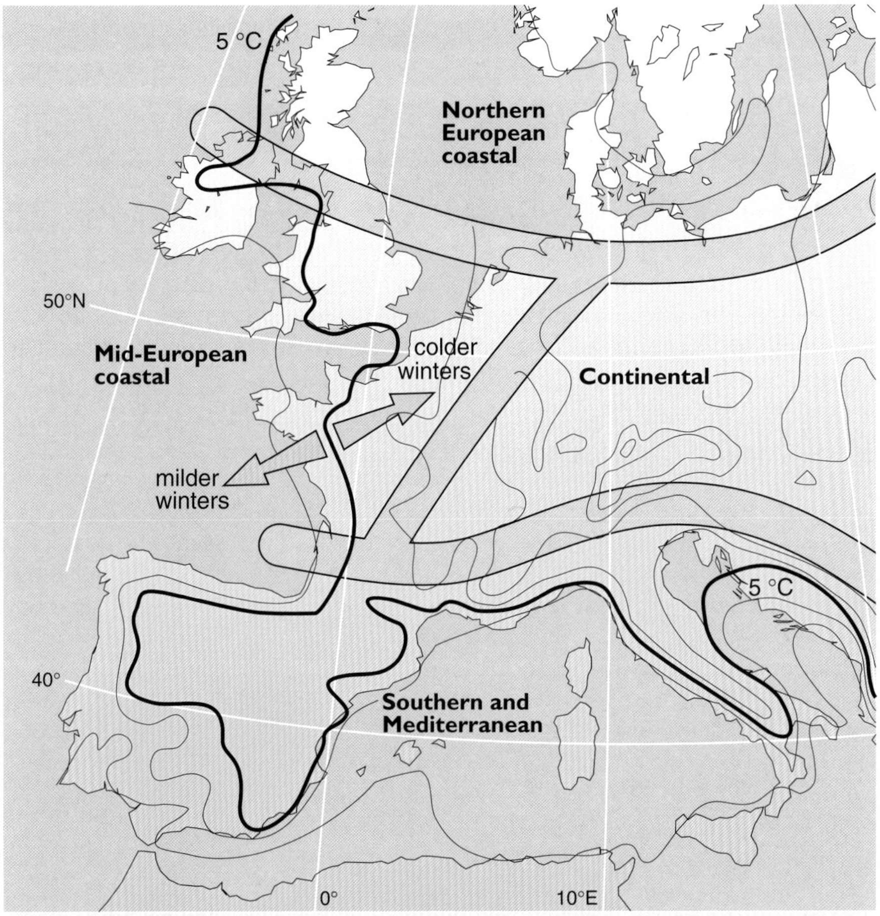

Figure 2.17 The different climatic zones of Europe. UK winters are mild compared with much of the rest of Europe. The 5 °C contour is for January

Broadly speaking, the climate of Western Europe can be split into four regions (Figure 2.17).

1. Northern European coastal zone: cold winters with little solar radiation in mid-winter; mild summers.
2. Mid-European coastal zone: cool winters with modest amounts of solar radiation; mild summers.
3. Continental zone: very cold winters with modest amounts of solar radiation; hot summers.
4. Southern and Mediterranean zone: mild winters with high solar radiation; hot sunny summers.

It is no coincidence that many solar experimental projects have been built on the boundaries of regions 3 and 4, in the Pyrenees and the area around the Alps. This kind of climate is also typical of Colorado in the central US, another area where solar-heated houses abound.

The broad view of passive solar heating is really about the subtle influence of climate on building design. Without this appreciation, it is all too easy to design buildings that are inappropriate to their surroundings.

As the Roman architect Vitruvius said in the first century BC.

> We must begin by taking note of the countries and climates in which homes are to be built if our designs for them are to be correct. One type of house seems appropriate for Egypt, another for Spain ... one still different for Rome, and so on with lands of varying characteristics. This is because one part of the earth is directly under the sun's course, another is far away from it ... It is obvious that designs for homes ought to conform to the diversities of climate.
>
> Cited in Butti and Perlin, 1980

Varieties of solar heating system

In practice, the categories 'active' and 'passive' are not clear cut: they blend into each other, with a whole range of possibilities in between. The following examples illustrate the range of solar heating systems available in addition to the roof-mounted solar water heaters that we have already considered.

Swimming pool heating

For swimming pool heating, the solar system can be extremely simple. Pool water is pumped through a large area of collector, usually unglazed. Typically, the collector will be about half the area of the pool itself. The best results are achieved with pools that do not have other forms of heating and are consequently at relatively low temperatures (under 20 °C). The aim may not necessarily be to save energy as much as to make the pool temperature more acceptable to bathers.

Conservatory (or 'sunspace')

A conservatory or greenhouse on the south side of a building can be thought of as a kind of habitable solar collector (see Figure 2.18(a)). Air is the heat transfer fluid, carrying energy into the building behind. The energy store is the building itself, especially the wall at the back of the conservatory.

Trombe wall

With a Trombe wall (named after its French inventor, Felix Trombe), the conservatory is replaced by a thin air space in front of a storage wall. This is a solar collector with the storage immediately behind. Solar radiation warms the store and is radiated into the house in an even fashion from its inner side. In addition, on sunny days, air is circulated through the air space into the house behind. At night and on cold days, the air flow is cut off.

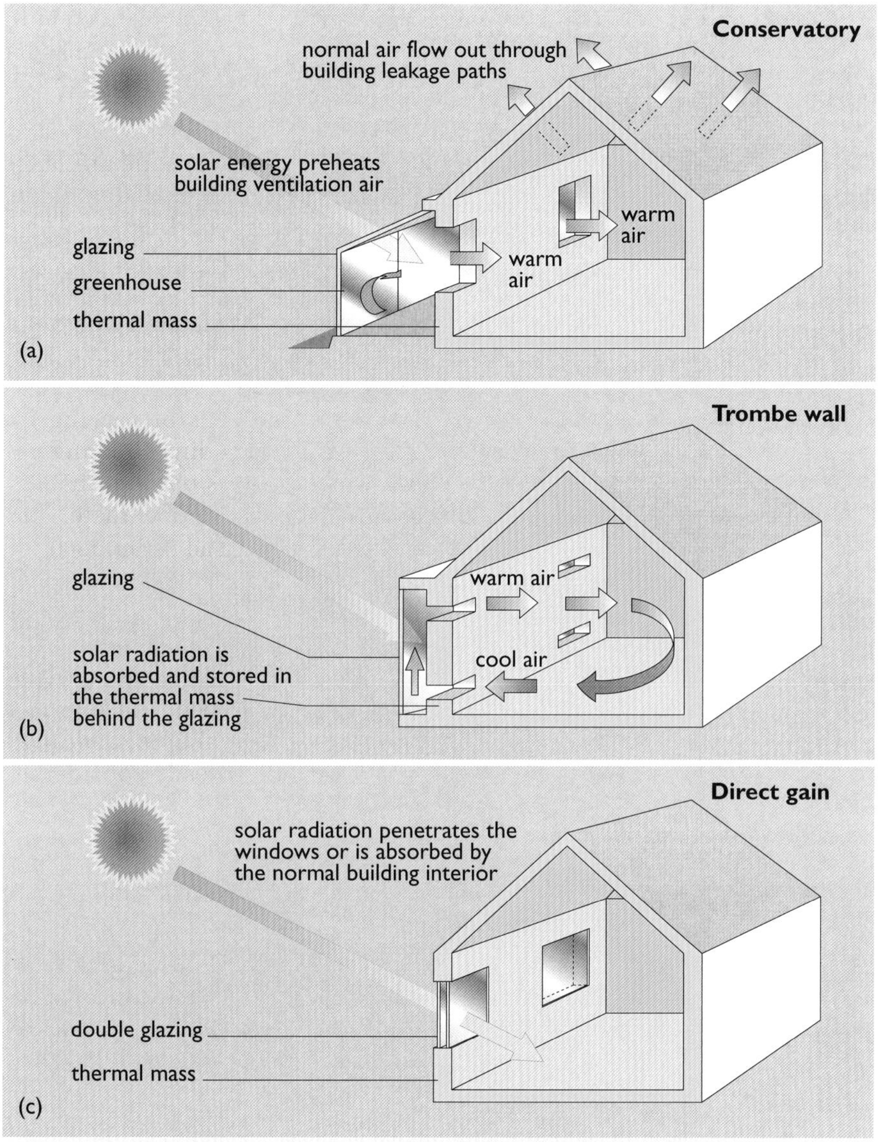

Figures 2.18 Different types of passive solar heating system: (a) Conservatory, (b) Trombe Wall, (c) Direct gain

This concept can take many forms. Small collector panels can be built directly on to the existing walls of buildings. In the extreme, the air path can be omitted, and walls simply covered with 'transparent insulation'.

Direct gain

Direct gain is the simplest and most common of all passive solar heating systems. All glazed buildings make use of this to some degree. The sun's rays simply penetrate the windows and are absorbed into the interior. If the building is 'thermally massive' enough, and the heating system responsive, the gains are likely to be useful. If the building is too 'thermally lightweight', it may overheat on sunny days and the occupants will perceive the effect as a nuisance.

2.6 Active solar heating

History

A solar water heater could be made simply by placing a tank of water behind a normal window. Indeed, many of the first systems produced in the US in the 1890s were little more than this.

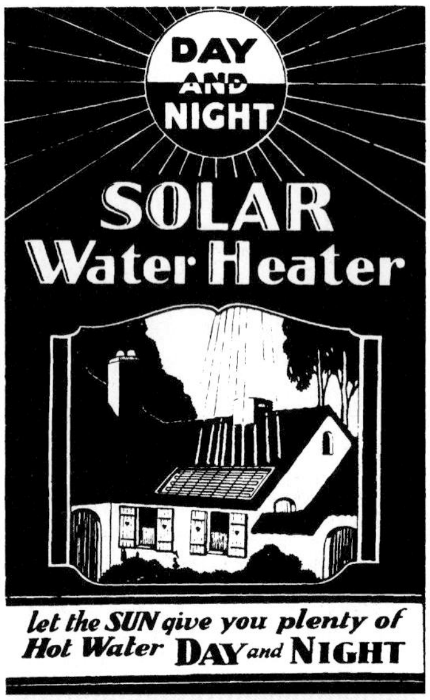

Figure 2.19 An advertisement for Bailey's thermosyphon solar water heaters circa 1915

The thermosyphon solar water heater as we know it was patented in 1909 by William J. Bailey in California. Since the system had an insulated tank, which could keep water hot overnight, Bailey called his business the 'Day and Night' Solar Water Heater Company. He sold approximately 4000 systems before the local discovery of cheap natural gas in the 1920s virtually closed his business.

In Florida, the solar water heating business flourished until the 1940s. Eighty per cent of new homes built in Miami between 1935 and 1941 had solar systems. Possibly as many as 60 000 were sold over this period in this area alone. Yet by 1950 the US solar heating industry had completely succumbed to cheap fossil fuel (Butti and Perlin, 1980).

It was not until the oil price rises of the 1970s that the commercial solar collector reappeared. By 2001 there were at least 57 million m^2 of solar collectors installed world-wide, 11 million m^2 of them in the US, but only 208 000 m^2 in the UK(IEA, 2003, ESTIF, 2003).

We have already considered the basic form of a solar water heating system, but what about the choices involved in selecting the components?

Solar collectors

Just as solar energy systems can have many variants, so can solar collectors. Figures 2.20 and 2.21 summarize most of the possibilities.

Unglazed panels. These are most suitable for swimming pool heating, where it is only necessary for the water temperature to rise by a few degrees above ambient air temperature, so heat losses are relatively unimportant.

Flat plate water collectors. World-wide, these are the mainstay of domestic solar water heating. Usually they are only single glazed but may have an additional second glazing layer, sometimes of plastic. The more elaborate the glazing system, the higher the temperature difference that can be sustained between the absorber and the external air.

The absorber plate usually has a very black surface that absorbs nearly all of the incident solar radiation, i.e. it has a high **absorptivity**. Most normal black paints still reflect approximately 10% of the incident radiation (a white surface, by way of comparison might reflect back 70–80%). Some panels use a selective surface that has both high absorptivity in the visible region and low emissivity in the long-wave infrared, to cut heat losses.

Many designs of absorber plate have been tried with success in recent years, including pressed steel central heating radiators, specially made pressed aluminium panels and small-bore copper pipes soldered to thick copper or steel sheet. Generally, an absorber plate must have high thermal conductivity, to transfer the collected energy to the water with minimum temperature loss.

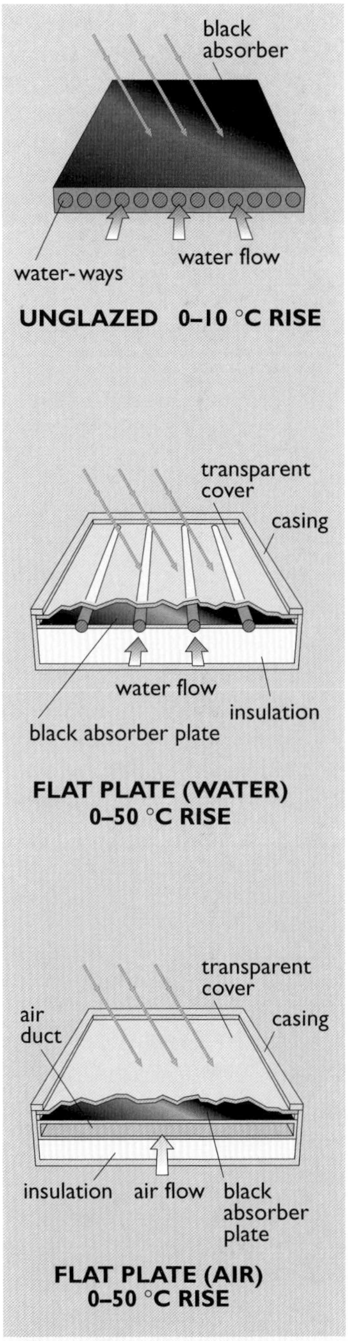

Figure 2.20 Solar collectors for low temperature collection

Flat plate air collectors. These are not so common as water collectors and are mainly used for space heating only. An interesting variant is to combine this type of collector with a photovoltaic panel, producing both heat and electricity.

Evacuated tube collectors. The example shown takes the form of a set of modular tubes similar to fluorescent lamps. The absorber plate is a metal strip down the centre of each tube. Convective heat losses are suppressed by virtue of a vacuum in the tube. The absorber plate uses a special 'heat pipe' to carry the collected energy to the water, which circulates along a header pipe at the top of the array.

A **heat pipe** is a device that takes advantage of the thermal properties of a boiling fluid to carry large amounts of heat. A hollow tube is filled with a liquid at a pressure chosen so that it can be made to boil at the 'hot' end, but the vapour will condense at the 'cold' end. The tube in effect has a thermal conductivity many times greater than if it had been made of solid metal, and is capable of transferring large amounts of heat for a small temperature rise.

Line focus collectors. These focus the sun on to a pipe running down the centre of a trough. They are mainly used for generating steam for electricity generation. The trough can be pivoted to track the sun up and down or east to west. A line focus collector can be oriented with its axis in either a horizontal or a vertical plane.

Point focus collectors. These are also used for steam generation or driving Stirling engines, but need to track the sun in two dimensions.

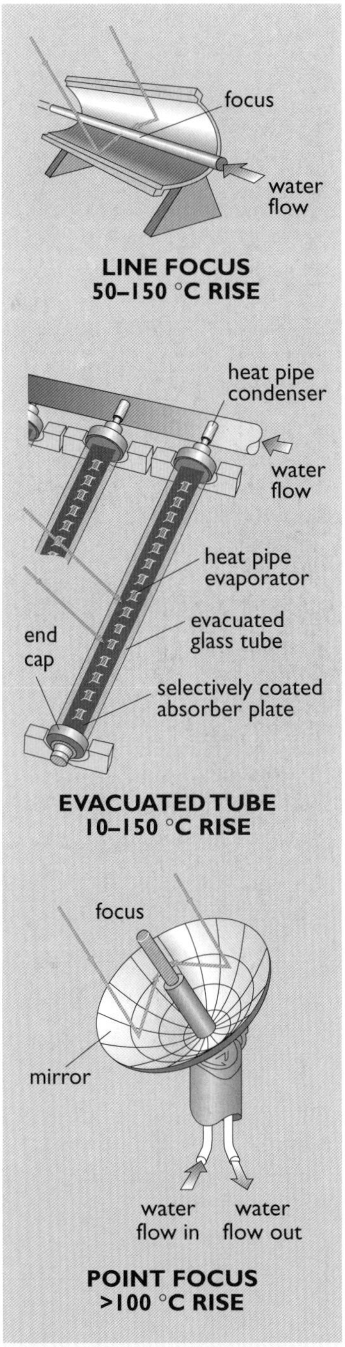

Figure 2.21 Solar collectors for medium to high temperature collection

Robustness, mounting and orientation

Solar collectors are usually roof mounted and once installed are usually difficult to reach for maintenance and repairs. They must be firmly attached to the roof in a leak-proof manner and then must withstand everything that nature can throw at them – frost, wind, acid rain, sea spray and hailstones. They also have to be proof against internal corrosion and very large temperature swings. A double-glazed collector is potentially capable of producing boiling water in high summer if the heat is not carried away fast enough. It is quite an achievement to make something that can survive up to 20 or more years of this treatment. However, a 1995 survey concluded that 85% of the 49 000 systems that had been installed in the UK by that date were still in working order (Sadler, 1996).

Fortunately, as we have seen, panels do not have to be installed to a precise tilt or orientation for acceptable performance. This in turn means that a large portion of the current building stock, possibly 50% or more, could support a solar collector.

Active solar space heating

Figure 2.22 Experimental active solar house in Milton Keynes seen in 1979

So far, we have looked in detail at domestic solar water heaters with only a few square metres of collector. If a far larger collector together with a much larger storage tank were fitted, solar energy should, in theory, be able to supply far more of the annual low-temperature needs of a building. In the 1970s, a number of experimental systems were built with this aim around the world. One example (shown in Figure 2.22) was constructed in Milton Keynes in 1975 and operated until mid-1994, experiencing a climate typical of central England. Thirty-six square metres of single-glazed collector were fitted to the roof of a relatively poorly insulated house. The energy store took the form of two insulated water tanks of 4.5 m^3 total capacity.

Monitoring showed that the system supplied half of the low-temperature heating needs, but that the bulk of the energy was used for domestic hot water. Only a small proportion of the space heating energy came from collected active solar energy, and that only at the sunnier ends of the heating season. In fact, almost as much came from passive solar gains through the windows. Although the system worked largely as designed, it highlighted some basic problems.

1 In order to collect enough solar energy to supply the winter demand, the collector would have to be very large. This would mean that over much of the summer its potential output would not be used because the demand would not be there, and the capital expenditure would effectively be wasted.

2 If the house had been better insulated, it would not have required so much space heating energy, and what it did consume could have been better met by passive solar means.

Here we have the key problem: whether to use renewable energy to heat a poorly insulated house, or use energy conservation to cut demands so that it is not necessary to supply so much energy. Subsequent analysis and experiment showed that the same saving in fossil fuel could be obtained from a well-insulated passive solar-heated house for a fraction of the cost.

Similar calculations have been made for solar-heated houses in Germany and France. Energy conservation and renewable energy supply must be treated on an equal footing.

Interseasonal storage and solar district heating

It is tempting to think that if the storage tank of a solar-heated house was made large enough, the summer sun could be saved through to the winter. This is known as **interseasonal** storage. However, the difficulties of this should not be underestimated. The volume of hot water storage needed to supply a house is almost the same size as the house itself. Also such a storage tank might need insulation half a metre thick to retain most of its heat from summer to winter. In order to reduce the ratio of surface area (and hence heat loss) to volume it pays to make the storage tank very large. This implies that such schemes are likely to be most useful for large buildings or district heating schemes. Even on this scale, though, payback times appear to be very long.

However, there are considerable economies of scale for large projects where solar collectors can be purchased and erected in bulk. Since the 1980s there has been a steady stream of construction of large arrays of solar collectors for district heating systems in mainland Europe, mainly in Denmark, Sweden and Germany. By late 2002, there were a total of 65 schemes having more than 500 m^2 of collector area. These systems had a total area of 110 000 m^2 and an estimated total peak thermal output of 50 MW (Dalenbäck, 2002). Most of these schemes serve residential buildings.

Although the arrays can be very large (such as the 8000 m^2 array shown in Figure 2.23) they usually only supply a small percentage of the total annual heat load. Essentially they can be thought of as shared domestic solar water heating systems.

Figure 2.23 An 8000 m^2 array of collectors feeding a district heating system at Marstal in Denmark (photo courtesy Arcon)

2.7 Passive solar heating

History

All glazed buildings are already to some extent passively solar heated – effectively they are live-in solar collectors. The art of making the best use of this dates back to the Romans, who put glass to good use in their favourite communal meeting place, the bath house. Window openings 2 m wide and 3 m high have been found at Pompeii.

After the fall of the Roman Empire, the ability to make really large sheets of glass vanished for over a millennium. It was not until the end of the seventeenth century that the plate glass process reappeared in France, allowing sheets 2 m square to be made.

Even so, cities of the eighteenth and nineteenth centuries were overcrowded and the houses ill-lit. It was not until the late nineteenth century that pioneering urban planners set out to design better conditions. They became obsessed with the medical benefits of sunlight after it was discovered that ultraviolet light killed bacteria. Sunshine and fresh air became the watchwords of 'new towns' in the UK like Port Sunlight near Liverpool, built to accommodate the workers of a soap factory.

The planners then did not realise that ultraviolet light does not penetrate windows, but the tradition of allowing access for plenty of sunlight continues, reinforced by findings that exposure to bright light in winter is essential to maintain human hormone balances. Without it, people are likely to develop mid-winter depression.

Given the UK's plentiful supply of coal, there was little interest in using solar energy to cut fuel bills until recent years. The construction of the Wallasey School building in Cheshire in 1961, inspired by earlier US and French buildings, was thus something of a novelty (see Figures 2.24 and 2.25).

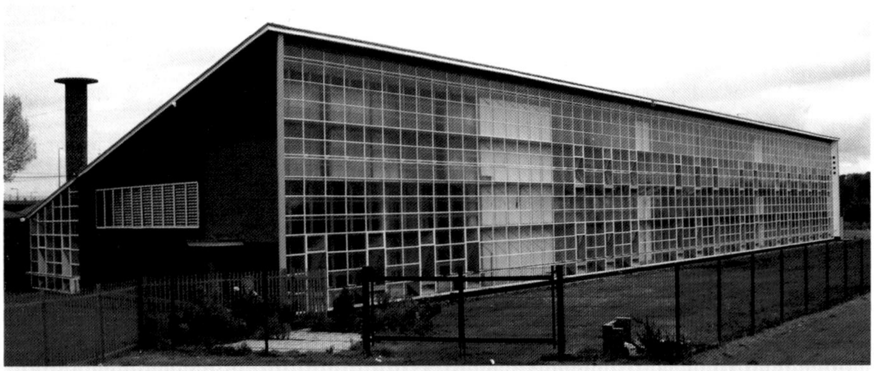

Figure 2.24 Wallasey School, Cheshire, UK – built in 1961

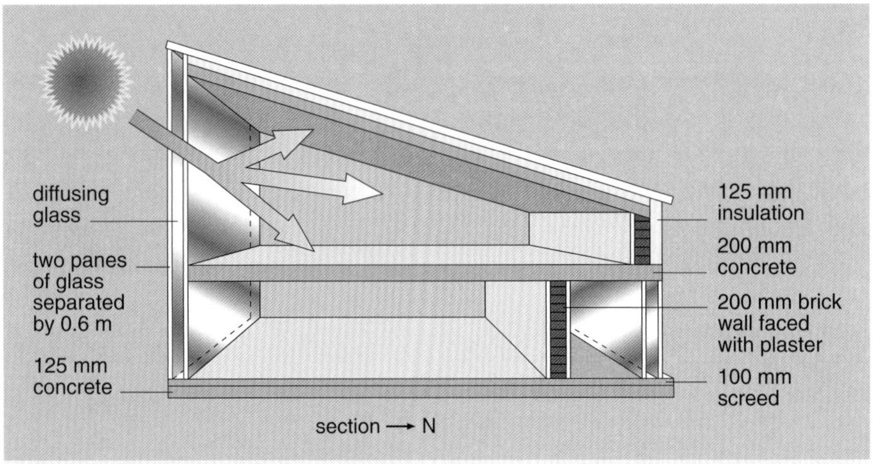

Figure 2.25 Wallasey School – section

Direct gain buildings as solar collectors

The Wallasey School building is a classic direct gain design. It has the essential features required for passive solar heating:

1 a large area of south-facing glazing to capture the sunlight;

2 thermally heavyweight construction (dense concrete or brickwork). This stores the thermal energy through the day and into the night;

3 thick insulation on the outside of the structure to retain the heat.

After its construction, the oil-fired heating system originally installed was found to be unnecessary and was for a time removed, leaving the building totally heated by a mixture of solar energy, heat from incandescent lights and the body heat of the students.

Passive solar heating versus superinsulation

Although the Wallasey School building is one style of low-energy building, there are others. The Wates house, built at Machynlleth in Wales in 1975 (Figure 2.26) was one of the first 'superinsulated' buildings in the UK. It features 450 mm of wall insulation and small quadruple-glazed windows. This was a radical design, given that at this time normal UK houses were built with single glazing, no wall insulation, and new Building Regulations requiring a mere 25 mm of loft insulation were only just being introduced.

Figure 2.26 The superinsulated Wates House at Machynlleth

Situated low in a mountain valley, the Wates house is certainly not well placed for passive solar heating. In fact, it was intended to be heated and lit by electricity from a wind turbine.

Which of the two design approaches – passive solar or superinsulation – is better? There are no easy answers to this question. The art of design for passive solar heating is to understand the energy flows in a building and make the most of them. There need to be sufficient solar gains to meet winter heating needs. These can be reduced by good levels of insulation. Solar energy is also needed to provide adequate lighting, but not so much in summer that there is overheating.

Window energy balance

We can think of a south-facing window as a kind of passive solar heating element. Solar radiation enters during the day, and, if the building's internal temperature is higher than that outside, heat will be conducted, convected and radiated back out.

The question is whether more heat flows in than out, so that the window provides a net energy benefit. The answer depends on several things:

1 the building's average internal temperature;

2 the average external temperature;

3 the available solar radiation;

4 the transmittance characteristics of the window, its orientation and shading;

5 the U-value of the window which is in turn dependent on whether it is single or double glazed (or even better insulated).

Figure 2.27 shows the average monthly 'energy balance' of a south-facing window in the vicinity of London for a building with an average internal temperature of 18 °C. In the dull, cold months of December and January, both single- and double-glazed windows can be net energy losers. However, in the autumn and spring months, November and March, a double-glazed window becomes a positive contributor to space heating needs. Its performance can be further improved by insulating it at night.

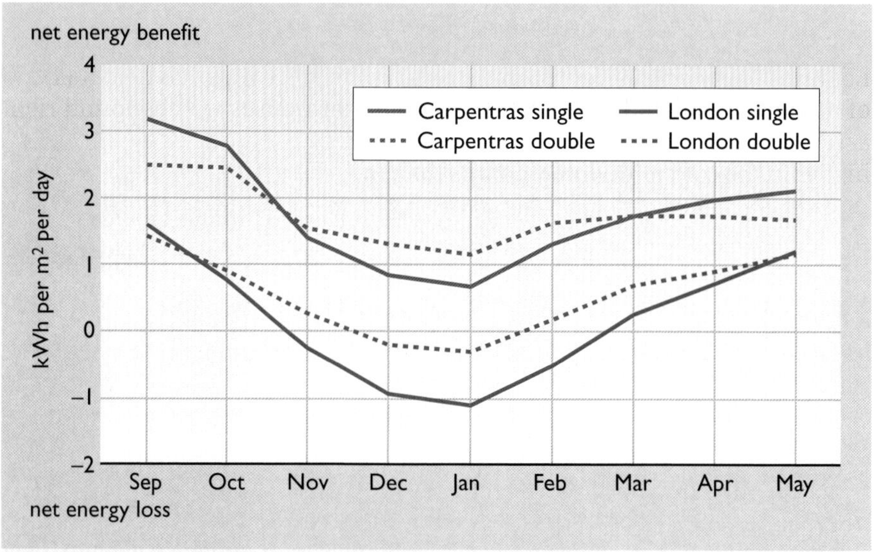

Figure 2.27 Window energy balance London and Carpentras

We can compare this with a similar energy balance for Carpentras, near Avignon, in the south of France (also shown in Figure 2.27). Although the mid-winter months there are almost as cold as in London, they are far sunnier, being at a lower latitude. The incoming solar radiation is far greater than the heat flowing out, even in mid-winter, and the energy balance is markedly positive.

To return to the UK, we need to consider how best use can be made of the solar energy available.

With a long heating season, a south-facing double-glazed window is a good thing. It can perhaps supply extra heat during October and November, March and April. On the other hand, with a very short heating season confined to the dullest months, say just December and January, it is not really much use at all.

How long is the heating season?

In order to answer this question, we must consider the rest of the building, its insulation standards and its so-called 'free' heat gains.

In a typical house, to keep the inside warmer than the outside air temperature, it is necessary to inject heat. The greater the temperature difference between the inside and the outside, the more heat needs to be supplied. In summer it may not be necessary to supply any heat at all, but in mid-winter large amounts will be needed. The total amount of heat that needs to be supplied over the year can be called the **gross heating demand.**

This will have to be supplied from three sources:

1 'free heat gains', which are those energy contributions to the space heating load of the building from the normal activities that take place in it: the body heat of people, and heat from cooking, washing, lighting and appliances. Taken individually, these are quite small. In total, they make a significant contribution to the total heating needs. In a typical UK house, this can amount to 15 kWh per day;

2 passive solar gains, mainly through the windows;

3 fossil fuel energy, from the normal heating system.

Let us now consider, for example, the monthly average gross heat demand of a poorly insulated 1970s UK house (similar houses will be found right across Northern and Central Europe). As shown in Figure 2.28, this will be higher in the cold midwinter months than in the warmer spring and autumn. In summer, when the outside air temperature is high, this heating requirement almost drops to zero.

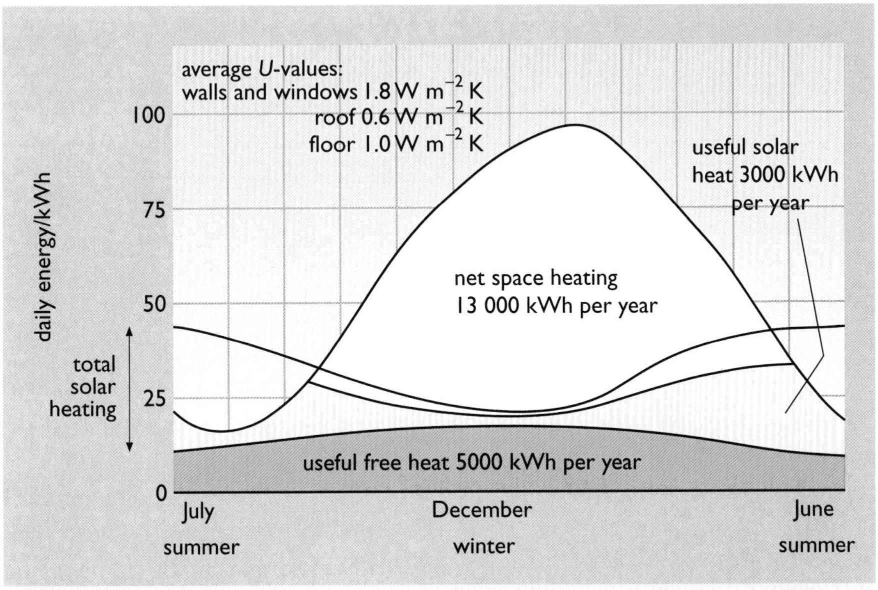

Figure 2.28 Contribution to the net space heating demand in a typical poorly insulated UK house of the 1970s

As shown in Figure 2.28, for this particular house, over the whole year, out of a total gross heating demand of 21 000 kWh, 5000 kWh come from free heat gains and 3000 kWh from solar gains (we have assumed to count free heat gains before solar gains).

Put another way, a perfectly ordinary house is already 14% passive solar heated. The **net heating demand**, to be supplied by the normal fossil fuel heating system, is simply the outstanding heat requirement, namely 13 000 kWh. This will have to be supplied from mid-September to the end of May.

It is possible to cut the house heat demand by putting in cavity wall and loft insulation and double instead of single glazing. This will reduce the gross heating demand and, as shown in Figure 2.29, allow the free heat gains and normal solar gains to maintain the internal temperature of the house for a longer period of the year. (The insulation levels chosen are typical of new Danish houses of the early 1980s, a standard eventually roughly equalled by the UK Building Regulations of 2002!)

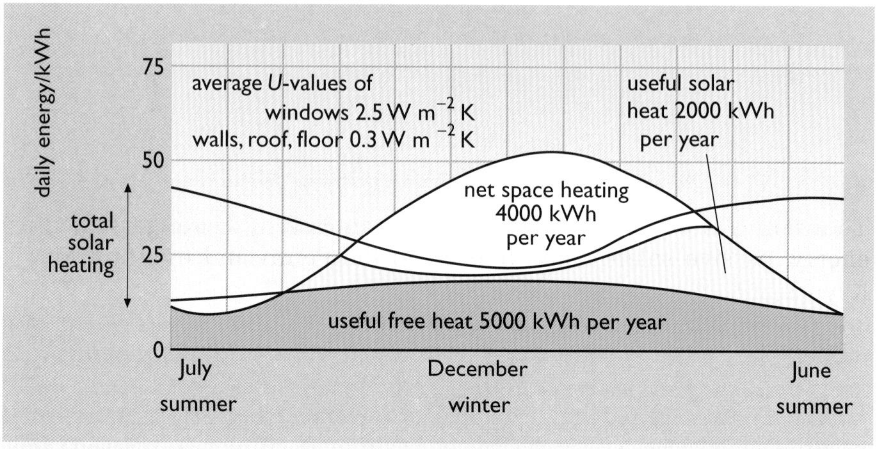

Figure 2.29 Contribution to net space heating demand in a house of normal design but well insulated

The heating season will then be reduced to between October and the end of April. Out of a total gross heat demand of 11 000 kWh, 5000 kWh still come from free heat gains, but as a result of the improved insulation, it will be possible to utilize only 2000 kWh of solar gains. Finally, 4000 kWh will remain to be supplied from the normal heating system.

By insulating the house, 9000 kWh per year will have been saved in fossil fuel heating, but the solar contribution (using our slightly arbitrary accounting system) will have fallen from 3000 to 2000 kWh.

There are two ways in which the space heating demand could be cut further.

1 By providing extra insulation. If the house was 'superinsulated', using insulation of 200 mm or more thick, the space heating load might disappear almost completely, leaving just a small need on the coldest, dullest days. Solar gains might not be needed.

2 By providing appropriate glazing to ensure that the best use is made of the mid-winter sun.

Which of these methods is chosen will depend on the local climate and the relative expense of insulation materials and glazing. Depending on the precise circumstances, it may be a lot easier to collect an extra 100 kWh of solar energy than to save 100 kWh with extra insulation. It also depends on the desired aesthetics of the building and the need for natural daylight inside.

General passive solar heating techniques

There are some basic general rules for optimizing the use of passive solar heating in buildings.

1 They should be well insulated to keep down the overall heat losses.

2 They should have a responsive, efficient heating system.

3 They should face south (anywhere from south-east to south-west is fine). The glazing should be concentrated on the south side, as should the main living rooms, with little-used rooms such as bathrooms on the north.

4 They should avoid overshading by other buildings in order to benefit from the essential mid-winter sun.

5 They should be 'thermally massive' to avoid overheating in summer.

These rules were used, broadly in the order above, to design some low-energy, passive solar-heated houses on the Pennyland estate in Milton Keynes in central England in the late 1970s. The design steps (see Figure 2.30) were carefully costed and the energy effects evaluated by computer model.

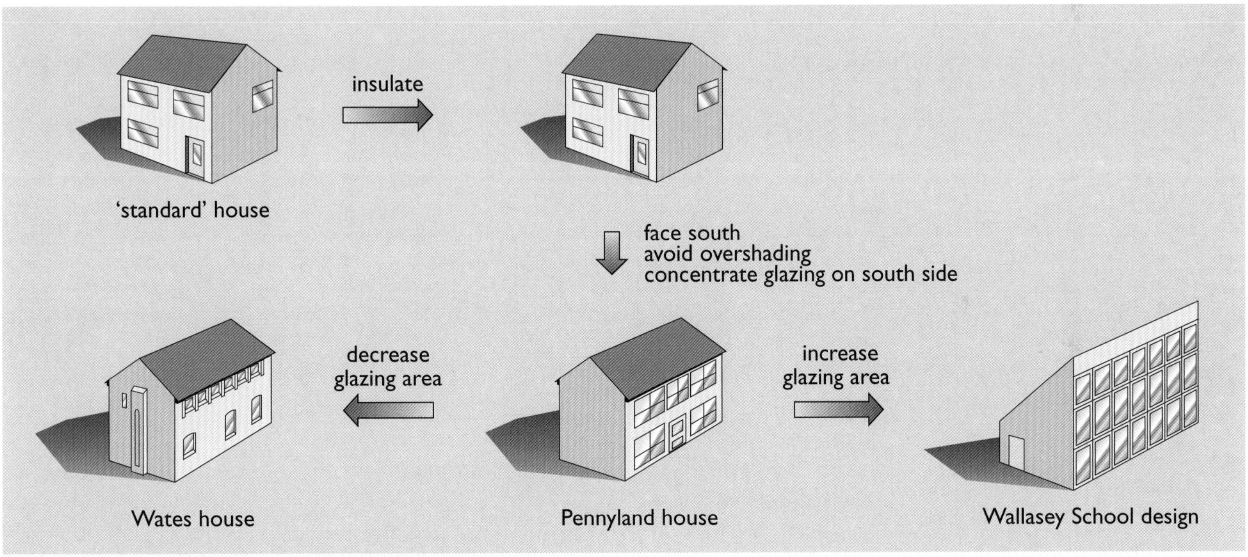

Figure 2.30 Design steps in low-energy housing

The resulting houses had a form that was somewhere between the Wallasey School building and the Wates house. There was not too much glazing, but not too little (see Figures 2.31 and 2.32 overleaf).

An entire estate of these houses was built and the final product carefully monitored. At the end of the exercise it was found that the steps 1–5 listed above produced houses that used only half as much gas for low-temperature heating as 'normal' houses built in the preceding year. The extra cost was 2.5% of the total construction cost and the payback time was four years.

Figure 2.31 Passive solar housing at Pennyland – south elevation – the main living rooms have large windows and face south

Figure 2.32 Passive solar housing at Pennyland – the north side has smaller windows

Figure 2.33 Passive solar housing at Pennyland – plans

Here we come back to the difference between the 'broad' and 'narrow' definitions of passive solar heating. In its broad sense, it encompasses all the energy-saving ideas (1–5 above) put into these houses. In its narrow sense, it covers only the points that are rigidly solar based (3–5).

In this project, insulation and efficient heating saved the vast bulk of the energy, but approximately 500 kWh per year of useful space heating energy came from applying points 3–5 (Everett, *et al.*, 1986).

Put another way, this 500 kWh is the difference in energy consumption between a solar and a non-solar house of the same insulation standard. We can call this figure the **marginal passive solar gain**. As we saw in Figures 2.28 and 2.29 above, even non-solar houses have some solar gains. What we are doing is trying to maximize them.

It is rather difficult to calculate the extra cost involved in producing marginal passive solar gains. After all, the passive solar 'heater' is an integral part of the building, not a bolt-on extra. Careful costing studies of different building designs and layouts have shown that modest marginal solar gains can be had at minimal extra cost.

Essentially, in its narrow sense, passive solar heating is largely free, being simply the result of good practice. In the 'wider' sense of integrated low energy house design, the energy savings have to be balanced against the cost of a whole host of energy conservation measures, some of which involve glazing and perhaps have a solar element, and others which do not.

The balance between insulation savings and passive solar gains is also highly dependent on the local climate. In practice the most ambitious passive solar buildings are built in climates that have high levels of sunshine during cold winters.

Conservatories, greenhouses and atria

Direct gain design is really for new buildings: it cannot do much for existing ones. However, for many old buildings, conservatories or greenhouses could be added on to the south sides, just as they can be incorporated into new buildings.

Figure 2.34 shows a typical example of a conservatory added to a poorly insulated nineteenth century terraced house also in Milton Keynes. Monitoring of its performance suggested that it would save 800 kWh per year of space heating energy. However, not all of this is 'solar' energy:

- 15% is due to the thermal buffering of the house, because the conservatory acts as extra insulation to the south side of the house;
- 55% is due to preheating the ventilation air to the house. Fresh air entering the house via the conservatory will be warmer than air entering another way;
- 30% is from normal solar gains entering the house via the greenhouse, principally by conduction through the intervening wall.

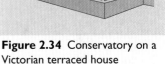

Figure 2.34 Conservatory on a Victorian terraced house

An extra 150 kWh per year could be obtained by using a fan to pump warm air into the house when the greenhouse became warm enough (Ford, 1982).

Add-on conservatories and greenhouses are expensive and cannot normally be justified on energy savings alone. Rather, they are built as extra areas of unheated habitable space. A strong word of caution is necessary here. A conservatory only saves energy if it is not heated like other areas of the house. There is a danger that it will be looked on as just another room and equipped with radiators connected to the central heating system. One house built like this can easily negate the energy savings of 10 others with unheated conservatories.

The costs can be reduced for new buildings if they are integrated into the design. The new Hockerton housing estate in Nottinghamshire (see Figure 2.35 overleaf) combines a 'Wallasey school' type design with a full-width conservatory (see EST, 2003). Not only do the walls and roof have 300 mm of insulation, but the rear of the houses is also built up with earth, giving extra thermal mass and protection from the worst winter weather. This is known as **earth sheltering.**

Glazed atria are also becoming increasingly common. At their simplest, they are just glazed-over light wells in the centre of office buildings. At the other extreme, entire shopping streets can be given a glazed roof, creating an unheated but well-lit circulation space.

Figure 2.35 The Hockerton houses feature a full width conservatory and earth sheltering

Figure 2.36 Trombe wall bungalows at Bebington near Liverpool

Trombe walls

As we saw in Figure 2.18(b) above, in a Trombe wall the conservatory or greenhouse is replaced by a thin, glazed, air space with the thermal storage immediately behind. The original designs were built in the 1950s in the south of France, but the idea has been tested in the UK (see Figure 2.36)

This is perhaps a technology that works best in sunnier climates, since the bulk of the building is hidden from the sun behind the storage wall, and without careful design internal lighting can be poor and direct solar gains blocked out. Other experimental variants of this are wall- or roof-mounted air collectors. As with water-based active solar space heating, we have to ask whether or not better overall cost-effectiveness could be achieved by investing in simple insulation measures.

Avoiding overshading

One important aspect of design for passive solar heating is to make sure that the mid-winter sun can penetrate to the main living spaces without being obstructed by other buildings. This will require careful spacing of the buildings.

There are many design aids to doing this, but a useful tool is the **sunpath diagram** (see Figure 2.37). For a given latitude, this shows the apparent path of the sun through the sky as seen from the ground.

In practice, the contours of surrounding trees and buildings can be plotted on it to see at what times of day during which months the sun will be obscured. The Pennyland houses were laid out so that the midday sun in mid-December just appeared over the roof-tops of the houses immediately to the south.

However, we need to ask whether it is advisable to cut down all offending overshading trees in the area to let the sun through. To obtain maximum benefit from passive solar heating in its broad sense, it is necessary to follow another rule:

> Houses should be sheltered from strong winter winds.

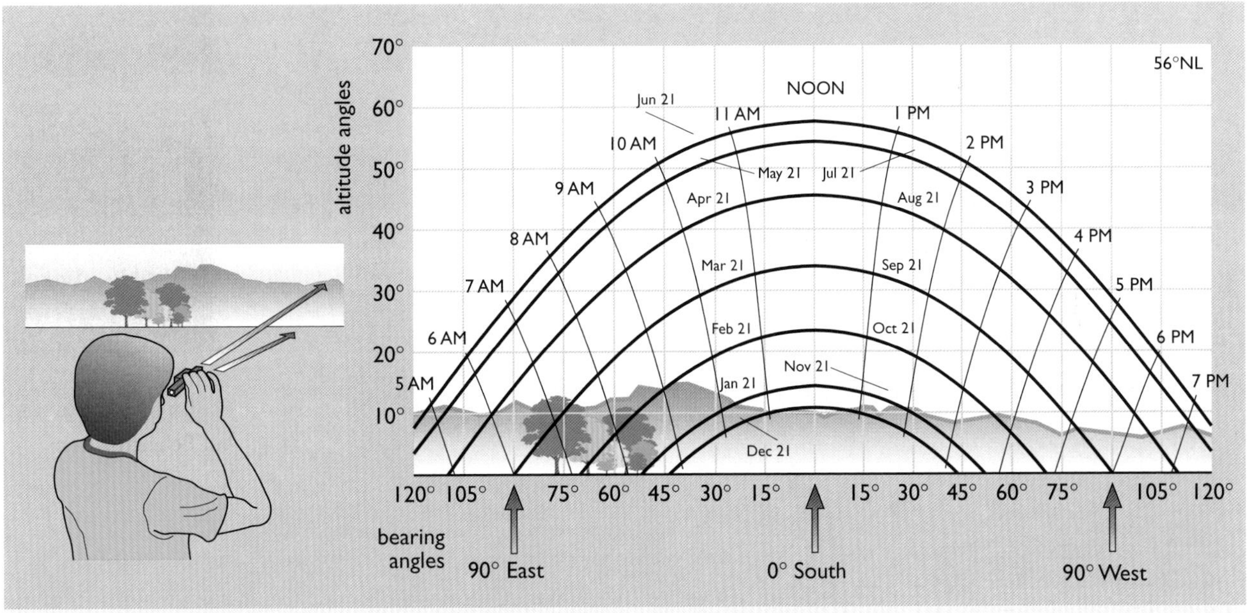

Figure 2.37 Plotting the skyline on a sunpath diagram can give important information about overshading. The sunpath diagram shown is for 56° N, which is approximately the latitude of Glasgow or Edinburgh

Computer modelling suggests that, in houses such as those built at Pennyland, sheltering can produce energy savings of the same order of magnitude as marginal passive solar gains, approximately 500 kWh per year per house.

Where do the winter winds come from? This is immensely dependent on the local micro-climate of the site. In large parts of the UK, the prevailing wind is from the south-west. It would thus be ideal if every house could have a big row of trees on its south-west side. But is it possible to provide shelter from the wind without blocking out the winter sun? This is where housing layout becomes an art. Every site is different and needs solutions appropriate to it.

2.8 Daylighting

Daylight is a commodity that we all take for granted. Replacing it with artificial light was, before the middle of the twentieth century, very expensive (a topic discussed in the companion text). With the coming of cheap electricity, daylight has been neglected and most modern office buildings are designed to rely heavily on electric light.

Houses are traditionally well designed to make use of natural daylight. Indeed, most of those that were not have long ago been designated slums and duly demolished. In the UK in 2000, domestic lighting accounted for only approximately 3% of the delivered energy use, and even this could be cut by the use of low-energy fluorescent lamps.

In some commercial offices, however, lighting can account for up to 30% of the delivered energy use. Modern factory units and hypermarket buildings are built with barely any windows. In 'deep-plan' office buildings, such as

Figure 2.38 Mirrors used to catch valuable daylight in narrow London streets before World War II

Figure 2.39 Modern deep-plan office buildings, such as Canary Wharf Tower above, require continuous artificial lighting in the centre, which may create overheating in summer

Canary Wharf Tower in London (Figure 2.39), there are many offices, central corridors and stairwells that require continuous lighting, even when the sun is shining brightly outside.

Although in winter the heat from lights can usefully contribute to space heating energy, in summer (when there is most light available) it can cause overheating, especially in well-insulated buildings. Making the best use of natural light saves both on energy and on the need for air conditioning.

Daylighting is a combination of energy conservation and passive solar design. It aims to make the most of the natural daylight that is available. Many of the design details will be found in the better quality nineteenth century buildings. Traditional techniques include:

- shallow-plan design, allowing daylight to penetrate all rooms and corridors
- light wells in the centre of buildings;
- roof lights;
- tall windows, which allow light to penetrate deep inside rooms;
- the use of task lighting directly over the workplace, rather than lighting the whole building interior

Many of these techniques have been used in the new Wansbeck Hospital, Northumberland, shown in Figure 2.40

Figure 2.40 These shallow-plan buildings of Wansbeck Hospital in Northumberland have been laid out to allow natural light to penetrate into most of the wards

Other experimental techniques include the use of steerable mirrors to direct light into light wells, and the use of optical fibres and light ducts.

When artificial light has to be used, it is important to make sure that it is used efficiently and is turned off as soon as natural lighting is available. Control systems can be installed that reduce artificial lighting levels when photoelectric cells detect sufficient natural light. Payback times on these energy conservation techniques can be very short and savings of 50% or more are feasible.

In designing new buildings, there is a conflict between lighting design and thermal design. Deep-plan office buildings have a smaller surface area per unit volume than shallow-plan ones. They will need less heating in winter. As with all architecture, there are seldom any simple answers and compromises usually have to be made.

2.9 Solar thermal engines and electricity generation

So far, we have considered only low-temperature applications for solar energy. If the sun's rays are concentrated using mirrors, high enough temperatures can be generated to boil water to drive steam engines. These can produce mechanical work for water pumping or, more commonly nowadays, for driving an electric generator.

The systems used have a long history and many modern plants differ little from the prototypes built 100 years ago. Indeed, if cheap oil and gas had not appeared in the 1920s, solar engines might have developed to be commonplace in sunny countries.

Legend has it that in 212 BC Archimedes used the reflective power of the polished bronze shields of Greek warriors to set fire to Roman ships besieging the fortress of Syracuse. Although long derided as myth, Greek navy experiments in 1973 showed that 60 men each armed with a mirror 1 m by 1.5 m could indeed ignite a wooden boat at 50 m.

If each mirror perfectly reflected all its incident direct beam radiation squarely on to the same target location as the other 59 mirrors, the system could be said to have a **concentration ratio** of 60. Given an incident direct beam intensity of, say, 800 W m^{-2}, the target would receive 48 kW m^{-2}, roughly equivalent to the energy density of a boiling ring on an electric cooker.

The most common method of concentrating solar energy is to use a parabolic mirror. All rays of light that enter parallel to the axis of a mirror formed in this particular shape will be reflected to one point, the focus.

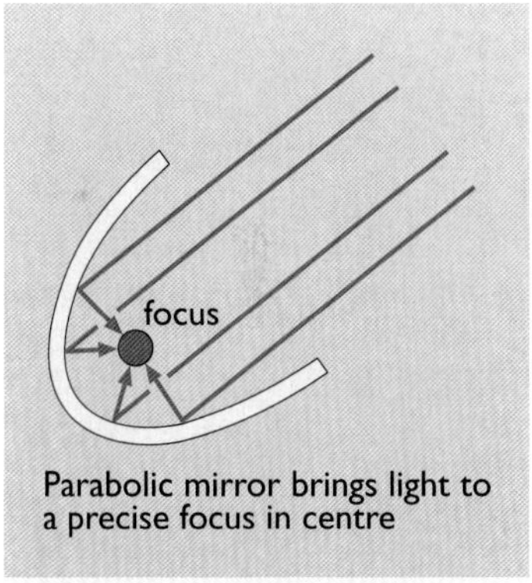
focus
Parabolic mirror brings light to a precise focus in centre

Figure 2.41 A parabolic mirror brings light to a precise focus in its centre

However, if the rays enter slightly off-axis, they will not pass through the focus. It is therefore essential that the mirror tracks the sun.

As shown in Figures 2.20 and 2.21 earlier, these mirrors can be made in either line focus or point focus forms. In the line focus form (sometimes called a trough collector), the image of the sun can be concentrated on a small region running along the length of the mirror. In order to keep the sun focused on this, the collector normally faces south and needs only to track the sun in **elevation**, that is, up and down.

In the point focus form, the image is concentrated on a boiler in the centre of the mirror. For optimum performance, the axis must be pointed directly at the sun at all times, so it needs to track the sun both in elevation and in **azimuth** (that is, side to side).

Most mirrors are assembled from sheets of curved or flat glass fixed to a framework.

There are trade-offs between the complexity of design of a concentrating system and its concentration ratio. A well-built and well-aimed parabolic collector can achieve a concentration ratio of over 1000. A line focus parabolic trough collector may achieve a concentration ratio of 50, but this is adequate for most power plant systems. The ratio required depends on the desired target temperature.

Unless the incident solar energy is carried away by some means, the target, be it a boat or a boiler, will settle at an equilibrium temperature where the incoming radiation balances heat losses to the surrounding air. The latter will be mainly by convection and re-radiation of infrared energy and will be dependent on the surface area of the target and its exposure to wind. A line focus parabolic trough collector can produce a temperature of 200–400 °C. A dish system can produce a temperature of over 1500 °C.

What is important to appreciate is that no concentrator can deliver in total any more energy than falls on it, but what it does receive is all concentrated into one small area.

The first solar engine age

The process of converting the concentrated power of the sun into useful mechanical work started in the nineteenth century. When, in the 1860s, France lacked a supply of cheap coal, Augustin Mouchot, a mathematics professor from Tours, had the answer: solar-powered steam engines. In the 1870s and 1880s, Mouchot and his assistant, Abel Pifre, produced a series of machines ranging from the solar printing press shown in Figure 2.42 to solar wine stills, solar cookers and even solar engines driving refrigerators.

Figure 2.42 Abel Pifre's solar-powered printing press

Their basic collector design was a parabolic concentrator with a steam boiler mounted at the focus. Steam pipes ran down to a reciprocating engine (like a steam railway engine) on the ground.

Although these systems were widely acclaimed, they suffered from the fundamental low power density of solar radiation and low overall efficiency.

In order to understand some of the problems of engines powered by solar energy, it is necessary to consider the Second Law of Thermodynamics (see Box 2.4).

BOX 2.4 Heat engines, Carnot efficiency and ORCs

The steam engine is familiar enough. It works by boiling water to produce a high-pressure vapour. This then goes to an 'expander', which extracts energy and from which low-pressure vapour is exhausted. The expander can be a reciprocating engine or a turbine. Such systems are known as **heat engines**.

All heat engines are subject to fundamental limits on their efficiency set by the **Second Law of Thermodynamics** (a topic discussed in more detail in the companion text). They all produce work by taking in heat at a high temperature, T_{in}, and ejecting it at a lower one, T_{out}. In the ideal case, the maximum efficiency they could hope to achieve is given by:

$$\text{maximum efficiency} = 1 - \frac{T_{out}}{T_{in}}$$

where T_{in} and T_{out} are expressed in degrees Kelvin (or degrees Celsius plus 273). This ideal efficiency is known as the **Carnot efficiency**, after the nineteenth-century French scientist Sadi Carnot.

For example, a turbine fed from parabolic trough collectors might take steam at 350 °C and eject heat to cooling towers at 30 °C. Its theoretical efficiency would therefore be

$$1 - (30 + 273)/(350 + 273) = 0.51, \text{ i.e. } 51\%.$$

Its practical efficiency is more likely to be about 25%, due to various losses.

Systems that use turbines are often referred to as **Rankine cycles**, after another pioneer of thermodynamics, William Rankine.

Normally, to boil water, its temperature must be raised to at least 100 °C. This may be difficult to achieve with simple solar collectors. It would be more convenient to work with a fluid with a lower boiling point. In order to do this, a 'closed cycle' system must be adopted, with a condenser that changes the exhaust vapour back to a liquid and allows it to be returned to the boiler.

Systems have been developed that use stable organic chemicals with suitably low boiling points, similar to the refrigerants used in heat pumps (see Box 2.3). One that uses an organic fluid and a turbine is known as an **organic Rankine cycle** or **ORC**. These are commonly used with solar ponds, OTEC systems and some types of geothermal plant as described in Chapter 9.

These low temperature systems are likely to have poor efficiencies. For example, the theoretical Carnot efficiency of a heat engine that was fed with relatively low-temperature vapour at 85 °C, say from a flat plate solar collector, and exhausted heat at 35 °C would be only 14%.

The early French solar steam engines were not capable of producing steam at really high temperatures and as a result their thermal efficiencies were poor. It required a machine that occupied 40 m^2 of land just to drive a one-half horsepower engine (less than the power of a modern domestic vacuum cleaner!).

By the 1890s, it was clear that this was not going to compete with the new supplies of coal in France, which were appearing as a result of increased investment in mines and railways.

At the beginning of the twentieth century, in the USA, an entrepreneur named Frank Shuman applied the principle again, this time with large parabolic trough collectors. He realised that the best potential would be in really sunny climates. After building a number of prototypes, he raised enough financial backing for a large project at Meadi in Egypt. This used five parabolic trough collectors, each 80 m long and 4 m wide. At the focus, a finned cast iron pipe carried away steam to an engine.

In 1913, his system, producing 55 horsepower, was demonstrated to a number of VIPs, including, the British government's Lord Kitchener. Given the expense of coal in Egypt at the time (it had to be imported from the UK), the payback time would have been only four years.

By 1914, Shuman was talking of building 20 000 square miles of collector in the Sahara, which would 'in perpetuity produce the 270 million horsepower required to equal all the fuel mined in 1909' (see Butti and

Perlin, 1980, for the full story). Then came World War I and immediately afterwards the era of cheap oil. Interest in solar steam engines collapsed and lay dormant for virtually half a century.

The new solar age

Solar engines revived with the coming of the space age. When, in 1945, a UK scientist and writer, Arthur C. Clarke, described a possible future 'geostationary satellite', which would broadcast television to the world, it was to be powered by a solar steam engine. In fact, by the time such satellites materialized, 25 years later, photovoltaics (see Chapter 3) had been developed as a reliable source of electricity.

Elsewhere, space rockets, guided missiles and nuclear reactors needed facilities where components could be tested at high temperatures without contamination from the burning of fuel needed to achieve them. The French solved this problem in 1969 by building an eight-storey-high parabolic mirror at Odeillo in the Pyrenees. This faces north towards a large field of heliostats: flat mirrors, which, like those held by Archimedes' warriors, track the sun. This huge mirror could produce temperatures of 3800 °C at its focus, but only in an area of 50 cm^2.

Power towers

In the early 1980s, the first serious, large, experimental electricity generation schemes were built to make use of high temperatures. Several were of the central receiver system or 'power tower' type. The 10 megawatt (MW) Solar One system at Barstow, California, shown in Figure 2.43, is a good example. This uses a field of tracking heliostats, which reflect the sun's rays onto a boiler at the top of a central tower.

Figure 2.43 Barstow central receiver system – heliostat field

Initially the Barstow plant used special high-temperature synthetic oils to carry away the heat to a steam boiler. In the 1990s, it was rebuilt as Solar Two, and between 1996 and 1999 it operated including heat storage using molten salt at over 500 °C. This allowed it to produce electricity potentially on a 24-hour basis. A new tower project is being built near Seville in Spain to explore the use of super-heated air as a heat transfer medium. This will transfer heat at 700 °C from the collector at the top to the tower to an insulated heat store filled with wire mesh and ceramic pellets, and then to a boiler to raise steam for use in a conventional turbine (Osana *et al.*, 2001).

Parabolic trough concentrator systems

Most of the world's solar-generated electricity is produced at a large solar power station developed by Luz International, at Kramer Junction in the Mojave desert in California. Between 1984 and 1990, Luz brought on-line nine Solar Electricity Generating Systems (SEGS) of between 13 and 80 MW rating. These are essentially massively-uprated versions of Shuman's 1913 design, using large fields of parabolic trough collectors (see Figures 2.44 and 2.45). Each successive project has concentrated on increasing economies of scale in purchasing mirror glass and the use of commercially available steam turbines. The latest 80 MW SEGS has 464 000 m^2 of collector area.

Figure 2.44 SEGS solar collector field at Kramer Junction in southern California

Figure 2.45 SEGS solar collector field – aerial view

The collectors heat synthetic oil to 390 °C, which can then produce high-temperature steam via a heat exchanger. On 1 July 1997, the SEGS VI plant reached its all time single day performance record, averaging 18% efficiency over the day and 20% between 9 a.m. and 5 p.m. This is competitive with commercially available photovoltaic systems.

The SEGS plants were intended to compete with fossil fuel generated electricity to feed the peak afternoon air-conditioning demands, and for several years this objective was successfully achieved. In 1992, reductions in the price of gas, to which the price paid for electricity from the plant was tied, brought financial difficulties. However they continue to operate reliably and cheaply, although plans for new plants have been delayed.

Current thinking is that a solar thermal plant should be combined with fossil generation and/or some kind of storage. This would allow generation at night, making better use of the capital investment in the steam turbine of the plant. Also, in order to maximize the thermal efficiency of the turbine, it is desirable to feed it with very high temperature steam. The best way to do this may be for the solar mirrors to act in conjunction with a gas-fired or even coal-fired generation plant. This kind of thinking may offend purists who would like a genuine 100% solar plant, but it would improve the economics of the overall system.

Parabolic dish concentrator systems

Instead of conveying the solar heat from the collector down to a separate engine, an alternative approach is to put the engine itself at the focus of a mirror (see Figure 2.46). This has been tried both with small steam engines and with Stirling engines.

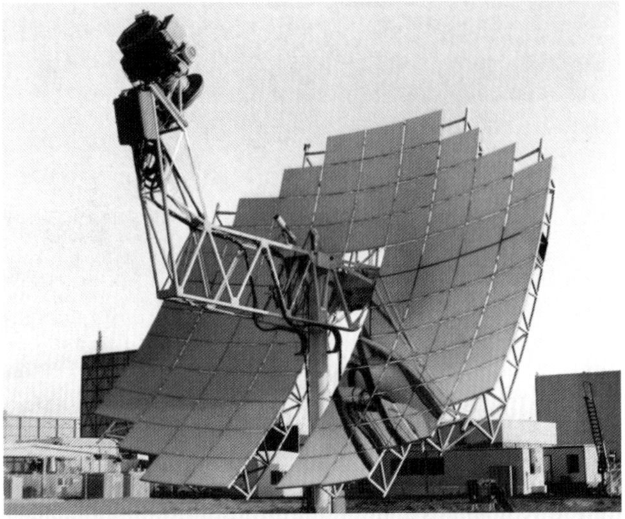

Figure 2.46 Boeing solar-powered Stirling engine

Stirling engines are described in the companion text and have a long history (they were invented in 1816). Although steam engines have fundamental difficulties when operating with input temperatures above 700 °C, Stirling engines, given the right materials, can be made to operate at temperatures of up to 1000 °C, with consequent higher efficiencies. Current experimental solar systems using these have managed very high overall conversion efficiencies, approaching 30% on average over the day.

The parabolic mirrors themselves are also undergoing interesting experimental developments in Germany. Instead of heavy glass mirrors, circular sheets of aluminized plastic film are being used. By creating a partial vacuum behind the film, it can be bent into a parabolic shape. This creates a very lightweight mirror, which in turn only requires a lightweight structure to support and track it.

Solar ponds

A totally different approach to solar thermal electricity production is the solar pond, which uses a large, salty lake as a kind of flat plate collector (Figure 2.47). If the lake has the right gradient of salt concentration (salty water at the bottom and fresh water at the top) and the water is clear enough, solar energy is absorbed on the bottom of the pond.

The hot, salty water cannot rise, because it is heavier than the fresh water on top. The upper layers of water effectively act as an insulating blanket and the temperature at the bottom of the pond can reach 90 °C. This is a high enough temperature to run an organic rankine cycle (ORC) engine.

However, the thermodynamic limitations of the relatively low temperatures mean low solar-to-electricity conversion efficiencies, typically less than 2%. Nevertheless, systems of 50 MW electrical output, fed from a lake of 20 hectares, have been demonstrated.

One advantage of this system is that the large thermal mass of the pond acts as a heat store, and electricity generation can go on day or night, as required.

In practice, the system has disadvantages. Large amounts of fresh water are required to maintain the salt gradient. These can be hard to find in the solar pond's natural location, the desert. Indeed, the best use for solar ponds may be to generate heat for water desalination plants, creating enough fresh water to maintain themselves and also supply drinking water.

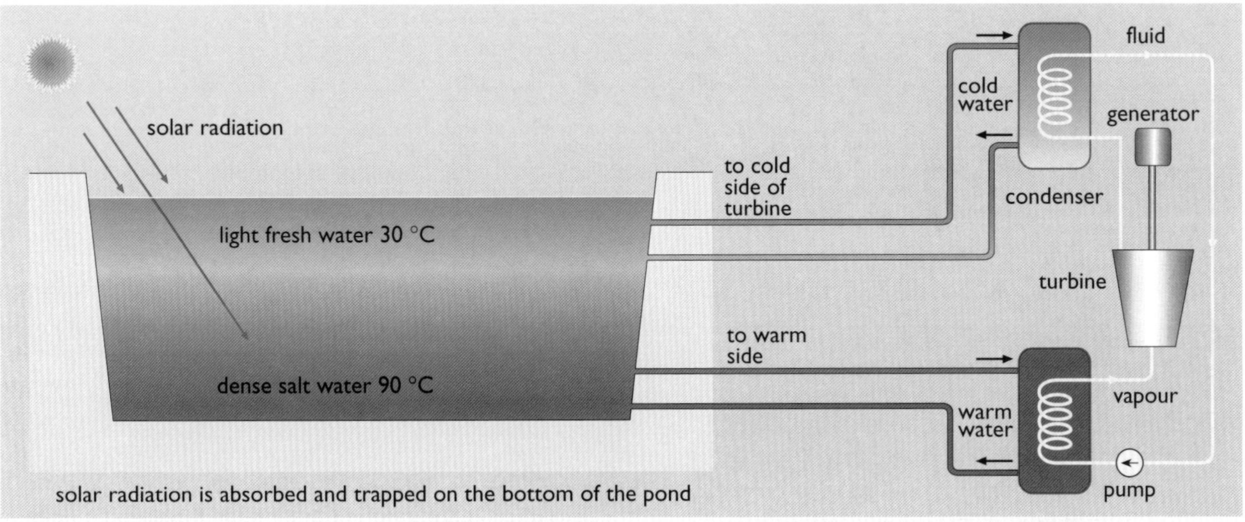

Figure 2.47 A solar pond used to drive a vapour cycle turbine

Although initially developed in Israel by the solar pioneer Harry Tabor, commercial systems have now been tried in the US and Saudi Arabia.

Solar ponds are not really viable at high latitudes, since the collection surface is, by its very nature, horizontal, and cannot be tilted. Their best location is in the large areas of the world where natural flat salt deserts occur.

Ocean thermal energy conversion (OTEC)

Ocean thermal energy conversion essentially uses the sea as a solar collector. It exploits the small temperature difference between the warm surface of the sea and the cold water at the bottom (Figure 2.48). In deep waters, 1000 m or more, this can amount to 20 °C.

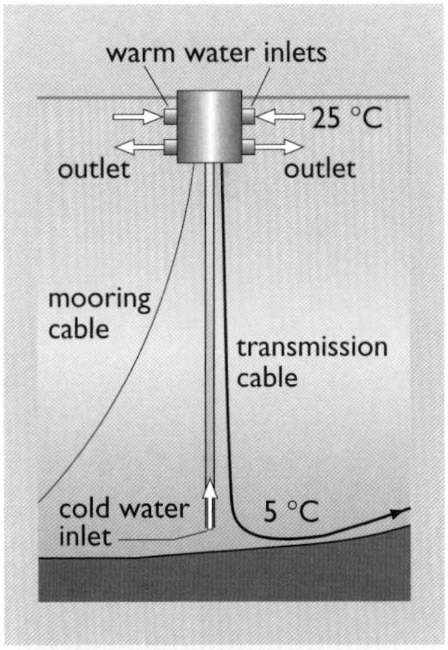

Figure 2.48 OTEC floating platform

Although the theoretical efficiency is likely to be very small and the organic rankine cycle system used needs to be finely tuned to boil at just the right temperature, there is an extremely large amount of water available.

Initial experiments made on a ship in the Caribbean in the 1930s were only marginally successful. Water had to be pumped from a great depth to obtain a significant temperature difference, and the whole system barely produced more energy than it used in pumping.

More recently, large-scale experiments have been carried out in the Pacific with more success, and a large number of experimental schemes have been conceived. The engineering difficulties are enormous. An OTEC station producing 10 MW of electricity would need to pump nearly 500 cubic metres per second of both warm and cold water through its heat exchangers, whilst remaining moored in sea 1000 metres deep.

Solar chimneys

The solar chimney exploits warm air produced in a very large greenhouse which is allowed to rise through a tall chimney. The updraught is used to turn an air turbine at the base of the chimney, driving a generator to produce electricity. This sounds simple enough. What is not so simple is the scale

Figure 2.49 Solar chimney at Manzanares in Spain

of construction. A 50 kW prototype built in Manzanares in Spain in 1981 used a greenhouse collector 240 m in diameter feeding warm air to a chimney 195 m high.

There are large economies of scale to be had. In a very sunny region of the world, a cost-optimum plant might have an output of 100 MW using a collector 3.6 km in diameter and feeding a chimney 950 m tall (Schlaich, 1995). However, because such design only produces warm air (a 35 °C temperature rise has been assumed in the calculations) the overall generation efficiency would be low, around 1.3%. Such a system, would thus require considerably more land area than one of equivalent output using high temperature concentrating collectors.

2.10 Economics, potential and environmental impact

The motto of Bailey's 'Day and Night' Solar Water Heater company was: 'Solar energy, like salvation, is free'. Even so, the company still had to charge its customers money for the collection equipment. Fortunately, the assessment of economics of solar thermal systems doesn't pose any great problems. Most solar thermal heating systems can be regarded in the same manner as conventional heating plant or building components and simple notions of 'payback times' often give an adequate assessment (see Appendix A1). Solar thermal electricity plants have similar ratings and life expectancies to small gas or diesel power stations. The difference is that they have no fuel costs.

Domestic active solar water heating

Domestic water heating for residential buildings makes up about 80% of the world solar thermal market. In the UK it has been estimated that there was a **solar park** (or total installed area) of approximately 120 000 m^2 of glazed and evacuated tube collectors installed at the end of 2001 (ESTIF, 2003). This contrasts with the 3.6 million m^2 installed in Germany.

It is likely that 50% of existing UK dwellings are suitable to have a solar water heater fitted. If this potential were to be fully taken up by 2025, this would require the deployment of approximately 12 million systems (allowing for a rise in the UK housing stock), or about 50 million m^2 of collectors. This could save an estimated 34.6 PJ (9.6 TWh) per year of oil and gas and 2.4 TWh per year of electricity, resulting in reductions of about 1% of current UK CO_2 emissions (DTI, 1999).

At present (2003) in the UK, sales of solar water heaters are low and prices are high. The DTI report suggested that a typical system of 3–4 m^2 area would cost between £1000 and £6000 including taxes. This range spans DIY systems and those installed in new houses at the bottom end, up to professionally installed retrofit systems at the top end. Such a system might have a life expectancy of 25 years and produce between 1000 and 1500 kWh of heat per year. The payback times are likely to be 10 years or more for a DIY system or over 20 years for a professionally installed one. In order to promote sales, the UK government has recently introduced a system of grants.

BOX 2.4 Do-it-yourself in Austria

The rapid growth in solar heating installations in Austria has largely been a product of home construction, starting in the province of Styria in 1983. By 1986, self-construction groups in Styria alone produced as much collector area as all the commercial suppliers in Austria together.

The Austrian Society for Renewable Energy (Arbeitsgemeinschaft Erneuerbare Energie, AEE), founded in 1988, has developed collector designs for home construction, runs training seminars and lends the tools for building collectors. Initially the commercial suppliers of systems were somewhat disapproving of self-construction groups, but eventually came to realise that there was a market for good quality DIY kits not just for end users, but also for local plumbers and builders. By the end of 1999 Austria had 2 million m^2 of solar collectors installed of which about 400 000 m^2 were self-build systems.

The most important marketing tool of the Austrian solar industry is said to be word-of-mouth promotion by satisfied owners of solar systems. About 60% of those working in DIY groups became interested because of the existing systems of friends and neighbours.

Similar schemes, such as that at Heeley City Farm near Sheffield (see Figure 2.50) are now being tried in the UK.

Looking further afield, prospects are more promising. In southern Europe, there is more sunshine, so a system may produce twice as much energy per square metre as in the UK. It is perhaps not surprising that Greece has a high level of solar water heater ownership. But who would have guessed that they would be overtaken in 1999 in terms of collector area per head of population by Austria? Between 1995 and 1998 the rate of installation there was running at around 200 000 m^2 per year. Much of the popularity of solar water heating is attributable to the success of 'do-it-yourself' schemes (see Box 2.4 and Weiss, 2000, 2001).

Figure 2.50 DIY solar water heating classes in Sheffield

Solar water heating is also popular in Germany, where collector sales reached 900 000 m^2 in 2001 before falling back to 540 000 m^2 in 2002, reflecting uncertainties in the German economy.

This has implications elsewhere. Although installations in the UK remain at a low level of around 15 000 m^2 per year, manufacturing production has been increasing, with exports (mainly of evacuated tube collectors) currently (2003) running at about 40 000 m^2 per year. A large proportion of these are sold on under the brand names of German suppliers and the UK export trade has followed the fortunes of the German market.

Given the experience of countries like Austria, in 1997 the European Commission suggested that promoting solar water heating could help cut EU CO_2 emissions (CEC, 1997). They suggested that an EU-wide annual installation growth rate of 20% per year would lead to a total installed capacity of 100 million m^2 by 2010. Although by 1999 the total installed figure had risen, 75% of this rise was in the same three countries, Germany, Austria and Greece. Although the rate of installations in Spain and the Netherlands increased significantly between 1994 and 1999, it seems doubtful that the 2010 target could be achieved without serious promotion in the remaining EU countries.

Looking still further, China represents the biggest solar market worldwide. In 2001 the market was estimated at 5.5 million m^2, mostly vacuum collectors, supplied from 1000 manufacturers.

As for environmental impact, that of solar water heating schemes in the UK would be very small. The materials used are those of everyday building and plumbing. Pumped solar collectors can be installed to be visually almost indistinguishable from normal roof lights, with storage tanks hidden inside the roof space. In Mediterranean countries, though, the use of free-standing thermosyphon systems on flat roofs can be visually intrusive. It is not so much the collector that is the problem but the storage tank above it. (Bailey's 'Day and Night' Solar Water Heater Company also had to face these problems. It offered to disguise the storage tank as a chimney!)

Swimming pool water heating

At the end of 2001 the UK had an estimated 89 000 m^2 of unglazed collectors installed, mainly used for swimming pools. Such systems can be simpler and cheaper than those for domestic water heating. The DTI's 1999 assessment suggested that in the UK, a 20 m^2 system might cost between £950 and £2700 (including taxes). This could be expected to produce approximately 300 kWh per square metre of collector. Payback times would be of the same order as the expected lifetime, about 15 years. There are only about 100 000 swimming pools in the UK, and even if this number were to double by 2025, the total potential energy saving would only amount an extra 7% over and above that for domestic water heating.

Active solar space heating and district heating

Although individual house solar space heating is technically feasible, it is likely to be far more cost effective to invest in insulation to cut space heating demands. A 1992 study (Long, 1992) concluded that with favourable collector and store prices the payback time for a system using interseasonal storage in the UK could be brought down to 30 years, but only assuming that it is an addition to an existing district heating system.

A district heating system is highly likely to be distributing heat from another source that deserves encouragement: waste heat from power stations (possibly biofuel-fired). On the one hand, this could be seen as discouraging the inclusion of solar energy. On the other, such communal systems offer economies of scale in the purchase and installation of collectors.

A Danish 1999 survey of the performance of 26 large systems between 1990 and 1999 found an average solar collection of 384 kWh per m^2 per year (kWh m^{-2} yr^{-1}) (CADDET, 2001). This perhaps sets the benchmark of technical performance. The problem is to be able to purchase and install collectors at an acceptable price. Analysis of a recent German programme of large-scale demonstration projects incorporating short-term storage suggests a 60–70% improvement in the ratio of costs to benefits compared to using individual solar water heaters (see Dalenbäck, 1999).

As for environmental impact, visually, where large solar collectors for space heating are roof mounted, they are unlikely to be a problem. However, if there is not sufficient roof space, they may take up valuable urban ground area. More practically, they may be difficult to protect from vandalism.

Passive solar heating and daylighting

In its narrow sense of increasing the amount of solar energy directly used in providing useful space heating, passive solar heating is highly economic, indeed possibly free. Passive solar-heated buildings have been generally well received by their occupants and are of interest to architects.

However, the potential is limited by the low rate of replacement of the building stock. The Department of Trade and Industry has estimated the potential in the UK by 2010 as a primary energy saving of 10 TWh yr^{-1} This has to be compared with the potentials for other common energy efficiency measures that could be applied to the existing housing stock, e.g. loft insulation: 6 TWh yr^{-1}, cavity insulation: 31 TWh yr^{-1}, and low energy lighting: 6 TWh yr^{-1} (DTI, 1999).

Designing buildings to take advantage of daylighting involves both energy conservation and passive solar heating considerations. A Building Research Establishment study estimated the national potential for exploiting daylight in the UK as worth between 5 TWh yr^{-1} and 9 TWh yr^{-1} by 2020. In warmer countries, daylighting may be far more important, since cutting down on summer electricity use for lighting can also save on air conditioning costs.

Designing and laying out buildings to make the best use of sunlight is generally seen as environmentally beneficial and has already shaped many towns and cities. For example, when in 1904 the city council of Boston, Massachusetts, USA, was faced with proposals for a 100 m high skyscraper, it commissioned an analysis of the shading of other buildings that this would cause. It was not pleased with the results and imposed strict limits on building heights.

However, a word of caution is necessary. In the UK, the tradition of new town development has been based partly on Victorian notions of the health aspects of 'light and air' in contrast to the overcrowded squalor of existing cities. This has been beneficial in terms of better penetration of solar energy into buildings. On the other hand, the encouragement of low building densities has led to vast tracts of sprawling suburbs and the consumption of enormous quantities of energy in transportation.

Solar thermal engines and electricity generation

As the original pioneers realised, it pays to build solar thermal power systems in really sunny places. In order to generate the high temperatures necessary for thermodynamically efficient operation, the local climate has to have plenty of direct solar radiation – diffuse radiation will not do. Low fossil fuel prices around the world have dampened interest in solar thermal electricity generation, in contrast to the continued enthusiasm for photovoltaics. A SolarPACES paper gives a good summary of the recent state of the technology (Tyner *et al.*, 2001).

In the US, the proven electricity price for the SEGS plants is considered to be around 12 to 14 US cents per kWh (8–9 per kWh). Even though the largest existing ones are 80 MW each, it is likely that the true optimum size is about 150–200 MW. This is set by the availability of commercial power turbines and economies of scale in the ordering of the glass for mirrors.

Currently, in sunny desert locations, solar thermal electricity is cheaper than photovoltaic power at current prices. It is suggested that it would still be so in these sunny areas even if PV panel prices fell by a factor of two (Quashning and Blanco, 2001).

The overall potential for such systems is enormous. Back in 1914, Frank Shuman was talking of building 20 000 square miles (50 000 square km) of collector in the Sahara desert. In terms of today's plants, this would imply a peak electricity generation capacity of some 2500 GW – nearly 50 times the peak demand on the UK electricity grid. The problem would be to find a suitable way of conveying the output to the loads. The manufacture and distribution of hydrogen would be one possibility. This is discussed in Chapter 10 and also in the companion text.

The environmental consequences of solar thermal power stations are somewhat mixed. A major problem is the sheer quantity of land required. A new optimum-sized SEGS plant would occupy 3–4 square kilometres. Although, typically, the collectors only take up one-third of the land area, it may be physically difficult to use it for anything else. This is unlike windfarms where the turbines are very widely spaced and crops can grow underneath. Even so, when the full fuel cycle land requirements of other energy resources are taken into account, (including mining and waste disposal), it is claimed that SEGS plants use no more land than other conventional plants.

Solar ponds and solar chimney projects would need even larger areas of flat land because of their low thermodynamic efficiency.

For all of these kinds of system, sunny deserts, within striking distance of large urban electricity demands, are needed. In California, the Mojave desert is ideal. In Europe, parts of central Spain and other southern Mediterranean countries are interesting possibilities.

The environmental consequences of OTEC systems may be mixed. On the one hand, it is claimed that the vast amounts of water being pumped circulate nutrients and can increase the amount of fish life. On the other, dissolved carbon dioxide can be released from the deep sea water, thereby negating some of the benefits of renewable energy generation. Only further experiments will resolve these issues.

Although Northern Europe is probably too cloudy to support economic solar thermal electricity schemes, this is not to say that they should not interest industry. The massive areas of mirror glass used in the SEGS schemes were not made in the US but mainly imported from Germany (see Benemann, 1994). As with many other renewable energy sources, sunshine is free, but the hardware has to be manufactured by someone somewhere.

Conclusions

Solar energy is a resource that is there for the taking. All that is needed is to produce the necessary hardware. We already make plenty of use of passive solar energy and daylight, but we take it for granted. It only requires a little more care in the design and layout of our buildings to make the best use of it.

Similarly, techniques of active solar heating and solar thermal power generation are technically feasible and in many countries well proven and regarded as cost effective. For example, 80% of residential buildings in Israel have solar heating systems. Whether solar systems can be promoted in less sunny countries is another matter. The perceived economics of systems are highly dependent not just on the particular local climate and energy needs, but also, as shown in Austria, on the attitudes on individuals.

References

Achard, P. and Gicquel, R. (eds) (1986) *European Passive Solar Handbook* (preliminary edition), Commission of the European Communities, DG XII.

Baker, N. *et al.* (eds) (1993) *Daylighting in Architecture – A European Reference Book*, European Commission Handbook EUR 15006 EN, Luxembourg.

Benemann, J. (1994) *The use of glass in solar applications*, Sun at Work in Europe, vol. 9, no. 3, Sept. 1994. [An interesting account of the production of large-scale mirrors.]

Butti, K. and Perlin, J. (1980) *A Golden Thread: 2500 Years of Solar Architecture and Technology*, Marion Boyars.

CADDET (2001) *Solar Collectors Supplement District Heating System*, Centre for the Dissemination of Demonstrated Energy Technologies. Online at: http://www.caddet-re.org [accessed 3/10/2003].

Cavanagh, J. E., Clarke, J. and Price, R. (1992) *'Ocean energy systems'*, in Johannson, T. (ed.) (1992) Renewable Energy, Island Press.

Commission of the European Communities (1997*) Energy for the Future: Renewable Sources of Energy – White Paper*, COM(97) 599 final, ISBN 92-78-28533-1, Luxembourg, C.E.C.

Dalenbäck, J-O (1999) *Information Brochure on Large-Scale Solar Heating,* European Large Scale Solar Heating Network. Downloadable from main.hvac.chalmers.se/cshp [accessed 3/10/2002].

Dalenbäck, J-O (2002) *European Large-Scale Solar Heating Plants,* European Large Scale Solar Heating Network. Downloadable from main.hvac.chalmers.se/cshp [accessed 3/10/2002].

DTI (1999) *New and Renewable Energy: Prospects in the UK for the 21st Century – Supporting Analysis,* ETSU-R122, Department of Trade and Industry.

DTI (2003) *Energy Consumption in the United Kingdom*, Department of Trade and Industry. Downloadable from www.dti.gov.uk [accessed 29/09/2003].

Duffie, J. A. and Beckman, W. A. (1980) *Solar Engineering of Thermal Processes*, John Wiley. [A classic textbook on the physics and engineering of solar thermal energy systems.]

EST (2003) *The Hockerton Housing Project,* Energy Efficiency Best Practice in Housing New Practice Profile 119, Energy Saving Trust. Downloadable from www.est.org.uk [accessed 2/10/2003].

ESTIF (2003) *Sun in Action II – A Solar Thermal Strategy for Europe*, European Solar Thermal Industry Federation, April 2003. Downloadable from www.estif.org [accessed 15/11/2003].

ETSU (eds) (1985) *Active Solar Heating in the UK*, Energy Technology Support Unit, Report R 25, HMSO.

Everett, R. *et al.* (1986) *'Pennyland and Linford Low Energy Housing Projects'*, Journal of Ambient Energy, Vol. 7, No. 2.

Ford, B. (1982) *Thermal Performance Modelling of a Terrace House with Conservatory*, ETSU Report 5-1056b.

Goulding, J. R., Lewis, J. O. and Steemers, T. C. (eds) (1992) *Energy in Architecture – The European Passive Solar Handbook*, Batsford. [A mine of technical information, highly recommended.]

Granqvist, C. G. (1989) 'Energy efficient windows: options with present and forthcoming technology', in Johansson, R. B., Bodlund, B. and Williams, R. H. (eds) *Electricity – Efficient End Use,* Lund University Press.

Hutchins, M. (1997) *Glazing Materials for Advanced Thermal Performance and Solar Gain Control,* Proceedings of Conference C69 of the UK Solar Energy Society, Using Advanced Glazing to Improve Daylighting and Thermal Performance in Buildings, May 1997.

IEA (2003) *Solar Update Newsletter No.39*, International Energy Agency, Feb. 2003. Downloadable from www.iea-shc.org [accessed 15th November 2003].

Long, G. (1992) *Solar Aided District Heating in the UK*, Energy Technology Support Unit, Report S 1190.

Mazria, E. (1979) *Passive Solar Energy Book*, Rodale Press. [A classic beautifully illustrated introduction to design for passive solar heating.]

Norton, B. (2000) *Heating Water by the Sun – a layperson's guide to the use of solar energy for providing domestic hot water and for heating swimming pools,* Solar Energy Society.

Osuna, R. *et al.* (2001) PS10, *A 10 MW Solar Tower Power Plant for Southern Spain*. Downloadable from http://www.solarpaces.org [accessed 14 March 2003].

Quaschning, V. Blanco, M. (2001) *Solar power – Photovoltaics or Solar Thermal Plants?*, Proceedings of VGB Congress Power Plants 2001. Downloadable from http://www.dlr.de/psa [accessed 14 March 2003].

Sadler, R. (1996) *The 1995 UK National Survey of Solar Water Heating,* Proceedings of Conference C67 of the UK Solar Energy Society, Solar Water Heating – Opportunities Today.

Schlaich, J. (1995) *The Solar Chimney – Electricity from the Sun*, Edition Axel Menges, Stuttgart.

Smith, P.F. (2001) *Architecture in a Climate of Change*, Architectural Press. [An up-to-date book on low energy building design for the UK.]

Tyner. C. E. *et al.* (2001) *Concentrating Solar Power in 2001*, SolarPACES. Downloadable from http://www.solarpaces.org [accessed 14 March 2003].

Weiss, W. (2000) *Successful Dissemination of 400 000 m² of Solar Systems by Do-It-Yourself Groups in Austria*, Arbeitsgemeinschaft Erneuerbare Energie, http://www.aee.at/verz/english/self01.html [accessed 14 March 2003].

Weiss, W. (2001) *Current development of the market and the potential of solar thermal systems in the medium-term*, Arbeitsgemeinschaft Erneuerbare Energie, http://www.aee.at/verz/english/therm11e.html [accessed 14 March 2003].

Chapter 3

Solar Photovoltaics

by Godfrey Boyle

3.1 Introduction

In Chapter 2 we saw how solar energy can be used to generate electricity by producing high temperature heat to power an engine which then produces mechanical work to drive an electrical generator.

This chapter is concerned with a more direct method of generating electricity from solar radiation, namely **photovoltaics**: the conversion of solar energy directly into electricity in a solid-state device.

The chapter starts with a brief look at the history and basic principles of photovoltaic energy conversion, concentrating initially on devices using monocrystalline silicon. We then review various ways of reducing the currently high costs of energy from photovoltaics.

The electrical characteristics of photovoltaic cells and modules are then described, followed by a discussion of the various roles of photovoltaic energy systems in supplying power in remote locations and in feeding power into local or national electricity grids.

The concluding sections review the economics and the environmental impact of photovoltaic electricity, examine the resources available from photovoltaics and how it might be increasingly integrated into electricity supplies, and look at the world market and future prospects for photovoltaics.

3.2 Introducing photovoltaics

If you were asked to design the ideal energy conversion system, you would probably find it difficult to come up with something better than the solar photovoltaic (PV) cell.

In this we have a device which harnesses an energy source that is by far the most abundant of those available on the planet. As earlier chapters have emphasised, the net solar power input to the earth is more than 10 000 times humanity's current rate of use of fossil and nuclear fuels.

The PV cell itself is, in its most common form, made almost entirely from silicon, the second most abundant element in the earth's crust. It has no moving parts and can therefore in principle, if not yet in practice, operate for an indefinite period without wearing out. And its output is electricity, probably the most useful of all energy forms.

A brief history of PV

The term 'photovoltaic' is derived by combining the Greek word for light, *photos*, with *volt*, the name of the unit of electromotive force – the force that causes the motion of electrons (i.e. an electric current). The volt was named after the Italian physicist Count Alessandro Volta, the inventor of the battery. Photovoltaics thus describes the generation of electricity from light.

Figure 3.1 Edmond Becquerel, who discovered the photovoltaic effect

The discovery of the **photovoltaic effect** is generally credited to the French physicist Edmond Becquerel (Figure 3.1), who in 1839 published a paper (Becquerel, 1839) describing his experiments with a 'wet cell' battery, in the course of which he found that the battery voltage increased when its silver plates were exposed to sunlight.

The first report of the PV effect in a solid substance appeared in 1877 when two Cambridge scientists, W. G. Adams and R. E. Day, described in a paper to the Royal Society the variations they observed in the electrical properties of selenium when exposed to light (Adams and Day, 1877).

In 1883 Charles Edgar Fritts, a New York electrician, constructed a selenium solar cell that was in some respects similar to the silicon solar cells of today (Figure 3.2). It consisted of a thin wafer of selenium covered with a grid of very thin gold wires and a protective sheet of glass. But his cell was very inefficient. The *efficiency* of a solar cell is defined as the percentage of the solar energy falling on its surface that is converted into electrical energy. Less than 1% of the solar energy falling on these early cells was converted to electricity. Nevertheless, selenium cells eventually came into widespread use in photographic exposure meters.

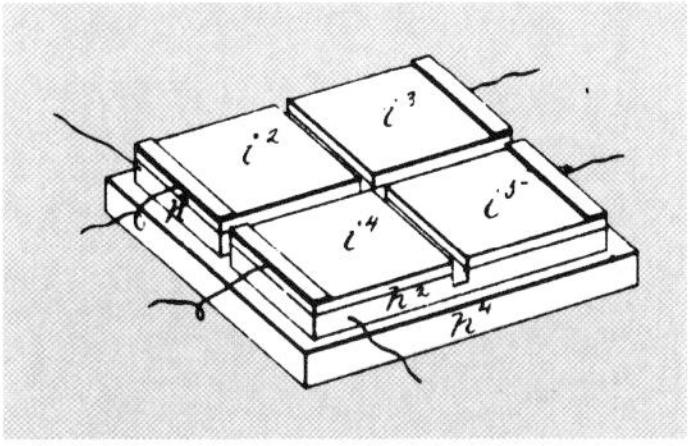

Figure 3.2 Diagram from Charles Edgar Fritts' 1884 US patent application for a solar cell

The underlying reasons for the inefficiency of these early devices were only to become apparent many years later, during the first half of the twentieth century, when physicists such as Planck and Einstein provided new insights into the nature of radiation and the fundamental properties of materials (see Section 3.3 below).

It was not until the 1950s that the breakthrough occurred that set in motion the development of modern, high-efficiency solar cells. It took place at the Bell Telephone Laboratories (Bell Labs) in New Jersey, USA, where a number of scientists, including Darryl Chapin, Calvin Fuller and Gerald Pearson (Figure 3.3), were researching the effects of light on **semiconductors**. These are non-metallic materials, such as germanium and silicon, whose electrical characteristics lie between those of conductors, which offer little resistance to the flow of electric current, and insulators, which block the flow of current almost completely. Hence the term *semi*conductor.

A few years before, in 1948, two other Bell Labs researchers, Bardeen and Brattain, had produced another revolutionary device using semiconductors – the transistor. Transistors are made from semiconductors (usually silicon) in extremely pure crystalline form, into which tiny quantities of carefully selected impurities, such as boron or phosphorus, have been deliberately diffused. This process, known as **doping**, dramatically alters the electrical behaviour of the semiconductor in a very useful manner that will be described in detail later.

In 1953 the Chapin–Fuller–Pearson team, building on earlier Bell Labs research on the PV effect in silicon (Ohl, 1941), produced 'doped' silicon slices that were much more efficient than earlier devices in producing electricity from light.

Figure 3.3 Bell Laboratories' pioneering PV researchers Pearson, Chapin and Fuller measure the response of an early solar cell to light

By the following year they had produced a paper on their work (Chapin *et al.*, 1954) and had succeeded in increasing the conversion efficiency of their silicon solar cells to 6%. Bell Labs went on to demonstrate the practical uses of solar cells, for example in powering rural telephone amplifiers, but at that time they were too expensive to be an economic source of power in most applications.

In 1958, however, solar cells were used to power a small radio transmitter in the second US space satellite, Vanguard I. Following this first successful demonstration, the use of PV as a power source for spacecraft has become almost universal (Figure 3.4).

Figure 3.4 The International Space Station is powered by large arrays of photovoltaic panels with a total output of 110 kilowatts

Rapid progress in increasing the efficiency and reducing the cost of PV cells has been made over the past few decades. Their terrestrial uses are now widespread, particularly in providing power for telecommunications, lighting and other electrical appliances in remote locations where a more conventional electricity supply would be too costly.

A single conventional PV cell produces only about 1.5 watts, so to obtain more power, groups of cells are normally connected together to form rectangular *modules*. To obtain even more power, modules are in turn mounted side by side and connected together to form *arrays*.

A growing number of domestic, commercial and industrial buildings now have PV arrays providing a substantial proportion of their energy needs. And a number of large, megawatt-sized PV power stations connected to electricity grids are now in operation in the USA, Germany, Italy, Spain and Switzerland.

The efficiency of the best single-junction silicon solar *cells* has now reached 24% in laboratory test conditions (see Box 3.1 and Box 3.3). The best silicon PV *modules* now available commercially have an efficiency of over 17%, and it is expected that in about 10 years' time module efficiencies will have risen to over 20% (Appleyard, 2003). Over the decade to 2002, the total installed capacity of PV systems increased approximately ten-fold, module costs dropped to below $4 per peak watt and overall system costs fell to around $7 per peak watt (see Figures 3.5 and 3.6). As we shall see, improvements in the cost-effectiveness of PV are likely to continue.

3.3 PV in silicon: basic principles

Semiconductors and 'doping'

PV cells consist, in essence, of a junction between two thin layers of dissimilar semiconducting materials, known respectively as 'p' (positive)-type semiconductors, and 'n' (negative)-type semiconductors. These semiconductors are usually made from silicon, so for simplicity we shall

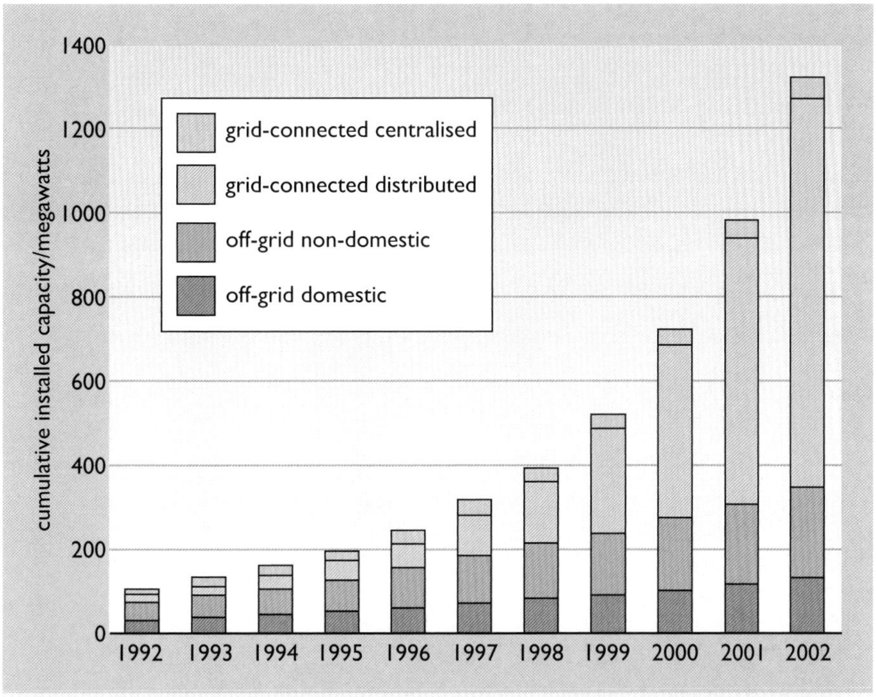

Figure 3.5 Cumulative PV power installed by type of application in IEA reporting countries, 1992–2002 (source: International Energy Agency, 2003). Note: data in this figure are from 20 International Energy Agency reporting countries, mainly 'developed' nations. There is additional significant PV module production and installed capacity in a number of other countries, particularly in the 'developing' world

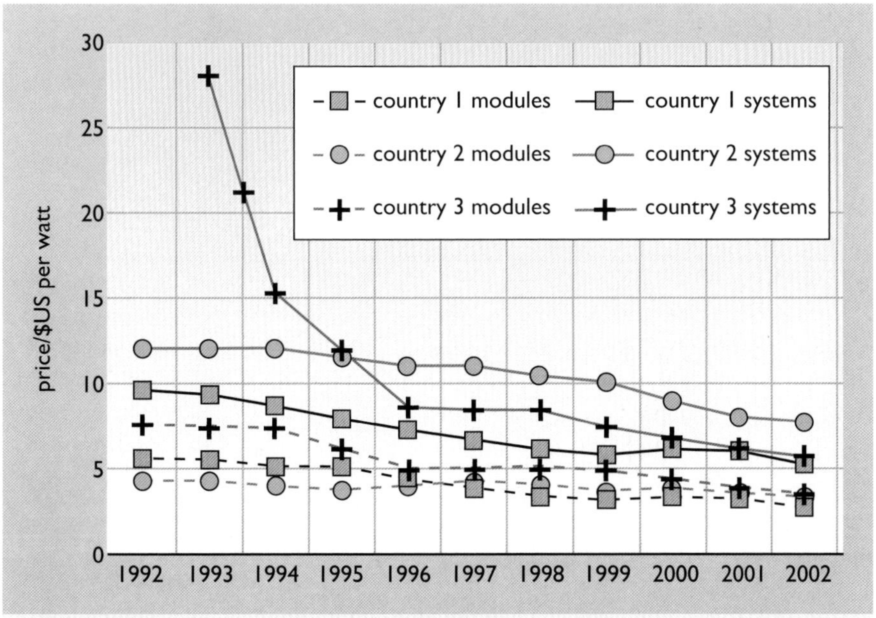

Figure 3.6 PV system and module price trends in three selected IEA reporting countries (source: International Energy Agency, 2003)

BOX 3.1 Standard test conditions for PV cells and modules

There is widespread international agreement that the performance of PV cells and modules should be measured under a set of standard test conditions.

Essentially, these specify that the temperature of the cell or module should be 25 °C and that the solar radiation incident on the cell should have a total power density of 1000 watts per square metre (W m^{-2}), with a spectral power distribution known as **Air Mass 1.5 (AM 1.5)**.

The spectral power distribution is a graph describing the way in which the power contained in the solar radiation varies across the spectrum of wavelengths.

The concept of 'Air Mass' describes the way in which the spectral power distribution of radiation from the sun is affected by the distance the sun's rays have to travel though the atmosphere before reaching an observer (or a PV module).

Just outside the earth's atmosphere, the sun's radiation has a power density of approximately 1365 W m^{-2}. The characteristic spectral power distribution of solar radiation as measured before it enters the atmosphere is described as the **Air Mass 0 (AM0)** distribution.

At the earth's surface, the various gases of which the atmosphere is composed (oxygen, nitrogen, ozone, water vapour, carbon dioxide, etc.) attenuate the solar radiation selectively at different wavelengths. This attenuation increases as the distance over which the sun's rays have to travel through the atmosphere increases.

When the sun is at its zenith (i.e. directly overhead), the distance over which the sun's rays have to travel through the atmosphere to a PV module is at a minimum. The characteristic spectral power distribution of solar radiation that is observed under these conditions is known as the **Air Mass 1 (AM1)** distribution.

When the sun is at a given angle θ to the zenith (as perceived by an observer at sea level), the Air Mass is defined as the ratio of the path length of the sun's rays under these conditions to the path length when the sun is at its zenith. By simple trigonometry (see Figure 3.7), this leads to the definition:

$$\text{Air Mass} \approx \frac{1}{\cos\theta}$$

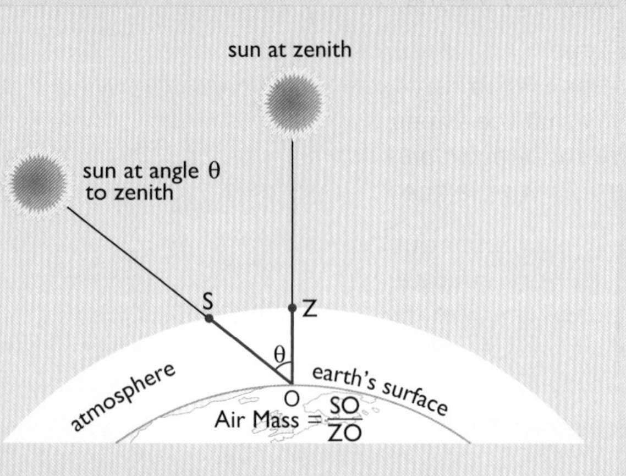

Figure 3.7 Air Mass is the ratio of the path length of the sun's rays through the atmosphere when the sun is at a given angle (θ) to the zenith, to the path length when the sun is at its zenith

An Air Mass distribution of 1.5, as specified in the standard test conditions, therefore corresponds to the spectral power distribution observed when the sun's radiation is coming from an angle to overhead of about 48 degrees, since cos 48° = 0.67 and the reciprocal of this is 1.5.

The approximate spectral power distributions for Air Masses 0 and 1.5 are shown in Figure 3.8.

In practice, the power rating in **peak watts (Wp)** of a cell or module is determined by measuring the maximum power it will supply when exposed to radiation from lamps designed to reproduce the AM1.5 spectral distribution at a total power density of 1000 W m^{-2} .

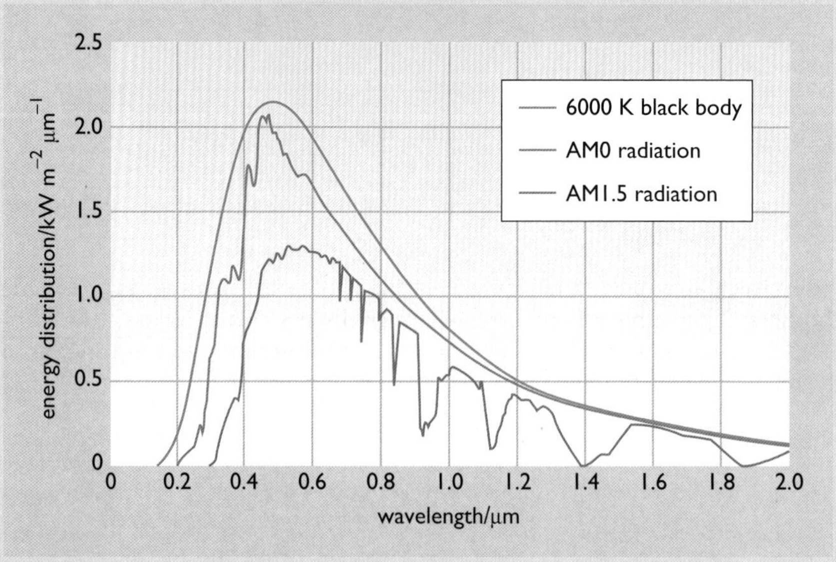

Figure 3.8 The spectral power distributions of solar radiation corresponding to Air Mass 0 and Air Mass 1.5. Also shown is the theoretical spectral power distribution that would be expected, in space, if the sun were a perfect radiator (a 'black body') at 6000 °C

initially consider only silicon-based semiconductors – although, as we shall see later, PV cells can be made from other materials.

n-type semiconductors are made from crystalline silicon that has been 'doped' with tiny quantities of an impurity (usually phosphorus) in such a way that the doped material possesses a *surplus of free electrons*. **Electrons** are sub-atomic particles with a negative electrical charge, so silicon doped in this way is known as an **n (negative)-type** semiconductor.

p-type semiconductors are also made from crystalline silicon, but are doped with very small amounts of a different impurity (usually boron) which causes the material to have a *deficit of free electrons*. These 'missing' electrons are called **holes**. Since the absence of a negatively charged electron can be considered equivalent to a positively charged particle, silicon doped in this way is known as a **p (positive)-type** semiconductor (see Box 3.2).

The p–n junction

We can create what is known as a **p–n junction** by joining these dissimilar semiconductors. This sets up an **electric field** in the region of the junction. This electric field is like the electrostatic field you can generate by rubbing a plastic comb against a sweater. It will cause negatively charged particles to move in one direction, and positively charged particles to move in the opposite direction. However, a p–n junction in practice is not a simple mechanical junction: the characteristics change from 'p' to 'n' gradually, not abruptly, across the junction.

The PV effect

What happens when light falls on the p–n junction at the heart of a solar cell?

Light can be considered to consist of a stream of tiny particles of energy, called **photons**. When photons from light of a suitable wavelength fall within the p–n junction, they can transfer their energy to some of the electrons in the material, so 'promoting' them to a higher energy level. Normally, these electrons help to hold the material together by forming so-called 'valence' bonds with adjoining atoms, and cannot move. In their 'excited' state, however, the electrons become free to conduct electric current by moving through the material. In addition, when electrons move they leave behind holes in the material, which can also move (Figure 3.9 in Box 3.2). The 'car parking' analogy shown in Figure 3.10 may be helpful in visualizing the processes involved.

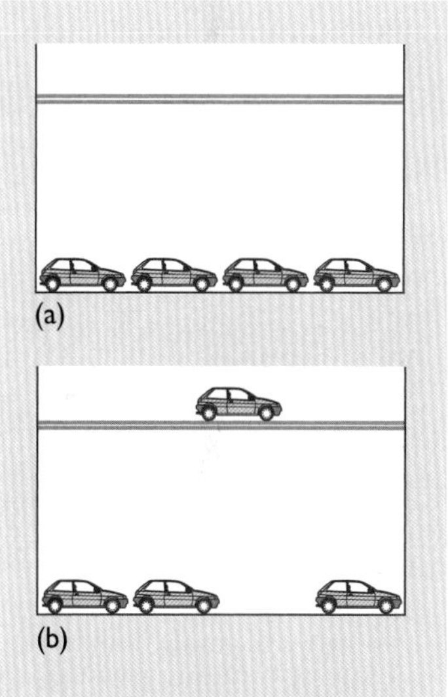

Figure 3.10 'Car parking' analogy of conduction processes in a semiconductor. (a) The ground floor of the car park is full – the cars there (representing electrons in the 'valence band') cannot move around. The first floor is empty. (b) A car (electron) is 'promoted' to the first floor (representing the 'conduction band'), where it can move around freely. This leaves behind a 'hole' that also allows cars on the ground floor (valence band) to move around (source: Green, 1982)

When the p–n junction is formed, some of the electrons in the immediate vicinity of the junction are attracted from the n-side to combine with holes on the nearby p-side. Similarly, holes on the p-side near the junction are attracted to combine with electrons on the nearby n-side.

The net effect of this is to set up around the junction a layer on the n-side that is more positively charged than it would otherwise be, and, on the p-side, a layer that is more negatively charged than it would otherwise be. In effect, this means that a *reverse* electric field is set up around the junction: negative on the p-side and positive on the n-side. The region around the junction is also depleted of charge carriers (electrons and holes) and is therefore known as the **depletion region**.

BOX 3.2 Crystalline silicon, doping, p–n junctions and PV cells

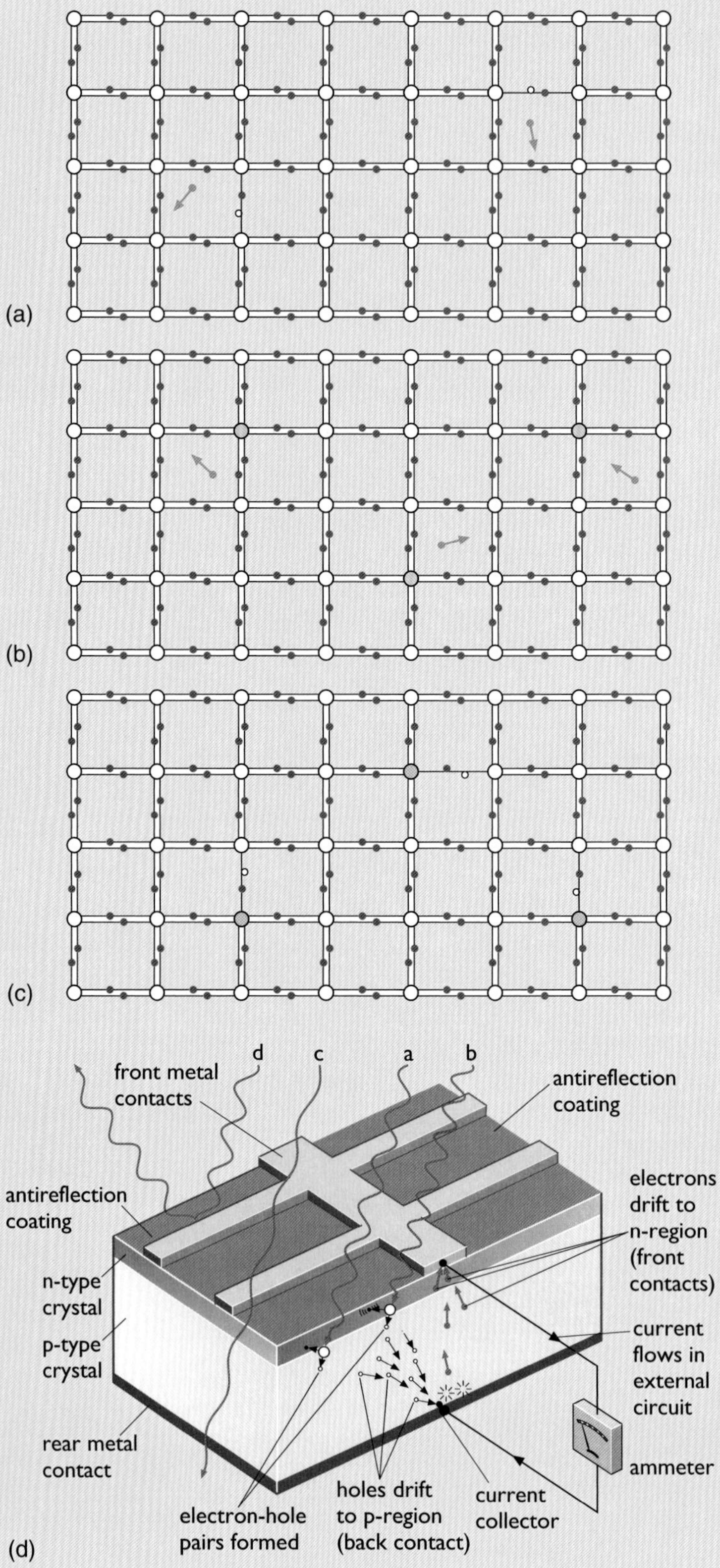

Figure 3.9 (a) Crystal of *pure* silicon has a cubic structure, shown here in two dimensions for simplicity. The silicon atom has four 'valence' electrons. Each atom is firmly held in the crystal lattice by sharing two electrons (small dots) with each of four neighbours at equal distances from it. Occasionally thermal vibrations or a photon of light will spontaneously provide enough energy to promote one of the electrons into the energy level known as the conduction band, where the electron (small dot with arrow) is free to travel through the crystal and conduct electricity. When the electron moves from its bonding site, it leaves a 'hole' (small open circle), a local region of net positive charge

(b) Crystal of *n-type* silicon can be created by doping the silicon with trace amounts of phosphorus. Each phosphorus atom (large coloured circle) has five valence electrons, so that not all of them are taken up in the crystal lattice. Hence n-type crystal has an excess of free electrons (small dots with arrows)

(c) Crystal of *p-type* silicon can be created by doping the silicon with trace amounts of boron. Each boron atom (large coloured circle) has only three valence electrons, so that it shares two electrons with three of its silicon neighbours and one electron with the fourth. Hence the p-type crystal contains more holes than conduction electrons

(d) A silicon solar cell is a wafer of p-type silicon with a thin layer of n-type silicon on one side. When a photon of light with the appropriate amount of energy (labelled a) penetrates the cell near the junction of the two types of crystal and encounters a silicon atom, it dislodges one of the electrons, which leaves behind a hole. The energy required to promote the electron into the conduction band is known as the band gap. The electron thus promoted tends to migrate into the layer of n-type silicon, and the hole tends to migrate into the layer of p-type silicon. The electron then travels to a current collector on the front surface of the cell, generates an electric current in the external circuit and then reappears in the layer of p-type silicon, where it can recombine with waiting holes. If a photon with an amount of energy greater than the band gap (labelled b) strikes a silicon atom, it again gives rise to an electron–hole pair, and the excess energy is converted into heat. A photon with an amount of energy smaller than the band gap (c) will pass right through the cell, so that it gives up virtually no energy along the way. Moreover, some photons (d) are reflected from the front surface of the cell even when it has an antireflection coating. Still other photons are lost because they are blocked from reaching the crystal by the current collectors that cover part of the front surface (source for text and figures: Chalmers, 1976)

When an electron in the junction region is stimulated by an incoming photon to 'jump' into the conduction band, it leaves behind a hole in the valence band. Two charge carriers (an **electron–hole pair**) are thus generated. Under the influence of the reverse electric field around the junction, the electrons will tend to move into the n-region and the holes into the p-region.

The process can be envisaged (Figure 3.11) in terms of the energy levels in the material. The electrons that have been stimulated by incoming photons to enter the conduction band can be thought of as 'rolling downwards', under the influence of the electric field at the junction, into the n-region; similarly, the holes can be thought of as 'floating upwards', under the influence of the junction field, into the p-region.

The flow of electrons to the n-region is, by definition, an electric current. If there is an external circuit for the current to flow through, the moving electrons will eventually flow out of the semiconductor via one of the metallic contacts on the top of the cell. The holes, meanwhile, will flow in the opposite direction through the material until they reach another metallic contact on the bottom of the cell, where they are then 'filled' by electrons entering from the other half of the external circuit.

In order to produce power, the PV cell must generate voltage as well as the current provided by the flow of electrons. This voltage is, in effect, provided

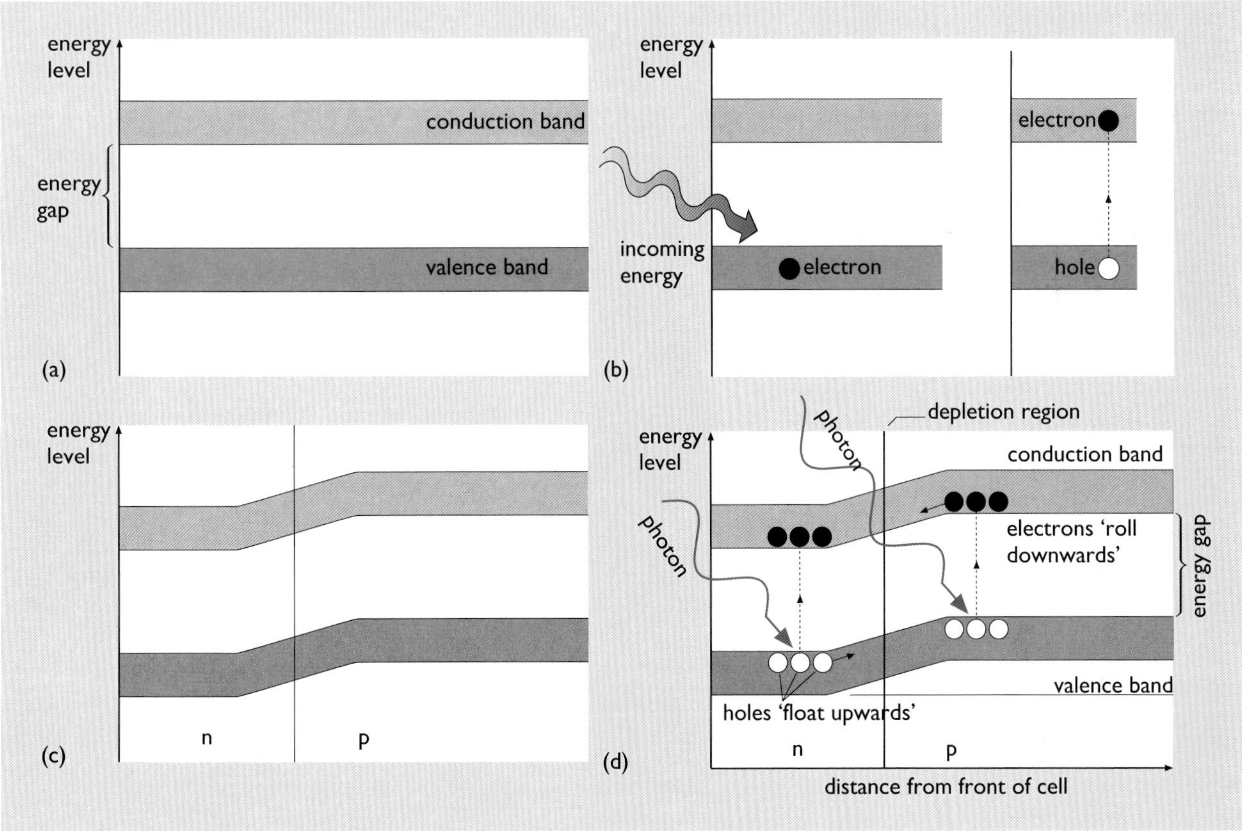

Figure 3.11 (a) Energy bands in a normal ('intrinsic') semiconductor; (b) An electron can be 'promoted' to the conduction band when it absorbs energy from light (or heat), leaving behind a 'hole' in the valence band; (c) When the n-type and p-type semiconductors are combined into a p–n junction, their different energy bands combine to give a new distribution, as shown, and a built-in electric field is created; (d) In the p–n junction, photons of light can excite electrons from the valence band to the conduction band. The electrons 'roll downwards' to the n-region, and the holes 'float upwards' to the p-region

BOX 3.3 Band gaps and PV cell efficiency

According to the quantum theory of matter, the quantity of energy possessed by any given electron in a material will lie within one of several levels or 'bands'. Those electrons that normally hold the atoms of a material together (by being 'shared' between adjoining atoms, as we saw in Figure 3.9) are described by physicists as occupying a lower-energy state known as the **valence band**.

In certain circumstances, some electrons may acquire enough energy to move into a higher-energy state, known as the **conduction band**, in which they can move around within the material and thus conduct electricity (Figure 3.11). There is a so-called **energy gap** or **band gap** between these bands, the magnitude of which varies from material to material, and which is measured using an extremely small energy unit: the *electron volt* (eV) (see Appendix A2).

Metals, which conduct electricity well, have many electrons in the conduction band. Insulators, which hardly conduct electricity at all, have virtually no electrons in the conduction band. Pure (or 'intrinsic') semiconductors have some electrons in the conduction band, but not as many as in a metal. But 'doping' pure semiconductors with very small quantities of certain impurities can greatly improve their conductivity.

If a photon incident on a doped, n-type semiconductor in a PV cell is to succeed in transferring its energy to an electron and 'exciting' it from the valence band to the conduction band, it must possess an energy at least equal to the band gap. Photons with energy less than the band gap do not excite valence electrons to enter the conduction band and are 'wasted'. Photons with energies significantly greater than the band gap do succeed in 'promoting' an electron into the conduction band, but any excess energy is dissipated as heat. This wasted energy is one of the reasons why PV cells are not 100% efficient in converting solar radiation into electricity. (Another is that not all photons incident on a cell are absorbed: a small proportion are reflected.)

Because the energy of a photon is directly proportional to the frequency of the light associated with it, photons associated with shorter wavelengths (i.e. higher frequencies) of light, near the blue end of the visible spectrum, have a greater energy than those of longer wavelength near the red end of the visible spectrum.

The spectral distribution of sunlight varies considerably according to weather conditions and the elevation of the sun in the sky (see Box 3.1). For maximum efficiency of conversion of light into electric power, it is clearly important that the band gap energy of the material used for a PV cell is reasonably well matched to the spectrum of the light incident upon it. For example, if the majority of the energy in the incoming solar spectrum is in the yellow–green range (corresponding to photons with energy of around 1.5 eV), then a semiconductor with a band gap of around 1.5 eV will be most efficient. In general, semiconductor materials with band gaps between 1.0 and 1.5 eV are reasonably well suited to PV use. Silicon has a band gap of 1.1 eV.

The maximum theoretical conversion efficiency attainable in a single-junction silicon PV cell has been calculated to be about 30%, if full advantage is taken of 'light trapping' techniques to ensure that as many of the photons as possible are usefully absorbed (Green, 1993). However, *multi-junction* cells have also been designed in which each junction is tailored to absorb a particular portion of the incident spectrum. Theoretically, such cells should have a much higher efficiency, possibly as high as 66% for an infinite number of junctions – though the efficiencies so far achieved by multi-junction cells in practice have been very much lower than this (see Section 3.6).

In practice, the highest efficiency achieved in commercially available single-junction monocrystalline silicon PV *modules* (as distinct from individual PV *cells*) is currently around 17%. The efficiency of PV modules is usually lower than that achieved by cells in the laboratory because:

- it is difficult to achieve as high an efficiency consistently in mass-produced devices as in one-off laboratory cells under optimum conditions;
- laboratory cells are not usually glazed or encapsulated;
- in a PV module there are usually inactive areas, both between cells and due to the surrounding module frame, that are not available to produce power;
- there are small resistive losses in the wiring between cells and in the diodes used to protect cells from short circuiting;
- there are losses due to mismatching between cells of slightly differing electrical characteristics connected in series.

by the internal electric field set up at the p–n junction. As we shall see, a single crystalline silicon PV cell typically produces a voltage of about 0.5 V at a current of up to around 3 A – that is, a peak power of about 1.5 W. (Depending on their detailed design, some PV cells produce more power than this, some less.)

Monocrystalline silicon cells

Until fairly recently, the majority of solar cells were made from extremely pure **monocrystalline** silicon – that is, silicon with a single, continuous crystal lattice structure (Figure 3.9) having virtually no defects or impurities. Monocrystalline silicon is usually grown from a small seed crystal that is slowly pulled out of a molten mass, or 'melt', of polycrystalline silicon (see below), in the sophisticated but expensive **Czochralski process** developed initially for the electronics industry. The entire process of

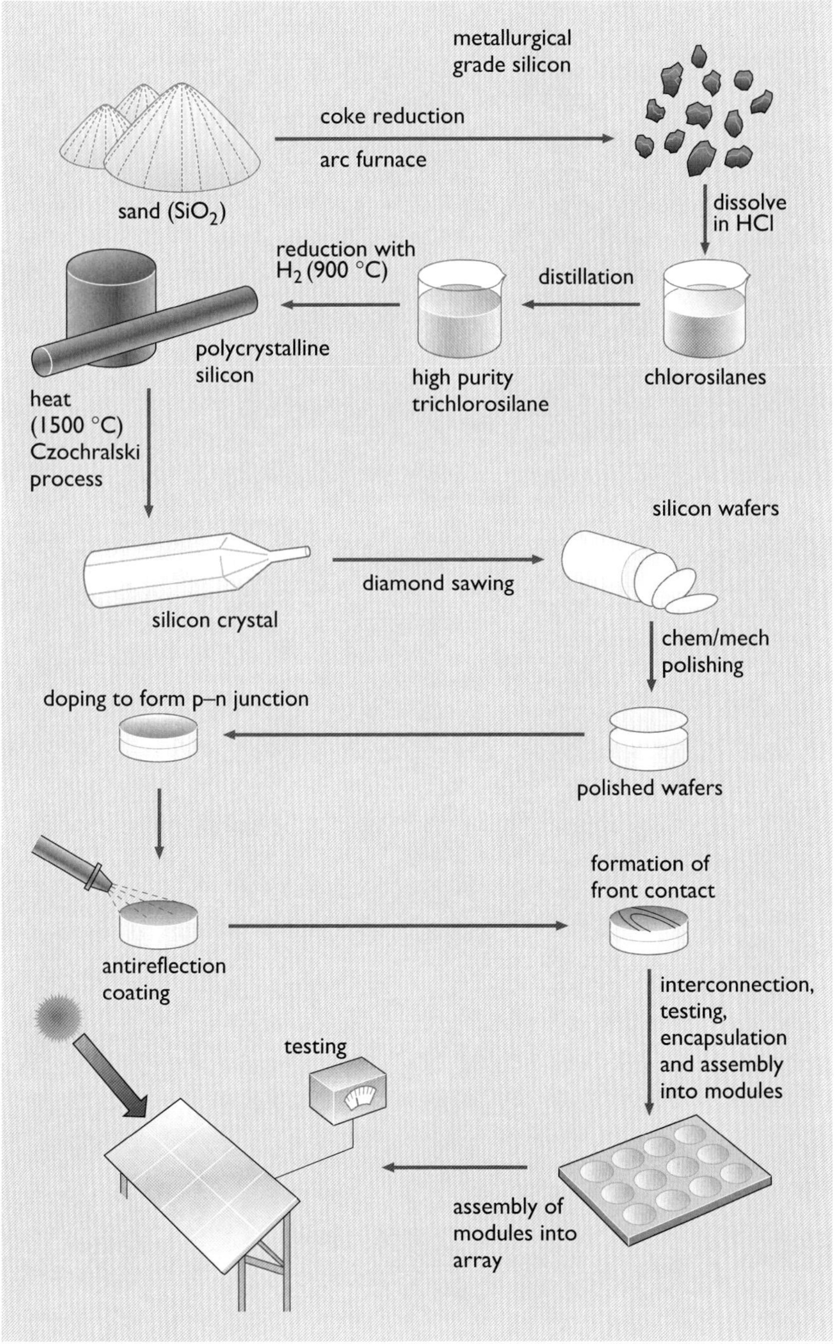

Figure 3.12 The overall process of monocrystalline silicon solar cell and module production

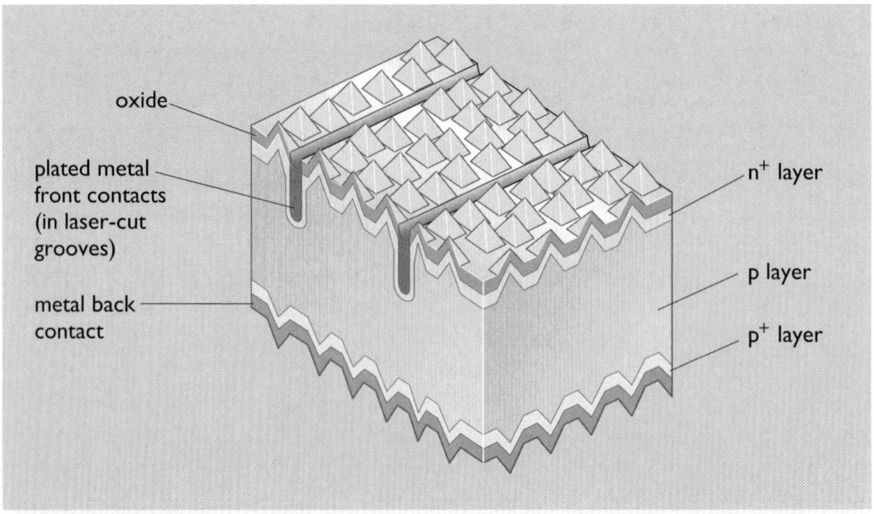

Figure 3.13 Main features of the 'laser-grooved buried-grid' monocrystalline PV cell structure, developed at the University of New South Wales (Green, 1993) and used in many high-efficiency PV modules. A pyramid-shaped texture on the top surface increases the amount of light 'trapped'. Buried electrical contacts give very low electrical resistance whilst minimizing losses due to overshadowing. The symbols p^+ and n^+ denote heavily doped layers that reduce electrical resistance in the contact areas

monocrystalline silicon solar cell and module production is summarized in Figure 3.12. Many of the most efficient single-junction monocrystalline PV modules currently available use a 'laser-grooved buried-grid' cell structure as shown in Figure 3.13.

3.4 Crystalline PV: reducing costs and raising efficiency

Although the latest monocrystalline silicon PV modules are highly efficient, they are also expensive because the manufacturing processes are slow, require highly skilled operators, and are labour- and energy-intensive. Another reason for their high cost is that most high-efficiency cells are fabricated from extremely pure 'electronic-grade' silicon. However, PV cells can be made from slightly less pure, but less costly, 'solar-grade' silicon, with only a small reduction in conversion efficiency.

A number of approaches to reducing the cost of crystalline PV cells and modules, or increasing their efficiency, have been under development during the past 20 years or so. These include cells using polycrystalline rather than single-crystal material; growing silicon in ribbon or sheet form; and the use of other crystalline PV materials such as gallium arsenide.

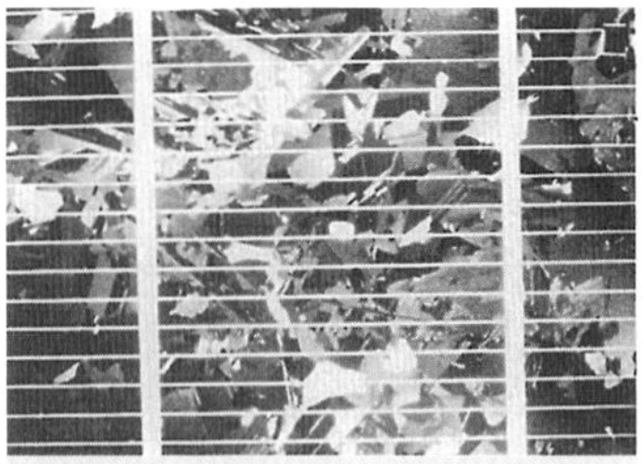

Figure 3.14 Polycrystalline silicon consists of randomly packed 'grains' of monocrystalline silicon

Polycrystalline silicon

Polycrystalline silicon essentially consists of small grains of monocrystalline silicon (Figure 3.14). Solar cell wafers can be made directly from polycrystalline silicon in various ways. These include the controlled casting of molten polycrystalline silicon into cube-shaped ingots which are then cut, using fine wire saws, into thin square wafers and fabricated into complete cells in the same way as monocrystalline cells.

Polycrystalline PV cells are easier and cheaper to manufacture than their monocrystalline counterparts. But they tend to be less efficient because light-generated charge carriers (i.e. electrons and holes) can recombine at the boundaries between the grains within polycrystalline silicon. However, it has been found that by processing the material in such a way that the grains are relatively large in size, and oriented in a top-to-bottom direction to allow light to penetrate deeply into each grain, their efficiency can be substantially increased. These and other improvements have enabled commercially available polycrystalline PV modules (sometimes called 'multi-crystalline' or 'semi-crystalline') to reach efficiencies of over 14%.

Companies such as Pacific Solar in Australia and Astropower in the USA are developing ways of depositing **polycrystalline films** on to ceramic or glass substrates, forming PV modules with an efficiency of around 10%. The silicon films used are somewhat thicker than in other 'thin film' PV cells (see below), so cells made in this way are sometimes known as '*thick* film' polycrystalline cells.

Silicon ribbons and sheets

This approach involves drawing thin 'ribbons' or 'sheets' of multicrystalline silicon from a silicon 'melt'. One of the main processes is known as edge-defined, film-fed growth (EFG), and was originally developed by the US company Mobil Solar. It is described in Figure 3.15. EFG cells are manufactured by the German company RWE Schott Solar (under the brand name ASE) and silicon ribbon cells by Evergreen Solar in the USA.

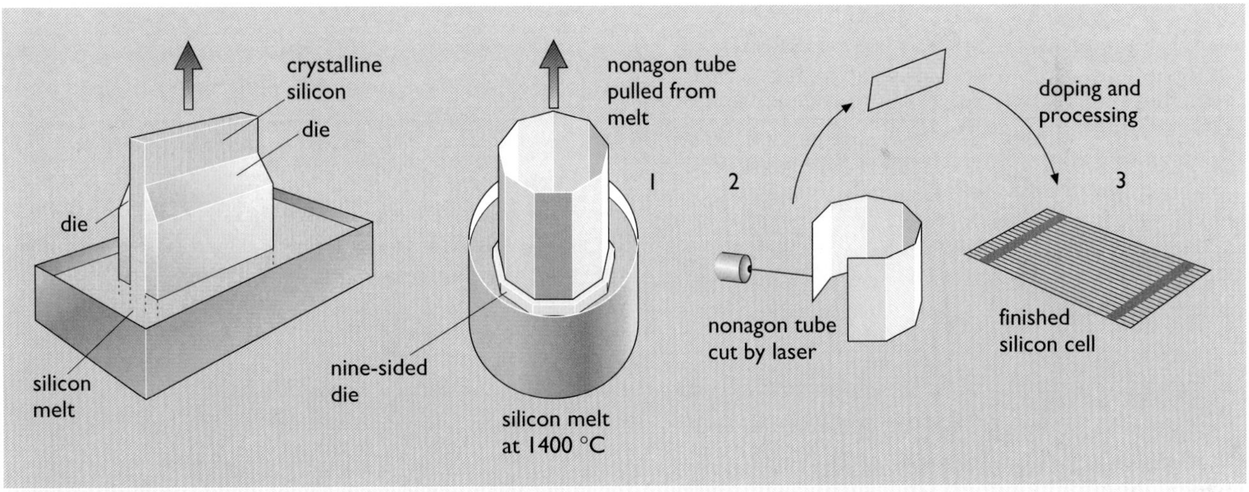

Figure 3.15 Edge-defined, film-fed growth process for PV production. Thin, hollow polygonal tubes of polycrystalline silicon up to 6 m long are slowly pulled through a die from a 'melt' of pure silicon, then cut by laser into individual cells

Gallium arsenide

Silicon is not the only crystalline material suitable for PV. Another is gallium arsenide (GaAs), a so-called **compound semiconductor**. GaAs has a crystal structure similar to that of silicon (see Figure 3.9), but consisting of

Figure 3.16 The Dutch Nuna solar car, winner of the 2001 World Solar Challenge, arriving at Adelaide, South Australia after completing the 3000 km journey from Darwin in just over 32 hours at an average speed (during daylight hours) of 91 km per hour. The ultra-lightweight car was produced by students from the universities of Delft and Rotterdam, sponsored by the utility Nuon. It is driven by electric motors powered by arrays of high-efficiency, dual-junction and triple-junction gallium arsenide PV cells developed by the European Space Agency for satellite use. (The 2003 solar challenge winner was Nuna II, also from the Netherlands, which completed the journey at an average speed of 97 km per hour)

alternating gallium and arsenic atoms. In principle it is highly suitable for use in PV applications because it has a high light absorption coefficient, so only a thin layer of material is required. GaAs cells also have a band gap wider than that of silicon, one close to the theoretical optimum for absorbing the energy in the terrestrial solar spectrum (see Box 3.3). Cells made from GaAs are therefore more efficient than those made from monocrystalline silicon. They can also operate at relatively high temperatures without the appreciable reduction in efficiency that affects PV cells made from silicon when their temperatures rise. This makes them well suited to use in *concentrating* PV systems (see Section 3.6 below).

On the other hand, cells made from GaAs are substantially more expensive than silicon cells, partly because the production process is not so well developed, and partly because gallium and arsenic are not abundant materials. GaAs cells have often been used when very high efficiency, regardless of cost, is required – as in many space applications. They have also powered many of the winning cars in solar car races (see Figure 3.16).

3.5 Thin film PV

Amorphous silicon

Solar cells can be made from very thin films of silicon in a form known as **amorphous silicon (a-Si)**, in which the silicon atoms are much less ordered than in the crystalline forms described above. In a-Si, not every silicon atom is fully bonded to its neighbours, which leaves so-called 'dangling bonds' that can absorb any additional electrons introduced by doping, so rendering any p–n junction ineffective.

However, this problem is largely overcome in the process by which a-Si cells are normally manufactured. A gas containing silicon and hydrogen (such as silane, SiH_4), and a small quantity of dopant (such as boron), is decomposed electrically in such a way that it deposits a thin film of amorphous silicon on a suitable substrate (backing material). The hydrogen in the gas has the effect of providing additional electrons that combine with the dangling silicon bonds to form, in effect, an alloy of silicon and hydrogen. The dopant that is also present in the gas can then have its usual effect of contributing charge carriers to enhance the conductivity of the material.

Solar cells using a-Si have a somewhat different form of junction between the p- and the n-type material. A so-called 'p–i–n' junction is usually formed, consisting of an extremely thin layer of p-type a-Si on top, followed by a thicker 'intrinsic' (i) layer made of undoped a-Si, and then a very thin layer of n-type a-Si. The structure is as shown in Figure 3.17. The operation of the PV effect in a-Si is generally similar to that in crystalline silicon, except that in a-Si the band gap, although wider, is less clearly defined.

Amorphous silicon cells are much cheaper to produce than those made from crystalline silicon. a-Si is also a much better absorber of light, so much thinner (and therefore cheaper) films can be used. The manufacturing process operates at a much lower temperature than that for crystalline silicon, so less energy is required; it is suited to continuous production; and it allows quite large areas of cell to be deposited on to a wide variety of both rigid and flexible substrates, including steel, glass and plastics.

But a-Si cells are currently much less efficient than their single-crystal or polycrystalline silicon counterparts: maximum efficiencies achieved with small, single-junction cells in the laboratory are currently around 12%. Moreover, the efficiency of many currently available single-junction a-Si modules degrades, within a few months of exposure to sunlight, from an initial 6–10%, stabilizing at around 4–8%.

Strenuous attempts have been made by manufacturers to improve the efficiency of a-Si cells and to solve the degradation problem. One approach involves the development of multiple-junction a-Si devices (see Section 3.6 below.)

Amorphous silicon cells are already widely used as power sources for a variety of consumer products such as calculators, where the requirement is not so much for high efficiency as for low cost.

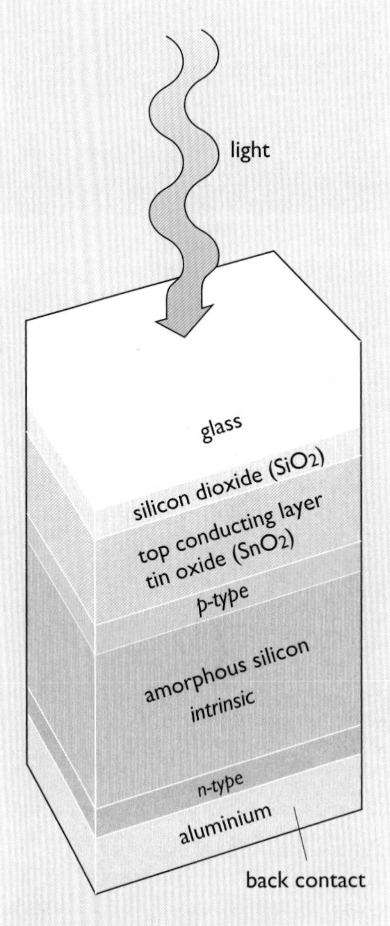

Figure 3.17 Structure of an amorphous silicon cell. The top electrical contact is made of an electrically conducting, but transparent, layer of tin oxide deposited on the glass. Silicon dioxide forms a thin 'barrier layer' between the glass and the tin oxide. The bottom contact is made of aluminium. In between are layers of p-type, intrinsic and n-type amorphous silicon

Other thin film PV technologies

Amorphous silicon is by no means the only material suited to thin film PV, however. Among the many other possible thin film technologies some of the most promising are those based on compound semiconductors, in particular copper indium diselenide ($CuInSe_2$, usually abbreviated to CIS), copper indium gallium diselenide (CIGS) and cadmium telluride (CdTe). Modules based on all of these technologies have reached the production stage, but production volumes are small.

Thin film CIGS cells have attained the highest laboratory efficiencies of all thin film devices, around 17%, and CIGS modules with stable efficiencies of around 10% are produced by Shell Solar in the USA and Würth Solar in Germany.

Cadmium telluride modules can be made using a relatively simple and inexpensive electroplating-type process. The band gap of CdTe is close to the optimum, and module efficiencies of over 10% are claimed, without the initial performance degradation that occurs in a-Si cells. However, the modules contain cadmium, a highly toxic substance, so stringent precautions need to be taken during their manufacture, use and eventual disposal (see Section 3.11). BP Solar, a subsidiary of the oil company BP, was involved in CdTe module manufacture but withdrew from production in 2002, explaining that although it did not consider the presence of cadmium in its modules to be a problem, customers appeared to believe otherwise. The Japanese firm Matsushita also withdrew from CdTe production in 2002, but First Solar Inc. in the USA continues to produce CdTe modules.

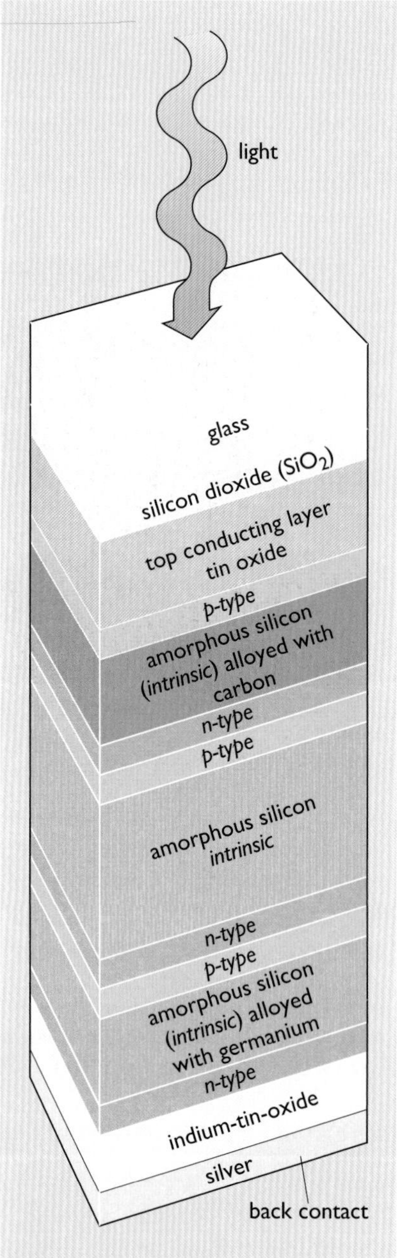

Figure 3.18 Structure of a multi-junction amorphous silicon cell

3.6 Other innovative PV technologies

Multi-junction PV cells

One way of improving the overall conversion efficiency of PV cells and modules is the 'stacked' or **multi-junction** approach, in which two (or more) PV junctions are layered one on top of the other, each layer extracting energy from a particular portion of the spectrum of the incoming light. A cell with two layers is often called a 'tandem' device.

The band gap of amorphous silicon, for example, can be increased by alloying the material with carbon, so that the resulting material responds better to light at the blue end of the spectrum. Alloying with germanium, on the other hand, decreases the band gap so the material responds better to light at the red end of the spectrum.

Typically, a wide band gap a-Si junction would be on top, absorbing the higher-energy photons at the blue end of the spectrum, followed by other thin film a-Si junctions, each having a band gap designed to absorb a portion of the lower light frequencies, nearer the red end of the spectrum (Figure 3.18). Multi-junction modules using amorphous silicon are available with stable efficiencies of around 8% from companies such as Unisolar and RWE.

Cells of different types can also be used, as in the Sanyo 'Hybrid HIT' module, in which a thin monocrystalline layer is sandwiched between two amorphous silicon layers. This enables very high conversion efficiencies to be achieved with low material and manufacturing energy requirements. Another example is the tandem module produced by Kaneka of Japan, which has a layer of amorphous silicon on top of a thin layer of 'microcrystalline' silicon (i.e. silicon in the form of extremely small crystals less than one micrometre in diameter).

Concentrating PV systems

Another way of getting more energy out of a given number of PV cells is to use mirrors or lenses to concentrate the incoming solar radiation on to the cells, an approach similar to that described in Chapter 2, Section 2.10, on solar thermal engines. This has the advantage that substantially fewer cells are required – to an extent depending on the concentration ratio, which can vary from as little as two to several hundred or even thousand times. The concentrating system must have an aperture equal to that of an equivalent flat plate array to collect the same amount of incoming energy. In concentrating PV systems the cells usually need to be cooled, either passively or actively, to prevent overheating.

Systems with the highest concentration ratios use complex sensors, motors and controls to allow them to track the sun on two axes – azimuth (horizontal orientation) and elevation (tilt) – ensuring that the cells always receive the maximum amount of solar radiation. Systems with lower concentration ratios often track the sun on only one axis and can have simpler tracking mechanisms.

Most concentrators can only utilize direct solar radiation. This is a problem in countries like the UK where nearly half the solar radiation is diffuse. However, some unconventional designs of concentrator allow some of the diffuse radiation, as well as direct radiation, to be concentrated (see Boes and Luque, 1993).

Silicon spheres

The US firm Texas Instruments has developed an ingenious way of making PV cells using tiny, millimetre-sized, spheres of polycrystalline silicon embedded at regular intervals between thin sheets of aluminium foil. Among the advantages of this approach are that impurities in the silicon tend to diffuse out to the surface of the spheres, where they can be 'ground off' as part of the manufacturing process. This allows relatively cheap, low-grade silicon to be used as a starting material. The resulting sheets of PV material are very flexible (see Figure 3.19), which can be an advantage in some applications.

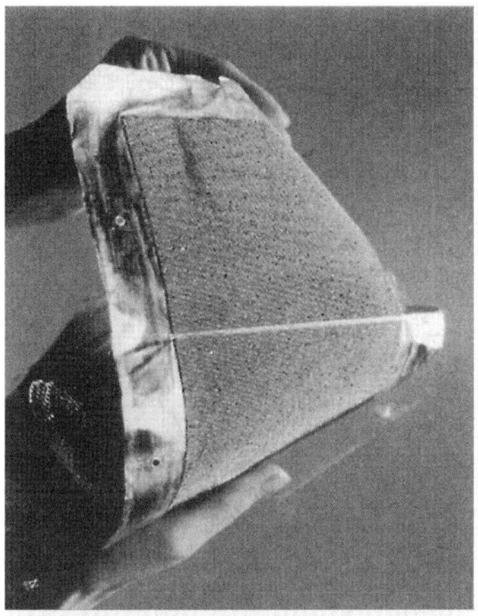

Figure 3.19 'Silicon spheres' PV technology

The technology is being commercialized by Automation Tooling Systems Inc. in Canada, where a 20 MWp (megawatts-peak) plant is under construction.

Photoelectrochemical cells

A radically different, photoelectrochemical, approach to producing cheap electricity from solar energy has been pioneered by researchers at the Swiss Federal Institute of Technology in Lausanne. (Strictly, photoelectrochemical devices are not photovoltaic: this term implies a solid-state device. Photoelectrochemical devices, however, use liquids.) The idea of harnessing photoelectrochemical effects to produce electricity from sunlight is not new – indeed, Becquerel's pioneering PV experiments were with liquid-based devices.

The Swiss researchers, led by Professor Michael Grätzel, have achieved much higher efficiencies than before, in a device that could be extremely cheap to manufacture.

It consists essentially of two thin glass plates, both covered with a thin, transparent, electrically conducting tin oxide layer (Figure 3.20, overleaf). To one plate is added a thin layer of titanium dioxide (TiO_2), which is a semiconductor. The surface of the TiO_2 has been treated to give it exceptionally high roughness, enhancing its light-absorbing properties.

Immediately next to the roughened surface of the TiO_2 is a layer of 'sensitizer' dye only one molecule thick, made of a proprietary 'transition metal complex' based on ruthenium or osmium. Between this 'sensitized' TiO_2 and the other glass plate is a thicker layer of iodine-based electrolyte.

On absorption of a photon of suitable wavelength, the sensitizer layer injects an electron into the conduction band of the titanium dioxide. Electrons so generated then move to the bottom electrically conducting layer (electrode) and pass out into an external circuit where they can do work. They then re-enter through the top electrode, where they drive a reduction–oxidation process in the iodine solution. This then supplies electrons to the sensitized TiO_2 layer in order to allow the process to continue.

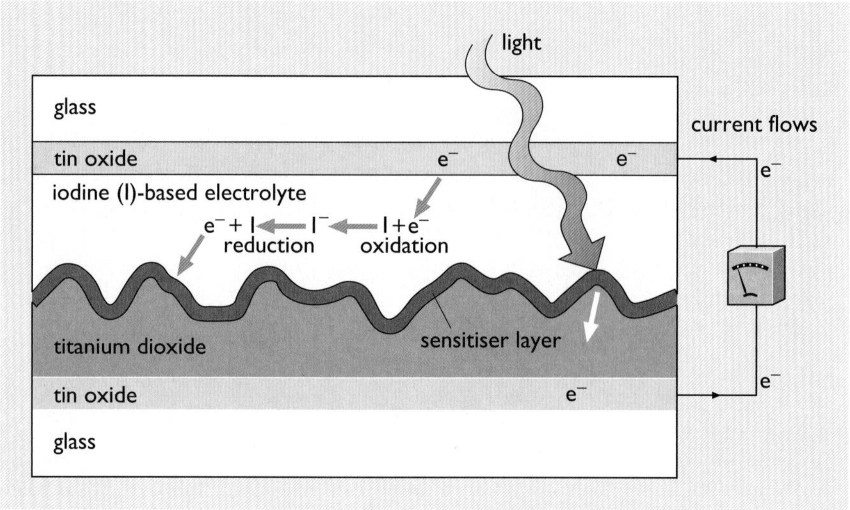

Figure 3.20 Principles of operation of the photoelectrochemical 'Grätzel Cell', developed at the Swiss Federal Institute of Technology, Lausanne

The Swiss team claims to have achieved efficiencies of 10% in full (AM 1.5) sunlight and that its cells are stable over long periods, though some researchers are not fully convinced of this. PV cells based on this technology are being manufactured on a small scale by STI in Australia and by Greatcell Solar SA in Switzerland. (See Grätzel *et al.*, 1989, Grätzel, 2001 and O'Regan *et al.*, 1991, 1993.)

'Third generation' PV cells

Still at the frontiers of research is a range of new photovoltaic technologies known as 'third generation' PV (crystalline PV is considered 'first generation' and thin film PV 'second generation'). These devices are generally based on **nanotechnology** – that is, technology which aims to manipulate molecules and atoms at extremely small scales, measured in billionths of a metre, or nanometres (nm). These tiny particles and structures are called 'nanoparticles' and 'nanostructures'; crystals of such sizes are termed 'nanocrystals'. Nanoparticles consisting of extremely small collections of atoms of semiconducting material are called 'quantum dots'.

As Dresselhaus and Thomas (2001) observe, a whole new class of materials is emerging,

> such as films containing nanocrystalline structures, quantum dots and nanostructured conducting polymers. These films typically contain nanoparticles with a size distribution showing a wide range of electronic band gaps (it is this gap that determines which wavelengths can be absorbed) so that much of the solar spectrum can be absorbed by the cell. If such films can be made cheaply, with an optimised distribution of nanoparticle diameters, and can be properly aligned, one might have an ideal solar collector. This field is still very young, and is moving very rapidly.

Eventually, if research on third generation PV proves successful, it could lead to PV cells made, for example, from extremely thin stacked plastic sheets, converting solar energy to electricity with very high efficiency at very low cost.

3.7 Electrical characteristics of silicon PV cells and modules

One very simple way of envisaging a typical 100 cm^2 silicon PV cell is as a solar powered battery, one that produces a voltage of around 0.5 V and delivers a current proportional to the sunlight intensity, up to a maximum of about 3 A in full sunlight.

In order to use PV cells efficiently we need to know a little more about how they behave when connected to various electrical loads. Figure 3.21 shows a single silicon PV cell connected to a variable electrical resistance R, together with an ammeter to measure the current (I) in the circuit and a voltmeter to measure the voltage (V) developed across the cell terminals. Let us assume the cell is being tested under standard conditions (see Box 3.1).

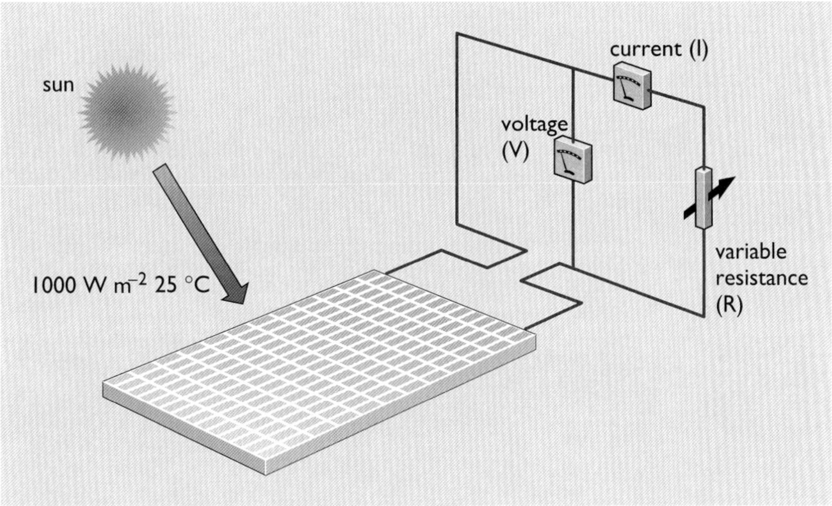

Figure 3.21 PV cell connected to variable resistance, with ammeter and voltmeter to measure variations in current and voltage as resistance varies

When the resistance is infinite (i.e. when the cell is, in effect, not connected to any resistance, or 'open circuited') the current in the circuit is at its minimum (zero) and the voltage across the cell is at its maximum, known as the **open circuit voltage** (V_{oc}). At the other extreme, when the resistance is zero, the cell is in effect 'short circuited' and the current in the circuit then reaches its maximum, known as the **short circuit current** (I_{sc}).

If we vary the load resistance between zero and infinity, the current (I) and voltage (V) will be found to vary as shown in Figure 3.22, which is known as the 'I–V characteristic' or 'I–V curve' of the cell. The *power output* (the product of voltage and current) is zero at V_{oc} (because $I = 0$) and zero again at I_{sc} (because $V = 0$). Between these points it rises and then falls, so there is one point at which the cell delivers maximum power. This is the **maximum power point** (MPP) on the I–V curve.

At lower levels of solar radiation than the standard 1000 W m^{-2} assumed in Figure 3.22, the general shape of the I–V characteristic stays the same, but the short circuit current decreases in proportion to the radiation

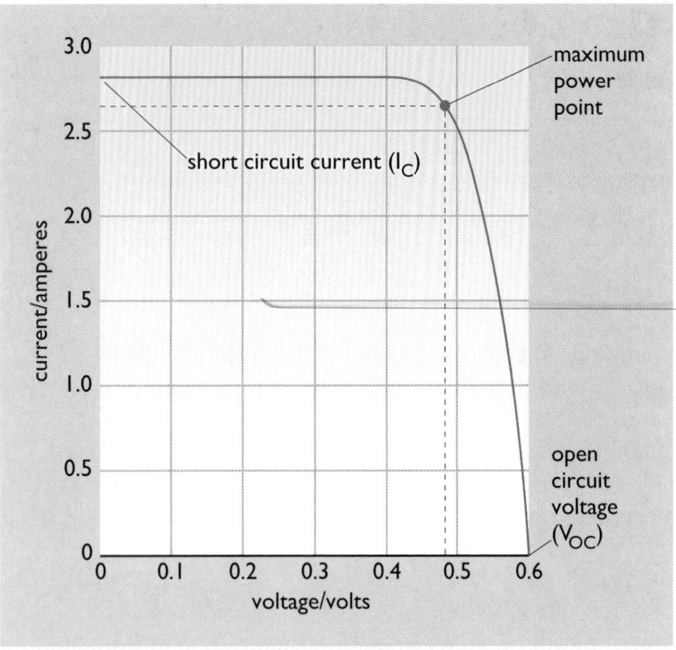

Figure 3.22 Current–voltage (I–V) characteristics of a typical silicon PV cell under standard test conditions

intensity, whilst the open circuit voltage falls less sharply. However, the open circuit voltage and the maximum power both decrease linearly as cell temperature increases.

When PV cells are delivering power to electrical loads in real-world conditions, the intensity of solar radiation usually varies substantially over time. Many PV systems therefore incorporate a so-called 'maximum power point tracking' device, a specialized electronic circuit that automatically varies the load 'seen' by the PV cell in such a way that it is always operating around the maximum power point.

As we have seen, a typical 100 cm^2 silicon PV cell produces a voltage of around 0.5 V. Since many PV applications involve charging lead–acid batteries, which have a typical nominal voltage of 12 V, PV modules often consist of around 36 individual cells wired in series to ensure that, even on overcast days, the total voltage is usually above 13 V and sufficient to charge a 12 V battery.

3.8 PV systems for remote power

PV modules are now widely used in developed countries to provide electrical power in locations where it would be inconvenient or expensive to use conventional grid supplies. They often charge batteries to ensure continuity of power. Some examples are shown in Figure 3.23.

Figure 3.23 (left) PV parking meter; (middle) navigation buoy; (right) telemetry system

BOX 3.4 PV system sizing

In order to be able to specify accurately how many PV modules would be required for a particular application, and in a remote location what the battery capacity should be, a PV energy system designer needs to know:

- What are the daily, weekly and seasonal variations in electrical demand?
- What are the daily, weekly and seasonal variations in the amount of solar radiation in the area where the system is situated?
- What is the proposed orientation and tilt angle of the PV array?
- For how many sunless days will the battery need to be able to provide backup electricity?

System sizing and other PV system design considerations are described in detail in Imamura *et al.* (1992), Treble (1991), Roberts (1991) and Lasnier and Ang (1990). An easy to follow step-by-step procedure is also given in Treble (1999).

Computer programs have also been developed to help engineers calculate the size and cost of PV systems to meet specified energy requirements in given locations and climatic conditions (see Further Information).

In many 'developing' countries, however, electricity grids are often non-existent or rudimentary, particularly in rural areas, and all forms of energy are usually very expensive. Here photovoltaics can be highly competitive with other forms of energy supply – especially in countries with high solar radiation levels – and its use is growing very rapidly. Applications include PV water pumping; PV refrigerators to help keep vaccines stored safely in health centres; PV systems for homes and community centres, providing energy for lights, radios, and audio and video systems; PV-powered telecommunications systems and PV-powered street lighting.

The European Union has recognized the huge potential contribution that photovoltaics could make towards improving the living standards of the 1–2 billion people who are classified by the World Bank as 'poor'. The EU's 1997 'Plan for Takeoff' for renewable energy, in addition to setting a target of 500 000 roof- and façade-mounted PV systems in Europe by 2010, also proposed an export initiative to install 500 000 village-scale PV systems by the end of the decade 'to kick-start decentralized electrification in developing countries' (see European Commission, 1997 and Palz, 1994).

This theme has been reiterated in more recent reports, including *Power to Tackle Poverty,* presented to the 2002 United Nations World Summit on Sustainable Development, which emphasized that PV and other renewables 'have the potential to meet the needs of the world's poorest people at an affordable price. Clean, affordable and friendly to the local environment, these technologies are better suited than those dependent on fossil fuels to meet the needs of people living in remote areas of poorer countries.' The report urged Governments to accept the challenging target of installing up to 4.5 GW of PV systems in developing countries over the next decade. This would require an annual production rate reaching over 1 GW by 2012, creating many tens of thousands of jobs (IT Power, 2001).

3.9 Grid-connected PV systems

PV systems for homes

In most parts of the developed world, grid electricity is easily accessible as a convenient backup to PV or other fluctuating renewable energy supplies. Here it makes sense for PV energy systems to use the grid as a giant 'battery'.

The grid can absorb PV power that is surplus to current needs (say, on sunny summer afternoons), making it available for use by other customers and reducing the amount that has to be generated by conventional means; and at night or on cloudy days, when the output of the PV system is insufficient, the grid can provide backup energy from conventional sources.

In these grid-connected PV systems, a so-called 'grid-commutated inverter' (or 'synchronous inverter') transforms the DC power from the PV arrays into AC power at a voltage and frequency that can be accepted by the grid, while 'debit' and 'credit' meters measure the amount of power bought from or sold to the utility.

In the UK, the electricity demand of a typical household is currently around 4000 kWh a year, but much of this is for heating and other uses that could be supplied by non-electrical energy forms such as gas. The demand of a small UK household for energy in forms that *necessitate* the use of electricity (such as lighting, domestic appliances, radio, TV, hi-fi and computers) is probably nearer 1000 kWh a year, if the most energy-efficient lighting and appliances are used.

What area of PV array would be required in the UK to supply 1000 kWh per year?

The annual energy 'yield' of PV arrays varies widely according to the type of module and various other factors. PV array yields can be assessed in two ways: in terms of their annual energy output *per peak kilowatt of 'rated' power* (kWh kWp^{-1} yr^{-1}); or in terms of their annual energy output *per square metre of module area* (kWh m^{-2} yr^{-1}).

In recent UK tests, as explained in Box 3.5, measured module yields *per peak kilowatt* ranged from just over 1000 kWh kWp^{-1} yr^{-1} to below 700 kWh kWp^{-1} yr^{-1}. On a *per square metre* basis, yields varied from nearly 120 kWh m^{-2} yr^{-1} to as low as 22 kWh m^{-2} yr^{-1}. These test figures, however, were for unshaded arrays with close-to-optimal orientation. In practice, UK PV installers assume annual yields of around 750 kWh per peak kilowatt for crystalline PV systems, and somewhat higher yields per peak kilowatt for arrays using thin film modules. To produce 1000 kWh per year using crystalline modules would therefore require an installed capacity of around 1.3 kWp with an array area of around 9 square metres.

Figure 3. 24 In this solar house in Oxford, UK, a 4 kWp grid-linked array of monocrystalline PV modules forms an integral part of the roof, alongside solar water heating panels. The PV array supplies the house's annual electricity requirements, plus a small surplus used to provide some of the power for a small electric car

The roofs of many UK houses could accommodate a PV array of this size, and it is likely that about half of UK roofs are oriented in a direction sufficiently close to south to enable them to be used for solar collection purposes. Conventional PV modules can more easily be integrated into the roofs of new dwellings, where they can replace all or part of the conventional roof (Figure 3.24). 'Solar slates' have also been developed that look very similar to conventional roofing materials (Figure 3.25).

Figure 3.25 This small terraced house in Richmond, London (on the right of the photo), has a 1.3 kWp array of 'solar slates' that supply its annual electricity requirements. It is equipped with the latest energy-efficient electrical appliances

In Germany, the government has been strongly encouraging the installation of domestic PV systems and the development of its PV industry (see Section 3.10). In Japan, subsidies have successfully encouraged the installation of tens of thousands of PV systems, mainly on houses, and the government provides generous long-term R&D funding to its PV industry. Total installed PV capacity in Japan reached 600 MW by the end of 2002. Switzerland, Italy and the Netherlands also have significant programmes supporting domestic PV installation (Figure 3.26).

The UK has lagged considerably behind many other European countries in stimulating PV development. However, in 2002 the UK Department of Trade and Industry launched a Major Photovoltaics Demonstration Programme, which should result in 3000 domestic roofs and 140 larger non-residential buildings having PV systems installed by 2006. This is in addition to an earlier programme of 'Domestic and Large Scale Field Trials' of PV systems, which should result in at least 500 installations on domestic roofs and 18 on large public buildings.

Figure 3.26 This large development at Niewland, a suburb of Amersfoort in the Netherlands, has a total of 1.3 MWp of PV capacity installed on the roofs of houses, schools and community buildings. Some of the PV arrays are owned by the local electricity company, ENECO

BOX 3.5 Energy yields from PV systems

The annual amount of energy that will be produced by a photovoltaic system depends on various conditions, including:

- the annual total amount of solar radiation available at the site. This can be estimated from meteorological data giving the measured number of kWh (or GJ) per square metre per year incident on a horizontal surface at the nearest meteorological station. In typical UK conditions, this is around 1000 kWh m^{-2} yr^{-1} but in very sunny countries the figure can rise to well over 2000 kWh m^{-2} yr^{-1};
- the orientation (azimuth) and tilt (elevation) of the PV arrays;
- the peak power rating of the arrays (or, alternatively, the area of the arrays in square metres);
- the energy conversion efficiency of the PV modules;
- the extent to which the efficiency of the modules is affected by temperature;
- the extent to which module efficiency is affected by the spectral distribution of the solar radiation. This varies according to the elevation of the sun and the extent of clouds and water vapour in the atmosphere;
- the efficiency of the inverter used to convert the DC power from the PV arrays into AC, and any losses that occur in the wiring between the PV system and the final consumer.

The effects of these variables in practice were measured by researchers at Oxford University in their 'PV-Compare' project (Jardine and Lane, 2003). Two arrays, each including 11 different types of PV module, were tested side-by-side under identical conditions at two different sites, one at Oxford in the UK and one at Mallorca in Spain.

PV modules are usually 'rated' by their manufacturers in terms of their peak output in watts, measured under standard test conditions (see Box 3.1), and are usually sold on the basis of their price per peak watt (Wp) – although this can give a misleading impression of their potential annual energy yield. Another way of quantifying a module's output is in terms of its annual output per square metre of module area.

The annual amount of solar radiation in Mallorca was 1700 kWh m^{-2} compared with 1022 kWh m^{-2} for Oxford. The measured annual energy yields, both in terms of kWh per kWp and in terms of kWh per square metre of module area, for both sites are given in Table 3.1. (These yields are net, allowing for losses in wiring, inverters etc.)

In terms of kWh *per peak kilowatt*, for Mallorca, the best annual energy yields were from the CIS and double-junction amorphous silicon modules. Monocrystalline and polycrystalline modules gave somewhat lower but roughly similar yields, as did triple-junction amorphous silicon. The lowest yields were given by CdTe and single-junction a-Si modules. The authors concluded that the high yields of the CIS and double-junction a-Si modules were mainly due to their better performance in the high temperature conditions found in Mallorca.

For Oxford, the rankings were very similar. The highest-yielding modules in kWh per kWp terms were

Table 3. 1 Comparison of annual energy yields of different PV module technologies in Mallorca, Spain and Oxford, UK

PV technology	Mallorca		Oxford	
	Yield/kWh per kWp per year	**Yield**/kWh m^{-2} per year	**Yield**/kWh per kWp per year	**Yield**/kWh m^{-2} per year
a-Si triple junction	1380.4	87.3	858.6	54.3
a-Si double junction (manufacturer A)	1655.3	88.3	991.8	52.9
a-Si double junction (manufacturer B)	1515.5	79.8	926.6	48.8
a-Si single junction	887.4	38.9	557.3	22.3
Monocrystalline Si	1389.2	187.9	871.8	117.2
Multicrystalline Si (ribbon)	1283.3	94.6	824.8	60.8
Multicrystalline Si (Si film)	1352.9	88.1	821.8	61.2
Multicrystalline Si	1368.0	143.1	842.0	96.2
Multicrystalline Si (EFG)	1340.4	155.9	875.1	101.8
CIS	1553.3	150.5	1025.3	99.2
CdTe	958.5	64.8	673.7	48.9

Source: Jardine and Lane, 2003

once again CIS and double-junction a-Si, with crystalline silicon modules and triple-junction a-Si producing somewhat lower yields, and CdTe and single-junction a-Si the least productive. The researchers concluded that the reason for the better performance of the thin film CIS and double-junction a-Si modules was their better response to the blue light wavelengths found in the often overcast UK weather conditions.

In terms of energy yield *per square metre*, however, monocrystalline PV modules, owing to their high overall energy conversion efficiency, outperformed their thin film competitors, yielding 188 kWh m^{-2} yr^{-1} and 117 kWh m^{-2} yr^{-1} in Mallorca and Oxford respectively. CIS modules also gave high yields per unit area, but most polycrystalline silicon PV modules gave significantly lower area-related yields than their monocrystalline counterparts. The lowest yields per unit area were given by CdTe and single-junction a-Si. (It is worth noting, however, that there were problems with the single-junction a-Si modules used in the tests and that some other single-junction a-Si modules would probably have given considerably better performance.)

PV systems for non-domestic buildings

PV arrays can also be integrated into the roofs and walls of commercial, institutional and industrial buildings, replacing some of the conventional wall cladding or roofing materials that would otherwise have been needed and reducing the net costs of the PV system. In the case of some prestige office buildings, the cost of conventional cladding materials can exceed the cost of cladding with PV.

Commercial and industrial buildings are normally occupied mainly during daylight hours, which correlates well with the availability of solar radiation. Thus the power generated by PV can significantly reduce a company's need to purchase power from the grid at the 'retail' price. Surplus power is usually sold to the grid at a lower 'wholesale' price, so it is economically advantageous to use as much PV power as possible on-site rather than selling it to the grid. A few UK utilities, however, operate 'net metering' schemes where the buying and selling prices are effectively the same and the customer only pays for the net number of units used. Although net metering is unusual in the UK, it is widely available in other countries such as Japan, Germany and the Netherlands.

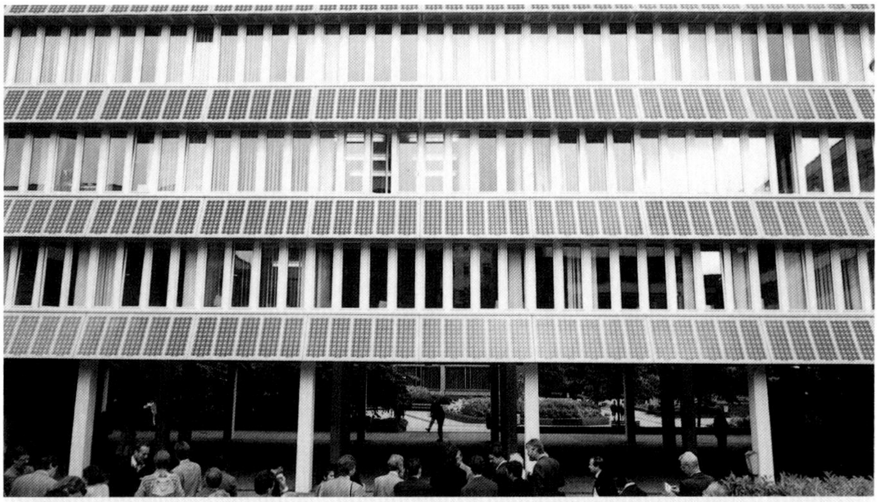

Figure 3.27 Britain's first building with PV cladding, a 40 kWp system installed in 1995 on the façade of a refurbished computer centre at the University of Northumbria, Newcastle

Figure 3.28 The roof of this petrol station in London has a large PV array, providing enough electricity to power the lights and petrol pumps. BP has installed several hundred such arrays on petrol stations around the world

Figure 3.29 This solar office at Doxford, near Sunderland in the UK, has a 73 kWp PV array integrated into its south-facing façade. The building also incorporates energy-efficient and passive solar design features to minimize its need for heating and lighting

There are now many examples of non-domestic buildings incorporating grid-connected PV systems in countries like Germany, Japan, the Netherlands, Italy, the UK and the USA (see Figures 3.27, 3.28 and 3.29).

Large, grid-connected PV power plants

Large, centralized PV power systems, mostly at the multi-megawatt scale, have also been built to supply power for local or regional electricity grids in a number of countries, including Germany, Switzerland, Italy and the USA (see Figure 3.30).

Compared with building-integrated PV systems, large stand-alone PV plants can take advantage of economies of scale in purchasing and installing large numbers of PV modules and associated equipment, and can be located on sites that are optimal in terms of solar radiation. On the other hand, the electricity they produce is not used on-site and has to be distributed by the grid. This involves transmission losses, and the price paid for the power by a local electricity utility will usually only be the 'wholesale' price at which it can buy power from other sources (although in some countries such as Germany, PV power is purchased at premium prices).

Large plants also require substantial areas of land, which has to be purchased or leased, adding to costs – although low-value 'waste' land, for example alongside motorways or railways, can be used (see Figures 3.31 and 3.33). The land can often be used for other purposes as well as PV generation. At Kobern-Gondorf in Germany (Figure 3.30) the site is used as a nature reserve for endangered species of flora and fauna. In some large PV plants, such as

Figure 3.30 The 340 kW PV power plant at Kobern-Gondorf, Germany, one of several large grid-connected PV installations operated by the electricity utility RWE

the 4 MW installation at Sonnen in Germany, the arrays have been mounted at least one metre above the ground, minimizing shading to the vegetation beneath and even allowing sheep to graze beneath the panels. It would also, in principle, be possible to have other forms of renewable energy generation alongside a large PV system, such as wind turbines – depending on wind conditions.

Large PV power plants are more economically attractive in those areas of the world that have substantially greater annual total solar radiation than northern Europe. Areas such as north Africa or southern California not only have annual solar radiation totals more than twice those in Britain, but also have clearer skies. This means that the majority of the radiation is direct, making tracking and concentrating systems effective and further increasing the annual energy output. The price of electricity from such PV installations is likely to be less than half of that from comparable non-tracking installations.

Figure 3.31 PV arrays can be installed on low-value land, as here beside a motorway in Switzerland

Satellite solar power

Probably the most ambitious – and some would say the most fanciful – proposal for a 'grid-connected' PV plant is the Satellite Solar Power System (SSPS) concept, first suggested more than 30 years ago (see Glaser, 1972 and 1992). The basic idea was to construct huge PV arrays, each over 50 km^2 in area and producing several GW of power, in geostationary orbit around the earth (Figure 3.32). The power would be converted to microwave radiation and beamed from 1 km-diameter transmitting antennas in space to 10 km^2 receiving antennas on earth. The received power would then be converted to alternating current and fed into a national electricity grid.

The advantages of the SSPS are, in theory, very substantial. In space, the PV arrays would receive a full 1365 W m^{-2} of solar power, instead of the 1000 W m^{-2} that is the maximum available

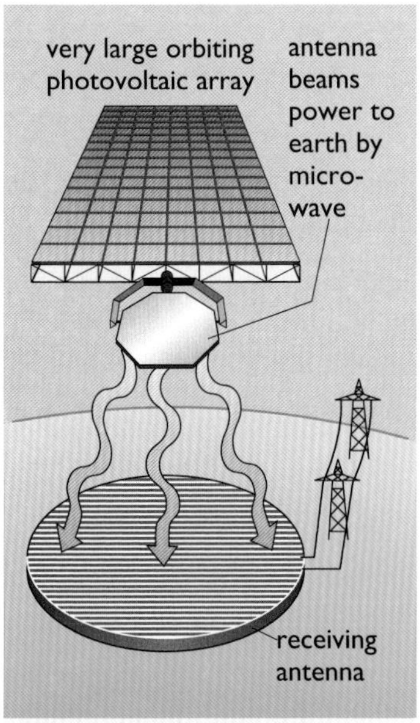

Figure 3.32 The satellite solar power station (SSPS) concept

at the earth's surface. Moreover, this high power would be available virtually constantly (except for occasional eclipses). And in the weightless and airless space environment, it should be possible to construct extremely large but very light PV arrays, without having to worry about the effects of wind and weather – though meteorites would be a problem.

On the other hand, the engineering challenges in constructing an SSPS, and the associated capital costs, would be enormous. One early US study estimated that a prototype would cost $79 billion at 1979 prices and would take 30 years to complete.

In 2001 the US space agency NASA reviewed the SSPS concept and found that, whilst great progress had been made in both photovoltaics and space technologies since the 1970s, further dramatic cost reductions and technological progress, involving greatly increased R&D funding, would be needed to make the concept viable.

The Japanese space agency NASDA has also recently expressed interest in satellite solar power systems. It envisages launching an experimental 500 kW SSPS in 2005–7, followed by 1 MW and 25 MW prototypes in 2010 and 2015, culminating in a 1 GW 'commercial' SSPS around 2020. NASDA is also investigating the use of laser beams rather than microwaves to transmit the power to earth.

However, huge amounts of capital would need to be raised to fund satellite solar power systems, restricting their use to a few rich nations. Building such systems also implies centralizing an energy source that many consider attractive precisely because of its decentralized nature. Concerns also persist about the health effects of the microwave (or laser) beams that would be employed – especially if there was a malfunction in the 'fail safe' control system that is meant to ensure that the beam always points at the receiving antenna. Interference with communications and radio astronomy could also be a problem. (For more details, see Further Information.)

3.10 Costs of energy from PV

As with any energy source, the cost per kilowatt-hour of energy from PV cells consists essentially of a combination of the capital cost and the running cost, as outlined in Appendix A1 and explained in more detail in Chapter 12 of the companion text.

The capital cost of a PV energy system includes not only the cost of the PV modules themselves, but also the so-called 'balance of system' (BoS) costs, i.e. the costs of the interconnection of modules to form arrays, the array support structure, land and foundations (if the array is not roof mounted), the costs of cabling, charge regulators, switching, inverters and metering, plus the cost of either storage batteries or connection to the grid.

Although the initial capital costs of PV systems are currently high, their running costs should be extremely low in comparison with those of other renewable or non-renewable energy systems. Not only does a PV system not require any fuel but also, unlike most other renewable energy systems, it has no moving parts (except in the case of tracking systems) and should require far less maintenance than, say, a wind turbine. (PV systems including batteries, however, have additional maintenance requirements.)

A report by the engineering consultancy Halcrow Gilbert Associates (HGA, 1996) examined in detail the costs of grid-connected PV for homes in the UK. It assumed a 2 kWp crystalline PV system delivering some 1500 kWh per year and supplying about 40% of the electricity needs of a four-bedroom home. It concluded that in UK conditions the maximum price buyers should have to pay for a 2 kW system was some £6.60 per peak watt, but that an achievable cost for a straightforward installation was between £5 and £5.50 per watt. With bulk purchase of systems for a number of houses, costs should reduce to around £4.40 per watt.

The HGA report examined three scenarios:

1. In the 'worst case' scenario, the local utility charges for connecting and commissioning the system and there are substantial operation and maintenance costs; the system is surface mounted so no savings can be made in roofing costs; and 45% of the electricity generated is used within the house, with 55% exported.
2. In the 'best case' scenario, there are no charges for connection and commissioning; the system replaces roof tiles costing £450; there are low operation and maintenance costs; and there is a small shift in consumption patterns by householders (in line with German experience of PV households) so that 60% of the electricity is consumed within the home and only 40% exported.
3. This scenario is the same as scenario 2, except that the utility permits a 'net metering' arrangement, increasing the value of power sold to the grid to the same level as the cost of power purchased from it.

For a one-off system, cost per kWh of electricity calculated by HGA ranged from 76p kWh^{-1} in the worst case scenario (assuming high interest on the capital costs) to 34p kWh^{-1} in the best case. For bulk-purchased systems, these costs reduced to 58p kWh^{-1} in the worst case and 30p kWh^{-1} in the best case. Because of the high capital costs and low value of electricity generated, even in the best case scenarios the payback times on the capital investment in the system were very long – over 100 years.

Since this 1996 study, as Figure 3.6 in Section 3.2 shows, PV system prices have reduced a little, to between about $5 and $7 per peak watt in the reporting countries shown, roughly equivalent (at 2003 exchange rates) to £3.50–£5 per watt – marginally lower than the prices assumed by HGA. PV system prices in the UK and a number of other countries are often substantially higher than these figures would suggest, however, owing to the relatively small number of installations, lack of installer experience and a relatively uncompetitive environment. (Small systems also tend to be more expensive per peak watt than larger systems.) In UK conditions, then, roof-top PV systems are not yet an attractive investment – at least in narrow financial terms.

In other countries, however, the financial climate is more attractive. The most favourable conditions are probably in Germany, where the REFIT (Renewable Energy Feed-In Tariff) scheme enables much higher prices to be paid for PV electricity sold to the grid – around 50 Euro cents per kWh, roughly equivalent to US$0.60 or £0.35. In addition, very low interest loans are offered to householders purchasing PV systems.

These attractive terms are intended to reflect the high environmental benefits of PV in comparison with other, more polluting energy sources, and to stimulate a strong PV industry sector. The price paid for PV electricity in Germany reduces by 5% per annum to encourage industry to lower its prices as its production and installation volumes build up. These measures enabled the German government's '100 000 roofs' target, involving an installed capacity of 300 MW of domestic PV, to be achieved in 2003. From 2004 onwards, a new scheme broadly similar to REFIT is due to come into operation.

In the UK, the Government in 2003 introduced a system of grants of between 40% and 60% of the costs of PV systems and has declared its intention to stimulate the growth of a strong PV sector comparable to that of Germany and Japan. But the 50% contribution, whilst welcomed by the industry, is unlikely to be sufficient to make investment in PV attractive to large numbers of UK home owners.

Reducing the costs of power from PV

How, then, might the price of PV power be made more competitive?

Firstly, the installed cost per peak watt needs to drop substantially. The key to this lies in mass production. Historically, PV production costs have dropped by more than 20% for every doubling of production quantity. Most PV production plants are still relatively small, producing modules with a total peak output of the order of 10 MWp per year. A report in 1999 by the consultancy KPMG, quoting widely accepted industry studies, concluded that if much larger plants of around 500 MWp per annum production capacity were built, module prices could be reduced by a factor of about four, to around $1 per peak watt. The price of PV electricity should then become competitive with average retail electricity prices in the developed OECD countries. Such plants have not yet been built because a large enough market for them does not exist; and a large market does not yet exist because large plants have not been built to bring down prices. This impasse could be broken by a major investment in such plants by either industry or Government, which would massively accelerate the take-up of PV systems. Otherwise, the unaided market could take decades to reduce prices and increase volumes to similar levels (KPMG, 1999).

Secondly, the overall annual conversion efficiency of the PV arrays needs to increase substantially. As mentioned above, the leading crystalline PV modules currently have an efficiency of over 17%. Japan's Sharp Corporation, the world's largest PV manufacturer, foresees module efficiencies rising to perhaps 23–24% in the next decade (Appleyard, 2003).

Thirdly, the 'balance-of-system' (BoS) costs need to be substantially reduced. In existing PV systems the BoS costs are roughly three-quarters of the module costs, so to keep this ratio these costs also need to be reduced to below $1 per watt. This should be feasible, given volume production of BoS components and the likelihood of substantially reduced installation costs when the industry has gained more experience.

Recent studies (PIU, 2002) suggest that, with increased production volume and continued technological improvements, the price of electricity from PV systems in the UK is likely to fall to around 10–16p per kWh by 2020. Such prices would still be relatively high, around twice the current retail price of domestic electricity in the UK, but could be seen as reflecting the high environmental value of PV-generated electricity (see Hohmeyer, 1988).

3.11 Environmental impact and safety

Environmental impact and safety of PV systems

The environmental impact of PV is probably lower than that of any other renewable or non-renewable electricity generating system.

In normal operation, PV energy systems emit no gaseous or liquid pollutants, and no radioactive substances. However, in the case of CIS or CdTe modules, which include very small quantities of toxic substances, there is a slight risk that a fire in an array might cause small amounts of these chemicals to be released into the environment.

PV modules have no moving parts, so they are also safe in the mechanical sense, and they emit no noise. However, as with other electrical equipment, there are some risks of electric shock – especially in larger systems operating at voltages substantially higher than the 12–48 V employed in most small PV installations. But the electrical hazards of a well-engineered PV system are, at worst, no greater than those of other comparable electrical installations.

PV arrays do, of course, have some visual impact. Roof-top arrays will normally be visible to neighbours, and may or may not be regarded as attractive, according to aesthetic tastes. Several companies have produced special PV modules in the form of roof tiles that blend into roof structures more unobtrusively than conventional module designs (see Figure 3.25).

As already mentioned, PV arrays on buildings require no additional land, but large, multi-megawatt PV arrays will usually be installed on land specially designated for the purpose, and this will entail some visual impact. One study (Gagnon *et al.*, 2002; see Chapter 13 of the companion text) calculates a net land requirement of 45 $km^2\ TWh^{-1}\ yr^{-1}$ for photovoltaic electricity production, but this assumes a large-scale PV power plant on a dedicated site. In some countries such as Switzerland, the authorities are installing large PV arrays as noise barriers alongside motorways and railways (Figure 3.33). Arguably PV is here *reducing* the overall environmental impact.

Figure 3.33 PV arrays installed beside a railway in Switzerland

Environmental impact and safety of PV production

The environmental impact of manufacturing silicon PV cells is unlikely to be significant – except in the unlikely event of a major accident at a manufacturing plant. The basic material from which the vast majority of PV cells are made, silicon, is not intrinsically harmful. However, small amounts of toxic chemicals are used in the manufacture of some PV modules. Cadmium is obviously used in the manufacture of cadmium telluride modules. Small amounts of cadmium are also currently used in manufacturing CIS and CIGS modules – although new processes now available allow this to be eliminated.

As in any chemical process, careful attention must be paid to plant design and operation to ensure the containment of any harmful chemicals in the event of an accident or plant malfunction.

Even though PV arrays are potentially very long-lived devices, eventually they will come to the end of their useful life and will have to be disposed of or, preferably, recycled. Some manufacturers are already recycling PV modules, and draft EU regulations on module recycling are in preparation.

Energy balance of PV systems

A common misconception about PV cells is that almost as much energy is used in their manufacture as they generate during their lifetime. This may well have been true in the early days of PV, when the refining of monocrystalline silicon and the Czochralski process were very energy intensive, and the efficiency of the cells produced was relatively low, leading to low lifetime energy output.

However, with more modern PV production processes introduced in recent years, and the improved efficiency of modules, the energy balance of PV is now more favourable. A recent study (Alsema and Niewlaar, 2000) found the energy payback time for PV modules (including frames and support structures) to be between 2 and 5 years in European conditions, and that with future improvements this should reduce to 1.5–2 years. (See also the companion text, Chapter 13, Box 13.11.) The use of materials with low embodied energy (such as wood) in PV array support structures can also improve the overall energy payback time of PV systems.

3.12 PV integration, resources and future prospects

Integration

If PV energy systems continue to improve in cost-effectiveness compared with more conventional sources, as seems likely, in what way would national energy systems need to be modified to cope with the long-, medium- and short-term fluctuations in the output of PV arrays?

In the UK, most PV power would be produced in summer, when electricity demand is relatively low; much less would be produced in winter when demand is high. And although PV power is quite reliable (during daylight hours) in climates with mainly clear skies, it can be highly intermittent in countries like the UK, where passing clouds can reduce output dramatically within seconds.

But as long as the capacity of variable output power sources such as PV is fairly small in relation to the overall capacity of the grid (most studies suggest between 10 and 20%), there should not be a major problem in coping with their fluctuating output. The grid is, after all, designed to cope with massive fluctuations in *demand*, and similarly fluctuating sources of *supply* like PV can be considered equivalent to 'negative loads'. Such fluctuations would also, of course, be substantially smoothed out if PV power plants were situated in many different locations subject to widely varying solar radiation and weather patterns.

However, if PV power stations, and other fluctuating renewable energy sources such as wind power, were in future to contribute more than about 20% of electricity supplies, then the 'generating mix' supplying the grid would have to be changed to include a greater proportion of 'fast-response' power plants, such as hydro or gas turbines, and increased amounts of short-term storage and 'spinning reserve'.

These considerations lead some analysts to suggest that without large quantities of cheap electrical energy storage, intermittent renewable energy sources like PV cannot make a major contribution. Whilst this would seem to be an exaggeration, at least for small to medium levels of 'penetration' of PV and other renewables into the system, it is certainly true that cheap storage in large amounts would make their integration easier.

This is one reason for the recent revival of interest in the use of **hydrogen** as a medium for energy storage and distribution. Hydrogen would be produced by the electrolysis of water, using PV (or other renewables) as the electricity source. The hydrogen would be stored and transported to wherever it was needed, then converted back to electricity using fuel cells (see Figure 3.34) . This topic is discussed further in Chapter 10: Integration.

Figure 3.34 This solar house in Freiburg, southern Germany, is not connected to the grid. PV arrays on the roof provide all its electricity requirements. When PV electricity is available but not being used in the house, it is used to produce hydrogen, which is stored in a large tank in the garden. When there is a deficit of PV electricity, the stored hydrogen is used to produce electricity via a fuel cell in the basement, which also produces hot water. The house also has very high levels of insulation, so its space heating requirements are very low. The house was constructed as a demonstration project in the 1990s and is now used as a laboratory

PV resources

The overall resource available in principle from PV is enormous. A simple calculation shows that if PV modules of 10% average efficiency were installed on 0.1% of the earth's surface (some 500 000 km^2, equivalent to about 1.3% of the earth's total desert area) they would produce enough electricity to supply all of the world's current energy requirements.

A similar calculation for Britain shows that if PV modules of 10% annual average conversion efficiency were located on about 1.4% of the country's land area (i.e. occupying 35 000 km^2) they could produce some 350 TWh annually, equivalent to current UK electricity consumption.

In practice, of course, there are many constraints that reduce the resource available from PV to a fraction of the amounts calculated above. (Box 10.1 in Chapter 10: Integration gives a consistent set of terms to describe the various types of resource, ranging from 'available resource' to 'economic potential'.)

A detailed study in 2002 by the International Energy Agency covering 14 countries, including most of Europe, Japan, Australia, Canada and the USA, concluded that the potential contribution to national electricity production from PV integrated into the built environment ranges from 15% (for Japan) to nearly 60% (USA). For the UK, the figure is 30%, implying a potential of around 100 TWh per year. The IEA calculations excluded any surfaces that presented installation problems and any that would yield less than 80% of the output of an optimal system because of poor orientation, inclination or shading (IEA, 2002).

A recent study for the UK Prime Minister's Strategy Unit (Chapman and Goss, 2002), echoing an earlier study by the Energy Technology Support Unit (ETSU, 1999), estimated the 'technical potential' for building-integrated PV in the UK by 2025 to be around 37 TWh per year (see Table 10.1 in Chapter 10: Integration). This estimate is rather lower than the c. 100 TWh per year implied by the IEA study, mainly because it takes into account the rate of construction of new buildings. It also concluded that if the maximum acceptable price of power from renewables were set at 7p kWh^{-1}, then the 'economic potential' resource available from PV on buildings would be dramatically reduced, to only around 0.5 TWh per year. However, a price of 7p kWh^{-1} (around the current UK level of domestic electricity prices) may be a somewhat arbitrary figure when looking as far ahead as 2025. In other countries such as Japan, electricity prices are much higher, and some analysts expect conventional UK electricity prices to rise substantially in real terms in the future, not least as a reflection of the environmental and social costs of conventional electricity production (see Hohmeyer, 1988). So the 'economic potential' of PV in the UK could eventually be substantially greater than the 0.5 TWh quoted in the Strategy Unit and ETSU studies.

The growing world photovoltaics market

In each of the years 2001 and 2002, world PV production (in both IEA and non-IEA countries) grew by just under 40% per annum, bringing total world PV production in 2002 to an estimated 560 MWp (Schmela, 2003). This growth rate implies a doubling of world PV production every two years.

The phenomenal expansion of recent years has been accompanied by major changes in the PV industry. In 1997, US manufacturers had the largest share of the world market, at 41%, while Japan's share was 25%, Europe's was 23% and the rest of the world's 11%. But by 2002, Japanese PV manufacturers had surged ahead to capture the largest market share, around 44%. The European manufacturers' share remained at around 25%, the USA's dropped to around 20% and that of the rest of the world (including India, the rest of Asia and Australia) stayed at about 10%.

In 2002 the world's largest PV manufacturer was Japan's Sharp Corporation, which expanded its production of PV cells by 66% in that year alone, reaching 123 MW. In 2003 Sharp aimed to boost production to 200 MW including a new PV plant for the European market, located in Wales. Other leading Japanese manufacturers were Sanyo, Kyocera and Mitsubishi.

The leading European manufacturers were BP Solar, which in 2002 produced an estimated 71 MW, and Shell Solar, which produced 55 MW. Since the merger of the oil companies BP and Amoco, BP Solar now incorporates the US firm Solarex, which was owned by Amoco. As noted earlier, BP has closed its amorphous silicon and cadmium telluride thin film PV plants in the USA but is expanding its crystalline PV manufacturing capacity in Spain, India, Australia and the USA. Another major oil company, Shell, which already had substantial involvement in PV, has entered into a partnership with the PV manufacturer Siemens Solar, forming a joint venture company known as Shell Solar.

Crystalline silicon still dominates PV technology in the marketplace, with monocrystalline and polycrystalline modules accounting for some 88% of world production in 2002. Perhaps surprisingly, given the optimism expressed by many about the prospects for thin-film PV modules when they were first introduced, the market share of thin-film amorphous silicon, cadmium telluride and copper indium diselenide modules has declined in recent years, from about 12% in 1999 to around 6% in 2002. This is probably due in part to silicon PV manufacturers improving the performance and reducing the price of their modules, making it more difficult for thin films to compete. The market share of PV modules using crystalline silicon ribbon and silicon sheet technology stayed roughly constant at around 5% between 1999 and 2002 (Schmela, 2003).

Future prospects: national and international PV research, development and demonstration programmes

If PV is to realize its ultimate potential to become one of the world's leading energy sources, there will need to be sustained programmes of R&D and market stimulation. Here we highlight very briefly a few of the main programmes in progress in some of the key countries and continents, and how each views the future prospects for PV.

USA

In 1998 the then US President Clinton announced a 'One million solar roofs' programme, aimed at stimulating the installation of a million solar roofs, including solar thermal as well as PV systems, in the United States, to be achieved by a variety of measures at both federal and state level (see http://www.millionsolarroofs.org/index.html).

In 1999, the US PV industry published a 20-year 'Industry Roadmap' to provide a strategic framework for PV development. It envisages a 25% per annum growth in US PV manufacturing capacity, with the aim of delivering 15% (equivalent to 3.2 GW) of the new peak generating capacity expected to be required in the USA in 2020. It predicts home PV prices falling to around $1.5 per peak watt by 2020 (see NCPV, 1999).

Europe

Germany has its ambitious 100 000 roofs programme, as mentioned above, and several other European countries also have substantial PV development programmes. The European PV Industry Association (EPIA) published in 2003 its 'Industrial Roadmap for Solar Electricity' (see Lysen, 2003), which stresses the need for major investments in PV during the decade to 2010 and beyond. It envisages 3 GW of installed PV capacity in Europe by 2010 and module manufacturing prices dropping to below 2 Euros per peak watt.

Japan

Over the past decade, Japan has undertaken a major investment in both photovoltaics R&D and in manufacturing capacity, taking advantage of a market stimulated by government subsidies of around 50% of the capital costs of residential PV systems, with the aim of encouraging the installation of some 70 000 roof-top PV systems. The Japanese PV industry is now the world's largest, and is growing very rapidly. Japan released its 'Vision of a Self-Sustainable PV Industry' in June 2002. This envisages some 4.8 GW of PV capacity installed in Japan by 2010, with production in 2010 reaching over 1200 MW, two thirds of it destined for the housing sector. By 2020, it foresees production rising to 4.3 GW per year, with a further rise to 10 GW per year by 2030. By then, the country's total installed capacity would be over 80 GW with a market value of over US$18 billion (see also DTI, 2003).

Realizing the global potential

To underline the enormous contribution that PV could make on a world scale, the European PV Industry Association and Greenpeace published in 2001 a report: *Solar Generation: Solar Electricity for over 1 Billion People and 2 Million Jobs by 2020.*

By 2020, it envisages just over 200 GW of installed PV capacity worldwide, supplying 1 billion off-grid and 82 million grid-connected customers, 30 million of them in Europe. By then, some 60% of PV production would be located in the non-industrialized countries, especially south Asia and Africa, and the PV industry would be supporting over 2 million full-time jobs. Looking even further ahead, to 2040, the report foresees a world solar electricity output of over 9000 TWh, meeting just over one quarter of global electricity demand by that date.

If this ambitious vision is successfully realized several decades hence, then the technology of photovoltaics, currently still in its adolescence, will truly have reached maturity.

References

Adams, W. G. and Day, R. E. (1877) 'The action of light on selenium', *Proceedings of the Royal Society*, London, Series A, 25, p. 113.

Alsema, E. A. and Niewlaar, E. (2000) 'Energy viability of photovoltaic systems', *Energy Policy*, vol 28, pp. 999–1010.

Appleyard, D. (2003) Interview with Takashi Tomita of Sharp Corporation, *Regen,* June–July, pp. 32–34.

Becquerel, A. E. (1839) 'Recherches sur les effets de la radiation chimique de la lumière solaire au moyen des courants électriques' and 'Mémoire sur les effets électriques produit sous l'influence des rayons solaires', *Comptes Rendus de l'Académie des Sciences*, vol 9, pp. 145–149 and 561–567.

Boes, E. C. and Luque, A. (1993) 'Photovoltaic concentrator technology', in Johansson, T. B. *et al.* (eds) *Renewable Energy Sources for Fuels and Electricity*, Washington DC, Island Press, pp. 369–370.

Chalmers, R. (1976) 'The photovoltaic generation of electricity', *Scientific American*, October, pp. 34–43.

Chapin, D. M., Fuller, C. S. and Pearson, G. L. (1954) 'A new silicon p–n junction photocell for converting solar radiation into electrical power', *Journal of Applied Physics*, vol 25, pp. 676–677.

Chapman, J. and Goss, R. (2002) *Technical and economic potential of renewable energy generating technologies: Potentials and cost reductions to 2020*, Working Paper for UK Cabinet Office Performance and Innovation Unit (now Strategy Unit) Energy Review, downloadable from http://www.strategy.gov.uk/2002/energy/workingpapers.shtml [accessed 26 June 2003].

Derrick, A., Francis, C. and Bokalders, V. (1991) *Solar Photovoltaic Products – A Guide for Development Workers*, IT Publications and IT Power.

Dresselhaus, M. S. and Thomas, I. L. (2001) 'Alternative energy technologies', *Nature*, vol 414, pp. 332–337.

DTI (Department of Trade and Industry) (2003) *Developments in Solar Photovoltaics in Japan*, Global Watch Mission Report, Published by Pera Innovation Ltd on behalf of the DTI, November, p. 40.

EPIA (European Photovoltaics Industry Association) (2001) *Solar Generation: Solar Electricity for over 1 Billion People and 2 Million Jobs by 2020*, published in association with Greenpeace, http://www.cleanenergynow.org/resources/solargenback.pdf.

ETSU (Energy Technology Support Unit) (1999) *New and Renewable Energy: Prospects for the 21st Century: Supporting Analysis.* Report R 122.

European Commission (1997) *White Paper for a Community Strategy and Action Plan*. Available from http://europa.eu.int/ [accessed 12 December 2003].

Glaser, P. (1972) 'The case for solar energy', paper presented at the annual meeting of the Society for Social Responsibility in Science, Queen Mary College, London, September.

Glaser, P. (1992) ‘An overview of the solar power satellite option’, *IEEE Transactions on Microwave Theory and Techniques*, vol. 40, no. 6, June, pp. 1230–1238.

Grätzel, M. *et al.* (1989) *The Artificial Leaf: Molecular Photovoltaics Achieve Efficient Generation of Electricity from Sunlight,* Research Report, Ecole Polytechnique Fédérale de Lausanne, Switzerland.

Grätzel, M. (2001) ‘Photoelectrochemical cells’, *Nature*, vol 414, 15 November, pp. 338–344.

Green, M. (1982) *Solar Cells*, Englewood Cliffs, Prentice-Hall.

Green, M. (1993) ‘Crystalline and polycrystalline silicon solar cells’, in Johansson, T. B. *et al.* (eds) *Renewable Energy Sources for Fuels and Electricity*, Washington DC, Island Press, pp. 337–360.

HGA (Halcrow Gilbert Associates) (1996) *Report on grid-connected PV.*

Hohmeyer, O. (1988) *Social Costs of Energy*, Berlin, Springer-Verlag.

IEA (International Energy Agency) (2003) *Trends in Photovoltaic Applications. Survey report of selected IEA countries between 1992 and 2002.* Available from http://www.oja-services.nl/iea-pvps/products/index.htm [accessed 12 December 2003].

Imamura, M. S., Helm, P. and Palz, W. (1992) *Photovoltaic System Technology: A European Handbook*, Bedfordshire, UK, W. H. Stephens, for Commission of European Communities.

IT Power (2001) *Power to Tackle Poverty,* report presented to the 2002 United Nations World Summit on Sustainable Development, Johannesburg, South Africa, Published by IT Power and Greenpeace, 20pp.

Jardine, C. N. and Lane, K. (2003) ‘PV-compare: energy yields of photovoltaic technologies in northern and southern Europe’, *Photovoltaic Science, Applications and Technology*, Proceedings of Joint Meeting of UK Solar Energy Society and PVNET, Loughborough University, UK, April, pp. 42–47.

Johansson, T. B., Kelly, H., Reddy, A. K. N. and Williams, R. H. (eds) (1993) *Renewable Energy Sources for Fuels and Electricity*, Washington DC, Island Press.

KPMG (1999) *Solar Energy: from Perennial Promise to Competitive Alternative*, report by KPMG Bureau voor Economische Argumentatie, Hoofdorp, Netherlands, commissioned by Greenpeace Netherlands, p. 52.

Lasnier, F. and Ang, T. G. (1990) *Photovoltaic Engineering Handbook*, Bristol, Adam Hilger.

Lysen, E. (2003) ‘Photovoltaics: an outlook for the 21st century’, *Renewable Energy World*, January–February, pp. 43–53.

McNelis, B. and Jesch, L. F. (eds) (1994) *Proceedings of UK International Solar Energy Society 20th Anniversary Conference*, January, London, ISES.

McVeigh, C. (1983) *Sun Power*, Oxford, Pergamon.

NCPV (National Center for Photovoltaics) (1999) *Report of the National Photovoltaic (PV) Industry Workshop*, Chicago, USA, September, p. 73.

Available from NCPV website: http://www.nrel.gov.ncpv [accessed December 2003].

Ohl, R. S. (1941) *Light Sensitive Device*, US Patent No. 2402622; and *Light Sensitive Device Including Silicon*, US Patent No. 2443542, both filed 27 May.

O'Regan, B. and Grätzel, M. (1991) 'A low cost, high efficiency solar cell based on dye-sensitised colloidal TiO_2 films', *Nature*, vol 235, pp. 737–740.

O'Regan, B., Nazeruddin, M. K. and Grätzel, M. (1993) 'A very low cost, 10% efficient solar cell based on the sensitisation of colloidal titanium dioxide films', *Proceedings of 11th European Photovoltaics Conference*, Montreux, Switzerland, Gordon and Breach.

Palz, W. (1994) 'Power for the world: a global photovoltaic action plan', in McNelis, B. and Jesch, L. F. (eds) *Proceedings of UK International Solar Energy Society 20th Anniversary Conference*, January, London, ISES, pp. 7–41.

PIU (Performance and Innovation Unit) (2002) *The Energy Review*, Chapter 6. Downloadable from: http://www.number-10.gov.uk/su/energy/9.html [accessed December 2003].

Roberts, S. (1991) *Solar Electricity: A Practical Guide to Designing and Installing Small Photovoltaic Systems*, London, Prentice Hall.

Schmela, M. (2003) 'A bullish PV year: market survey on world cell production in 2002', *Photon International*, vol 3, March, pp. 42–48.

Treble, F. C. (ed.) (1991) *Generating Electricity from the Sun*, Oxford, Pergamon Press.

Treble, F. C. (1999) *Solar Electricity: a Lay Guide to the Generation of Electricity by the Direct Conversion of Solar Energy*, 2nd edition, The Solar Energy Society, Oxford Brookes University, Oxford OX3 0BP.

Further information

An excellent introduction to the physical principles underlying photovoltaic solar energy conversion and the detailed design of solar cells is available on the CD ROM *Photovoltaics: Devices, Systems and Applications* produced by the Key Centre for Photovoltaic Engineering at the University of New South Wales (UNSW) in Australia. Details from http://www.pv.unsw.edu.au. Complementing the CD ROM are three companion textbooks:

Green, M. A. (1982) *Solar Cells: Operating Principles, Technology and System Applications*, Englewood Cliffs, Prentice Hall;

Green, M. A. (1995) *Silicon Solar Cells: Advanced Principles and Practice*, Sydney, Centre for Photovoltaic Devices and Systems, UNSW;

Wenham, S. R., Green, M. A. and Watt M. (n.d.) *Applied Photovoltaics*, Sydney, Centre for Photovoltaic Devices and Systems, UNSW.

A detailed appraisal of the current status and future prospects of photovoltaics can be found in Archer, M. D. and Hill, R. (eds) (2001) *Clean*

Energy from Photovoltaics (series on *Photoconversion of Solar Energy, Volume 1*), London, Imperial College Press.

A good concise introductory text on PV is *Solar Electricity* (Treble, 1999). The territory is covered in more detail in the same author's *Generating Electricity from the Sun* (Treble, 1991).

Engineers and others wishing to design PV energy systems will find *Photovoltaic Systems Technology: A European Handbook* (Imamura *et al.*, 1992) and the *Photovoltaic Engineering Handbook* (Lasnier and Ang, 1990) particularly useful. Equally useful, but pitched at a less advanced technical level, is *Solar Electricity: A Practical Guide to Designing and Installing Small Photovoltaic Systems* (Roberts, 1991).

Software to assist in designing PV systems (and many other renewable energy systems) has been developed by Natural Resources Canada. Called *RETScreen*, it can be downloaded free of charge from http://www.retscreen.net/ang/ [accessed 8 December 2003].

Details of US proposals for satellite solar power systems (SSPS) are available from the NASA website at: http://spacesolarpower.nasa.gov/ [accessed 8 December 2003]; and from the US Office of Energy Efficiency and Renewable Energy website at: http://www.eere.energy.gov/consumerinfo/refbriefs/l123.html [accessed 8 December 2003].

Details of Japanese SSPS proposals can be found at: http://www.nasda.go.jp/lib/nasda-news/2002/07/front_line_e.html [accessed 8 December 2003].

The history of photovoltaics since the early 1950s is described in Loferski, J. (1993) 'The first forty years: a brief history of the modern photovoltaic age', *Progress in Photovoltaics*, vol 1, no. 1, pp. 67–78.

For further information on PV in the built environment see the International Energy Agency's PV Power Systems website: http://www.iea-pvps.org [accessed 14 March 2003]. The IEA PVPS project has also produced a very informative CD ROM, *Photovoltaics in the Built Environment*, that includes an extensive set of reports, documents, photographs and presentations on the subject.

Architects interested in PV will find *Photovoltaics in Architecture* (Thomas, R. (ed) (2001) London and New York, Spon Press, 156pp.) a valuable guide.

Further information on the DTI grants scheme for PV systems in the UK is available at http://www.solarpvgrants.co.uk [accessed 8 December 2003] and from the Energy Saving Trust at: http://www.est.org.uk/ [accessed 25 November 2003].

Academic journals covering the photovoltaics field include *Solar Energy Materials and Solar Cells* (Elsevier) and *Progress in Photovoltaics* (Wiley). There are also various industry newsletters, including *PV News* (PO Box 290, Casanova, VA 22017, USA) and a monthly magazine, *Photon*, published (in English) by Solar Verlag GmbH, Wilhelmstrasse 34, 52070 Aachen, Germany.

Chapter 4

Bioenergy

by Stephen Larkin, Janet Ramage and Jonathan Scurlock

4.1 Introduction

Bioenergy is the general term for energy derived from materials such as wood, straw or animal wastes, which were living matter relatively recently – in contrast to the fossil fuels. Such materials can be burned directly to produce heat or power, but can also be converted into **biofuels**. Charcoal and biodiesel, for example, are biofuels made from wood and plant seeds respectively.

All the earth's living matter, its **biomass**, exists in the thin surface layer called the biosphere. It represents only a tiny fraction of the total mass of the earth, but in human terms it is an enormous energy store. More significantly, it is a store which is continually replenished by the flow of energy from the sun, through the process of photosynthesis. Although only a small fraction of the solar energy reaching the earth each year is fixed in this way by organic matter on land (see Figure 1.8 in Chapter 1), it is nevertheless equivalent to almost seven times the world's total primary energy consumption.

It is important to appreciate the role of the biomass in maintaining the earth's atmosphere. If a cosmic hurricane were to sweep away all the plant life on earth, the resulting loss of mass would be no more than one part in a billion – like blowing the dust off a school globe. Yet the physical consequences of this infinitesimal change would be enormous. There would no longer be a supply of oxygen to the atmosphere, and it is the composition of the atmosphere – the particular mixture of nitrogen, oxygen and trace gases such as CO_2 – which in turn maintains the surface conditions on the earth. When we consider the possible effects of human actions on the environment it is essential to bear in mind the important fact that the biomass and the atmosphere are not two separate features of the surface layers of the earth: their interdependence is so strong that it is essential to treat them as one single system.

In nature, the energy stored in the carbohydrates in plants is dissipated through a series of conversions involving chemical and physical processes in the plant, the soil, the surrounding atmosphere and other living matter, until it is eventually radiated away from the earth as low-temperature heat. Some will be lost within a year, but biomass can accumulate over decades in the wood of trees. A small fraction may accumulate over centuries as peat, and a tiny proportion has become fossil fuel over periods measured in millions of years.

The significance of these processes is that if we can intervene and 'capture' some of the biomass at the stage where it is acting as a store of chemical energy, we have a fuel. Moreover, provided our consumption does not exceed the natural level of production, the combustion of biofuels *should generate no more heat and create no more carbon dioxide* than would have been formed in any case by natural processes. So it seems that here we have a truly sustainable energy source, with no deleterious global environmental effects at all. However, as we shall see later, this is not always the case in practice – and there may be other effects to consider.

Material such as firewood, rice husks and other plant or animal residues can simply be burned to produce heat, and in many developing countries this **traditional biomass** continues to account for a large part of energy consumption. In recent decades, however, terms such as **'new' biomass**

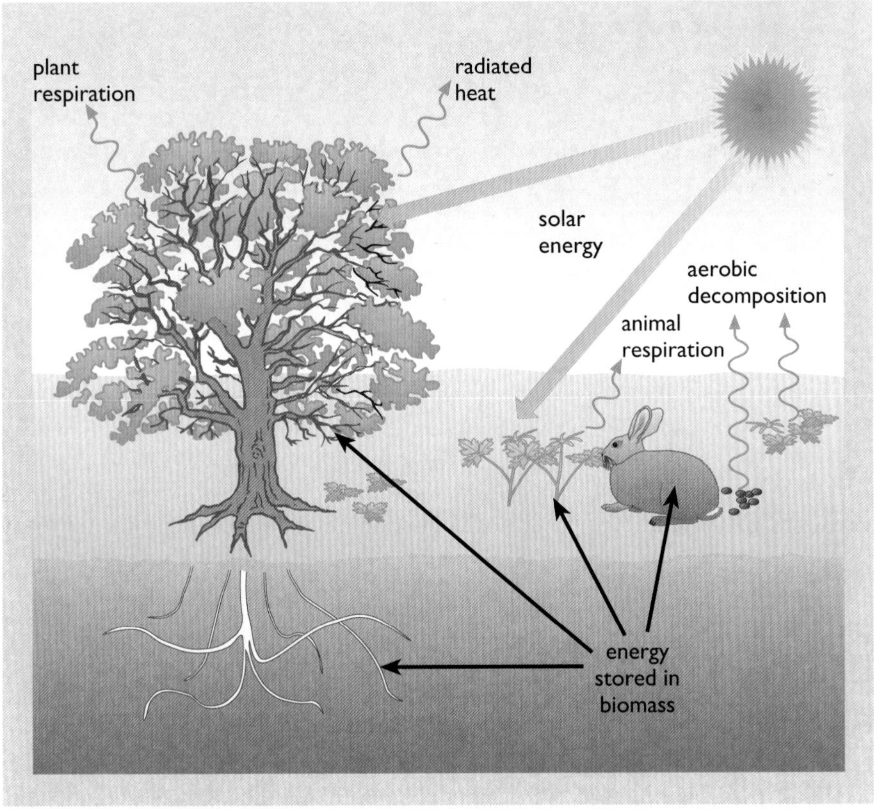

Figure 4.1 The bioenergy cycle on the local scale

have come into use to characterize materials that are processed on a large, commercial scale, usually in the more industrialized countries. The input to these processes may be purpose-grown **energy crops**, but often they are organic **wastes**. The output may be useful heat, or any of a wide range of solid, liquid or gaseous biofuels.

This chapter discusses all the above features of bioenergy, and other aspects such as the economics and the potential future for this world-wide renewable resource.

BOX 4.1 Biomass – some basic data

Note that almost all the data here are subject to considerable uncertainty (see Section 4.2).

World totals

Total mass of living matter (including moisture)	2000 billion tonnes
Total mass in land plants	1800 billion tonnes
Total mass in forests	1600 billion tonnes
World population (2002)	6.2 billion
Per capita terrestrial plant biomass	300 tonnes
Energy stored in terrestrial biomass	25 000 EJ
Net annual production of terrestrial biomass	400 000 Mt y^{-1}

World energy comparisons

Rate of energy storage by land biomass	3000 EJ y^{-1} (95 TW)
Total primary energy consumption (2002)	451 EJ y^{-1} (14.3 TW)
Biomass energy consumption	56 EJ y^{-1} (1.6 TW)
Food energy consumption	16 EJ y^{-1} (0.5 TW)

4.2 Bioenergy past and present

From wood to coal

Until recent times, the history of fuels was essentially the history of biofuels. Although there is evidence of coal-burning as early as 3000 years ago, its contribution remained relatively small until about 200 years ago. Indeed, bioenergy was still dominant in many areas of life well into the Industrial Revolution, with wood for heat, tallow candles made from animal fats for light, and grasses as 'fuel' for the only means of transport – the horse.

The move from biofuel to fossil fuel was a key feature of the Industrial Revolution. For many centuries, the high temperatures needed for iron smelting could be achieved only in furnaces using charcoal (see Section 4.6 below). The impurities and variable nature of coal made it unsuitable for smelting, and attempts to reduce it to a type of charcoal had little success initially. But in the early 1700s, an effective 'coal charcoal' was produced, and within a few decades, this new fuel, now called coke, was replacing charcoal throughout the growing industrial sector.

The increased demand for coal led to deeper mines, and the need to pump flood water from great depths led to the first steam engines – whose main fuel was of course coal. By the end of the nineteenth century, coal was dominant in the world's industrialized countries. The twentieth century saw the rise of oil and natural gas, but it is worth noting that coal consumption also increased five-fold between 1900 and 2000.

Will the twenty-first century see the reverse process: '*from coal to wood*'? Could biofuels completely replace fossil fuels? The data in Box 4.1 show that it is theoretically possible, and the case has been made for such a change to occur (see, for instance, Hall, 1991 and Hall *et al.*, 1993). But there will of course be circumstances where other sustainable resources are more appropriate. These possibilities for the future are discussed in Section 4.11 below, and in more detail in Chapter 10.

Present biomass contributions

Obtaining a reliable estimate of the total world-wide energy contribution from the many sources of bioenergy is a task fraught with difficulties. Unlike the fossil fuels, bioenergy has no global companies producing detailed reports on production and consumption. Indeed, the use of traditional biomass often involves no financial transaction at all, or the trading is local and unrecorded. In recent years, the International Energy Agency has endeavoured to collect national data based on agreed categories of renewable energy, but they warn that their figures are still subject to a great deal of uncertainty. Recent estimates of the annual contribution to world primary energy from traditional biomass fall in the range 40–60 EJ. Figure 1.1 in Chapter 1 is based on a contribution of 48 EJ, with 'new' biomass adding a further 9 EJ or so. (See UNDP, 2000: IEA, 2002: IEA, 2003).

The details may be uncertain, but there is little doubt that biomass is a major energy provider over much of the world. It accounts for about a third of total primary energy consumption in the developing countries (up to 90% in some of the poorest), and its total annual contribution continues to rise. Its *percentage* contribution to world primary energy is falling slightly as developing countries industrialize, but it remains important even in the more advanced of these, accounting for about 20% of primary energy in China and 40% in India.

Even in the industrialized world, the energy contribution from biomass can be significant, particularly in countries with large forestry industries or well-developed technologies for processing residues and wastes. In Sweden and Finland, where biomass contributes about 20% of primary energy, the use of residues in the pulp and paper industries is important. And advanced systems for domestic heating, district heating and CHP (combined heat and power) have helped Sweden towards a 5-fold increase in the use of bioenergy over the past decade or so, reaching nearly 40 GJ per head of population in 2001. In the UK, the bioenergy contribution rose at about 7% per year during the 1990s, and reached a total of 134 PJ, or 2.2 GJ per capita in 2002 (DTI, 2003a, see also Figure 1.5 in Chapter 1).

4.3 Biomass as a fuel

What are fuels?

It is common experience that some materials will burn whilst others won't. Why? What is it that makes wood a fuel and sand not? We might start with a few well-known facts about combustion:

- it needs air – or to be more precise, oxygen;
- the fuel disappears – or at least undergoes a major change;
- heat is produced, i.e. energy is released.

It would appear then that a fuel is a substance which interacts with oxygen and in doing so releases energy and changes into different chemical compounds – the combustion products.

We know enough about the composition of common fuels to be able to predict what these products will be. Consider, for instance, methane – a biofuel and also the principal component of natural gas. Each methane molecule consists of one carbon and four hydrogen atoms: CH_4. Oxygen is a diatomic gas, with molecules consisting of two atoms (O_2), so in full combustion each methane molecule reacts with two of these:

$$CH_4 + 2O_2 \rightarrow CO_2 + 2H_2O + \text{energy.}$$

The heat energy released in this process is the difference between the chemical energy of the original fuel and oxygen, and the chemical energy of the resulting carbon dioxide and water. In practice, it is usual to refer to it as the **energy content** (or *heat content*, or *heat value*) of the fuel – the methane in this case.

The reaction shown above contains the essential features of the burning of any common fuel: a compound containing carbon and hydrogen interacts with oxygen from the air to produce carbon dioxide and water (the latter usually as water vapour or steam). If we know the composition of the fuel and the relative masses of the chemical elements making up its molecules, we can predict how much carbon dioxide will be produced in burning a given amount of fuel (see Box 4.2).

BOX 4.2 CO_2 from fuel combustion

We take the combustion of methane (CH_4) as an example. The masses of the atoms of carbon and oxygen are respectively 12 times and 16 times the mass of a hydrogen atom, so we can associate masses with the items in the combustion equation:

$$CH_4 \quad + \quad 2O_2 \quad \rightarrow \quad CO_2 \quad + \quad 2H_2O$$
$$12 + (4 \times 1) + 2 \times (2 \times 16) \rightarrow 12 + (2 \times 16) + 2 \times (2 \times 1 + 16)$$

We see, therefore, that burning 16 tonnes of CH_4 releases 44 tonnes of CO_2.

The energy content of natural gas is 55 GJ t^{-1}. So burning one tonne of natural gas releases 2.75 tonnes (2750 kg) of CO_2 in producing 55 GJ of heat.

The other fossil fuels are more complex than methane, but their combustion is a similar process. The heat produced per tonne is rather less, however, and they also produce more CO_2 per tonne because they have a higher ratio of carbon to hydrogen atoms, so the CO_2 per unit of heat output is greater (see Table 4.1).

Table 4.1 Heat content and CO_2 emissions

Fuel	Heat content /GJ t^{-1}	CO_2 released /kg GJ^{-1}
coal	~30	~80
oil	42	70
natural gas	55	50
air-dry wood	~15	~80*

Note that the composition of coal, wood and to a lesser extent oil can vary significantly.

*If the wood is grown sustainably and combustion is complete, its *lifecycle* CO_2 emission should be close to zero.

The fossil fuels, the result of hundreds of millions of years of slow geological change acting on plant or animal matter, are examples of **hydrocarbons**, consisting almost entirely of carbon and hydrogen. The fossil fuels and their uses are discussed in more detail in the companion text, *Energy for a Sustainable Future* (Boyle *et al.* 2003).

Most of the biofuels, derived from living or recently dead biomass, contain significant amounts of oxygen as well. The molecules of biological materials are also much larger and more complex than methane, but we can typify their combustion by considering the relatively simple case of glucose, a sugar whose chemical formula is $C_6H_{12}O_6$:

$$C_6H_{12}O_6 + 6O_2 \rightarrow 6CO_2 + 6H_2O + \text{energy}.$$

The energy produced in burning one tonne or one cubic metre of various biological materials is shown in Table 4.2, with the main fossil fuels for comparison.

Table 4.2 Average heat energy content of fuels

Fuel	Energy content GJ t^{-1}	GJ m^{-3}	Fuel	Energy content GJ t^{-1}	GJ m^{-3}
Wood (green, 60% moisture)	6	7	Straw (as harvested, baled)	15	1.5
Wood (air-dried, 20% moisture)	15	9	Sugar cane residues	17	10
Wood (oven-dried, 0% moisture)	18	9	Domestic refuse (as collected)	9	1.5
Charcoal	30	*	Commercial wastes (UK average)	16	*
Paper (stacked newspapers)	17	9	Oil (petroleum)	42	34
Dung (dried)	16	4	Coal (UK average)	28	50
Grass (fresh-cut)	4	3	Natural gas (at supply pressure)	55	0.04

Note that the composition of coal and most biofuels is variable and the energy content per kg can differ significantly from the above averages. The energy *per cubic metre* depends on the density and can vary even more widely. (* Indicates dependence on specific types of material.)

BOX 4.3 Energy and moisture content

We have seen that water is an inevitable combustion product of any common fuel. It appears of course as steam, and some extra heat can be obtained by cooling and condensing this. The heat output per kilogram when this is not done is called the low heat value (LHV) of the fuel, whilst the high heat value (HHV) includes the extra contribution. The difference between LHV and HHV is often small enough to be ignored in rough calculations, but this may not be the case for biomass.

Unlike fossil fuels, plant matter often has a high water content, which adds to the mass but contributes no energy. It also increases the energy used in heating and evaporation. In general, each 10% increase in moisture content reduces the LHV by roughly 11%. But condensing all the water makes a great difference, and for green plants can increase the heat output by 50% or more.

Biomass as a solar energy store

In natural decomposition, plant material interacts with oxygen to produce carbon dioxide and water, just as in combustion. But in nature the process does not stop there. Solar energy completes the cycle, recreating fuel and oxygen. The mechanism is **photosynthesis** (from *photo*: to do with light and *synthesis*: putting together), in which plants take in carbon dioxide and water from their surroundings and use energy from sunlight to convert these into the sugars, starches, cellulose, etc. which make up 'vegetable matter'.

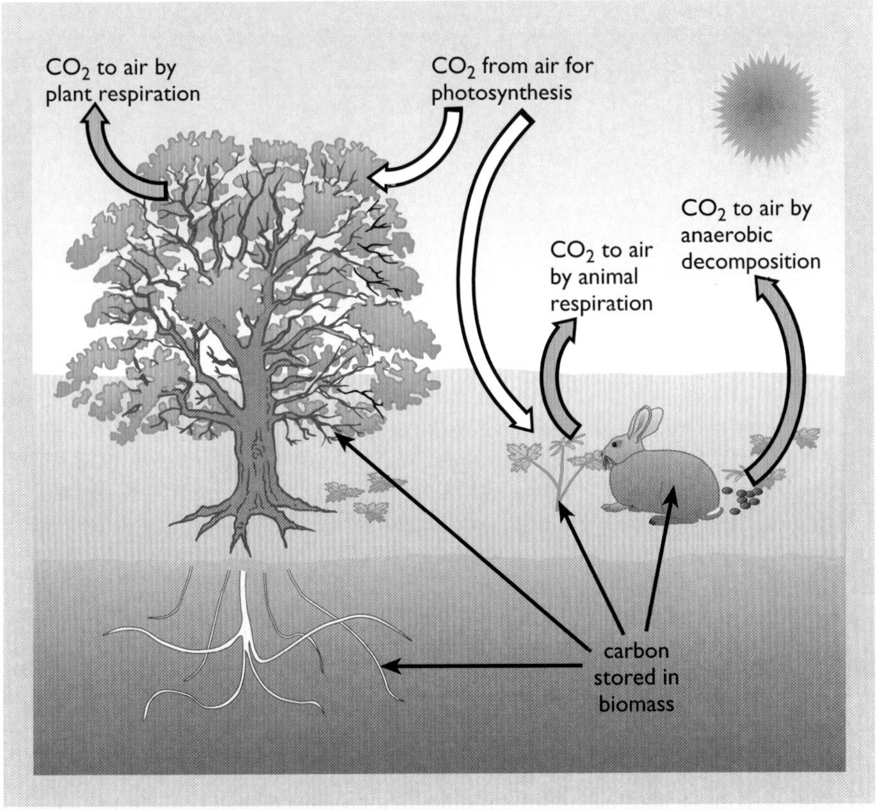

Figure 4.2 The carbon cycle on the local scale

The essential features of the process can be represented by an example:

$$6CO_2 + 6H_2O + \text{light energy} \rightarrow C_6H_{12}O_6 + 6O_2$$

Notice that the first item on the right, $C_6H_{12}O_6$, is again the formula for glucose. This is not necessarily the final 'vegetable matter', but as before it will serve for our simple example. The second product is of course oxygen, and it will by now be obvious that this process is exactly the reverse of the combustion/decomposition discussed above. The plant *grows* by using solar energy to convert carbon dioxide and water into carbohydrate or similar material, with a release of oxygen. When it *decays* – or we burn it – oxygen is used and energy is released as heat.

Photosynthesis is clearly dependent on both light and carbon dioxide, but the availability of light is very variable, and too much light for the available CO_2 can lead to damage to the photosynthetic systems. The process of *photorespiration* acts as a safety valve by reversing the process of photosynthesis, but without releasing energy in a form that the plant can use to grow (unlike the normal metabolic process of *respiration*). This apparently wasteful process has a valuable protective function in the **C3 plants** that form the majority, especially in temperate areas. (The name reflects the important role of molecules with three carbon atoms in the photosynthesis of these plants.)

Plants from tropical areas that are at greater risk from light-induced damage, and have evolved in areas of high temperatures and risk of drought, especially those in the grass family (e.g. maize, sugar cane, miscanthus), have an alternative to photorespiration. They concentrate CO_2 in the cells where photosynthesis is carried out, so there is never a shortage, and no risk of damage. These **C4 plants**, by avoiding wasteful photorespiration, can potentially produce higher yields of biomass than C3 plants. (For a more detailed treatment of photosynthesis, see Hall and Rao (1999).)

Conversion efficiencies

The **yield**, the tonnes of biomass produced per hectare per year, is obviously as important for energy crops as for food crops. Yields depend on many factors: the location, climate and weather, the nature of the soil, supplies of water, nutrients, etc., and the choice of plant. Even confining ourselves to **energy crops**, we find that the air-dry mass of plant matter produced annually on an area of one hectare can be as little as one tonne or, in favourable circumstances, as much as 30. In energy terms, this represents a range from perhaps 15 GJ to 300 GJ per hectare per year. (Table 4.3 in Section 4.4 below shows the gross yields in tonnes per hectare for various energy crops.)

It is easy to see that these yields imply an extremely low conversion efficiency. Consider, for instance, Northern Europe, where the average solar energy delivered in a year is about 1000 kWh per square metre (see Chapter 2). An energy crop in this region might yield perhaps 200 GJ per hectare per year: a solar-to-bioenergy conversion efficiency of about two-thirds of one percent.

BOX 4.4 Conversion of solar energy

Consider one hectare (ha) of land, in an area such as southern England where the annual energy delivered by solar radiation is 1000 kWh m^{-2} y^{-1}.

1000 kWh is 3.6 GJ and 1 ha is 10 000 m^2, so the total annual energy is	36 000 GJ

After losses (see the main text) about an eighth of this reaches the crop at the right time. Say...

12% of the annual energy reaches growing leaves	4320 GJ
50% of this is photosynthetically active radiation	2160 GJ
85% of which is captured by the growing leaves	1836 GJ
21% of which is converted into stored energy	386 GJ
40% of which is lost in photorespiration or consumed in respiration sustaining the plant, leaving	231 GJ

This is about 5.3% of the solar radiation reaching the growing plant, and only 0.64% percent of the original total annual energy.

Why is the solar-to-biomass conversion efficiency so low? Box 4.4 traces the losses, and the first significant fact is that most of the annual solar radiation may be ineffective: it misses the plant altogether, or arrives during the wrong season, or delivers more energy than the upper leaves can use but too little to the lower, shaded ones, or there may not be enough water or nutrients for the plant to use the solar energy that does reach it. Then there are diseases, pests (reducing leaf area) and weeds (competing for light or nutrients). The remaining losses shown in Box 4.4 are specific to the interaction of the plant and the sunlight. A leaf, like the human eye, responds only to a part of the solar spectrum (see Figure 2.5 in Chapter 2). The pattern of absorption is shown in Figure 4.3. Note the dip in the green region: a leaf looks green because it absorbs less green light than red or blue. Next, only about 85% of the 'useful' radiation is captured by the leaves. Then the actual process of photosynthesis converts only 21% the energy of absorbed light into the energy content of fixed carbohydrate. Finally, a further 40% is lost in photorespiration and the energy the plant requires to support itself, which comes from respiration.

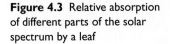

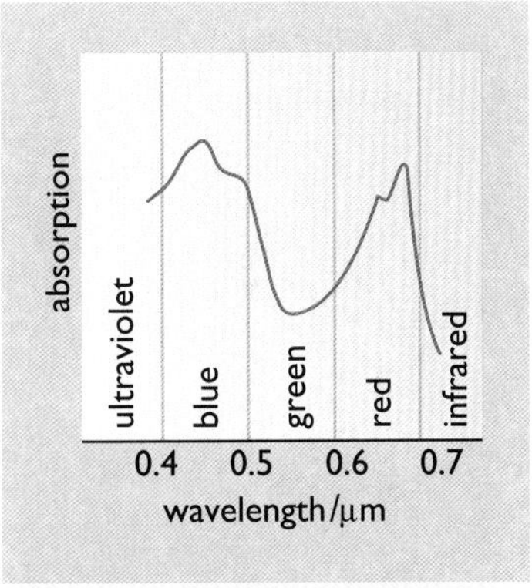

Figure 4.3 Relative absorption of different parts of the solar spectrum by a leaf

For C4 plants grown in tropical regions, the 5.3% of Box 4.4 might increase to perhaps 6.7%, giving an overall efficiency of 1–2%. And in general, good crop management can increase the useful fraction of solar radiation. Nevertheless, plants remain much less efficient solar energy converters than PV cells – but they are much cheaper.

4.4 Bioenergy sources I: Energy crops

The two main sources of bioenergy are the purpose-grown **energy crops** that are the subject of this section, and **'wastes'**, the otherwise unwanted products of human activities, to be discussed in Section 4.5. The technologies for using the resources and the extent of their contributions are briefly introduced in these two sections, but treated in more detail in Sections 4.6–4.8.

Table 4.3 Annual gross yields* of energy crops

Crop	Average yield* /t ha^{-1} y^{-1}
Sugar cane	35
Bagasse	10
Maize, wheat, rice, sorghum, miscanthus	15
Wood (temperate region)	10
Wood (tropics)	20

*Oven-dry mass, including all above-ground matter

Note that yields depend on growing conditions, and could be less than half or more than twice the average values shown here

Source: Johansson, 1993

We use the term *energy crop* here in its widest sense, to include any plants that are grown specifically for use as fuel or for conversion into other biofuels. It includes, for instance, wood for burning, plants for fermenting to ethanol and crops whose seeds are particularly rich in oils (but not of course those that are grown primarily for food).

Energy crops have attracted increasing attention in recent years, for several different reasons:

- the need for alternatives to fossil fuels, to reduce net CO_2 emissions
- the search for indigenous alternatives to imported oil
- the problem of surplus agricultural land.

The relative importance of these has been a major factor in determining the preferred crops in different countries or regions – subject of course to the constraints imposed by the local climate, soil, etc. We shall consider the crops under two headings: woody plants and others.

Woody crops

Forestry occupies an ambiguous position in environmental debates. A well-managed forest can be a sustainable fuel source, reducing atmospheric CO_2 as it grows and later providing a substitute for fossil fuel. In other cases, however, 'forestry' can mean the decimation of the world's major natural forests, a potential global environmental catastrophe. The wood in such cases is not cut primarily for fuel, so it does not really fall within the remit of this book – although one might ask whether the *residues* of unsustainable forestry should be included as 'renewables' in the woody wastes category. The past decade or so has seen a move away from the earlier view that forests in developing countries were being destroyed mainly by people cutting wood for domestic use. The evidence is that most of these small-scale users collect wood from local scattered areas, and that the large-scale disappearance of forest trees is the result of commercial forestry or charcoal production – or in some areas, clearance to establish other crops (UNDP, 2000).

The uncontrolled removal of wood can bring economic as well as environmental problems, reducing productive capacity in countries where it contributes appreciably to industrial energy consumption – in Brazil, for instance, where the steel industry uses over two million tonnes of charcoal per year. (Fourteenth-century England had a similar problem, but it was 'solved' temporarily when the Black Death killed a third of the population.)

In conventional forestry, any energy production is incidental, often using wastes (see Section 4.5). In **modified conventional forestry**, whilst energy is still not the main product, it is part of the plan. Coniferous trees are planted at higher than usual density and vigorously thinned after a few years, using integrated harvesting techniques to produce chipped wood. The remaining trees grow to maturity in the normal way. There have been trials of this system, but the main developments in recent years have been with woody crops planted and harvested entirely for energy production.

The general term for these is **short rotation forestry (SRF)**, the 'rotation' being periodic cutting of the wood every few years. In Europe, the term **short rotation coppice (SRC)** is more usual, referring to the centuries-old practice of coppicing willow or other fast-growing trees. (Some of the resulting 'copses' remain as attractive features of the countryside). In the modern version, cuttings are planted at 10–15 000 per hectare. Cut back close to the ground after a year, they re-grow, either with multiple stems or as single-stem trees. The crop is allowed to grow for 2 to 4 more years and then harvested by cutting the stems close to the soil level. The stumps re-grow and the cycle is repeated, for perhaps 30 years. Annual yields of 10 t per hectare should be achievable in Northern Europe. Modern harvesting machines (see Figure 4.20 below) can reduce the stems *in situ* to short lengths suitable for transporting, storage and future use. Adequate water, appropriate fertilizers and weed control are important during the establishment phase.

Figure 4.4 Short Rotation coppice

An obvious use of SRC is the provision of **heat**, and one of its most successful applications has been in Sweden where around 18 000 ha of willow coppice supplies energy for district heating. The generation of **electricity** using SRC, either as the sole fuel or co-fired with coal, has received considerable attention in Europe, where some 30 MW of capacity is fuelled by energy crops. In the UK, the planned ARBRE (Arable Biomass Renewable Energy) power station in North Yorkshire would be one of the world's most advanced energy crop systems, a biomass integrated gasification combined cycle (BIGCC) plant running on the gas from a wood gasifier (see Box 4.8). Fuel for the initial 8 MW pilot plant would come from some 2000 ha of SRC. Unfortunately, despite an EU grant, support under the UK Renewables Obligation, and set-aside payments (Box 4.5) for the SRC, the project is in abeyance for financial reasons at the time of writing.

There has been some development of SRF outside Europe, mainly in the Americas and Australia and New Zealand. Eucalyptus is commonly used, and there is interest in its potential for improving degraded land as well as its value as an energy source.

BOX 4.5 Set-aside

After the second World War, many countries established systems of subsidies to promote food production. In the EU, the institutionalization of these within the Common Agricultural Policy (CAP) eventually resulted in large food surpluses, and these in turn led to the introduction of the **set-aside** requirement. Farmers may set aside a certain percentage of their arable land in return for agricultural support payments on the rest of their arable area and payments for each hectare set aside. The set-aside land may remain fallow, or can be used to grow crops for other purposes. Since 2000, the compulsory level of set-aside has been 10%, but farmers may claim set-aside payments for up to 100% if energy crops are grown on the land. In the UK, there are also planting grants of up to £1600 per hectare for woody coppice and £920 per hectare for miscanthus.

Agricultural crops

Globally the most widely grown crops for bioenergy purposes are sugar cane and maize. Both are C4 crops, with the high yields needed for a favourable energy balance (see Section 4.9). The main interest lies in their potential for conversion to **liquid fuels**.

A completely different type of energy crop is grown for its 'oily' seeds. Sunflowers, oilseed rape, soya beans, etc. are grown for the oil in their seeds, which can be converted to a diesel substitute, known as biodiesel.

Both these potential replacements for crude oil products are discussed in Section 4.8.

Figure 4.5 Harvesting miscanthus using conventional agricultural machinery

Non-woody energy crops suitable for temperate climates have also received attention in Europe and the USA. These include **miscanthus**, a C4 grassy plant. It originated in Asia and Africa, but some forms will grow in northern Europe, and research in the 1990s suggested that it could yield up to 18 dry tonnes per hectare per year under UK conditions. The thick woody stems are suitable for direct combustion since they have a very low water content (20–30%) when harvested. It has the advantage over SRC that it can be grown using normal farming techniques, and its annual cycle allows more flexibility in land use. At the time of writing, Europe has no operational 'grassy' energy crop project, but the process is considered to be ready for commercial exploitation in the UK.

4.5 Bioenergy sources II: Wastes

We have already seen that the wastes resulting from our 'non-energy' uses of biomass are potential fuels. This is less evidently the case for urban and industrial waste, but much of it is organic (biological) material, and will release energy if burned. The question is whether this material should be regarded as a *renewable* resource. In recent years it has become customary to treat only **biodegradable** wastes as renewables, excluding, for instance, polymeric materials ('plastics') that do not degrade easily.

In this section we look first at the wastes arising from direct uses of biomass: forestry and agriculture, and animal husbandry. We then move on to household, or more generally, municipal wastes, and finally to the specific wastes associated with industrial processes.

Wood residues

Operations such as thinning plantations and trimming felled trees generate large volumes of **forestry residues**. At present these are often left to rot on site. This has the environmental merit of retaining nutrients, and in any case their bulk and inconvenient form makes transporting them for wider use uneconomic. However, with the development of integrated harvesting techniques, the use of some fraction of the residues for heat and/or power generation is increasing in many countries. Solid biomass, mainly forestry residues, fuels some 6% of Austria's generating capacity (two-thirds of it in CHP plant), and the USA has over 6 GW of biomass-fired plant, again mainly using forest residues.

The quantity of residues used as firewood in the UK is uncertain, but thought to be about a million dry tonnes per year. (ETSU, 1999). It is estimated that a further million or more tonnes is available for use in power plants or by similar large users, a potential input of some 20 PJ, sufficient for about 200 MW of power station capacity. However, the investment required for the harvesting and processing machinery means that the resource is unlikely to become economically viable until a market for the wood is fully established.

Temperate crop wastes

World-wide, residues from **wheat** and **maize** (corn), the two main temperate cereal crops, amount to more than a billion tonnes per year, with an estimated energy content of 15–20 EJ. They have uses as bedding, feed, etc., but in major cereal-growing regions more than half may remain unused. In East Anglia in the UK, for instance, about 80% of the nearly 2 million tonnes of **straw** produced annually is surplus to agricultural requirements. In the past most of the surplus was burned in the field, but air pollution concerns led to a ban on field burning from the end of 1992. There has been similar legislation across Europe, and farmers must now bear the costs of chopping and digging in the straw. China experienced a similar pollution problem in the 1990s, when residues that had been used for heating and cooking were replaced by 'modern' fuels, leaving a surplus that was burned in the fields. In this case, the solution was the introduction of biomass-fuelled village-scale gasifiers (see Section 4.7) distributing gas to households. Emissions were reduced and the conversion efficiency was better than with direct combustion.

The UK market for straw-burning systems for space heating has been relatively slow to develop. The straw must be baled, removed from the fields, stored in a dry atmosphere and transported to its point of use. Although straw has a reasonable energy density of about 15 GJ t^{-1}, it has a relatively low mass density. In bales (see Box 4.6), 1 tonne occupies a volume of some 6 cubic metres, which makes transport and storage expensive. There has been some commercialization of systems for producing high-density pelletted straw, typically in excess of 1 t m^{-3} (denser than wood). These allow automatic stoker feeding of boilers, and reduce transport and storage costs, but the systems are expensive.

The total UK rate of production of useful heat from straw is thought to be about 300 MW, corresponding to some 0.6 Mt (600 000 tonnes) of straw per year. Several European countries already have wide experience of straw burning, and Denmark has a programme to use 1.2 Mt per year in CHP plants with district heating. The first straw-fired power station in the UK, commissioned in 2000, is the world's largest single plant, using 0.2 Mt per year (Box 4.6).

BOX 4.6 Electricity from straw

The Elean straw-fired power station – the world's largest – began operation in 2000 at Sutton near Ely in Cambridgeshire. Its output capacity is 36 MW, and the capital cost was £60M.

The plant is optimally located, with minimum transport distance from the main cereal producing region of the country. 200 000 tonnes of straw per year are supplied, in the form of 550 kg Hesston bales, from large farms within 80 km of the plant. The straw is kept on the farm until required, and transported in covered lorries to the plant, where 2100 tonnes of straw, sufficient for 76 hours operation, can be stored.

The bales are shredded and the straw is burned on two-stage grates. The volatile matter (see Section 4.6) is released in an initial stationary phase, with the balance of combustion occurring on a secondary vibrating grate. Air injection is closely controlled to maintain optimum combustion conditions. Power generation is by a conventional steam turbine and generator (see Box 4.8), and the annual electrical output is over 270 GWh per year – sufficient for 80 000 households.

The ash, 5% of the mass of fuel input, is rich in potassium and phosphate and is used to manufacture agricultural fertilizers, returning the nutrients to the land. There are plans to use the waste heat in future, and also to take other baled crops such as miscanthus, allowing for agricultural diversification in the area.

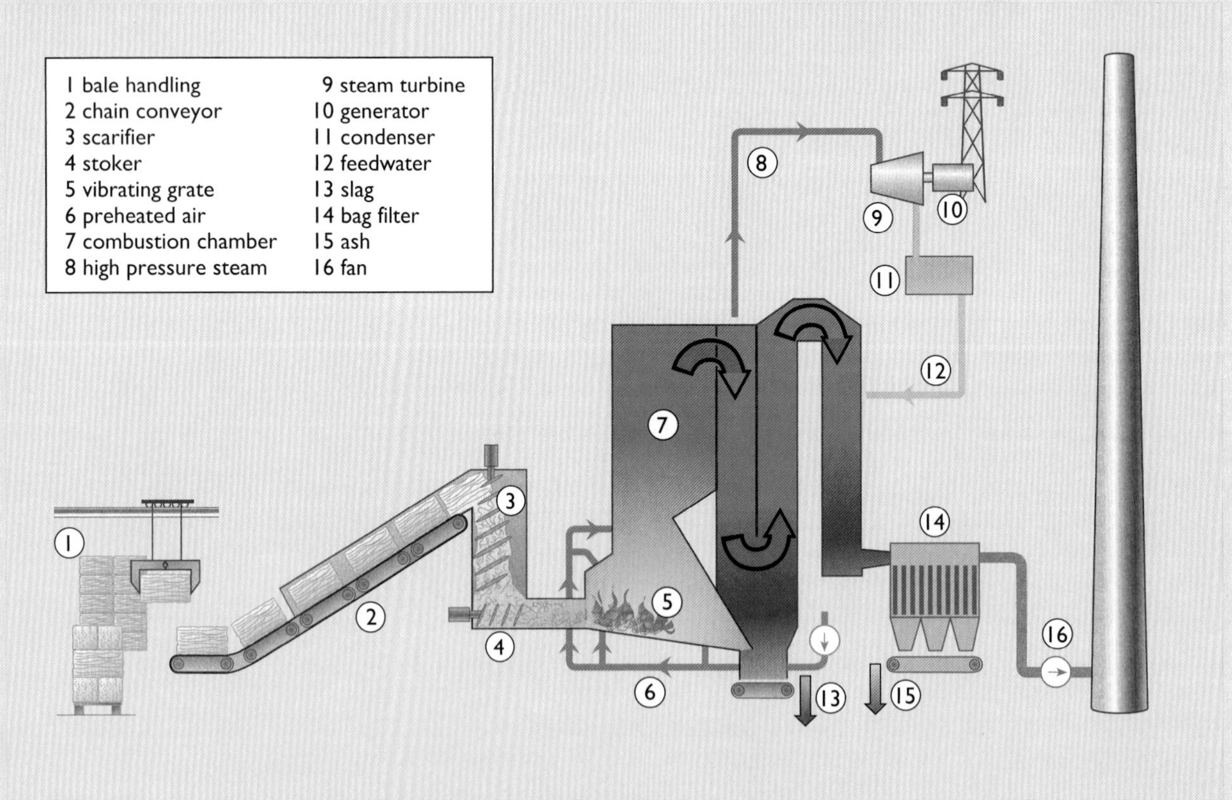

Figure 4.6 The Elean straw-fired power station: (top) the plant; (bottom) unloading Hesston straw bales

Tropical crop wastes

The total energy content of the annual residues of the world's two main tropical food crops, **sugar** and **rice**, is estimated as about 18 EJ – similar to the total for temperate crops. In this case, however, significant quantities are already being used as fuels.

Bagasse, the fibrous residue of sugar cane, is used in sugar factories as a fuel for raising steam, and to produce electricity for use in the plant. During the cane-crushing period there is often a surplus, and whilst transporting the bagasse may not be economic, selling surplus electricity could be. In the past, the problem of selling power that is available for only half the year has limited investment, but liberalization of energy markets in many countries has improved the prospects, and efficient boilers have now been installed in many production facilities. The potential for year-round generation using wood wastes in the non-crushing season has also created interest. Increased recovery of wastes, combined with improved efficiency of conversion to electricity, could result in up to 50 GW of generating capacity from the sugar industry world-wide. The use of bagasse to produce ethanol (see Section 4.8) is another possibility.

Rice husks are among the most common agricultural residues in the world, accounting for about one-fifth of unmilled rice dry weight. Although they have a high silica (ash) content compared with other biomass fuels, their uniform texture makes them suitable for technologies such as gasification (see Section 4.7). Rice husk gasifiers have been successfully operated in Indonesia, China and Mali (Manurung, 1990).

Animal wastes

Animal manure can be a major source of greenhouse gases. It is estimated, for instance, to account for 10% of methane emissions in the USA. When not correctly managed, farm slurries can also seriously pollute local watercourses. The combination of intensive animal rearing and stricter environmental controls on odour and water pollution is encouraging farmers to invest in anaerobic digestion as a means of waste management. As its name suggests, this is the decomposition of organic matter in the absence of air. The process, described in Section 4.7, generates useful biogas, and leaves an effluent that can be used (directly or dried) as a fertilizer.

Sewage sludge can be treated by anaerobic digestion, as has been done in the UK since the first large 'sewage farms' were built in the last century. Originally, much of the biogas was simply flared, but with 70% of all sewage now treated, an increasing proportion is used for heat and electricity production on-site. By mid-1994, 26 projects with a total generating capacity of 33 MW were supplying electricity under the NFFO scheme (see Chapter 10), but recent years have seen no new projects.

Another option for extracting energy from animal wastes, if the water content is low, is direct combustion for conventional power generation. **Poultry litter**, a mixture of chicken droppings and material such as straw, wood shavings etc., has an energy content in the range 9–15 GJ t^{-1}, depending on its moisture. The first UK power plant, with 13 MW output capacity, started operating in 1992 at Eye in Suffolk, using some 200 000 tonnes of litter per year from surrounding poultry farms. A second 13 MW plant was commissioned a year later, followed by a 39 MW plant in 1998, and others in 2000 and 2001. Almost all were NFFO projects in East Anglia, the exception being a 39 MW plant supported under the Scottish Renewables Obligation. Some of the larger plants are designed to accept forestry wastes together with poultry litter, and in recent years, the second 13 MW plant has used other animal wastes, including bone meal, as fuel (ETSU, 1999; DTI, 2003a).

Municipal solid waste

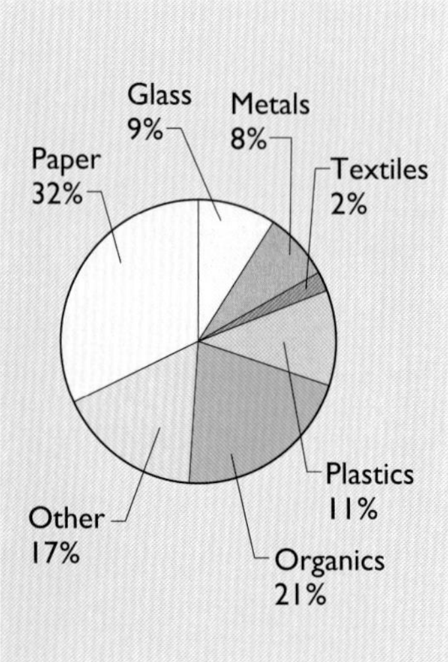

Figure 4.7 The composition of municipal solid wastes in the UK. Note that about 50% of the total is organic material (biomass)

The average household in the world's industrialized countries generates rather more than a tonne of solid waste per year, with an energy content of about 9 GJ per tonne. It appears therefore that the average UK household could in principle supply *one tenth* of its total annual energy consumption of about 90 GJ from its own wastes. There are however both technological and social impediments to this, as we shall see.

In practice, there are three main ways in which **municipal solid waste (MSW)** is treated at present:

- disposal in landfills
- combustion
- disposal in anaerobic digesters

Each of these processes may be preceded by some form of initial treatment, from simple removal of metallic items to recycling and other more radical methods described in the following sections. And each process may be accompanied by some system for extraction of useful energy.

Landfill, using suitable cavities such as old quarries, is the main disposal method in a number of countries, including the UK (with 80% of MSW currently landfilled), Germany and the USA. Most other European countries landfill a smaller fraction (or none) and incinerate much more – up to 60% or so (see Figure 4.9). Anaerobic digesters play a much smaller role at present, but this may change with increasing constraints on landfill and problems of acceptability of combustion plant.

Extraction of useful energy is a feature of many MSW disposal systems – and of virtually all newly constructed plants. In *anaerobic digesters* and *landfill* (which also involves anaerobic digestion) the energy is carried initially by a gas, which is then used either to produce heat or to drive an engine. In *combustion*, the energy is of course produced directly as heat. In all three cases, the heat may be used directly or for electric power generation.

Landfill gas is the subject of the next subsection. Combustion and anaerobic digestion of MSW are discussed in Sections 4.6 and 4.7 respectively.

Landfill gas

A large proportion of municipal solid waste is biological material (Figure 4.7), and its disposal in deep landfills furnishes suitable conditions for anaerobic digestion. It was known for decades that landfill sites produced methane, and systems were fitted to burn it off safely, but the idea of using this **landfill gas (LFG)** developed only in the 1970s.

The digestion in a landfill (Figure 4.8(a)) takes place over years, rather than the weeks of 'wet' systems. In developing a site, each area is covered with a layer of impervious material after it is filled, and the gas is collected by an array of interconnected perforated pipes at depths of up to 20 metres in the refuse (Figure 4.8(b)). In a large well-established landfill there can be several kilometres of pipes, with as much as 1000 m^3 per hour of gas being pumped out.

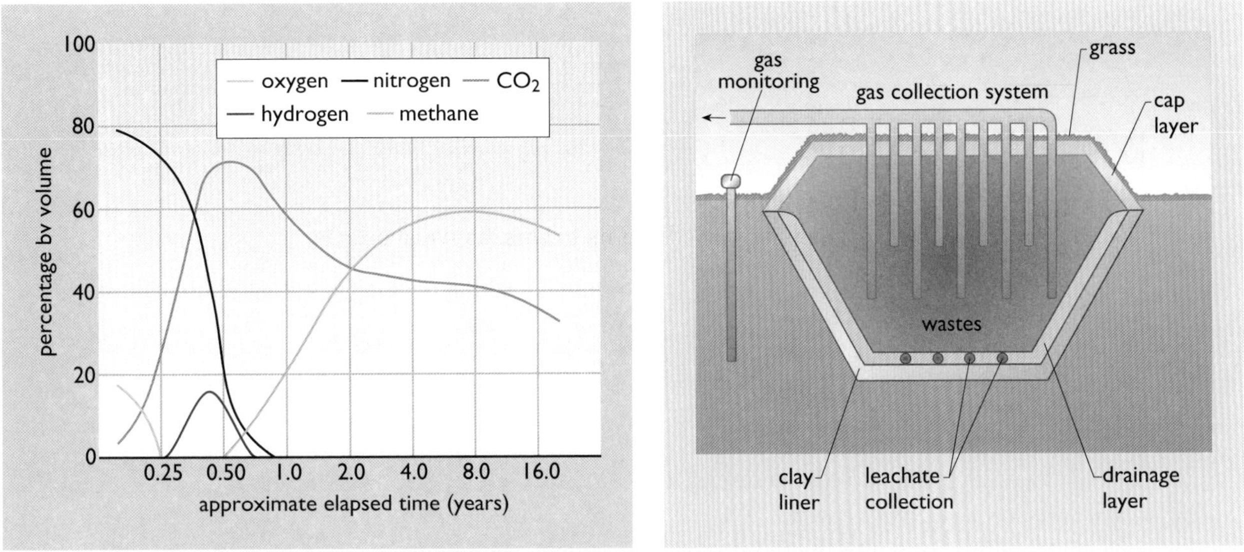

Figure 4.8 (a) The changing gas composition in a landfill site; (b) Extraction of landfill gas

In theory, the lifetime yield per tonne of wastes in a good site should lie in the range 150–300 m^3 of gas, with between 50% and 60% by volume of methane, which suggests a total energy of 5–6 GJ per tonne of refuse. In practice, at the average UK gas extraction rate, the heat energy output per tonne of wastes (as collected) is rather less than 2 GJ, but yields are extremely variable and often unpredictable – a factor that tends to deter potential investors.

The gas may be used directly, to fire kilns, furnaces or boilers, but there are rarely enough large users close to a landfill site, and the output is increasingly used to generate electricity for local use or for sale. The generators are driven either by large internal combustion engines (see Figure 4.14 below) or by gas turbines (see Section 4.7). Assuming a gas-to-electricity energy efficiency of perhaps 35%, this brings the overall energy efficiency of the system below 10%. A site containing a million tonnes of MSW might support an electrical capacity of perhaps 2 MW over a 15–20 year generating lifetime. Despite the low energy conversion efficiency, LFG plants have been amongst the most financially attractive of the systems receiving contracts under the NFFO scheme, with an annual electrical output in 2002 of some 2700 GWh from an installed capacity of about 400 MW. (See Section 4.10 for a cost calculation for LFG.)

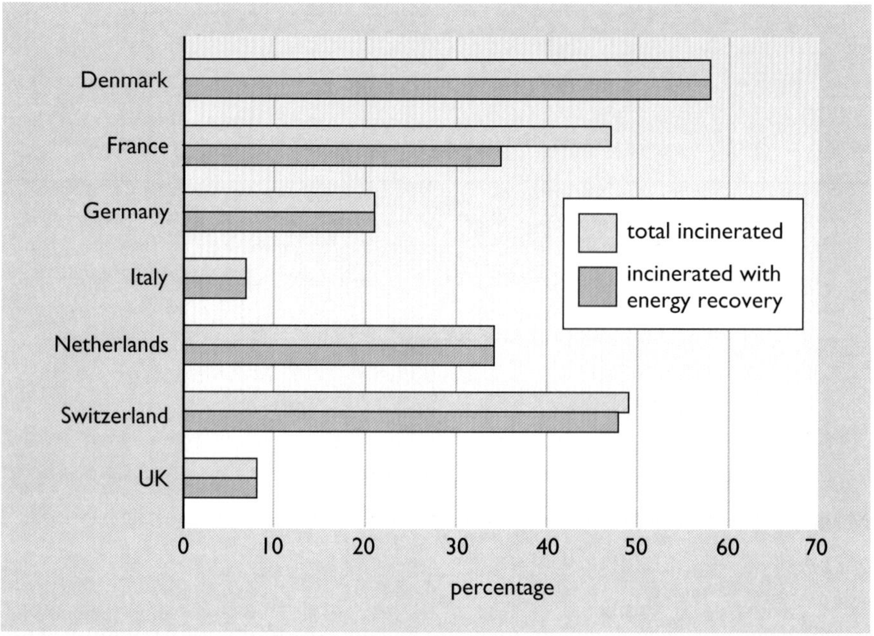

Figure 4.9 Incineration of MSW in EU countries

Commercial and industrial wastes

Commercial and industrial wastes of organic origin can be used as fuels. The UK generates about 36 Mt of specialized wastes each year, about two-thirds of which are combustible. The furniture industry, for instance, has been estimated to burn 35 000 tonnes of off-cuts and sawdust per year, one third of its production, providing 0.5 PJ of space and water heating and process heat (FOE, 1991). Some wastes are unsuitable for combining with domestic waste, for reasons of safety or cost, but energy recovery can help to reduce total costs where dedicated equipment is required in order to meet environmental and safety standards. For example, food processing wastes, which must be treated before discharge to reduce biological and chemical oxygen demand, may be anaerobically digested, with the resulting biogas used for process heat. Hospital wastes, all of which must be incinerated, are increasingly subject to energy recovery as health authorities upgrade their waste-handling equipment. Total production of hospital wastes in Britain is equivalent to 12 PJ y^{-1}.

The majority of the 40 million tyres discarded in Britain every year are unsuitable for reuse, but with an energy content of 32 GJ per tonne, they constitute a major fuel resource. Under EU legislation, whole tyres have been banned from landfill sites from 2003 and chipped tyres from 2006. A 20 MW 'tyre-fired' power station was commissioned in 1994 in Wolverhampton, and a cement factory uses tyres as a partial substitute for coal or coke.

4.6 Combustion of solid biomass

We now turn from the biomass sources to the ways in which they are used or might be used. If biofuels are to compete with our present fuels, they must be able to meet the demand for appropriate forms of energy at competitive prices. Two important criteria are the *availability* and the '*transportability*' of the supply. The premium fuels – oil and natural gas – are valued because their energy can be stored with little loss and made available where and when we need it. The biomass resource comes in a variety of forms, as we have seen. Many are relatively high in water content and decompose rather quickly, so few are good long-term energy stores. Relatively low energy densities mean that they are also likely to be bulky and expensive to transport over appreciable distances. Recent years have therefore seen considerable research effort devoted to means for converting biomass into more convenient forms of energy.

Most biomass is initially solid, and it can be burnt in this form to produce heat for use *in situ* or at not too great a distance. It may first require relatively simple *physical* processing, involving sorting, chipping, compressing and/or air-drying. These uses are discussed in this section. Alternatively, the biomass can be upgraded by *chemical* or *biological* processes to produce gaseous or liquid fuels. These options are discussed in Sections 4.7 and 4.8.

Combustion of wood and crop residues

Boiling a pan of water over a wood fire is a simple process. Unfortunately, it is also very inefficient, as Box 4.7 reveals.

BOX 4.7 Boiling a litre of water

How much wood is needed to bring one litre of water to the boil?

Data

Specific heat capacity of water = 4200 J kg^{-1} K^{-1}

Mass of 1 litre of water = 1 kg

Heat value of wood (Figure 5.13) = 15 MJ kg^{-1}

Density of wood = 600 kg m^{-3}

1 cubic centimetre (1 cm^3 = 10^{-6} m^3

Calculation

Heat energy needed to heat 1 litre of water from 20 °C to 100 °C = 80 × 4200 J = 336 kJ

Heat energy released in burning 1 cm^3 of wood = 15 × 600 × 10^{-6} MJ = 9.0 kJ

Volume of wood required = 336 ÷ 9.0 = **37 cm^3**.

Experience suggests that on an open fire much more than two thin 20 cm sticks would be needed. But a well-designed stove using small pieces of wood could boil the water with as little as four times this 'input' – an efficiency of 25%

Designing a stove or boiler that will make good use of valuable fuel requires an understanding of the series of processes involved in combustion. The first process, which consumes rather than produces energy, is the evaporation of any water in the fuel (see Box 4.3). Then, in the combustion process itself, there are always two stages, because any solid fuel contains two combustible constituents. The **volatile matter** is released as a mixture of vapours as the temperature of the fuel rises. The combustion of these produces the little spurts of flame seen around burning wood or coal. The solid which remains consists of the **char** together with any inert matter. The char, mainly carbon, burns to produce CO_2, whilst the inert matter becomes clinker, slag or ashes.

A feature of the biofuels is that three-quarters or more of their energy is in the volatile matter (unlike coal, where the fraction is usually less than half). The design of any stove, furnace or boiler should ensure that these vapours burn, and don't just disappear up the chimney. Air must also reach all the solid char, which is best achieved by burning the fuel in small particles. This can raise a problem, because finely divided fuel means finely divided ash – particulates that must be removed from the flue gases. The air flow should also be controlled: too little oxygen means incomplete combustion and leads to the production of poisonous carbon monoxide. Too much air is wasteful because it carries away heat in the flue gases.

Modern systems for burning biofuels are as varied as the fuels themselves, ranging in size from small stoves through domestic space and water heating systems to large boilers producing megawatts of heat (see for instance Box 4.6).

Wood burning produces a wide range of pollutants, and the past few decades have seen many programmes in developing countries for the design and dissemination of improved stoves. With the joint aims of reducing both fuel consumption and smoke emissions inside houses, these have ranged from small scale trials to major national programmes, as in China and India. A project to introduce a locally-produced charcoal-fired stove, the jiko, in Kenya is claimed to have been particularly successful. (For more detailed accounts, see UNDP, 2000 and Anderson *et al*, 1999.)

Charcoal

Charcoal is traditionally produced in the forests where the wood is cut. The 'kiln', consisting of stacked wood covered with an earth layer, is allowed to smoulder for a few days in the near absence of air, typically at 300–500 °C, a process called *pyrolysis*. The volatile matter is driven off, leaving the charcoal (the 'char' component mentioned above). Charcoal is almost pure carbon, with about twice the energy density of the original wood and burning at a much higher temperature, so it is much easier to design a simple and efficient stove for use with this high quality fuel. However, from 4 to 10 tonnes of wood are needed for each tonne of charcoal, and if no attempt is made to collect the volatile matter, up to three-quarters of the original energy content can be lost. The process also releases vaporized tars and oils and the products of incomplete combustion into the atmosphere, making this charcoal fuel cycle probably the most greenhouse gas intensive in the world. (See Section 4.8 for modern pyrolysis methods.)

Combustion of municipal solid wastes

In many countries in continental Europe and elsewhere, refuse incineration with heat recovery, or **energy-from-waste (EfW)**, is an important means of waste disposal. The heat may be used directly for district heating, or for power production (often in CHP plant). The inert ash can be used as hardcore. Countries with successful recycling and composting programs have often seen parallel growth in EfW, which accounts for 30–60% of MSW disposal in most Western European countries. World installed capacity is over 3 GW, about half of it in Europe.

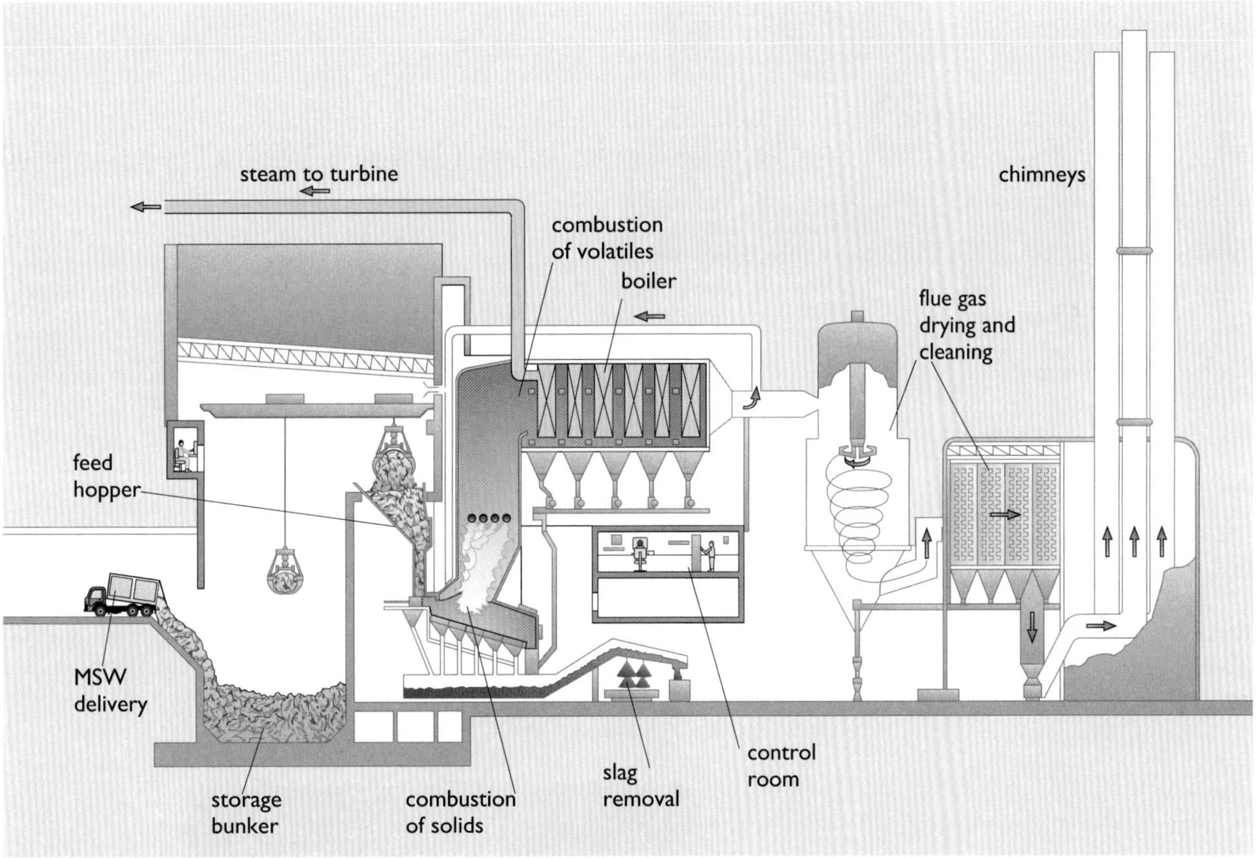

Figure 4.10 A large MSW combustion plant

EfW is becoming financially attractive in large cities in Britain because of the shortage of suitable landfill sites and the high costs of transporting wastes to distant sites (see Section 4.10). Encouraged by support under the NFFO scheme, UK capacity tripled during the 1990s, and all MSW combustion plants in the UK now make use of their heat output (see, for instance, Figure 4.11).

However, three-quarters of the UK EfW schemes proposed during the 1990s failed to obtain planning permission, mainly following public opposition. This often centred on air pollution (see Section 4.9), with further concerns about the delivery and storage of wastes. Many environmental groups are opposed to waste combustion, arguing for more waste reduction and

Figure 4.11 The SELCHP (South-East London Combined Heat and Power) plant, commissioned in 1984, is designed to incinerate 420 000 tonnes of MSW per year, producing steam for a 31 MW turbo-generator and heat for a local district heating scheme

recycling. Measures to encourage recycling and the reduction of packaging, making manufacturers responsible for the recycling of their products, could eventually lead to a reduction in the volume of MSW, but they may take years to become fully effective. Meanwhile, with ever more stringent restraints on landfill, EfW may be the only viable short-term option.

Pelletted fuel

Household refuse is hardly an ideal fuel. Its contents are variable, its moisture content tends to be high (20% or more) and its energy density is about a thirtieth of that of coal. So it is expensive to transport, and requires combustion plant designed specifically for this type of fuel. Methods for converting refuse into a fuel suitable for burning in conventional plant have therefore attracted considerable attention.

The term **refuse-derived fuel** (RDF) refers to a range of products resulting from separation of unwanted components, shredding, drying and otherwise treating the raw material. Relatively simple processing might involve separation of very large items, magnetic extraction of ferrous metals and perhaps rough shredding. The most fully processed product, known as **densified refuse-derived fuel** (or d-RDF), is the result of separating out the combustible part which is then pulverized, compressed and dried to produce solid fuel pellets with perhaps twenty times the energy density of the original material (Figure 4.12). The UK has six processing plants, mechanically separating non-combustibles such as metals and glass, and pelletizing the remaining organic matter. Their reduced ash content makes the pellets suitable for co-combustion with coal in conventional plant.

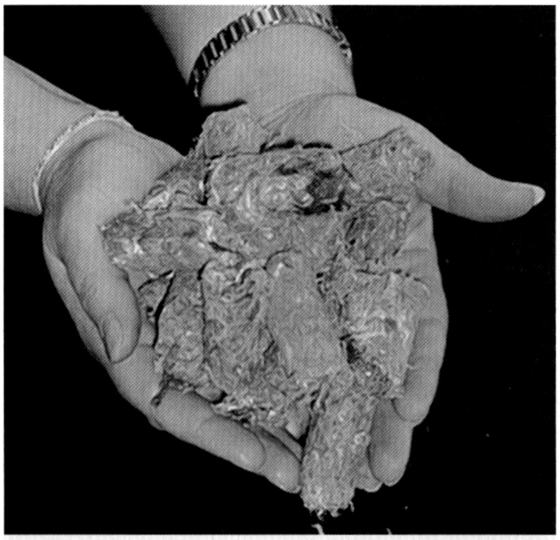

Figure 4.12 Refuse-derived fuel (d-RDF) pellets

4.7 **Production of gaseous fuels from biomass**

There are several reasons for growing interest in the production of **gaseous fuels** from biomass. The result is a more versatile fuel, suitable not only for burning but for use in internal combustion engines or gas turbines. It is easier to transport, and if undesirable pollutants and inert matter are removed during processing, it will be cleaner. It offers a route to electric power that could be more efficient than the direct combustion of biomass in a conventional power station. And finally, gasification under suitable conditions can produce **synthesis gas**, a mixture of carbon monoxide and hydrogen from which almost any hydrocarbon, synthetic petrol, or even pure hydrogen can be made.

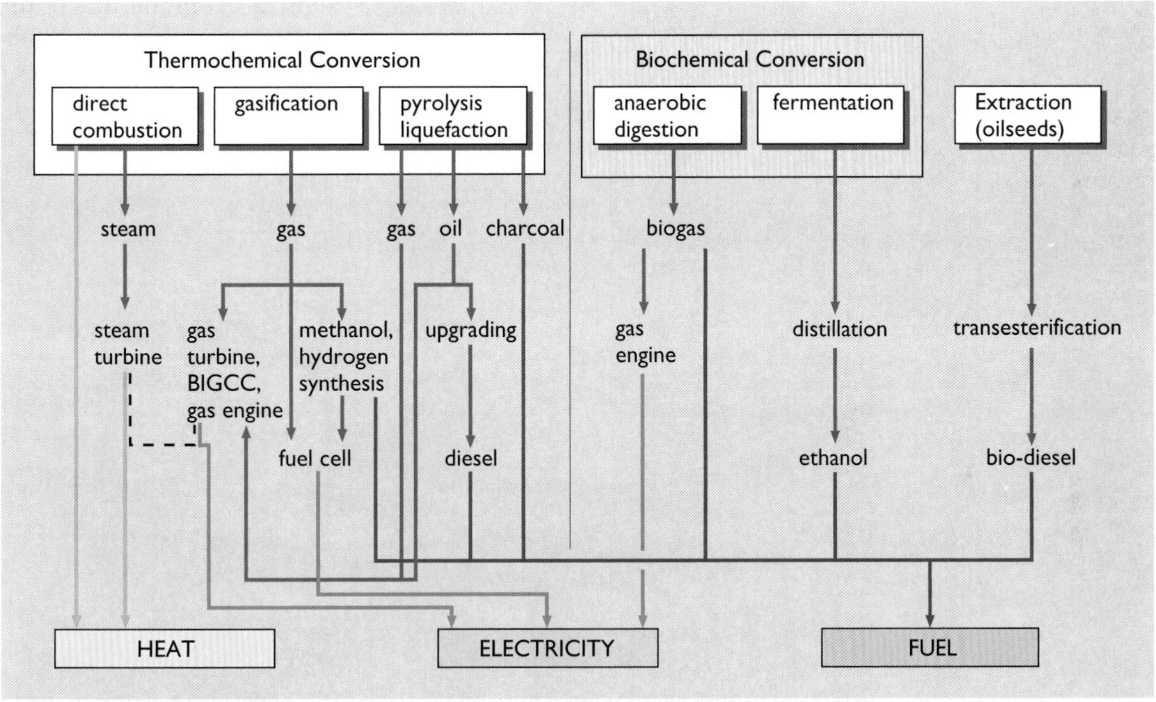

Figure 4.13 Main bioenergy conversion routes

Anaerobic digestion

We have seen two examples of anaerobic digestion of wastes: biogas and landfill gas. In the latter case, the 'digester' is the landfill itself, and the operator has only limited control of the processes. The digestion of 'wet wastes' is quite different. The feedstock, dung or sewage, is converted to a slurry with up to 95% water, and fed into a purpose-built **digester** whose temperature can be controlled. Digesters range in size from perhaps one cubic metre for a small 'household' unit (roughly 200 gallons) to some ten times this for a typical farm plant and more than 1000 m^3 for a large installation (Figure 4.15). The input may be continuous or in batches, and digestion is allowed to continue for a period of from ten days to a few weeks.

The process of anaerobic digestion is complex, but it appears that bacteria break down the organic material into sugars and then into various acids which are decomposed to produce the final gas, leaving a residue whose composition depends on the system and the original feedstock. The bacterial action itself generates heat, but in cooler climates additional heat is normally required to maintain the ideal process temperature of at least 35 °C, and this must be provided from the biogas. In extreme cases *all* the gas may be used for this purpose, but although the net energy output is then zero, the plant may still pay for itself through the saving in fossil fuel which would have been needed to process the wastes.

In a well-run digester, each dry tonne of input will produce 200–400 m^3 of biogas with 50% to 75% methane, an average energy output of perhaps 8 GJ per tonne of input. This is only about half the fuel energy of dry dung or sewage, but the process may be worthwhile in order to obtain a clean fuel and dispose of unpleasant wastes.

The biogas produced by a digester can be used to produce heat or electric power – or in many cases both. It can be used in large internal combustion engines (Figure 4.14) to drive electric generators, with the engine cooling water and exhaust gases providing heat to the digester. If it is scrubbed to remove the carbon dioxide and hydrogen sulphide, biogas is similar to natural gas and can be used as vehicle fuel. Most spark ignition engines can be converted to dual fuel operation, as some sewage companies have done for their own vehicles.

Figure 4.14 Sewage gas engine

The energy content of the annual wastes from housed animals in the UK is thought to exceed 100 PJ, and the estimated *accessible resource* in the form of biogas is about 10 PJ, enough to support an installed generating capacity of perhaps 100 MW. However, although the technology is well developed, with a range of digesters commercially available, the relatively high capital cost has tended to limit UK investment. A further disincentive has been the difficulty in maintaining the right operating conditions and blend of feedstock in small-scale plants. In the mid-1990s the total installed capacity in the UK remained under 1 MW (ETSU, 1999).

Manures from cattle, chickens and pigs are the most common wet wastes in Europe, especially in the Netherlands and Denmark, where limited land is available for the spreading of slurry. A Danish government programme investigating the economics of anaerobic digestion concluded after a six-year trial that a *large-scale* biogas plant could be profitable if three realistic conditions were met:

- the plant is operated in combined heat and power mode
- the gas is sold at a price comparable to that of natural gas
- credit is given for the disposal of other wastes

This proved to be the case. Central plants set up by farmers' co-operatives allowed better control of the processes, and dealing with wastes from other sources brought additional income. Such plants supply some 40 MW of heat output in Denmark and about 10 MW in the Netherlands. The first large-scale plant in Britain was commissioned in 2002 (Figure 4.15). Based on a design that has proved successful in Denmark and Germany, it uses an annual 146 000 t of slurry from 28 farms, together with wastes from food processors, to supply the heat input for a generating capacity of 1.43 MW.

Figure 4.15 Large-scale anaerobic digestion plant at Holsworthy in Devon

The developing world has seen many schemes for biogas plants during the past few decades. A major Chinese programme in the 1970s initially resulted in more than 7 million digesters, but suffered from many failures. A later drive, with better technology and supporting infrastructure, resulted in some 5 million domestic plants operating successfully by the mid-1990s.

In India 2.8 million biogas plants were installed by the end of 1998 and a potential for 12 million has been identified (UNDP, 2000). However, in many developing countries the capital cost of a digester is out of reach of the typical small farmer, and attempts to introduce community biogas plants have met with mixed fortunes, largely due to difficulties with balancing 'ownership' of animal dung against credit in the form of biogas consumption. As a result, many of the biogas plants in India are concentrated on wealthier farms with larger numbers of cattle.

Anaerobic digesters for MSW

As an alternative to recovery of biogas from landfills, MSW can be subjected to more carefully controlled processing in the type of large digester described above. Under these conditions gas yields are much higher and digestion is complete within a matter of weeks rather than years. Feedstock for the digester is the organic fraction of MSW diluted into a slurry – possibly mixed with sewage.

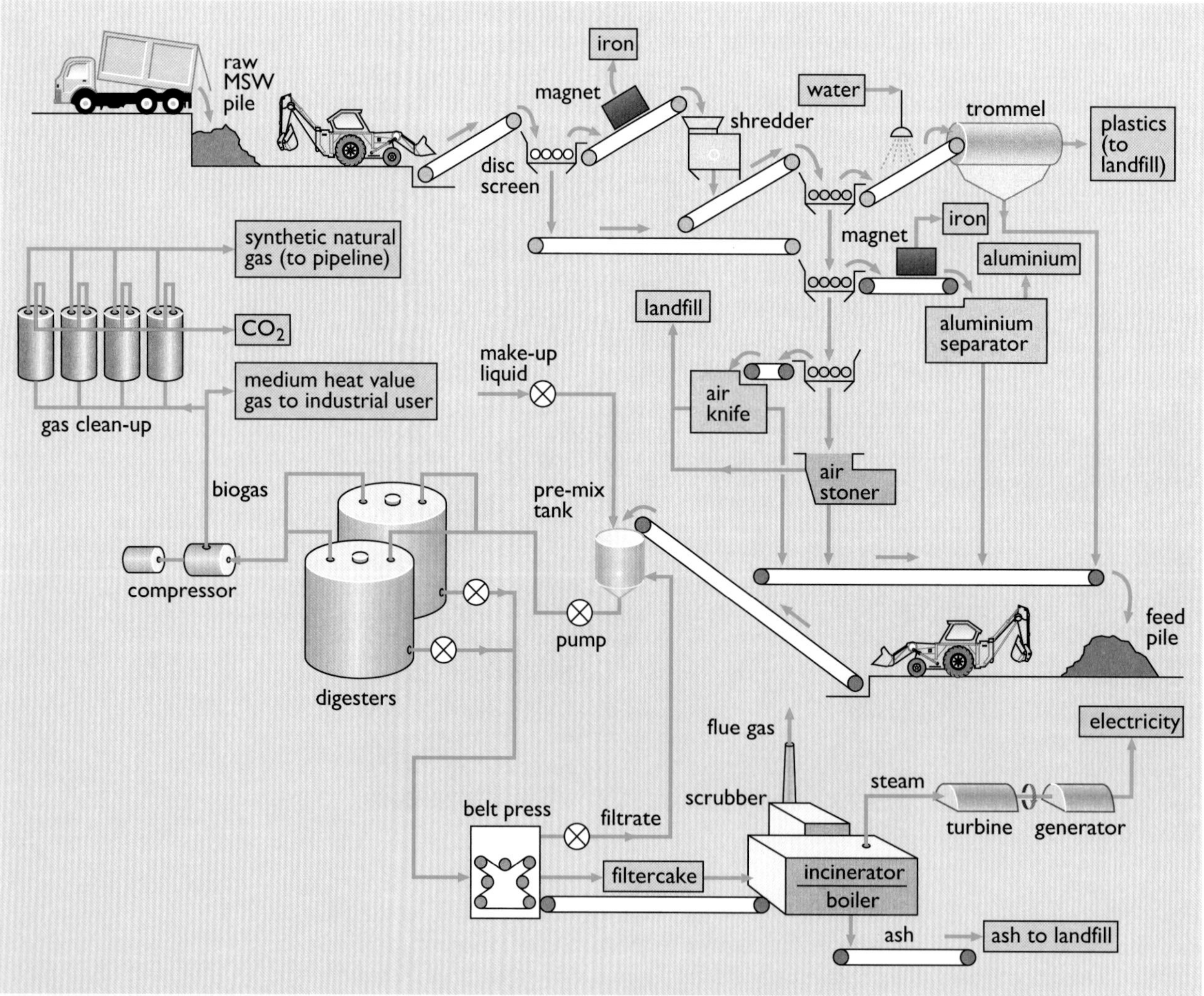

Figure 4.16 This integrated waste materials plant has facilities for recovery of metals and removal of plastics, followed by anaerobic digestion of the remainder. The solid residue from the digester serves as fuel for power production

An advantage over landfill is that these digesters can be closer to urban areas, reducing transport costs. They also require much less land. The greater capital and processing costs have, however, been a disincentive to investment. Plants already exist in several European countries, and although the UK has none yet, interest is growing due to the rising cost of landfill. Low-solids systems have been developed particularly in the USA, and Figure 4.16 shows the full complexity of a system which first recovers useful materials from the MSW, then produces methane by digestion and finally generates electric power using the combustion heat of the residual solids.

Gasification

In contrast to the mainly *biological* process of anaerobic digestion, the term **gasification** is used for *chemical* processes by which a gaseous fuel is produced from a solid fuel (see Figure 4.13 above.) Gasification is not a new process. 'Coal gas', the product of coal gasification, was widely used in the UK and elsewhere for many decades, and 'wood gas' was used for heating, lighting, and even as vehicle fuel. Both were superseded by natural gas – although wood gasifiers reappeared during the coal shortages of World War II.

There are many different designs of modern gasifier, but essentially one basic process: hot steam and oxygen interacting with the solid fuel. The gasification reactions do not occur easily, and need operating temperatures from a few hundred to over a thousand degrees Celsius, with pressures from a little above atmospheric pressure to 30 times this. The process begins with the release of the volatiles from the heated solid, leaving the char. These two components in turn undergo a reactions with steam and oxygen, resulting in **producer gas**, a mixture of combustible components (mainly carbon *monoxide* and hydrogen, with some methane, higher hydrocarbons and condensable tars) together with carbon dioxide and water. Further processing may break down some of the combustibles to give a cleaner gas. Nitrogen will also be present if air is used, rather than oxygen, and the energy content of the resulting gas is then only 3–5 MJ m^{-3}, about a tenth of that of natural gas. (Using pure oxygen gives a more valuable gas, but the additional cost of the oxygen plant makes this economic only for very large systems, and at the time of writing is used only in *coal* gasifiers.)

Small gasification plants (<300 kW) are available commercially, often combined with gas engines driving small generators, and demonstration plants in the range 10–30 MW have been in operation since the mid-1990s. The overall conversion efficiency from the energy of the solid fuel to that of the resulting gas varies widely, from as little as 40% or so in relatively simple systems, to 70% or more in the most sophisticated plants. The financial viability of biomass gasification on a large scale is not yet established; but the gas itself is not necessarily the desired end product. Much of its attraction stems from the future potential for three low-carbon biomass-based systems:

- electricity from integrated biomass-fuelled gas turbine plant
- liquid fuels as substitutes for petroleum products
- hydrogen or other fuel for fuel cells.

The first of these is briefly described in Box 4.8, the second is treated in Section 4.8 below; and fuels cells are discussed in Box 10.3 of Chapter 10.

BOX 4.8 Power station turbine systems

Most of the world's power stations use the heat from burning fuel to produce hot, high-pressure steam for the **steam turbines** that drive the generators (Figure 4.17(a)). Steam temperatures are limited to about 600 °C for technical reasons, so the maximum Carnot efficiency is about 65%, and the actual efficiency perhaps 45% (see Box 2.4 in Chapter 2).

Gas turbines, driven directly by the combustion products of a burning gas at 1000 °C or more, should have higher efficiencies, but the significant improvement is achieved in the **combined-cycle gas turbine** (**CCGT**) system. This makes use of the fact that the gases leaving the gas turbine are still hot enough to raise steam for a steam turbine. In order not to corrode or foul the turbine blades, the gases must be very clean – which is why nearly all present gas-turbine or CCGT plants burn natural gas.

However, if the output from a biomass gasifier can be cleaned and used to run the gas turbine, the possibility arises of a self-contained **biomass integrated gasification combined cycle** (**BIGCC**) system, for local power generation. Pilot demonstration plants using woody or bagasse fuel, with electrical outputs in the 5–10 MW range are being tested at the time of writing.

For more detailed accounts of steam and gas turbines and gasification processes in general, see the companion text. For descriptions of some biomass gasification plants, see Sims, 2002.

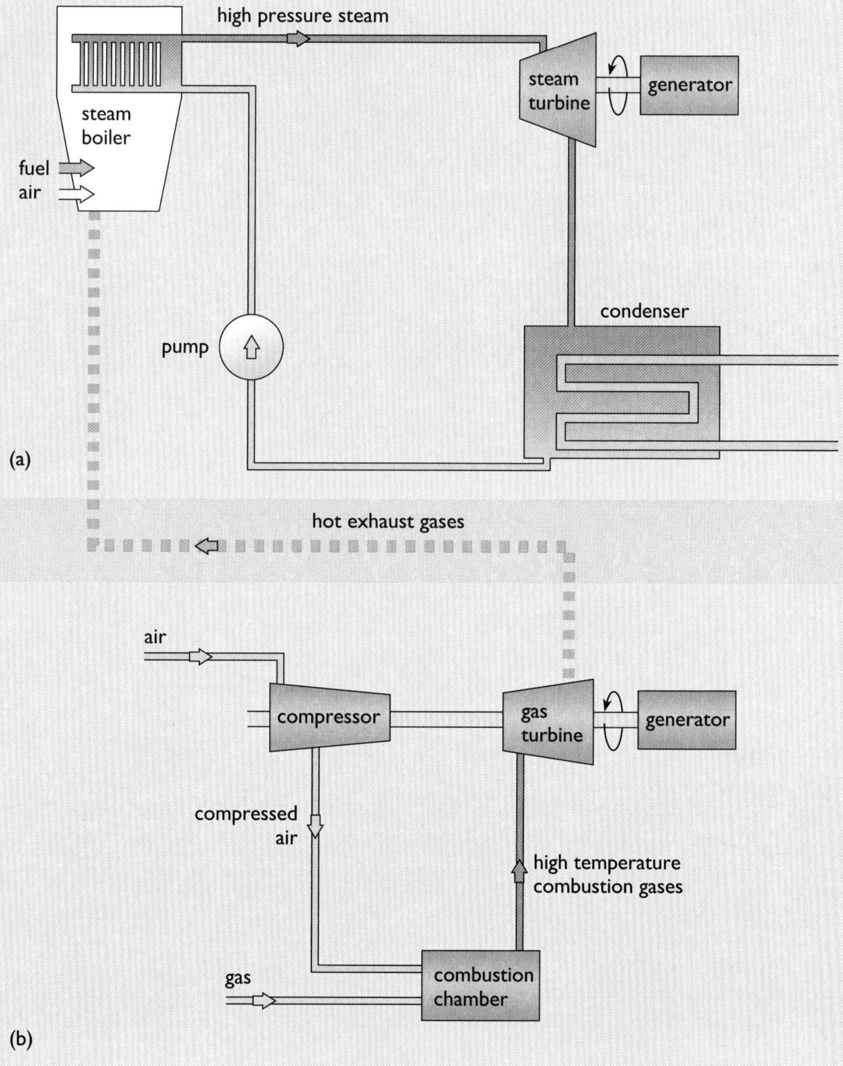

Figure 4.17 Types of generating system (a) conventional steam turbine: (b) simple gas turbine. In the CCGT plant, the gas turbine exhaust gases (dotted line) replace the fuel/air input as heat source for the boiler. For more details of turbines and gasification processes, see the companion text. For types of gasification plant, see Sims, 2002.

4.8 Production of liquid fuels from biomass

A major objective in bioenergy research is the production of liquid biofuels as substitutes for crude oil products. The three main approaches, treated in this section, could hardly be more different. The first is *thermochemistry*, which might be described as careful cookery. The second, *synthesis*, could be summarized as 'take it to pieces and start again'. The third method, *fermentation*, has of course been familiar to brewers and vintners for centuries.

Pyrolysis to produce bio-oil

Pyrolysis is the simplest and almost certainly the oldest method of processing one fuel in order to produce a better fuel. The traditional process that reduces wood to charcoal, today called **slow pyrolysis**, is very wasteful of energy, as we have seen. The term **pyrolysis** is now normally applied to processes where the aim is to collect the *volatile components* and condense them to produce a liquid fuel or **bio-oil**. As mentioned earlier, it is characteristic of biomass that the volatile matter carries more of the energy than the char, so this process should be more efficient.

The method involves heating the bio-material with a carefully controlled air supply. It must not burn, of course, and as the aim is a liquid product, gasification must be minimized. The resulting reactions are complex and hard to predict, giving a range of oils, acids, water, solid char and uncondensed gases, depending on the feedstock and operating conditions. The product, **bio-oil**, usually has about half the energy content of crude oil, and contains acid contaminants that must be removed. But it can be used as an oil substitute for heating or power generation, or could be refined to produce a range of chemicals and fuels.

Variations on the basic process include **solvolysis**, the use of organic solvents at 200–300 °C to dissolve the solids into an oil-like product, and **fast pyrolysis**, requiring temperatures of 500–1300 °C and high pressures (between 50 and 150 atmospheres). Fast pyrolysis, with capture of the volatiles, is used in the production of commercial charcoal, but at the time of writing, these methods for liquid fuels are at the pilot or demonstration stage (Sims, 2002).

Pyrolysis of wastes

Pyrolysis is an option for obtaining energy from waste materials, and pilot studies with MSW and plastics wastes have suggested relatively high energy efficiencies. But both pyrolysis and gasification (another option) seem likely to be considerably more expensive than incineration. Economies of scale, and perhaps greater acceptability than incineration, may mean better future prospects in more densely populated areas.

Synthesizing liquid fuels

One route from solid biomass to liquid biofuel starts with *gasification*. A gasifier using oxygen rather than air can produce a gas consisting mainly of H_2, CO and CO_2. Removal of the CO_2, and impurities such as tars, methane and traces of sulphur, leaves the highly active mixture of hydrogen and carbon monoxide called **synthesis gas**, or **syngas**, from which almost any hydrocarbon compound may be synthesized.

The first stage in the synthesis is a *shift reaction*, to adjust the proportions of H_2 and CO to the ratio required in the desired product. Methanol, for instance, is CH_3OH and therefore needs two H_2 molecules for each CO: a hydrogen-to-carbon monoxide ratio of 2:1. In the **Fischer-Tropsch** process, named for the chemists who developed it in the 1920s, the two components are passed over a suitable catalyst at high temperature and pressure, and the product, initially formed as a gas, is condensed. (A catalyst is a substance that influences a chemical reaction without itself being changed.)

The result, depending on the syngas composition and plant conditions, is a mixture of liquid and gaseous hydrocarbons. The gases can be recycled or used for heating, and the liquids can be up-graded and refined to produce vehicle fuels (Daey Ouwens and Küpers, 2003. For more on these processes, see also Chapter 7 of the companion text.)

Fermentation to produce ethanol

Fermentation is an **anaerobic biological** process in which sugars (e.g. $C_6H_{12}O_6$) are converted to alcohol by the action of micro-organisms, usually yeast. The required product, **ethanol** (C_2H_5OH), is separated from other components by distillation. Unlike methanol, ethanol cannot simply substitute for petrol, but it can be used as a gasoline extender in **gasohol**: petrol (gasoline) containing up to 26% ethanol – or possibly more in future. With suitable engine modifications (retuning, etc.) it can also be used directly.

Fermentation requires sugars, so the obvious source is sugar-cane, and this is the basis of Brazil's gasohol programme (see below). Plants whose main carbohydrate is starch (potatoes, maize, wheat) require initial processing to convert the starch to sugar, and this route is followed in the US, where the main source is maize. Even wood, with suitable pre-treatment, can be used as feedstock, but the processes are expensive, and the gasification/methanol route described above might be a better option.

The liquid resulting from fermentation contains about 10% ethanol, which is distilled off. The complete process requires a considerable heat input, usually supplied by crop residues. Table 4.4 shows the achievable yields of ethanol per tonne of biomass and per hectare of land for several crops. The energy content of ethanol is about 30 GJ t^{-1}, or 0.024 GJ per litre. Comparison of Tables 4.4 and 4.2 reveals that the conversion efficiency of fermentation is very poor indeed, but the technology is comparatively simple and plant costs low. And when the inputs are residues or surpluses and the output is a desirable product, low efficiency may be a price worth paying.

Table 4.4 Ethanol yields

Raw material	Litres per tonne[1]	Litres per hectare per year[2]
Sugar cane (harvested stalks)	70	400–12 000
Maize (grain)	360	250–2000
Cassava (roots)	180	500–4000
Sweet potatoes (roots)	120	1000–4500
Wood	160	160–4000[3]

1 This depends mainly on the proportion of the raw material that can be fermented
2 The ranges reflect world-wide differences in yield
3 The upper figure is a theoretical maximum

Figure 4.18 Sugar cane

Brazil's PRO-ALCOOL programme, producing ethanol from sugar residues, is the world's largest commercial biomass system. It was established in 1975, when oil prices were high and sugar prices low, and during its first 25 years, the avoided fossil fuel imports saved $40 billion directly in hard currency, with further saving from reduced interest on foreign debt. Production has reached 15 billion litres per year at times. Most vehicles currently run on gasohol with 26% ethanol content, and 1999 estimates suggested that the programme was reducing annual greenhouse gas emissions by almost 13 Mt of carbon.

The economics of ethanol production are very uncertain. Its viability depends critically on world prices of sugar and crude oil, both of which have varied widely – and often rapidly – over the past 30 years. Within Brazil, disputes between the growers and distillers, the government and the national oil company have had their effect. Production rose and production costs fell during the 1990s, but at the time of writing, continuing depressed oil prices and the growth in Brazilian domestic crude oil production, coupled with the continued failure of growers, distillers and government to agree on a fair price, threaten the future expansion of large-scale liquid fuel production.

The USA has a long history of bio-ethanol production from **maize** (encouraged by Henry Ford, amongst others). But as with other biofuels enterprises, cheap petroleum destroyed the market. In the later twentieth century, the threat of oil shortages and the problem of surplus maize led to a minor revival, centred on the main grain-growing states of the Midwest. The US output of liquid biofuels has continued to rise at a few percent per year, reaching some 5 million tonnes in 2001 (IEA, 2003). As with sugar ethanol in Brazil, the economic viability of maize ethanol depends on two volatile factors, in this case the world prices of oil and grain. The possibility of grain price rises has encouraged the establishment of research projects investigating the potential use of alternative feedstocks such as agricultural residues, MSW and even woody energy crops.

Sweet-stemmed varieties of the grass-like grain crop, **sorghum**, are already used as feedstock for ethanol production in the USA and Brazil, using the same equipment as for sugar cane. Two crops per year are possible, and sorghum has a much lower water requirement than sugar cane. There is genetic potential for improved yields, and plant breeders are developing hybrids suitable for the European climate. An EU directive already allows up to 7% ethanol in gasoline throughout Europe, and there are tax incentives for other liquid biofuels (see below). Current US and European programmes are based on producing a useful product from surplus food, but if all costs are taken into account, the product is far from being competitive with conventional vehicle fuels. At a wheat price of £113 per tonne, the cost of producing bio-ethanol has been estimated to be about 36p per litre, considerably higher than the current *untaxed* petrol and diesel prices of under 20p. However, as discussed in the following sub-section, there is scope to encourage these biofuels by changing the tax regime.

Vegetable oils to biodiesel

In 1911 Dr Rudolf Diesel wrote, '*The diesel engine can be fed with vegetable oils and would help considerably in the development of the agriculture of the countries which will use it.*' He demonstrated the use of a variety of vegetable oils, and more have been tried since, but as we have seen in other cases, cheap crude oil came to dominate the market.

Vegetable oils are quite different from the liquid fuels discussed above. They occur naturally in the seeds of many plants, and are extracted by crushing. Their energy content of 37–39 GJ t^{-1} is only a little less than that of diesel (about 42 GJ t^{-1}). They can be burned directly in diesel engines, either pure or blended with diesel fuel, but incomplete combustion is a problem, leading to carbon build-up in the cylinders, and conversion of the vegetable oils to **biodiesel** is preferred.

The oils are compounds called triglycerides, whose large molecules are effectively various organic acids combined with glycerol (an alcohol). The main components of diesel oil, usually called esters, are organic acids combined with other, lighter alcohols. The conversion process, called *transesterification*, involves adding methanol or ethanol to the vegetable oil. This converts the triglycerides into esters of methanol or ethanol, together with free glycerol. The glycerol, a valuable by-product, is removed, and the excess alcohol extracted for recycling, leaving the biodiesel. No engine modification other than tuning is required to use it, although its lower energy content means that fuel consumption is about 10% higher.

Global production of biodiesel, from a variety of plants, is around 1.5 million tonnes per year and growing. In Europe, the potential of **rape methyl ester (RME)** from rape grown on set-aside land (Box 4.5) has generated interest. In France all diesel contains 5% RME, and oil from rape in the UK is exported to France. In the USA, production is based on oil from **soya** beans and **recycled cooking oil**. The UK government is also interested in this option for waste oils from chip shops, etc.

The estimated cost of producing biodiesel from rape seed at £110 per tonne is approximately 26p per litre, including credits for the glycerol and cattle-feed by-products. However, EU member states have been allowed since 1993 to adjust fuel tax levels to make biofuels more attractive. The UK tax was cut by 20p per litre in 2001, and biodiesel is now available from over 100 filling stations. Germany imposes no tax, and has more than 400 outlets selling biodiesel at a little under the normal (taxed) price for diesel.

A European Commission directive has proposed a target of 5.75% for the share of biofuels in the transport sector by 2010.

Using waste cooking oil could reduce costs, but collection logistics will probably restrict this to catering operations rather than domestic use. A British supermarket showed interest in converting oil from staff catering facilities to fuel for its vehicle fleet in 2002. (There were also reports in 2002 of UK drivers mixing vegetable oils from their local supermarkets with their diesel fuel. This is illegal – unless they take care to pay the tax – and may harm the engine.)

In countries with warmer climates, blends of up to 30% vegetable oil with diesel are used often without transesterification. Coconut oil is used in tractors and lorries in the Philippines, palm and castor oil in Brazil and sunflower oil in South Africa. However, the food and cosmetics markets can usually pay a better price, so these applications are limited to places where diesel fuel is expensive and in short supply.

4.9 Environmental benefits and impacts

An account of the methods used in assessing the environmental effects of energy systems appears in the companion text, Chapter 13, *Penalties*, which also presents the detailed results of a number of comparative studies. (See Boyle *et al*, 2003).

As we saw at the start of this chapter, the world's biomass plays a very basic role in maintaining the environment, so it is important to consider not only the *benefits* of bioenergy but the possible *deleterious* effects, global or local, of our interference with these natural processes. Space does not allow a detailed account of every effect of every form of bioenergy, so we'll concentrate in the following on the most significant benefits and impacts, considering first atmospheric emissions and then other aspects.

Atmospheric emissions

Carbon dioxide

The concept of 'fixing' atmospheric CO_2 by planting trees on a very large scale has attracted much attention. There is little doubt that the halting of deforestation and the replanting of large areas of trees would bring many environmental benefits, but absorption of carbon dioxide by a new forest plantation is a once-and-for-all measure, 'buying time' by fixing atmospheric CO_2 while the trees mature, say for 40–60 years. A wider bioenergy strategy, concentrating on the substitution of biofuels for fossil fuels may be a more effective lasting solution.

To analyse the benefits of substitution, it is essential to assess all the effects in a **life cycle analysis**. We'll start with just one form of energy in one context: electricity generating plants that are either current or near to commercial implementation in the UK. Table 4.5 shows the emissions of carbon dioxide and of the two main sources of acid rain, sulphur dioxide and oxides of nitrogen. The data are *life cycle emissions per unit of electrical output*, taking into account all the processes involved. For instance, the totals for energy crops include emissions associated with fertilizer production and the use of fossil fuel in processing or transporting the fuel. But there is also 'credit' for the CO_2 *removed* from the atmosphere by the growing crop.

As can be seen, even the best systems are not carbon-neutral. But all the bioenergy systems, even MSW combustion, have lower CO_2 emissions than any of the fossil fuel plants. And it is easy to show that if the annual 270 GWh from the straw-fired plant described in Box 4.6 is replacing the same output from a coal-fired power station, it will be reducing UK annual CO_2 emissions by a quarter of a million tonnes.

Table 4.5 Net life cycle emissions from electricity generation in the UK

	Emissions[1] / g kWh^{-1}		
	CO_2	SO_2	NO_x
Combustion, steam turbine			
poultry litter	10	2.42	3.90
straw	13	0.88	1.55
forestry residues	29	0.11	1.95
MSW (EfW)	364	2.54	3.30
Anaerobic digestion, gas engine			
sewage gas	4	1.13	2.01
animal slurry	31	1.12	2.38
landfill gas	49	0.34	2.60
Gasification, BIGCC[2]			
energy crops	14	0.06	0.43
forestry residues	24	0.06	0.57
Fossil fuels			
natural gas: CCGT[2]	446	0.0	0.5
coal: 'best practice'	955	11.8	4.3
coal: FGD & low NO_x[3]	987	1.5	2.9

1 Note that 1 g kWh^{-1} is the same as 1 t GWh^{-1}.
2 See Box 4.8
3 Flue gas desulphurization and low NO_x burners.
Source: Adapted from ETSU, 1999

Such CO_2 reductions are the obvious benefits of bioenergy, whether used for power generation as here or directly for heat or for conversion to liquid fuel. But we must also look at other emissions. Nitrogen oxides (NO_x) are an inevitable product of the combustion of any fuel, because four-fifths of the air is nitrogen. High temperatures – in furnaces or internal combustion engines – increase NO_x production, and bioenergy systems will need to meet the same 'clean-up' requirements as those using fossil fuels. This also applies to the removal of particulates (see Section 4.6).

Methane

An important omission from Table 4.5 is *methane*. As we have seen, this powerful greenhouse gas is a product of the anaerobic digestion of biomass – whether naturally, as in a pond, or as a consequence of human activities. Its relationship to the use of bioenergy is rather complicated. Dung heaps are the result of our keeping animals for *food*, and landfills are the result of our accumulation of *wastes*. In neither case is the extraction of energy responsible for the methane emissions. Indeed, combustion of the gas is more nearly the solution than the problem. A molecule of CH_4 is nearly 30 times as effective as a molecule of CO_2 in trapping the earth's radiated heat, and full combustion effectively replaces each CH_4 molecule by a CO_2 molecule. To take one example, combustion of landfill gas is estimated to have reduced UK greenhouse gas emissions by the equivalent of some 20 Mt of carbon in 2002. Without this, total UK greenhouse gas emissions in that year would have been more than 10% higher.

In this context, there is controversy about the relative merits of landfill and MSW combustion (EfW). As Table 4.5 shows, life cycle CO_2 emissions are much higher for the EfW route, but with careful storage and efficient combustion, the methane emissions should be low. With landfill, it is never possible to collect all the gas, and there are inevitably methane emissions to the atmosphere. Depending on the collection efficiency, these could add the equivalent of another 100–200 g kWh^{-1} to the actual CO_2 emission shown in Table 4.5. However, as we shall see, criteria other than greenhouse emissions may determine the future roles of these two technologies.

Other emissions

There are also other important atmospheric emissions which are released at lower concentrations from the combustion of MSW and landfill gas – and to lesser degree in the combustion of any biomass. These include heavy metals and organic compounds (such as dioxins) that can potentially cause a wide range of health effects. Other sources of pollution include the fly ash residue from MSW combustion, which has a relatively high concentration of heavy metals and needs special disposal (e.g. in controlled or hazardous waste landfill sites). And there may be liquid effluents, from flue gas cleaning, for instance, that must be treated before release.

However, it has been estimated that EfW accounts for only 0.1% of UK dioxin emissions, and a Swiss study found that domestic bonfires were a greater source there than controlled MSW incineration. Both the UK and the EU are enforcing increasingly stringent emission standards and the installation of pollution control technology; but there are concerns that the standards are not always maintained – and history shows that, as data improve, the accepted 'safe' levels of such pollutants tend to become lower and lower.

Land use

Biomass is one of the most land-greedy energy sources, and it has been suggested that using land for other forms of renewable energy may do more to mitigate the impacts of CO_2 (Smith *et al.*, 2000). Comparing bioenergy yields with those of other 'low density' renewables can be illuminating. Consider, for instance, the land needed for an annual electrical output of ten million kWh – the equivalent of a 1.5 MW power station. An array of PV modules might need an area of some 40 ha to provide this, and a windfarm slightly more: perhaps 100 ha. With reasonable yields and conversion efficiencies, the land area of energy crops required to fuel this power plant would be in the range 300–1000 ha (3–10 km^2).

Area is of course not the only consideration, and in any case, the above three systems are unlikely to be competing for the same land. PV arrays are currently more likely to occupy rooftops than large areas of countryside, and wind turbines are often on high land that is also used for other purposes – grazing, or even woodland. The energy crop, as we have seen, may be on farm land that is surplus to food requirements.

In the case of energy crops, particularly oil-seed rape and short rotation coppice, there are concerns about the effect on the agricultural landscape, the reduction in biological diversity and the high inputs of fertilizers and pesticides. But proponents of bioenergy point out that coppices can use different tree species interspersed with indigenous vegetation, and that the life cycle fertilizer demand is perhaps one tenth of that of a cereal (food) crop. It is also claimed that diversity of animal life is greater, particularly for coppice. There is also interest in coppice as a biofilter, improving groundwater quality, and for land treatment of sewage sludge. In Denmark, it has been estimated that 30 000 ha of willow coppice could treat all the sewage produced by a population of over five million.

Energy balance

The terms **energy balance**, **energy payback ratio**, or sometimes just **energy ratio** are used to describe the relationship between the energy output of a system and the energy inputs needed to operate it (usually from fossil fuels). The concept came to the fore when doubts arose concerning some of the early fuel-from-biomass projects introduced following the oil price increases of the 1970s. There were claims that, when all energy inputs (fertilizers, harvesting, transport, processing, etc.) were taken into account, the fossil-fuel energy input for some schemes was actually greater than their bioenergy output.

The ratio of output to input will of course depend on the type of system, and the extent of the processing involved. In particular, ratios will normally be lower if the final 'output' is electricity, rather than the heat content of a biofuel. Over the full range of renewable sources, the ratio of output to input can vary from as little as 1:1 (see below) to as much as 300:1 (this for some hydroelectric plants). Woody energy crops perform well, with ratios between 10:1 and 20:1 on a heat output basis, but biodiesel may achieve only 3:1, whilst ethanol from grain barely breaks even at just over 1:1. When *wastes* are the input, the question arises of how much of the energy input to attribute to the energy extraction system. Where this is only a small fraction of the energy that would be used in any case, the result can be a high ratio. The value of 30:1 for electricity from woody sawmill wastes is an example.

Payback ratios can be improved by well-designed systems. Part of the biofuel output can, for instance, replace fossil fuels in supplying heat for the processes – as in the anaerobic digestion of wet wastes, or the use of bagasse instead of coal to provide process heat for ethanol production from sugar cane (Goldemberg, 1993).

The energy balance of a biomass energy system is also a reflection of its environmental impact. The greater the outputs, the greater the quantity of fossil fuel displaced. The lower the inputs, the lower the extra demands put upon the environment by the biomass system.

4.10 Economics

Bioenergy, as we have now seen, comes from many different sources and in many different forms. In the limited space available, it would be impossible to treat the economics of all of these in any detail, so we have chosen to limit the discussion here to the UK, and to the generation of electricity, which accounts for nearly 80% of UK use of bioenergy. (Note that there is brief discussion of the economics of *liquid* biofuels in Section 4.8 above.)

Energy Prices

The cost of any form of bioenergy must be seen in the context of the prices of similar forms of energy from present sources – usually the fossil fuels. Table 4.6 offers a summary of current (2003) UK energy prices. For ease of comparison, the normal £ per tonne or barrel, or pence per litre have been converted into prices per unit of energy.

Table 4.6 Average UK fuel prices(1), 2002

	Large industrial users (excluding tax)		Domestic consumers (including VAT)	
	/£ GJ^{-1}	/p kWh^{-1}	/£ GJ^{-1}	/p kWh^{-1}
coal	1.3	0.47	5.4 (2)	1.95
oil	3.0 (3)	1.1	4.0 (4)	1.4
natural gas	2.1	0.74	5.0	1.8
petrol, diesel (5)			~20	~6
electricity	7.4	2.7	19.4	7.0

1 Note that 1p per kWh = £2.78 per GJ

2 This is a figure adopted in modelling studies, as sales are too low to establish a reliable average (ETSU, 2002)

3 Heavy fuel oil

4 Gas oil, heating oil

5 Three-quarters of this price is tax

Source DTI, 2003b

The differences between industrial and domestic prices call for comment. The fuel prices paid by large consumers are more directly related to world prices of oil and coal and the European market price for natural gas. And fierce competition between UK electricity generators means that industrial consumers are able to negotiate contracts at low prices (see Chapter 10). Extra distribution costs account in part for the higher domestic prices, and domestic heating oil is also a more refined product than the heavy fuel oil used by industry. (The subject of comparative fuel prices is discussed in more detail in the companion text.)

Costing bioenergy

The four main factors that determine the cost of energy from any system are described in Appendix A1. For most renewable energy systems, the initial *capital cost* (including the cost of borrowing the money) is the main component. Unlike many other renewable energy technologies, bioenergy systems can have significant *fuel costs*. Energy crops, for example, must be planted, fertilized, protected against weeds and pests, harvested and transported. On the other hand, EfW may have *negative* fuel costs in the form of payment for disposal of wastes.

The remaining two factors are common to most energy systems. Operation and Maintenance costs (O & M) are usually proportional to the output of the plant, and will depend on the type of fuel – in particular, the nature of its emissions and the residues it leaves. It is usually assumed that the *decommissioning costs* will be covered by the scrap value of equipment.

It is also worth noting that for electrical plant, the cost per kWh of output depends on the annual output, so it is important to maximize the *load factor* (see Box 5.2 in Chapter 5).

Electricity from wastes

The economics of electric power from **landfill gas** are relatively simple. The gas itself is a waste product that must in any case be collected and flared off to protect the environment and prevent explosions. So the marginal additional cost of piping it to a gas engine is likely to be small. The principal costs are thus the capital cost of the engine and generator and connection to the grid, together with a modest allowance for O & M. Box 4.9 shows a cost calculation using data from a relatively recent assessment (ETSU, 1999).

However, with landfill sites in ever shorter supply, the cost of transport to distant sites is rising, with some councils paying more than £20 per tonne. Owners of sites can charge increasingly high **gate fees** to accept waste, particularly near large cities, and many countries also impose a **landfill tax**. Together, these charges can add a further £40–60 per tonne to the basic cost of local waste collection. Their deterrent effect can be appreciated by noting that the plant in Box 4.9 would need a landfill containing almost a million tonnes of wastes.

As we have seen, modern EfW generating plant using **municipal solid wastes** must meet strict requirements on emissions, residues and other environmentally sensitive issues. This requires complex and expensive

BOX 4.9 Costing landfill gas

We take as an example a landfill site large enough to support 2 MW of generating capacity over an operating life of 15 years with an average load factor of 88%. The following calculation assumes a **discount rate** of 8% (see Appendix A1).

Financial data
Plant cost: £750 per kW
O & M costs: 1 p kWh^{-1}
Discount rate: 8%

There are 8760 hours in a year, so with 88% load factor, the 2 MW plant will generate

2000 x 8760 x 0.88 = 15.42 million kWh per year.

Table A.1 in Appendix 1 shows that the annual repayment on £1000 at 8% over 15 years is £117. The total plant cost is £1 500 000, so the total annual repayment is

1500 x £117 = £175 500 = 17.55 million pence per year.

Expressed as a cost per kWh of output, this is 17.55 / 15.42 = 1.14 p kWh^{-1}.

Adding the O & M costs gives a final electricity unit cost of **2.14 p per kWh.**

Source: ETSU, 1999

plant, and UK estimates in recent years have suggested capital costs of over £4000 per kW and O & M costs as high as 4p per kWh. Using the method of Box 4.9, with the same plant life, load factor and discount rate, these figures would imply an electricity unit cost of about 10p per kWh. However this will be offset in part by the gate fees charged for accepting waste for disposal. In a plant generating 450 kWh per tonne of wastes, a gate fee of £25 per tonne would reduce the unit cost by about 5.5p per kWh.

The relative *financial* merits of LFG and EfW will depend on the factors discussed here, but other aspects are likely to enter into any choice of disposal method. The use of waste heat from the EfW plant for district heating may be a consideration. And public acceptability and the enthusiasm of local authorities for alternative waste strategies will certainly be relevant.

The capital costs of power plants based on conventional boiler technology for clean **forestry wastes** or **straw** range from £1200 to £1700 per kW. For **poultry litter** plants the costs are higher: £2240 per kW, for instance, for the plant at Eye described in Section 4.5. There are also 'fuel costs', for transportation, storage, etc., but economies of scale, including higher generation efficiencies, can compensate for any advantage of smaller plants closer to their fuel sources.

Electricity from Energy crops

It has been argued (ETSU, 1999) that two factors are critical for the future development of energy crops for large-scale electricity generation:

1 The technical and commercial development of large-scale high-efficiency plant. A full-scale 33 MW BIGCC plant based on the ARBRE pilot scheme described in Section 4.4 might, for instance, achieve an efficiency of 44% and cost under £1000 per kW.

2 A steady supply of fuel at a low price. If fuel for the above 33 MW plant, from SRC or other crops, could be supplied at £1.1 per GJ (£16.50 per tonne of air-dry wood), the unit cost of the electricity supplied should be less than 4p per kWh.

Unfortunately, the need for large-scale developments in two separate 'technologies' leads to a vicious circle. The investor in plant development will want to be assured of a firm fuel supply at suitable cost, and the farmer will only invest in SRC if assured of a firm contract for the sale of the wood at an acceptable price. Neither is able to give the other the necessary assurance. Despite set-aside and other grants (see Box 4.5), convincing farmers that SRC is a credible crop remains a problem.

Figure 4.20 Harvester chipping coppiced hazel

4.11 Future prospects

In the **industrialized countries** of the world, about 7% of total primary energy comes from renewables. Large-scale hydroelectricity accounts for some three-quarters of this, so geographical factors can mean major differences in the renewables totals of otherwise similar countries. The next main contributions, from the combustion and anaerobic digestion of wastes, account for all but a few percent of the remainder (with the exception of the countries with a significant geothermal resource).

Hydro power has little scope for expansion in most of the industrialized world (see Chapter 5), and the present baseline for solar, wind, tidal or wave energy is low. So the main growth in renewables consumption in the *near* future is seen as coming from increased direct combustion of residues and wastes for electricity generation. In the longer term, newer biomass technologies, such as electricity from BIGCC plants or fuel cells, or liquid biofuels for vehicles, are expected to play an increasing role – but biomass may by then be overtaken in the developed world by other renewables such as wind and solar power.

In the UK, the 1999 study discussed in the previous section investigated the *economic potential* (see Box 10.1 in Chapter 10) of the main biomass sources of electricity. At a discount rate of 8%, the potential annual contributions by 2025 were estimated as:

- 7.5 TWh from *landfill gas*, at a cost of under 3.0p per kWh
- 19 TWh from *straw, chicken litter* and *forestry wastes*, at under 5.0 p per kWh
- 13.5 TWh from *MSW* combustion, at under 7.0 p per kWh.

The total electrical output from these three sources in 2002 was 5.4 TWh, so an average annual growth rate of about 9% would be needed to reach the above total of 40 TWh.

The future contribution of electricity from *energy crops* is very uncertain, for the reasons discussed in the previous section. If the problems were resolved, the estimated additional UK contribution from this source could be 33 TWh from SRC, at under 4.0 p per kWh. This alone would represent almost 10% of current (2003) UK electricity demand.

In the **developing countries**, future demand for biomass energy is likely to be determined mainly by two opposing factors: the rise in overall energy demand due to increasing population, and the reduction in the demand for bioenergy due to a shift from traditional to 'modern' forms of energy. Most projections suggest that biomass consumption will continue to rise during the next few decades, but at a lower rate than renewables in general or total primary energy. Nevertheless, it is estimated that traditional biomass will remain the sole domestic fuel for over a quarter of the world's population in 2030 (IEA, 2002). In purely quantitative terms, therefore, it is arguable that the improvement of wood stoves should be the most important technological aim in the field of energy today.

In much of the world outside Europe and North America, the 'problem' of surplus agricultural land does not exist. Rather the reverse, with competition for land between local needs for energy and for food. Nevertheless, it has been suggested that, after allowing for increased food production for a growing world population, as much as 400–700 million additional hectares could be available for energy crops world-wide in the year 2050, without unacceptable loss of biodiversity (UNDP, 2000). If used, this could provide of the order of 100 EJ of bioenergy per year. As we have seen, traditional biomass currently contributes an estimated 48 EJ to world primary energy, with a further 9 EJ or so from 'new' bioenergy (mainly in the form of electricity). 100 EJ therefore represents a three-fold increase, and a new contribution equal to over a fifth of present world total primary energy consumption.

Table 4.7 Potential future world roles for bioenergy

Source[1], date	Time frame (year)	Bioenergy contribution	
		EJ per year	percent[2]
Johansson *et al*, 1993	2025	145	37%
	2050	206	37%
Shell, 1994	2060	220–222	15–22%
WEC, 1994	2050	94–157	~15%
	2100	132–215	11–15%
Greenpeace, 1993	2050	114	19%
	2100	181	18%
IPCC, 1996	2050	280	50%
	2100	325	46%

1 See *Energy Futures and Scenarios* in Chapter 10 References and Further information

2 Bioenergy as a percentage of total primary energy

Source: Adapted from UNDP, 2000

The future for renewable energy on the world scale is discussed in more detail in Section 10.10 of this book, and it is worth noting that each of the scenarios described there sees an increasing contribution from biomass. Table 4.7 summarizes the potential bioenergy contributions in future years according to a few of these scenarios. In detail, most foresee a diminishing contribution from traditional biomass, i.e. the direct combustion of wood, dung, etc., with the growth coming from increasing biomass-fuelled generating capacity, and in the longer term, from liquid biofuels, or even a biomass-based hydrogen economy.

References

Anderson, T., Doig, A., Rees, D. and Khennas, S. (1999) *Rural energy services*, London, IT Publications.

Boyle, G., Everett, B. and Ramage, J. (2003) *Energy Systems and Sustainability*, Oxford, Oxford University Press in association with The Open University.

Daey Ouwens C. and Küpers G. (2003) 'Lowering the cost of large-scale, biomass based, production of Fischer-Tropsch liquids', *Proceedings, Bioenergy 2003*, Helsinki, pp 384–389.

DTI (2003a) *Digest of UK Energy Statistics*, Chapter 7 Renewable sources of energy. Available as www.dti.gov.uk/energy/inform/dukes/dukes2003/07main.pdf [accessed 29 October 2003].

DTI (2003b), *Quarterly Energy Prices* – September 2003. Downloadable from www.dti.gov.uk/energy [accessed 24 August 2003].

ETSU (1999) *New and Renewable Energy: Prospects in the UK for the 21st Century: Supporting Analysis* (ETSU R-122) Available as www2.dti.gov.uk/renew/condoc/support.pdf [accessed 24 August 2003].

ETSU (2002), *Options for a Low Carbon Future* – Phase 1 (ED50099/1) 2002, Annex B. Downloadable at www.etsu.com/en env/ClimateChange AnnexB.pdf [accessed 24 August 2003].

FOE (1991) *Energy without End*, London, Friends of the Earth.

Goldemberg, J., Monaco, L. C. and Macedo, I. C. (1993) 'The Brazilian fuel-alcohol program' in Johansson *et al.* (eds) *Renewable Energy: Sources for Fuels and Electricity*, Washington DC, Island Press.

Hall, D. O. (1991) 'Biomass energy', *Energy Policy*, vol. 19, No.8, October 1991, pp. 711–733.

Hall, D. O., Rosillo-Calle, F., Williams, R. H. and Woods, J. (1993) 'Biomass for energy: supply prospects' in Johansson *et al.* (eds) *Renewable Energy: Sources for Fuels and Electricity*, Washington DC, Island Press.

Hall, D. O. and Rao, K. K. (1999) *Photosynthesis* (6th edn), Studies in Biology, Cambridge, Cambridge University Press.

IEA (2002) *Renewables in Global Energy Supply*, International Energy Agency, France. Available as www.iea.org/leaflet.pdf [accessed 19 October 2003].

IEA (2003) Available as www.iea.org/stats/files/renew2003.pdf [accessed 19 October 2003].

Johansson T. B., Kelly, H., Reddy, A. K. N. and Williams, R. H. (eds) (1993) *Renewable Energy: Sources for Fuels and Electricity*, Washington DC, Island Press.

Manurung, R. and Beenackers, A. A. C. M. (1990) 'Field test performance of open core downdraft rice husk gasifiers' in Grassi, G. *et al.* (eds) *International Conference on Biomass for Energy and Industry: proceedings*, Elsevier Applied Science.

Sims, R. E. H. (2002) *The Brilliance of Bioenergy: In Business and in Practice*, London, James and James.

Smith, P., Powlson, D., Smith, J., Fallon, P. and Coleman, K. (2000) 'Meeting Europe's climate change commitments: quantitative estimates of the potential for carbon mitigation by agriculture', *Global Change Biology*, vol. 6, pp. 525–39.

United Nations Development Programme (2002) *World Energy Assessment*, New York, United Nations.

Chapter 5

Hydroelectricity

by Janet Ramage

5.1 Introduction

Like most other renewables, water-power is indirect solar power. Unlike most of the others, however, it is already a major contributor to world energy supplies. Hydroelectricity is a well-established technology, which has been producing power reliably and at competitive prices for about a century. It provides about a sixth of the world's annual electrical output and over 90% of electricity from renewables (BP, 2003; IEA, 2003). Its power plants include some of the largest artificial structures in the world.

Our treatment of hydro power starts with a short account of a Scottish hydroelectric scheme, commissioned nearly 70 years ago and still operating with much of its original plant. The story of this medium-scale scheme exemplifies rather well both the technical and the economic aspects of hydroelectricity.

The chapter continues with a discussion of the nature of the resource and its present contribution to world energy. A summary of the basic science follows, together with a brief history of the development of water power, tracing the evolution of water-wheels and the turbines that succeeded them. Modern turbine systems are the subject of Sections 5.6–5.10, describing the main types and their uses.

The remaining sections are concerned with the problems and the potentialities of hydroelectricity. We find the familiar issues of cost, reliability of supply and integration, which arise for all the renewable sources; but for large-scale hydroelectricity, the questions are rather different: whether there are limits to growth, what determines these limits, and whether we are already reaching them.

5.2 The Galloway hydros

The Galloway Hydroelectric Scheme on the River Dee in south-west Scotland makes an interesting study for several reasons. Initially commissioned in 1935, it was the first major UK scheme designed specifically to provide extra power at times of peak demand. Its six power stations are controlled as one integrated system. Several of its dams are on major salmon-fishing rivers, raising environmental issues common to many hydro schemes. It is also technically interesting, because significant differences in site conditions at the power stations mean that the system includes several types of turbo-generator.

Origins

The Galloway Hydros owe their origin to local pride and individual enthusiasm – and to an Act of Parliament. The first proposals to use the rivers and lochs of south-west Scotland for hydro power appeared in the 1890s, but the scheme became feasible only with the establishment of a National Grid in the 1920s. This meant that the great industrial conurbation of Glasgow became a potential customer, and the need for plant with fast response to meet daily and seasonal peaks in demand favoured the inclusion of some hydroelectric power in the otherwise coal-dominated national system.

The scheme

The system (Figure 5.1) has three main elements. The first is Loch Doon, which provides the main long-term seasonal storage. Its natural outflow is not into the Dee at all but to the north. However, a dam now restricts this northerly stream and the main flow is diverted westwards through a 2 km tunnel into the upper Dee valley. An interesting feature is the Drumjohn Valve; when demand for power is low, this directs the flow from two eastern tributaries of the Dee through the tunnel in the 'reverse' direction, *into* Loch Doon, adding to the stored volume. The level in the loch can vary by 12 metres, releasing 80 million cubic metres of water. Falling through the 200 metre height difference down to the final outflow at Tongland, this represents a gross release of some 150 million MJ of energy – over 40 million kWh.

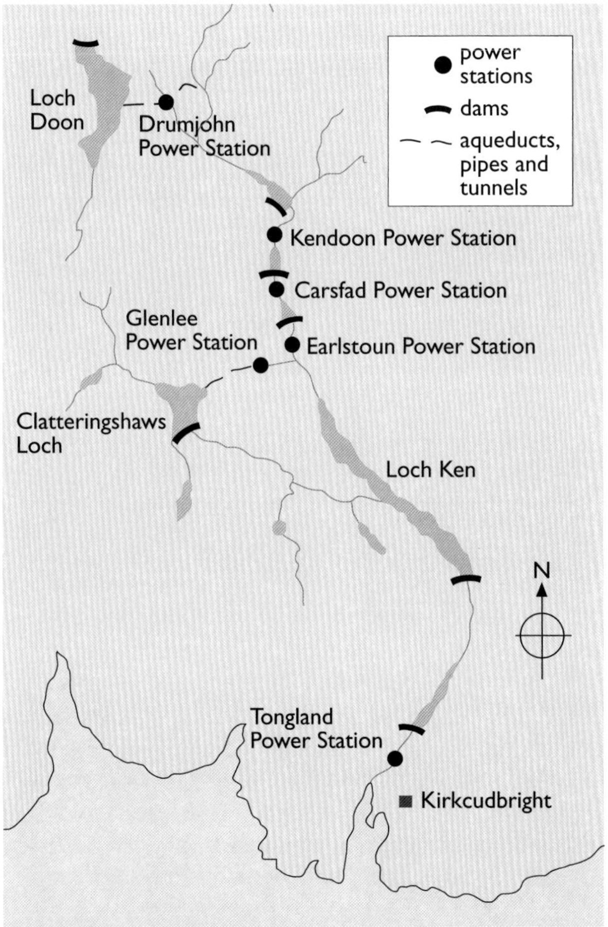

Figure 5.1 The Galloway Hydros
Source: Adapted from Hill, 1984

Clatteringshaws Loch, the only completely artificial reservoir in the system, is the second long-term storage element. Its outflow, through a tunnel nearly 6 km long and pipes with a fall of over 100 metres, supplies the 24 MW Glenlee power station before joining the Dee.

Figure 5.2 Carsfad power station and dam, Scotland

The third element, for fast response to short-term demand variations in the course of a day, is the series of dams and power stations along the course of the Dee: Kendon, Carsfad, Earlstoun and Tongland.

Power

The essential characteristics of a hydro site are the **effective head** (the height H through which the water falls) and the **flow rate** (the number of cubic metres of water per second, Q). As we shall see in Section 5.4, the power carried by the water is roughly ten times the product of these two quantities:

$$P\,(\text{kW}) = 10 \times Q \times H$$

The *electric* power output will of course be rather less than this input, as the data in Table 5.1 show.

The conversion of energy carried by water into electrical energy is carried out by the **turbo-generator**: a rotating turbine driven by the water and connected by a common shaft to the rotor of a generator.

Table 5.1 The Galloway power stations

Power station	Average head (m)	Maximum flow ($m^3\,s^{-1}$)	Output capacity (kW)	Number of turbines
Drumjohn	11	16	2000	1
Kendoon	46	55	24 000	2
Carsfad	20	73	12 000	2
Earlstoun	20	71	14 000	2
Glenlee	116	26	24 000	2
Tongland	32	127	33 000	3

Source: Scottish Power

Table 5.1 Data is available from: www.dti.gov.uk/energy/inform/energy_stats/electricity/dukes5_10.xls [accessed 28 October 2003]

The turbines

The head and the required power are critical in determining the most suitable type of turbine for a site. Glenlee's high head puts it at one extreme in the Galloway system, with Drumjohn's very low head and power rating at the other. Of the four river plants, Kendon and Tongland have intermediate heads and fairly high power ratings whilst Carsfad and Earlstoun have almost identical low heads and powers.

Any turbine consists of a set of curved blades designed to deflect the water in such a way that it gives up as much as possible of its energy. The blades and their support structure make up the turbine **runner**, and the water is directed on to this either by channels and guide vanes or through a jet, depending on the type of turbine. The Galloway plants include two types of runner: 'propellers' and Francis turbines (see Figures 5.15 and 5.16). 'Propeller' types are most suitable for large flows at low heads, and Francis turbines for medium to high heads. Comparison of Tables 5.1 and 5.2 shows that this is true for the Galloway scheme. (The choice of turbines for different sites is discussed in more detail in Section 5.10 below.)

Table 5.2 The turbines

Power station	Turbine rating (MW)	Turbine type	*Rate of rotation (rpm)
Drumjohn	2	Propeller	300
Kendoon	12	Francis	250
Carsfad	6	Propeller	214.3
Earlstoun	7	Propeller	214.3
Glenlee	12	Francis	428.6
Tongland	11	Francis	214.3

* See Box 5.1

Source: Scottish Power

BOX 5.1 **Rates of rotation**

The rate of rotation of a turbine (n) is its number of complete revolutions per minute (rpm). As Table 5.2 shows, this can vary appreciably, depending on the site and the turbine type. However, if the turbine drives the generator directly through a common shaft, as is usual, only certain rates of rotation are permitted, for the following reason.

The alternating voltages from all the power stations contributing to any grid system must have the same frequency. In Europe and many other countries, the agreed frequency is 50 Hz (hertz). So in a very simple generator, consisting of a magnet and a pair of coils, the magnet (or the coils) would need to spin at this rate: 50 cycles per second, or $50 \times 60 = 3000$ rpm. In the usual terminology, this simple system would be called a **two-pole** generator.

A large power-station generator will have more than one pair of coils on its spinning rotor (see Figure 5.20, right), and a **20-pole** machine, for instance, would need to rotate at only 300 rpm, a tenth of the above rate, to produce the required 50 Hz. A little arithmetic reveals that all the rates of rotation in Table 5.2 are sub-multiples of 3000 rpm.

Note that in the USA and other countries where the supply frequency is 60 Hz, the rates of rotation are sub-multiples of 3600 rpm.

(Generators are discussed in more detail in Chapter 12, *Electricity*, of the companion volume, (Boyle *et al*, 2003))

The salmon

The principal environmental issue raised during the approval process was the possible effect on salmon fishing. Several dams blocked rivers below their salmon spawning pools, and concern was expressed about the fate of adult salmon making their way upstream and young smolts on the reverse journey. The response was the incorporation of fish ladders at four dams. These are series of stepped pools with a constant downward flow of water to attract the fish, which leap up from pool to pool. The Doon dam had insufficient space for a long series of pools, so the fish ladder there is partly inside a round tower, but it is claimed that the fish find this spiral staircase no problem. The issue doesn't arise with Glenlee, as salmon do not use the man-made Clatteringshaws Loch. (It is worth noting that a much more detailed environmental impact statement would be required today.)

BOX 5.2 Capacity factor

There are 8760 hours in a year, so a 1 MW plant running constantly at its rated capacity would generate 8760 MWh, or 8.76 million kWh, in a year. The annual **capacity factor** of any plant is equal to its actual annual output divided by its maximum possible output (both in the same units, e.g. kWh, MWh, etc.). The result is usually expressed as a percentage.

If, for instance, Carsfad, with its capacity of 12 000 kW, generates 30 million kWh of electricity in a year, its capacity factor will be...

$$30 \times 10^6 / (12\,000 \times 8760) = 0.285, \text{ or } 28.5\%.$$

In practice, the annual capacity factor of a plant is determined by a combination of demand for its output and the power that it is able to produce at any time. The terms **load factor** and **plant factor** are often used as synonymous with capacity factor in the context of power plants.

Economics

The Galloway Scheme was built to supply power at times of peak demand. In other words, it was assumed that it would generate for only a few hours a day: an annual capacity factor, or load factor, of no more than perhaps 25%. This would suggest a poor return on the investment, which was in any case higher than for coal-fired plant of similar capacity. However, a number of circumstances made the scheme financially attractive.

- The company was able to assume a firm demand for the planned output.
- During its first three years the scheme received an annual treasury grant of £60 000 (close to a million pounds at present-day values), reducing the cost per unit by about 20%.
- The company also argued successfully that the local property taxes should be lower for hydro than coal-fired plant.

Figure 5.3 The fish ladder at Tongland

From the start, demand and the consequent economic performance exceeded expectations and only in a few years of serious drought did output fall below the planned level. This remains the case today. After nearly 70 years the original five plants are still generating power, joined in 1984 by the 2 MW plant at the Drumjohn Valve. The entire scheme is operated by the engineer at Glenlee, the only permanently manned plant – and the original construction costs were of course repaid many years ago.

5.3 Hydro: The resource

'The resource' for hydroelectricity, as for the other renewables, is not some finite quantity of stored energy, but a flow of power which, over a year, adds up to an *annual* energy contribution; in this case, the energy that becomes potentially available when rain (or snow) falls on high land. As we are concerned with electricity, it is customary to express the resource in kWh or TWh per year.

Nearly a quarter of the 1.5 billion TWh of solar energy incident on the Earth each year is consumed in the evaporation of water (see Figure 1.8), so the water vapour in the atmosphere represents an enormous store of renewable energy. Unfortunately most of this is not available to us. When the water vapour condenses, most of the energy is released into the atmosphere as heat and ultimately re-radiates into space. But a tiny fraction, about 0.06%, is retained by the precipitation that falls on hills and mountains.

The world resource

The natural route back to sea level for rain or snow falling on high land is by streams and rivers, and estimates of the energy carried by the world's flowing rivers might be regarded as the **total** or **available resource** (see Box 10.1, Chapter 10). Recent estimates have tended to suggest just over 40 000 TWh per year – about fifteen times the world's present hydroelectric output. (WEC, 2003b)

Estimates of the **technical potential** are, not surprisingly, appreciably lower, at 14 000–15 000 TWh per year. (It is characteristic of such predictions that 50 years ago the figure was only 6000 TWh.) As Table 5.3 reveals, the extent to which different regions of the world have developed their hydro potential varies considerably.

Table 5.3 Regional hydro potential and output

Region	Technical potential /TWh y^{-1}	Annual output* /TWh y^{-1}	Output as percent of technical. potential
Asia	5093	572	11%
S America	2792	507	18%
Europe	2706	729	27%
Africa	1888	80	4.2%
N America	1668	665	40%
Oceania	232	40	17%
World	14379	2593	18%

* Based on average output for the four years 1999–2002

Source: Adapted from WEC, 2003b and BP, 2003

For the UK, similar estimates, based on mean annual rainfall, land area and elevation, have suggested a technical potential of 40 TWh a year from 13 GW of installed capacity (ETSU, 1999). This is ten times the present output (see below) but would certainly be severely reduced if geographical and environmental factors were taken into account.

Two important questions remain, nationally and internationally. How much of the hydroelectric potential is it *practical* and *acceptable* to develop, and how much can we *afford* to develop? In other words, we need to consider the **practicable potential** and the **economic potential**. We return to these issues later.

World capacity and output

The world's installed capacity for large-scale hydroelectricity has increased every year for over a century, and by 2002 had reached about 740 GW. The contribution from small-scale hydro plants is very uncertain, but may increase the total by 5–10% (see Section 5.11).

Over the past few decades, the annual output from large-scale hydro has risen remarkably steadily, with an average yearly increase of about 50 TWh (Figure 5.4). In 2000, it exceeded 2700 TWh for the first time. However, there are inevitably annual fluctuations of a percent or so due to weather variations. In 2001, capacity rose but output fell by some 5%, due mainly to exceptionally dry conditions in the Americas. 2002 saw some recovery, despite a major drop of 70 TWh in European output. (In the UK, however, 2002 was a 'wet year' and output rose by nearly 20%.)

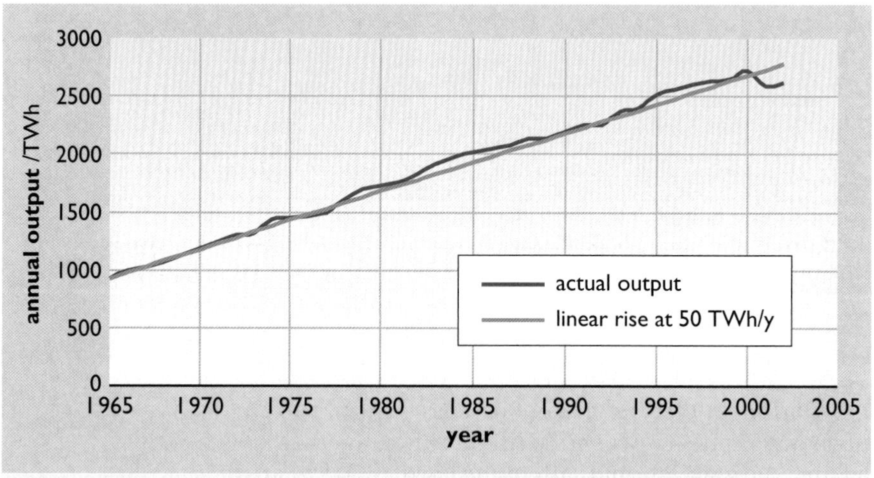

Figure 5.4 World annual hydroelectricity output, 1965–2002 (source: BP, 2003)

Despite its generally rising output, hydro has failed to keep pace with the growth in world total electricity production. Even before the recent fall, the decade 1991–2000 saw only a 24% increase in annual hydro output against a 30% increase in total electricity. In consequence the hydro contribution to the total fell from 18.5% to 17.6% in that decade, and by 2002 was only just over 16%.

The data on world capacity and output imply an average load factor (Box 5.2) for large-scale hydro of about 40%. (Some reasons for this relatively low value are discussed in Section 5.14.) On this basis, the annual increase in world output of 50 TWh corresponds to an average world-wide addition of some 14 GW of new capacity each year.

Tables 5.3 and 5.4 show the regional contributions in recent years, and the main national contributors. (Four-year averages are used, as better indicators than the fluctuating outputs of single years.) Norway obtains almost all her electricity from hydro, Brazil some 80%, and Canada and Sweden about half. Compared with these, the hydro resource of the UK is small, with an installed capacity in 2002 of 1.43 GW (90% of it in Scotland), and an annual output in recent years ranging from 3.3 to 5.1 TWh (DTI, 2003) – a wide spread reflecting year-to-year weather variations.

Table 5.4 National hydro contributions

country	output* ($TWh\ y^{-1}$)
Canada	345
Brazil	288
USA	264
China	231
Russia	167
Norway	129
Japan	91
India	76
Sweden	74
France	74
Venezuela	61
Italy	51
Austria	42
Switzerland	40
Spain	35
Rest of the World	626
World	2593

*Average over four years, 1999–2002

Source: BP, 2003

BOX 5.3 **Direct uses of water power**

There are many possible uses of water power which do not involve the generation of electricity. The kinetic energy of moving water (or the potential energy of water at a height) can be used directly to drive machines – and indeed these were the sole uses of water power until the mid-nineteenth century.

In the present day, direct use accounts for only a very small fraction of the total. Old water-mills still operate in a few places, grinding corn or sawing wood. And some mountain railways use a counter-balancing tank filled with water at the top and emptied at the foot of the hill. But these are rarities in the industrialized world.

In the developing countries, direct water power plays a slightly greater role. In Nepal, for instance, simple turbines which can be produced locally are used to drive machinery, and in the Middle East and Asia the use of a flowing stream to raise water for irrigation has by no means disappeared. Nevertheless, the world-wide energy contribution from direct uses is negligibly small compared with the hydroelectric power output.

5.4 Stored energy and available power

Stored potential energy

Water (or anything else) held at a height represents stored energy – the gravitational **potential energy** discussed in Section 1.3. About 9.81 joules of energy input are needed to lift one kilogram vertically through one metre against the gravitational pull of the Earth. If M kilograms are raised through H metres, the stored potential energy in joules is given by:

$$\text{Potential energy} = MgH \qquad (1)$$

The 'g' here is the acceleration due to gravity, whose value is about 9.81 m s^{-2}. (For rough calculations needing less than two percent precision, the approximation g = 10 m s^{-2} is often used.)

Equation 1 allows us to calculate the energy store represented by water held at a given height, if we know the stored mass and the height – the available head (see Box 5.4).

BOX 5.4 Stored energy

Suppose we want to find an absolute upper limit to the earth's hydro capacity. World-wide annual precipitation over land is about 10^{17} kg, and the mean height of the land above sea level is about 800 m. Equation (1) shows that the annual addition to the energy store is some 8×10^{20} joules, which is a little over 200 million million kWh, or 200 000 TWh, per year.

This is about twice the world's *total* primary energy consumption, and five times the 'total resource' mentioned in Section 5.3 – not surprisingly, as there is no way in which we could capture every drop of rain that falls. (And the above use of averages is in any case of doubtful validity.)

Power, head and flow rate

In estimating the resource, the *power* available at any time is at least as important as the total annual energy. The power supplied (P, in watts) is the rate at which energy is delivered – the *number of joules per second*. This will obviously depend on the flow rate of the falling water – the *number of cubic metres per second* (Q). The mass of a cubic metre of fresh water is 1000 kg, so the mass falling per second will be $1000 \times Q$, and it follows from Equation (1) that

$$P = 1000 \times Q \times g \times H$$

Resource estimates must take into account **energy losses**. In any real system the water will lose some energy due to frictional drag and turbulence, and the **effective head** will thus be less than the actual head. These flow losses vary greatly from system to system: in some cases the effective head is no more than 75% of the actual height difference, in others as much as 95%. Then there are energy losses in the plant itself. Under optimum conditions, a hydroelectric turbo-generator is one of the most efficient machines, converting all but a few percent of the input power into electrical output. Nevertheless, the **efficiency** – the ratio of output to input power – is always less than 100%. With these factors incorporated, the output power becomes

$$P = 1000 \times \eta \times Q \times g \times H \qquad (2a)$$

where H is now the effective head and η is the turbo-generator efficiency.

If we express P in kilowatts, and use the approximate value for g, we obtain a useful simple expression:

$$P\,(kW) = 10\eta QH \qquad (2b)$$

Box 5.5 gives examples of the use of this for rough calculation of power output.

BOX 5.5 Available power

As examples of power calculations, consider two systems, each with a plant efficiency of 83%, but of very different sizes.

The first site is a mountain stream with an effective head of 25 metres and a modest flow rate of 600 litres a minute, i.e. 0.01 cubic metres per second. Using Equation (2), we find that the power is approximately 2 kW.

In contrast, suppose that the effective head is 100 m and the flow rate is 6000 cubic metres per second – roughly the total flow over Niagara Falls. The power is now about 5 million kW, or 5 GW.

5.5 A brief history of water-power

The prime mover

Moving water was one of the earliest energy sources to be harnessed to reduce the work load of people and animals. No-one knows exactly when the water-wheel was invented, but irrigation systems existed at least 5000 years ago and it seems probable that the earliest water-power device was the **noria**, raising water for this purpose. This device (Figure 5.5) appears to have evolved over at least six centuries before the birth of Christ, perhaps independently in different regions of the Middle and Far East.

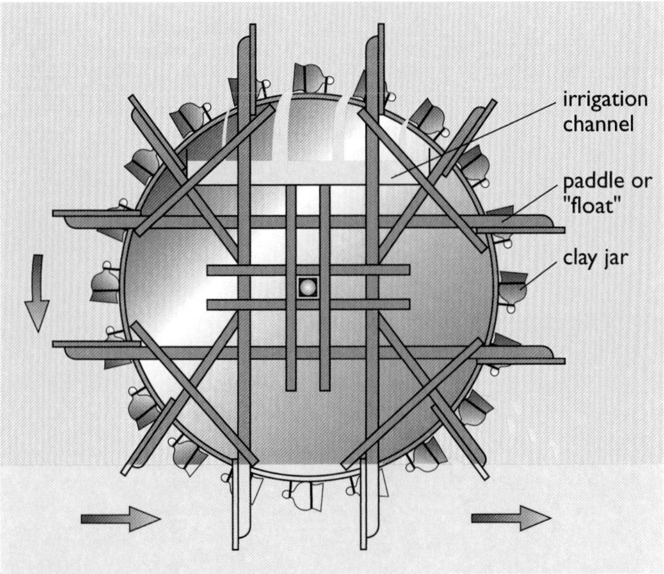

Figure 5.5 A noria. In this earliest water-wheel the paddles dip into the flowing stream and the rotating wheel lifts a series of jars, raising water for irrigation

The earliest **water mills** were probably vertical-axis corn mills, known as Norse or Greek mills (Figure 5.6), which seem to have appeared during the first or second century BC in the Middle East and a few centuries later in Scandinavia. In the following centuries, increasingly sophisticated water-mills were built throughout the Roman Empire and beyond its boundaries in the Middle East and Northern Europe (Figures 5.7 and 5.8). In England, the Saxons are thought to have used both horizontal- and vertical-axis wheels. The first documented mill was in the eighth century, but three centuries later the Domesday Book recorded about 5000, suggesting that every settlement of any size had its mill.

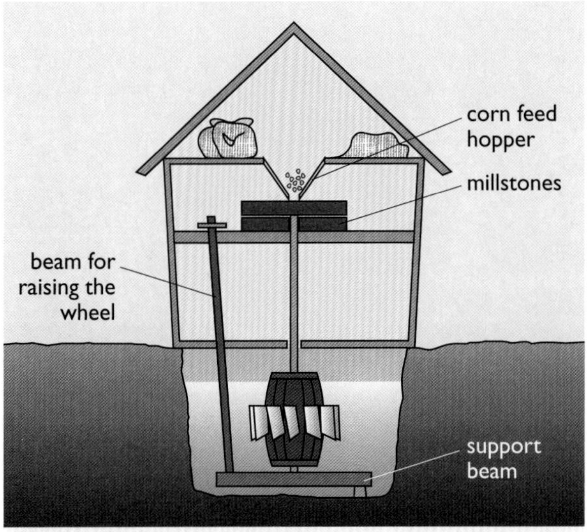

Figure 5.6 A Norse mill. Mills of this early vertical-axis type are still in use for mechanical power in remote mountainous regions

Raising water and grinding corn were by no means the only uses of the water mill, and during the following centuries the applications of this power source kept pace with the developing technologies

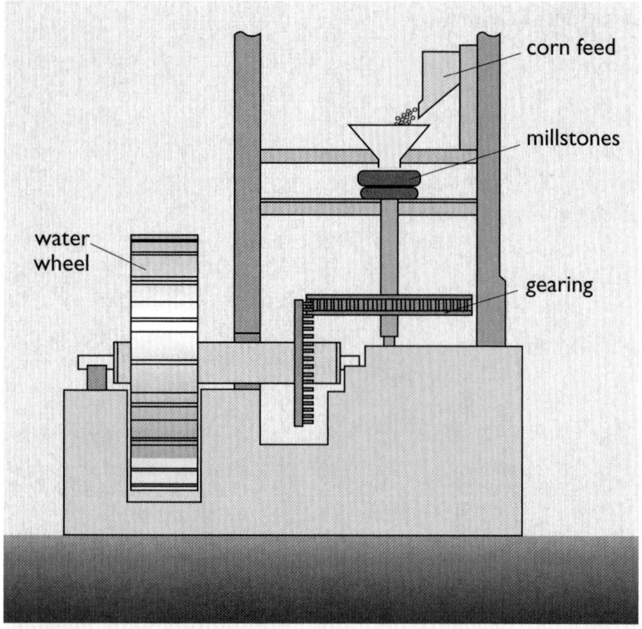

Figure 5.7 A Roman mill. This corn mill with its horizontal-axis wheel was described by Vitruvius in the first century BC. Note the use of gears

of mining, iron working, paper-making, and the wool and cotton industries. Water was the main source of mechanical power, and by the end of the seventeenth century England alone is thought to have had some 20 000 working mills.

There was much debate on the relative efficiencies of different types of water wheel (Box 5.6). The period from about 1650 until 1800 saw some excellent scientific and technical investigations of different designs. These revealed output powers ranging from about one horsepower to perhaps 60 for the largest wheels – in modern terms, from three-quarters of a kilowatt to 45 kW, and confirmed that for maximum efficiency the water should pass across the blades as smoothly as possible and fall away with minimal speed, having given up almost all its kinetic energy. (They also proved that, in principle, overshot wheels should win the efficiency competition.)

But then steam power entered the scene, putting the whole future of water-power in doubt.

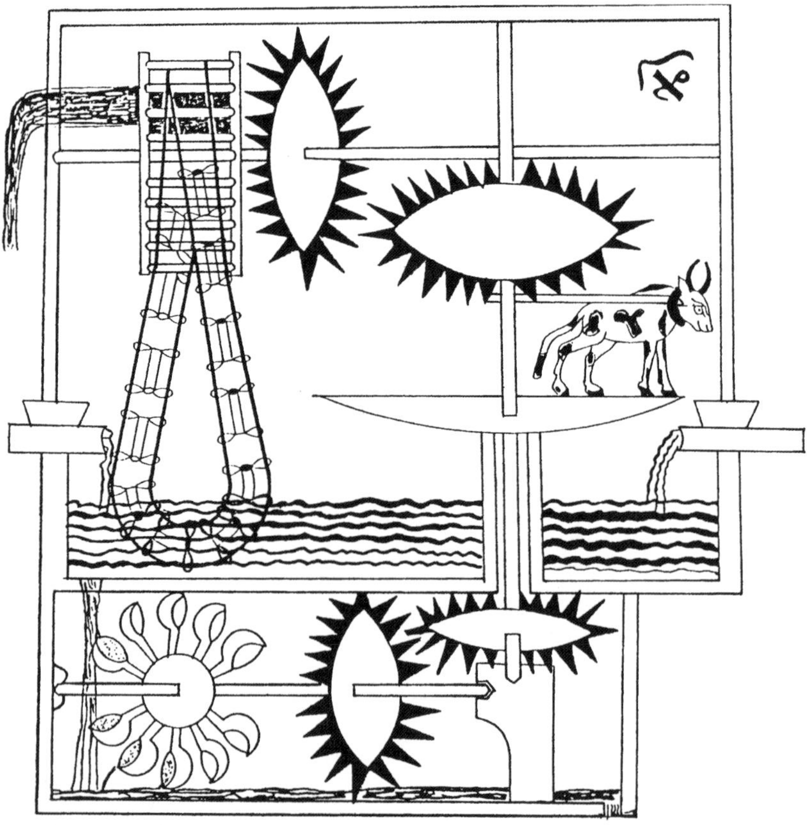

Figure 5.8 Medieval saqiya. The diagram comes from the Book of Knowledge of Ingenious Mechanical Devices of al-Jahazi, written in Mesopotamia 700 years ago. (The ox is a wooden cut-out designed to fool the public, who can't see the hidden water-wheel and gears below)

BOX 5.6 Types of water-wheel

By the end of the eighteenth century three main types of wheel were in use (Figure 5.9), two of which had remained virtually unchanged for well over 1000 years. (Vertical-axis wheels had also survived, but played only a minor role – a situation that was soon to change.) History is not our main concern here, but it is worth looking briefly at some of the features in order to understand the problems which later developments were designed to overcome.

- The **undershot** wheel is driven by the pressure of water against its lower blades which dip into the flowing stream. The advantage is that it can be used in almost any stream or channel; but it becomes very inefficient if the water backs up due to flooding, impeding the motion of the wheel.
- The **overshot** wheel is driven by water falling from above on blades with closed sides – effectively buckets. Overshot wheels don't suffer the flooding problem, but need a 'head' at least as high as the diameter of the wheel, making them unsuitable for streams and rivers with gentle gradients.
- The **breastshot** wheel, a later development, is a compromise. The water is channelled between parallel breast walls and strikes the paddles at about the level of the wheel axle. It has the advantage of overcoming the flooding problem without requiring the high head and massive construction of the overshot wheel.

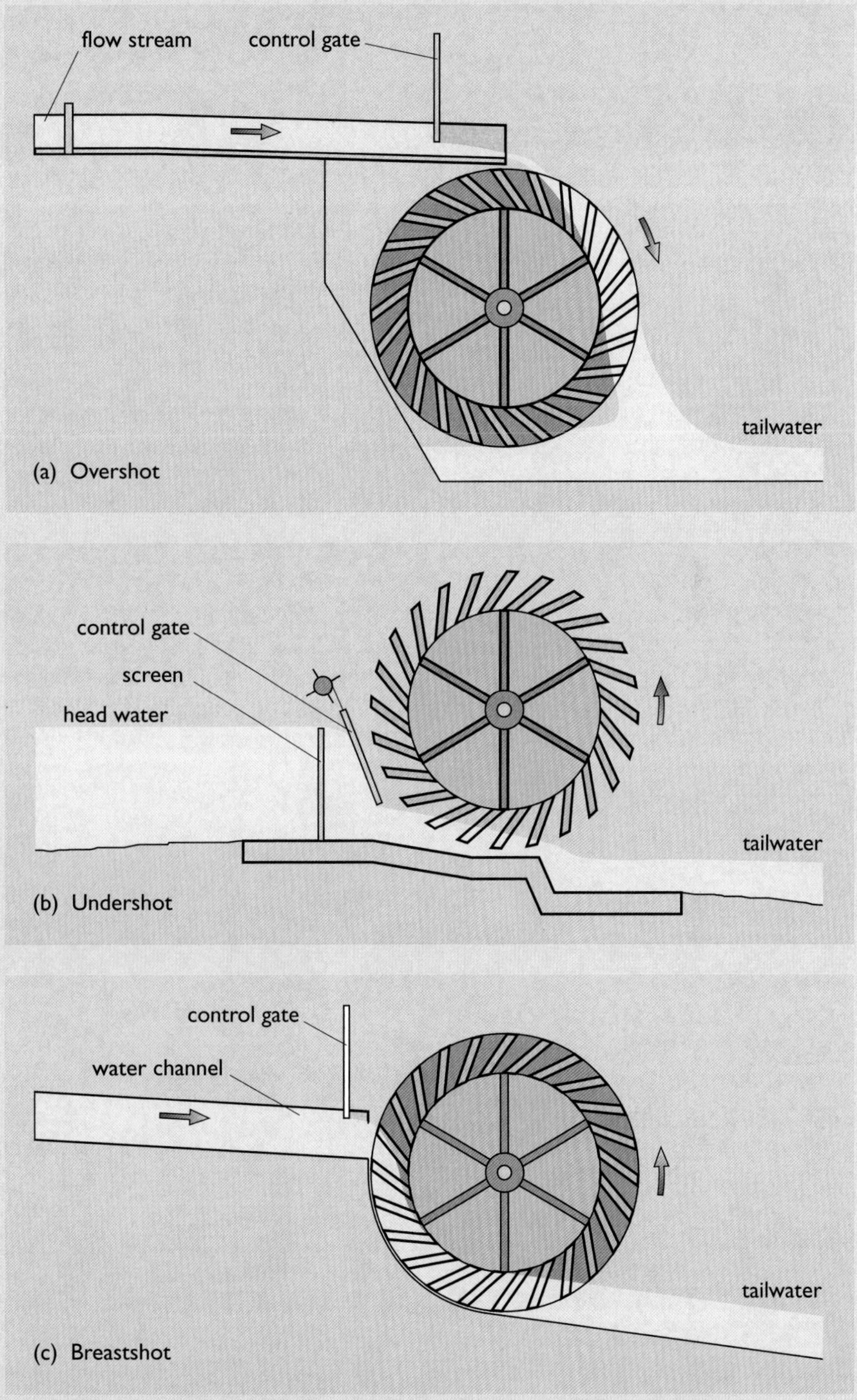

Figure 5.9 Types of traditional water-wheel: (a) overshot; (b) undershot; (c) breastshot

Nineteenth-century hydro technology

An energy analyst writing in the year 1800 would have painted a very pessimistic picture of the future for water-power. The coal-fired steam engine was taking over, and the water-wheel was fast becoming obsolete. However, like many later experts, this one would have suffered from an inability to see into the future. A century later the picture was completely different: the world now had an electrical industry and a quarter of its generating capacity was water-powered.

The growth of the power industry was the result of a remarkable series of scientific discoveries and developments in electro-technology during the nineteenth century; but significant changes in what we might now call *hydro-technology* also played their part. In 1832, the year of Faraday's discovery of electromagnetic induction, a young French engineer patented a new and more efficient water-wheel. His name was Benoit Fourneyron and his device was the first successful water **turbine**. (The name, which comes from the Latin *turbo*: something that spins, was coined by Claude Burdin, one of Fourneyron's teachers.) The water-wheel, essentially unaltered for nearly two thousand years, had finally been superseded.

Fourneyron's turbine (Figure 5.10) incorporates many new features. It is a vertical-axis machine, itself something of a novelty. But the most important innovations are the use of **guide vanes** to direct the water, on to the blades and the fact that the turbine runs *completely submerged*. These are the features that ensure the smooth flow of water, which is essential for high efficiency. The water enters at the centre of the turbine and is diverted

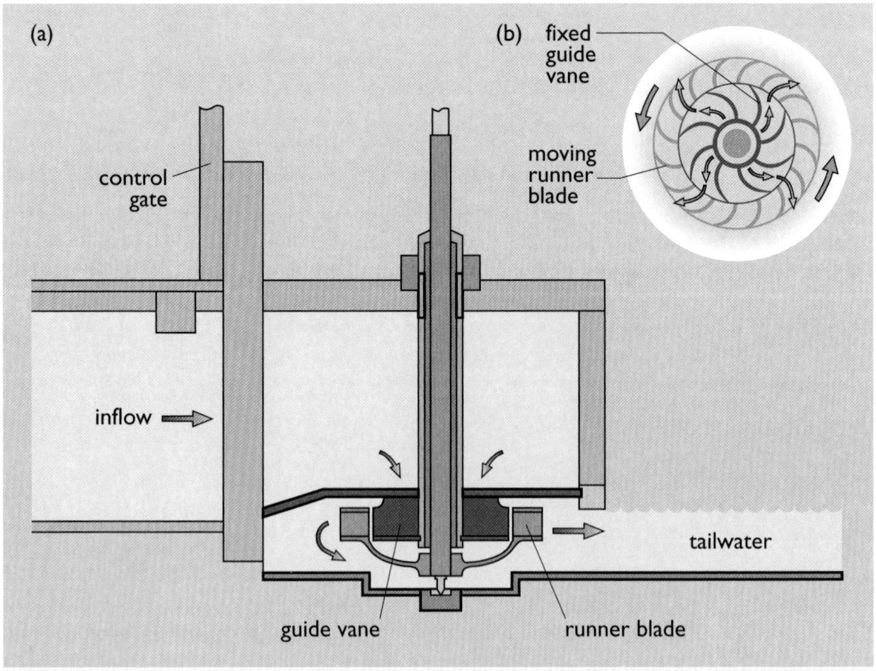

Figure 5.10 Fourneyron's turbine. The runner consists of a circular plate with curved blades around its rim and a central shaft. It spins under the force exerted by water flowing outwards between the fixed guide vanes and across its blades: (a) vertical section; (b) flow across guide vanes and runner

across the curved faces of the fixed guide vanes so that it is travelling horizontally outwards almost parallel to the curve of the runner blades as it reaches them. Deflected as it crosses the faces of the blades, it exerts a sideways pressure which transmits energy to the runner. Having given up its energy, it falls away into the outflow.

Tests showed that Fourneyron's turbine converted as much as 80% of the energy of the water into useful mechanical output – an efficiency previously equalled only by the best overshot wheels. The rotor could also spin much faster, an advantage in driving 'modern' machines. The first pair of these turbines to come into use were installed in 1837 in the small town of St Blasien in the Grand Duchy of Baden (now part of southern Germany).

Development did not stop there, and within a few years the American engineer James Francis started his experiments on **inward-flow radial** turbines which ultimately led to the modern machines known by his name (Section 5.7).

Half a century of development was needed before Faraday's discoveries in electricity were translated into full-scale power stations. Godalming in Surrey, UK, can claim the world's first public electricity supply, opened in 1881 – and the power source of this most modern technology was a traditional water-wheel. Unfortunately this early plant experienced the problem common to many forms of renewable energy: the flow in the River Wey was unreliable, and the water-wheel was soon replaced by a steam engine.

From this primitive start, the electrical industry grew during the final 20 years of the nineteenth century at a rate seldom if ever exceeded by any technology. The capacity of individual power stations, many of them hydro plants, rose from a few kilowatts to over a megawatt in less than a decade.

5.6 Types of hydroelectric plant

Figure 5.12 Nine 125 MW generators in one of the power houses of the Grand Coulee plant on the Columbia river, commissioned in the 1940s. The original 18 Francis turbines are currently being replaced after 60 years' service. A further six main turbines and a smaller pumped storage plant, both installed in the 1970s, bring the total generating capacity to 6.8 GW

Figure 5.11 The Hoover Dam on the Colorado River (originally the Boulder Dam) has a total height of 220 m, and its reservoir, Lake Mead, holds 35 billion cubic metres of water. The two power plants, first commissioned in 1936, now have 17 main turbo-generators with a total output of 2.1 GW

Present-day hydroelectric installations range in capacity from a few hundred watts to more than 10 000 megawatts – a factor of some hundred million between the smallest and the largest. We can classify installations in different ways:

- by the effective head of water
- by the capacity – the rated power output
- by the type of turbine used
- by the location and type of dam, reservoir, etc.

These categories are not of course independent of one another. The available head is an important determinant of the other factors, and the head and output largely determine the type of plant and installation. We start therefore with the customary classification in terms of head, but shall soon see that it is really the fourth criterion that matters.

Low, medium and high heads

Two hydroelectric plants with the same power output could be very different: one using a relatively low volume of high-speed water from a mountain reservoir and the other the huge volume flow of a slowly moving river. Sites, and the corresponding hydroelectric installations, can be classified as low-, medium- or high-head. The boundaries are fuzzy, and tend to depend on whether the subject of discussion is the civil engineering work or the choice of turbine; but high-head usually implies an effective head of appreciably more than 100 metres and low-head less than perhaps 10 metres. Figure 5.13 shows the main features of the three types.

The low dam or barrage of the installation in Figure 5.13(a) serves to maintain a head of water and also houses the plant. It may incorporate locks for ships

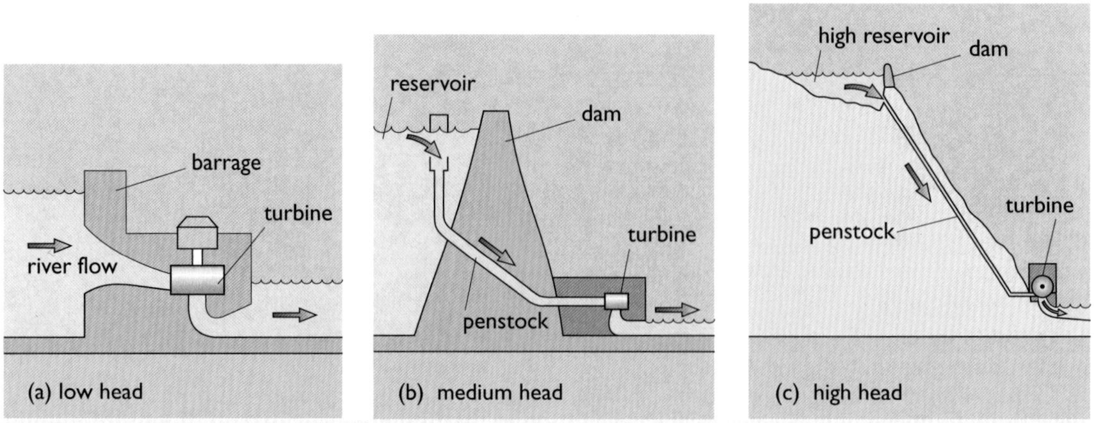

Figure 5.13 Types of hydroelectric installation

(or as we have seen, a fish-ladder for salmon). 'Run-of-river' power stations of the type shown, having little storage capacity, are dependent on the prevailing flow rate and can present problems of reliability if the flow varies greatly with the time of year or the weather. The large volume flow through a low-head plant means that the plant and the associated civil engineering works are large too, which means high capital cost, although this may be ameliorated where there is a second function such as flood control or irrigation.

The plant in Figures 5.11 and 5.13(b) is typical of the very large hydroelectric installations with a dam at a narrow point in a river valley. The large reservoir behind the dam provides sufficient storage to meet demand in all but exceptionally dry conditions. (It will also have flooded an extensive area and may not have been entirely welcomed by the population (see Section 5.12).) The USA has some of the largest dams of this type, including the Grand Coulee, 170 metres high and creating a reservoir that stores 9 billion cubic metres of water. On this scale, the civil engineering costs are obviously considerable, but the large reservoir normally ensures a reliable supply. Systems of this type don't of course have to be on a gigantic scale, and quite small reservoirs can provide power for hydroelectric plant located below their dams.

The 160 m effective head of the Francis turbines at the foot of the Hoover Dam might in other contexts be regarded as 'high' rather than 'medium', illustrating the fact that the distinction lies more in the type of installation. Figure 5.13(c) shows the difference. In the 'high-head' plant the entire reservoir lies well above the outflow level, and the **penstock** carrying the water may even pass through a mountain to reach the turbine. (The penstock was originally the wooden gate or 'stock' which controlled the flow of 'penned-up' water. It later came to mean the channel or pipe carrying the flow.)

With a high head, the flow needed for a given power is much smaller than for a low-head plant, so the turbines, generators and housing are more compact. But the long penstock adds to the cost, and the structure must be able to withstand the extremely high pressures below the great depth of water– as much as 100 atmospheres for a 1000-metre head (Box 5.7).

BOX 5.7 **Height and pressure.**

The pressure in a liquid (or gas) is the force with which it presses on each square metre of surface of anything immersed in it.

Atmospheric pressure is due to the weight of the air above us, equivalent at sea level to the weight of a 10–tonne mass acting on each square metre of any surface. As you move up through the atmosphere, the pressure decreases, initially by about 1% per 100 metres vertically.

As you move *down* through any body of water, the pressure *increases* due to the increasing weight of water above. Because water is several hundred times denser than air, the change is much more noticeable. At a depth of about 10 metres the pressure is twice that at the surface: a pressure of two atmospheres. And this increase of about one atmosphere per 10 metres continues as the depth becomes greater (Figure 5.14).

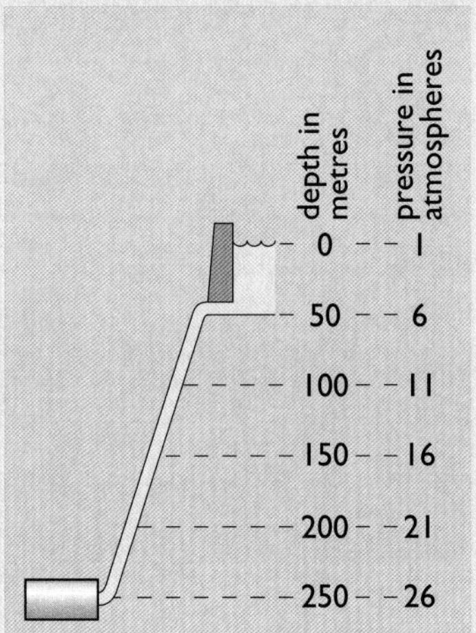

Figure 5.14 Water pressure and depth

Estimating the power

Reliable data on flow rates and, equally important, their variations, are essential for the assessment of the potential capacity of a site. Stopping the flow and catching the water for a measured time is hardly practicable for large flows or as a routine method. The preferred techniques depend on establishing empirical relationships between flow rate and either water depth or water speed at chosen points. Simple depth or speed monitoring then provides a record of flow rates. For many major rivers, particularly in developed countries, such data have been accumulated for years.

Where such records are not available, an entirely different approach is to determine the annual precipitation over the catchment area. This gives the total flow into the system and is particularly suitable for large systems. Allowance must be made, however, for losses due to processes such as re-evaporation, take-up by vegetation or leakage into the ground, and as these could account for as much as three-quarters of the original total they are hardly negligible corrections.

Dealing with time variations adds further problems. In most areas there will be seasonal changes, but these at least come at known times. The more serious problems are with changes over very long or very short periods. Year-to-year variations can be large: the average annual precipitation on the catchment area of the River Severn in the UK, for instance, is 900 mm but it can range from as little as 600 mm to as much as 1200 mm. For countries which depend heavily on hydroelectric power a succession of dry years can mean a serious supply shortage.

At the other extreme, the installation must be designed to survive the '100-year flood', the sudden rush of water following unusually heavy rain. As in any power system, the need to guard against rare but potentially catastrophic events adds to the cost (see also Section 5.12).

5.7 The Francis turbine

Present-day turbines come in a variety of shapes (Figure 5.15). They also vary considerably in size, with runner diameters ranging from as little as a third of a metre to some 20 times this. In the next four sections we look at how they work, the factors that determine their efficiency, and the site parameters that determine the most suitable turbine.

Francis turbines (Figures 5.16 to 5.18) are by far the most common type in present-day medium or large-scale plants. They are used in installations where the head is as low as two metres or as high as 300. These are radial-flow turbines, and although the water flow is inwards towards the centre instead of the outward flow of Fourneyron's turbine, the principle remains the same.

Action of the turbine

As the Francis turbine is completely submerged, it can run equally well with its axis horizontal (Figure 5.17) or vertical (Figure 5.18). In medium- or high-head turbines the flow is channelled in through a scroll case (also called the volute) a curved tube of diminishing size rather like a snail shell,

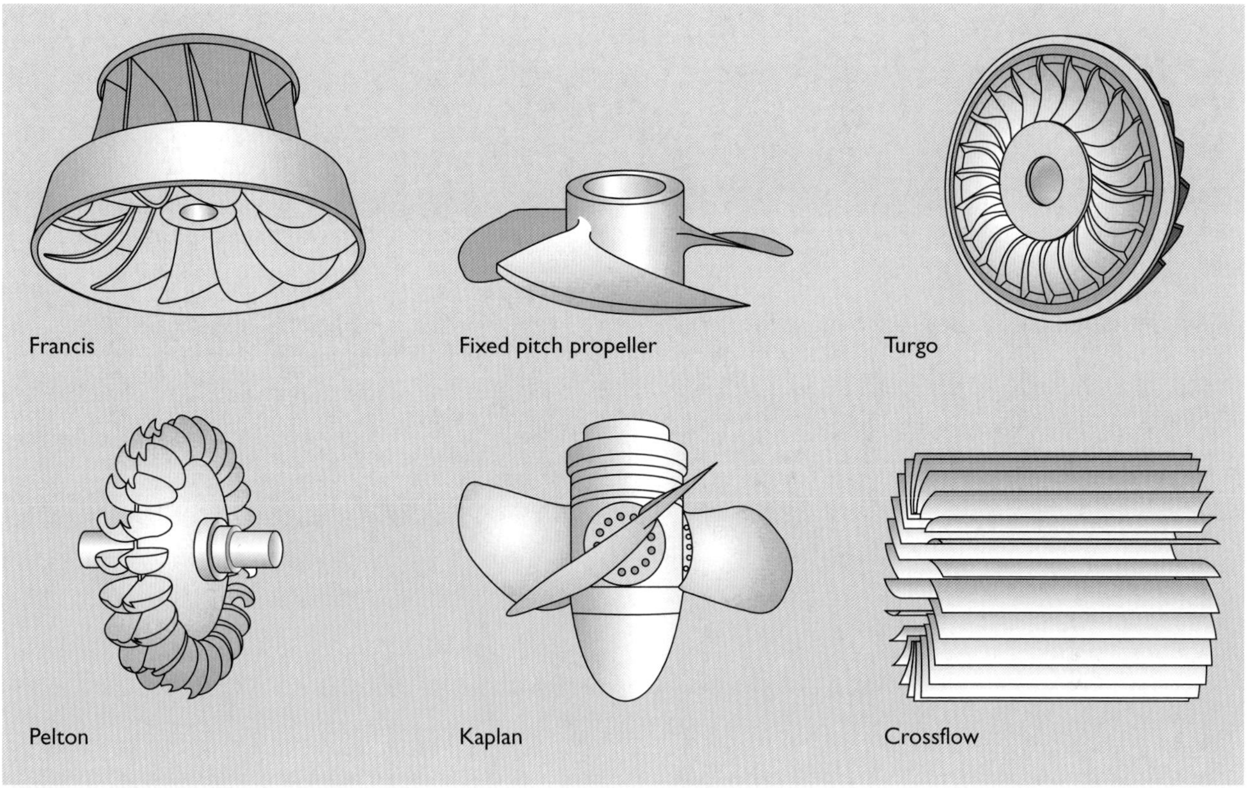

Figure 5.15 Types of turbine runner

Figure 5.16 Francis and Turgo runners. Front two rows, Francis, from left: pair, 2000 kW output, 80 m head; 600 kW, 80 m; 10.2 MW, 278 m; pair, 412 kW, 29 m. Back row, Turgos: left and right, 1575 kW, 190 m; centre, 428 kW, 175 m

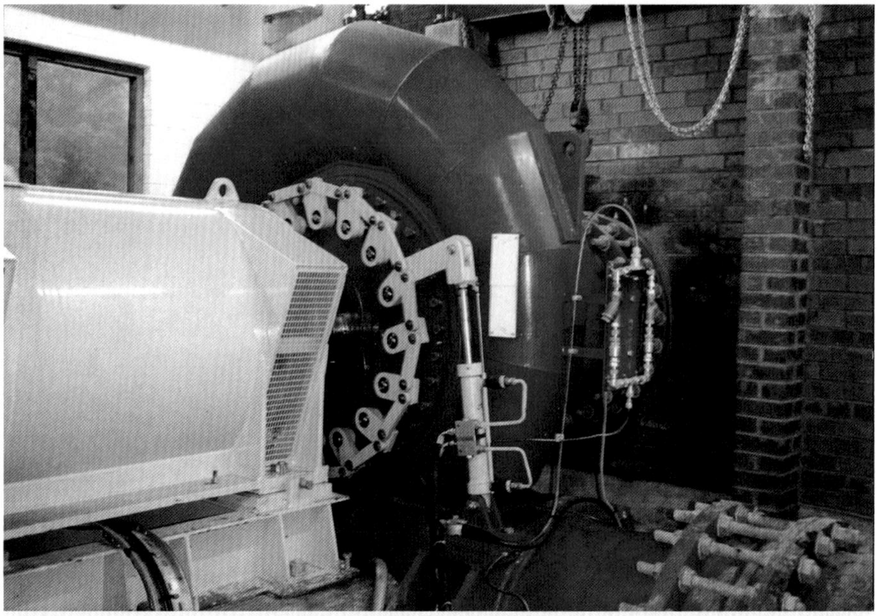

Figure 5.17 The 450 kW horizontal-axis Francis turbine of a small-scale plant in Scotland, commissioned in 1993. The inflow (at lower right) is 2.1 $m^3 s^{-1}$ at a head of 25 m. The turbine, rotating at 750 rpm, drives the generator whose casing can be seen on the left

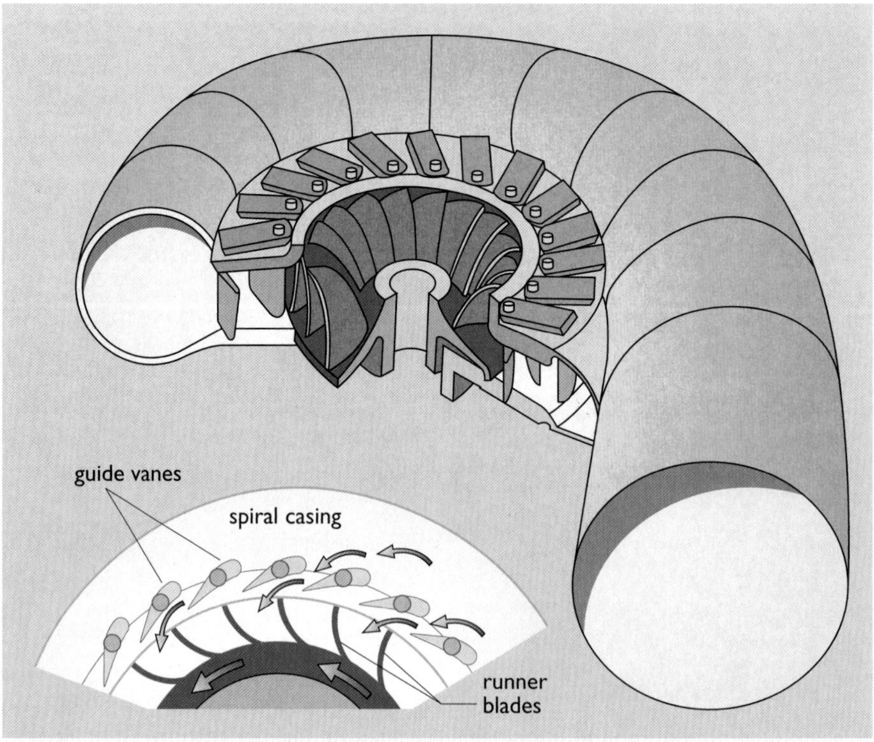

Figure 5.18 Francis turbine: (a) cut-away diagram; (b) flow across guide vanes and runner

with the guide vanes set in its inner surface. Directed by the guide vanes, the water flows in towards the runner. The shapes of the guide vanes and runner blades and the speed of the water are critical in producing the smooth flow that leads to high efficiency. Francis turbines run most efficiently when the blade speed is only slightly less than the speed of the water incident on the blades.

As it crosses the curved runner blades the water is deflected sideways, losing its whirl motion. It is also deflected into the axis direction, so that it finally flows out along the central draft tube to the tail race. That the water exerts a force on the blades is obvious because it has changed direction in passing through the turbine. In being deflected by the blades, it pushes on them in the opposite direction – the way they are travelling – and this reaction force transfers energy to the runner and maintains the rotation. For this reason, these are called reaction turbines. An important feature of this type is that the water arrives at the runner under pressure, and the pressure drop through the turbine accounts for a large part of the delivered energy.

Maximizing the efficiency

Although, as we saw above in Section 5.4, there are always energy losses, modern turbines can achieve efficiencies a high as 95% – but only under optimum conditions. Maintaining exactly the right speed and direction of the incoming water relative to the runner blades is important, and this leads to a problem. Suppose demand falls. The output power can be reduced by reducing the water flow, and in a Francis turbine this is done by turning the guide vanes; but this changes the angle at which the water hits the moving blades and the efficiency falls. This is a characteristic which must be accepted with this type of turbine. (As we'll see below, some 'propeller' types allow adjustment of the pitch, changing the angle of the blades to match the new conditions.)

A rather different cause of less than 100% efficiency is that the water flowing out carries away kinetic energy. A partial remedy is to flare the draft tube. If the tube becomes larger but the volume flow stays the same, the actual speed of the water must decrease, reducing the energy loss. It may seem strange that this change after the water has left the turbine makes any difference, but the effect of the deceleration is to reduce the pressure back at the exit from the turbine, increasing the pressure drop across it and therefore the energy it can extract.

Limits to the Francis turbine

The available head is an important factor in selecting the best turbine for a particular site. If the head is low a large volume flow is needed for a given power. But a low head also means a low water speed, and these two factors together mean that a much larger input area is required. Attempts to increase this area whilst adapting the blades to the reduced water speed and at the same time deflecting the large volume into the draft tube led to turbines with wide entry and blades which were increasingly twisted. Ultimately the whole thing began to look remarkably like a propeller in a tube, and this is indeed the type of turbine now commonly used in low-head situations (Section 5.8).

High heads bring problems too, because they mean high water speeds. As mentioned above, Francis turbines are most efficient when the blades are moving nearly as fast as the water, so high heads imply high speeds of rotation. A look at Tables 5.1 and 5.2 in Section 5.2 reveals that the turbine at Glenlee, with its much higher head, rotates at up to twice the speed of the other Francis turbines in the Galloway system. For sites with very high heads, the Francis turbine becomes unsuitable, and yet another type takes over (Section 5.9).

5.8 'Propellers'

In the 'propeller' or axial-flow turbines shown in Figures 5.15 and 5.19, the area through which the water enters is as large as it can be: it is the entire area swept by the blades. Axial-flow turbines are therefore suitable for very large volume flows and have become usual where the head is only a few metres. They have the advantage over radial-flow turbines that it is technically simpler to improve the efficiency by varying the angle of the blades when the power demand changes. Axial-flow turbines with this feature are called Kaplan turbines.

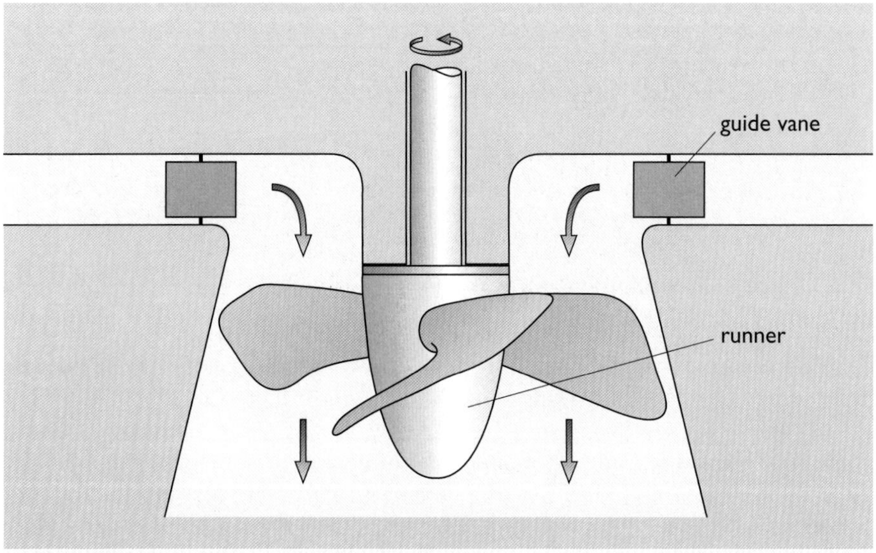

Figure 5.19 A 'propeller' or axial-flow turbine

As before, the entering water is swirling round, but an important feature is that the optimum blade speed is now appreciably greater than the water speed – as much as twice as fast. This allows a rapid rate of rotation even with relatively low water speeds. (Note that because the outer parts of the blade move faster than the more central part, the blade angle needs to increase with distance from the axis. This is why a propeller has its familiar twisted shape.)

Once you have axial flow there is no need to feed the water in from the side. It might appear to be better to let it flow in along the axis instead of deflecting it through a right angle; but if the water flows in along the axis of the turbine, the generator will get in the way and/or get wet! Some solutions to this problem appear in Chapter 6, Tidal Power.

5.9 Impulse turbines

Pelton wheels

For sites of the type shown in Figure 5.13(c), with heads above 250 metres or so (or lower for small-scale systems) the Pelton wheel is the preferred turbine. It evolved during the gold rush days of late nineteenth century California, was patented by Lester Pelton in 1880, and is entirely different from the types described above. It is essentially a wheel with a set of double cups or 'buckets' mounted around the rim (Figures 5.16 and 5.20). A high-speed jet of water, formed under the pressure of the high head, hits the splitting edge between each pair of cups in turn as the wheel spins (Figure 5.21). The water passes round the curved bowls, and under optimum conditions gives up almost all its kinetic energy. The power can be varied by adjusting the jet size to change the volume flow rate, or by deflecting the entire jet away from the wheel.

Figure 5.20 Finlarig power station, on the shores of Loch Tay, Scotland draws its water from Loch na Lairige at a gross head of 415 metres. Its average annual output is 64 million kWh. Left: the power station; Right: the original double twin-jet Pelton wheel and horizontal-axis 30 MW generator

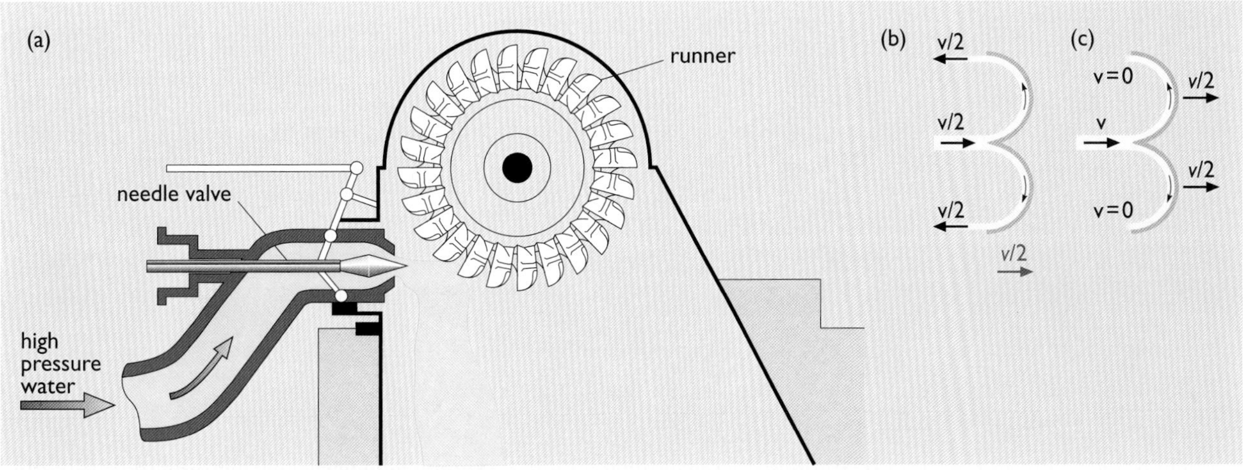

Figure 5.21 Pelton wheel turbine: (a) vertical section; (b) water flow as seen from moving cup; (c) actual motion of water and cup

The efficiency of a Pelton wheel is greatest when the speed of the cups is half the speed of the water jet (see Box 5.8). As the cup speed depends on the rate of rotation and the wheel diameter, and the water speed depends on the head, there is an optimum relationship between these three factors.

BOX 5.8 Optimum speed for a Pelton wheel

The following informal argument shows that a Pelton wheel extracts the maximum energy from the water if the cups move at half the speed of the water jet.

Consider the situation when water at speed v approaches a cup which is already moving in the same direction at half this speed ($v/2$). As seen from the cup, the water will be approaching at the difference between these two speeds, which in this case is again $v/2$ (Figure 5.21(b)).

Suppose now that the water passes smoothly round inside the curved cup until it leaves travelling in the opposite direction – as seen from the cup. You now have water moving backwards at speed $v/2$ relative to a cup that is moving forwards at just this speed. A person on the ground would see the cup moving on while the water simply falls vertically out of it (Figure 5.21(c)). The water has given up all its kinetic energy to the wheel. 100% efficiency, in principle!

In practice this is only approximately true, and the best cup speed is a little less than $v/2$.

The Pelton wheel is an impulse turbine, in contrast to the reaction turbines discussed previously. One important difference is that whereas a reaction turbine runs fully submerged and with a pressure difference across the runner, the impulse type is essentially operating in air at normal atmospheric pressure.

Input power

The power input to the Pelton wheel is determined as usual by the effective head and the flow rate of the water. Box 5.9 shows that, ideally, the volume rate of flow (Q) corresponding to an effective head H is:

$$Q = A \times \sqrt{(2gH)}$$

where A is the area of the jet.

In Section 5.4 we saw that the input power to a turbine is

$$P = 1000 \times Q \times g \times H$$

so we find that:

$$P = 1000 \times A \times \sqrt{(2gH)} \times g \times H$$

Using the approximate value of g, the power in kilowatts becomes

$$P(\text{kW}) = 45A\sqrt{(H^3)}$$

If adjacent cups are not to interfere with the flow, the wheel diameter needs to be about ten times the diameter of the jet. But two or even four jets can be spaced around the wheel to give greater output without increasing the size. If the number of jets is j, the power equation becomes:

$$P(\text{kW}) = 45jA\ \sqrt{(H^3)}$$

BOX 5.9 Effective head, water speed and flow rate.

Although there are in practice always energy losses in forming a jet, we'll assume here that the water leaves the jet at the speed that it would have gained in 'free fall' through the effective head.

We know from Equation (1) in Section 5.4 that the potential energy lost by M kg of water in falling through H metres is given by:

Potential energy = MgH

In Chapter 1 we saw that the kinetic energy of a moving object is proportional to its mass and the square of its speed:

Kinetic energy = $0.5\ M v^2$

So if all the lost potential energy is converted into kinetic energy, we have:

$0.5\ M v^2 = MgH$

so $v^2 = 2gH$ and

$$v = \sqrt{(2gH)} \quad (3)$$

If this water flows as a jet with a circular area of A square metres, the *volume* flowing out in each second (Q) will be equal to A times v. So the volume flow rate for an effective head H is given by

$Q = A \times \sqrt{(2gH)}$

Turgo and cross-flow turbines

A variant on the Pelton wheel is the **Turgo turbine** (Figures 5.15 and 5.22), developed in the 1920s. The double cups are replaced by single, shallower ones, with the water entering on one side and leaving on the other. The water enters as a jet, striking the cups in turn, so this is still an impulse turbine (and Box 5.9 still applies). However, its ability to handle a larger volume of water than a Pelton wheel of the same diameter gives it an advantage for power generation at medium heads.

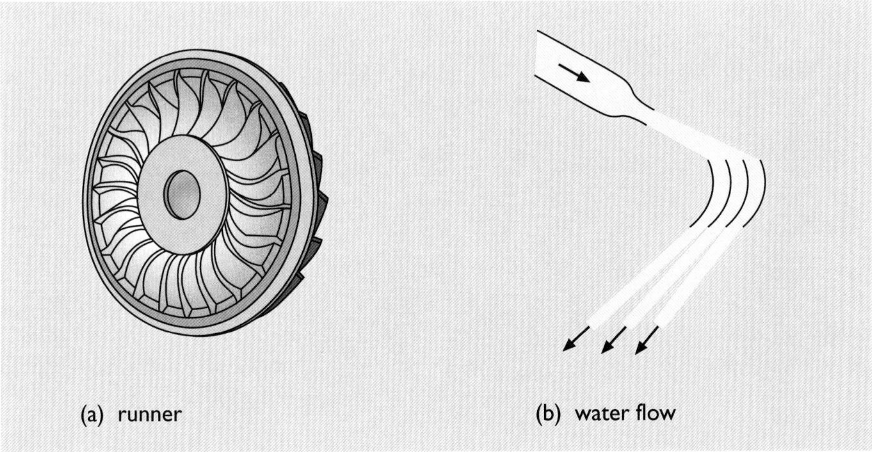

Figure 5.22 Turgo turbine: (a) runner; (b) water flow

The cross-flow turbine (**Mitchell-Banki**, or **Ossberger turbine**, see Figure 5.15) is yet another impulse type, but the water enters in a flat sheet rather than a round jet. It flows across the blades, around the central shaft, and across the blades on the opposite side as it leaves. Cross-flow turbines are often used instead of Francis turbines in small-scale plants with outputs below 100 kW or so. Some ingenious technological ideas have gone into the development of simple types which can be constructed (and maintained) without sophisticated engineering facilities and are therefore suitable for remote communities.

5.10 Ranges of application

We have seen that in general Pelton wheels are most suitable for high heads, propellers for low heads and Francis turbines for the intermediate ranges. But the effective head is not the only factor determining the most appropriate type for a given situation. The available power also matters.

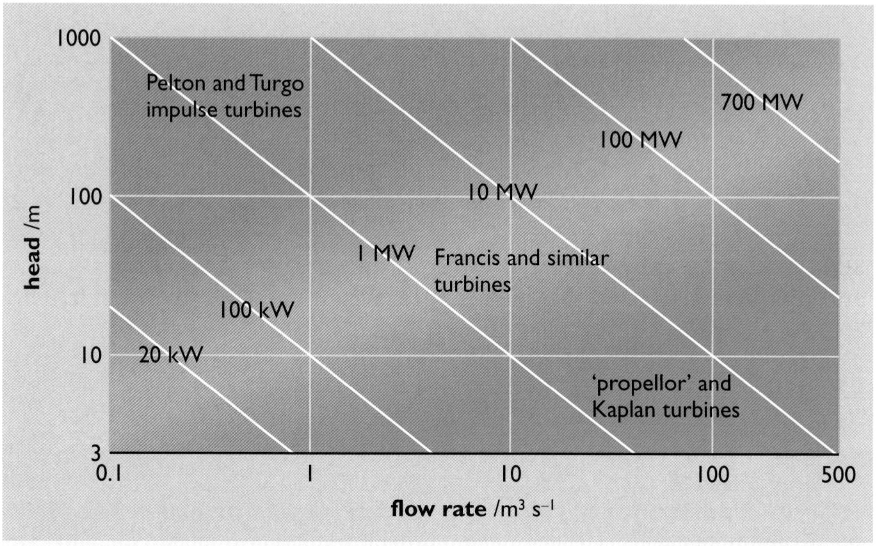

Figure 5.23 Ranges of application of different types of turbine. Note the overlap at the boundaries (see text)

Figure 5.23 represents one way to display the ranges of application of the different turbines. It shows the ranges of head, flow rate and corresponding power which best suit each type. It can be instructive to locate the data for the Galloway plants (and others in this chapter) on Figure 5.23 and compare the turbine types actually used with those suggested by the diagram.

Specific speed

One factor missing from diagrams such as Figure 5.23 is the *rate of rotation* (n) of the turbine, and as we saw in Box 5.1, this can take very different values. The rather strange quantity called the **specific speed** (N_S) takes n into account, allowing a more detailed assessment. N_S is calculated as follows

$$N_S = n \times \sqrt{\frac{P}{H^2 \times \sqrt{H}}} \qquad (4)$$

As can be seen, it depends on the rate of rotation n (in rpm), and the two main features of the site: the effective *head* H (in m) and the available *power* P (in kW).

Once the effective head H and the available water power P have been estimated for a particular site, the specific speeds at different possible rates of rotation can be calculated. The point of this is that each type of turbine, *regardless of size*, operates best within a certain range of values of N_S (Table

5.5). So a suitable turbine and rate of rotation can be found for the proposed site. It should be noted however that other technical factors may enter, and criteria such as cost, simplicity in manufacture and ease of maintenance are likely to influence the final choice.

The actual *size* of the turbine will of course depend on the power, but it is an intriguing feature of these machines that essentially the same *type* may be used for a giant 500-MW plant or a little system delivering only a ten-thousandth of this output.

Table 5.5 Specific speeds

Type of turbine	***Range of specific speeds**
Francis	70–500
'Propeller' or Kaplan	350–100
Pelton, Turgo, Cross-flow	10–80

*The ranges shown are approximate (see main text), and more detailed data would be used in practice.

5.11 Small-scale hydroelectricity

There is no agreed formal definition of the term **small-scale hydro (SSH)**. The prevailing view places the upper limit at 10 MW capacity (as in Switzerland, for instance) but the UK sets it at 5 MW and the USA at 30 MW – enough power for a small town. Other terms such as **mini-**, **micro-** and **pico-** are also used, for ranges of diminishing capacity down to a few hundred watts.

Small-scale installations can also be classified by the available head, but the ranges may be very different from those of large plants. Many SSH plants are run-of-river, with heads of only a few metres, and as little as 10 metres could be regarded as 'high head' for a very small plant. The corresponding choices of turbine type may vary appreciably from those indicated in Figure 5.23 and Table 5.5.

World-wide developments

In the earliest days of electric power, generators in the kW to MW range were installed on streams or rivers, often using the dams and sluices of old water-mills, and as late as the mid-twentieth century, many towns in Europe and elsewhere were still served by hydro or other plants that today would fall into the 'small' category. But with rising demand and the growth of national transmission networks, 500–1000 MW became the norm for modern power stations. The renewed interest in smaller systems appears to be a consequence of rather different factors in different regions of the world. In the *industrialized countries*, environmental issues are increasingly limiting the potential for further large-scale hydro development, and small-scale plants are encouraged. And some *developing countries* are seeing advantages in step-wise electrification by the establishment of local rather than nation-wide grid systems. The growing interest has resulted in technical improvements such as standardization of components and civil works, and advances in electronic controls, leading to off-the-shelf systems that should reduce costs and increase reliability.

World small-scale hydro data

It is generally agreed that world SSH capacity and output are rising. Just how rapidly is not easy to estimate with any precision. There are many privately owned small plants, and many in remote areas, or in countries whose data are unreliable or unavailable. The '100 000 SSH plants' in China, for instance, widely reported in the early 1990s, have become 43 000 in the more detailed accounts now available (see below). A World Energy Council survey (WEC, 2003b) reports an installed SSH (<10 MW) capacity of 18 GW in 38 selected countries at the end of 1999. These included the main contributors in the Americas and Europe, but not China (see below).

Estimates of the world-wide rate of increase in SSH capacity tend to lie between 1 and 2 GW a year. In summary, therefore, it seems probable that world operational capacity in 2003 lies in the range 50–60 GW. Taking the lower figure and assuming an average load factor of 35% suggests an annual output of about 150 TWh – about 6% of total hydro output, or 1% of world electricity generation.

China

About 300 million people in China derive their electricity from SSH (defined there as plants with less than 25 MW capacity). An intensive program of local electrification over the past few decades has led to a total installed capacity in early 2002 of over 26 GW (Tong, 2002).

China distinguishes between *micro* (<100 kW), *mini* (100–500 kW) and *small* (0.5–25 MW) plants. Of the 43 000 plants, about 90% are micro or mini, in roughly equal numbers; but three-quarters of the output comes from the remaining 10%, the 'small' installations. This is raising concern, as the environmental effects of a 25 MW installation (larger than all but one of the Galloway plants described in Section 5.2.) can have more in common with those of large-scale hydro than with a little 250 kW plant.

Rest of the world

The total SSH hydro capacity outside China is probably slightly greater than China's 26 GW, although different definitions can confuse the issue. Under the customary criterion (<10 MW), total Western European capacity is about 10 GW, but this would more than double under the Chinese definition (<25 MW). Within the 10 MW limit, the leading country outside China in 2002 was Japan, with about 3.5 GW of operational capacity, followed by Austria, France, Italy and the USA, each over 2 GW, and Brazil, Norway and Spain over 1 GW.

In most contexts, SSH remains more costly than electricity from conventional sources, but it is claimed that the technical improvements mentioned above are bringing costs to a level where in suitable locations these systems are competitive with other options (IEA, 2003). Nevertheless, in many European countries, investment in electricity from renewables during the past decade has concentrated on wind and solar PV rather than small-scale hydro power.

Elsewhere in the world, considerable SSH potential remains unused, particularly in mountainous regions. The concept of complete 'water-to-wire' systems for remote sites has attracted the attention of manufacturers in Europe and elsewhere (Maurer, 1997), noting that the annual gigawatt or so of new SSH capacity represents a world-wide market of up to 5 billion dollars.

Local manufacturers have also been encouraged. Extensive areas of Nepal, for instance, have no electricity supply and only mules or human porters for transport, but many mountain streams suitable for 'high-head' plants. This led to the development of a local industry producing extremely small-scale systems, transportable by a single person on foot. The Peltric turbo-generator set, for example, consists of a tiny Pelton wheel driving a simple generator. Operating under heads of 50–70 metres, it produces a kilowatt or so of output. With cheap and relatively simple 'civil works', these tiny systems proved sufficiently popular to be copied in other countries (Upadhyay, 1997). Unfortunately, social unrest in Nepal in the past few years has brought this encouraging development to a halt in many areas, and the total operational capacity has probably fallen below the 13 MW recorded at the turn of the century (WEC 2003b).

Small-scale hydro in the UK

With most of the potential UK sites for large-scale hydroelectric plant lying in areas of natural beauty, it has been recognized for some time that few if any major new developments are likely to prove acceptable (ETSU, 1994, 1999), and proposals for an increased UK hydro contribution have tended to centre on small-scale installations with outputs between a few kilowatts and a few megawatts.

At the end of 2002, the total operational small-scale hydro capacity in the UK was 70 MW, with an annual output of about 200 GWh (DTI, 2003). To place this in context, we might note that it represents about 4% of all UK hydroelectricity, 3% of 'new renewables' electricity – and one two-thousandth of total UK electricity.

However, capacity had been increasing during the 1990s at about 6% a year, due in part to encouragement under the various Non-Fossil Fuel Obligation (NFFO) schemes (see Chapter 10). Over the period 1990–99, contracts were awarded for 146 hydro systems with a total capacity of 95.4 MW. At the end of 2002, 48 were operational, with a total capacity of 30 MW and annual output approaching 90 GWh – contributing just under half the national SSH total (DTI, 2003; ETSU, 2003).

The Elan Valley scheme of 1998 (Box 5.10) is interesting as an example of two growing trends: the up-grading of old hydro plants and the use of existing dams for new plants (see also Section 5.14 below).

A detailed examination, was carried out by Salford University during 1987–88 to assess the UK SSH resource. It covered some 1300 sites with potential output in the range 25 kW–5 MW. On slightly optimistic assumptions about costs and load factor, the conclusion (ETSU, 1989) was that the total potential at the then tariff price and expected rate of return on capital was about 1300 GWh a year from a capacity of 320 MW.

More recent estimates, based in part on the original data, suggest a 'practically feasible' resource in the range 40 MW–110 MW (including sites currently contracted under the NFFO schemes). There may be an unexploited Scottish resource of up to 3 TWh y^{-1} from 1 GW, but this would require new reservoirs and is thought likely to be limited by environmental constraints. (ETSU, 1999)

BOX 5.10 Elan Valley

The Elan Valley scheme in the Cambrian Mountains in mid-Wales (Figure 5.24) is unusual in that one NFFO contract supported the development of five power stations. Like the Galloway Hydros (Section 5.2), they are in sequence along one river, but these plants use the dams and reservoirs of an existing water-supply system (for Birmingham, 70 miles away), and their *total* capacity is only 4.2 MW.

Table 5.6 gives some details. The contribution from Foel Tower, on the Garreg-Ddu reservoir, is worth noting: its Kaplan turbine sits inside one of the three 14-metre diameter pipes carrying the water supply for Birmingham.

All the reservoirs and dams date from the early twentieth century – the latter are listed historic monuments. Caban Coch was already a hydro plant. One of its old turbines was replaced some time ago and the other in 1998 as part of this scheme. In a region of great natural beauty and with 90% of its area designated as Sites of Special Scientific Interest, the four new plants have required careful siting. To meet environmental constraints, their turbines make use of existing discharge pipes, and their associated buildings are mainly below ground.

Table 5.6 The Elan Valley Plants

site	Craig Goch	Pen-y-Garreg	Caban Coch	Foel Tower	Claerwen
head (m)	36.5	37.5	37	13.5	56
capacity (kW)	480	810	950	300	1680
turbine	Francis	Francis	2 × Francis	Kaplan	Francis

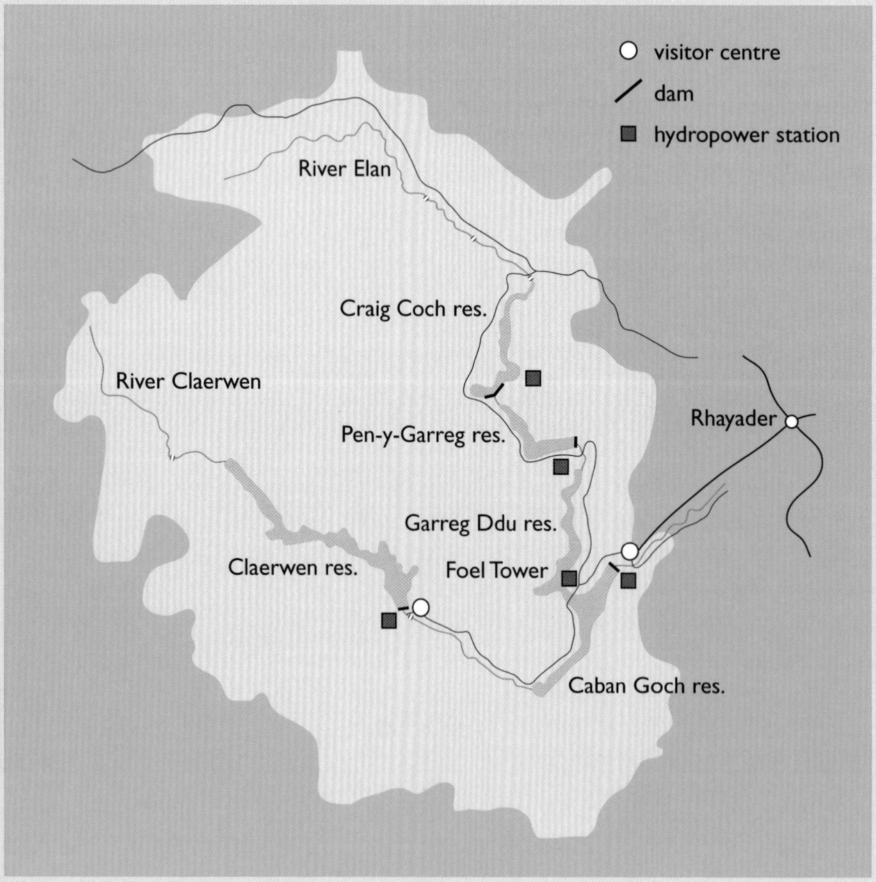

Figure 5.24 The Elan Valley Hydro Scheme
Sources: ETSU, 1998, 2000

5.12 Environmental considerations

> The environmental impacts of a hydroelectric project must be thoroughly analyzed since, after it is completed, they are essentially irreversible.
>
> Dorf, 1978

> The ecological damage per unit of energy produced is probably greater for hydroelectricity than for any other energy source.
>
> CONAES, 1979

> ...carefully planned hydropower development can, and does, make a great contribution to improving electrical system reliability and stability throughout the world. [It] will play an important role in the improvement of living standards in the developing world, [and] make a substantial contribution to the avoidance of greenhouse gas emissions and the related climate change issues.
>
> WEC, 2003a

As the quotations suggest, the environmental issues associated with hydroelectricity are no less controversial than those for other energy sources. We might usefully start by briefly summarizing the environmental *benefits* of hydroelectricity compared with other types of power plant. It releases no CO_2, and negligible quantities of the oxides of sulphur and nitrogen that lead to acid rain. It produces no particulates or chemical compounds such as dioxins that are directly harmful to human health. It emits no radioactivity. Dams may collapse, but they will not cause major explosions or fires. Moreover, hydroelectric plant is often associated with positive environmental effects such as flood control or irrigation, and in some cases, its development leads to a valued amenity or even a visual improvement to the landscape.

However, during the twentieth century, the construction of large dams has led to the displacement of many millions of people from their homes, and dam failures have killed many thousands. We'll consider these and other deleterious effects under three headings:

- hydrological effects – water flows, groundwater, water supply, irrigation, etc.
- other effects of large dams and reservoirs
- social effects.

Hydrological effects

The three categories of effect listed above are not of course independent. Any hydrological change will certainly affect the ecology and thus the local community. A hydroelectric scheme is not basically a *consumer* of water, but the installation does 'rearrange' the resource. Diverting a river into a canal, or a mountain stream into a pipe, may not greatly change the total flow, but it can have a marked effect on the environment (see for instance Box 5.11). And evaporation from the exposed surface of a large reservoir may appreciably reduce the available water supply.

BOX 5.11 Gabcikovo-Nagymaros

The River Danube is already used for hydroelectric power in several countries, but the development on the Slovak-Hungarian border (Figure 5.25) has been a controversial issue for over two decades. The 880 MW scheme, agreed in 1977, included a large reservoir downstream from Bratislava, a canal to carry diverted water to a power plant at Gabcikovo, and a second barrage and plant at Nagymaros.

Work proceeded for about a decade, but the political changes of the late 1980s brought increasingly vocal opposition on environmental grounds, particularly in Hungary. Construction on the Hungarian side effectively stopped in 1989, and in May 1992 the government announced cancellation of the 1977 agreement.

The newly established Slovak state, committed to the scheme, declared that the cancellation was illegal. In October 1992 Slovak engineers used the period when the river was at its lowest to complete the diversion into the new canal, 18 km long and with concrete walls rising to a height of 15 metres above the surroundings.

The resulting fall in the water table, with wells drying up, vegetation dying and unique forms of wild life in danger, reinforced calls for legal limits to the diversion of water. An artificial irrigation scheme only slightly alleviated the situation, and in 1993 the Hungarian and Slovak governments took the dispute to the International Court of Justice at the Hague. In 1995, before the Court ruled, Slovakia agreed to reduce the water diverted.

In September 1997, the Court ruled that *both* countries had acted illegally. The ruling demanded that they should jointly negotiate a new solution (still without the Nagymaros plant) and that this must '*accommodate both the economic operation of the system of electricity generation and the satisfaction of essential environmental concerns.*' The two governments met for negotiations but have yet (in mid-2003) to agree a solution. The present half-scheme, with only one power station to generate income, is a financial disaster.

Some sources: Fleischer, 1993; WWF, 1997, 1998; McGriff 1999; Slovakia 2000; Zinke 2002

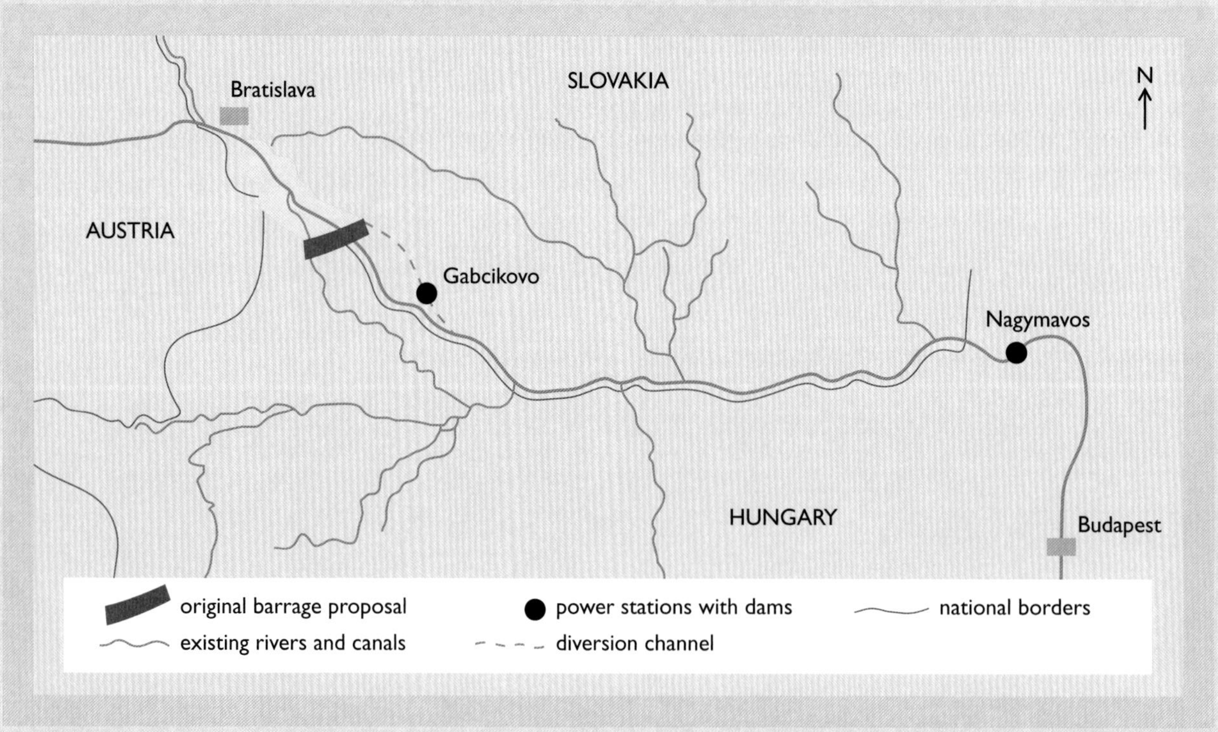

Figure 5.25 The proposed Danube scheme

Dams and reservoirs

Any structure on the scale of a major hydroelectric dam will affect its environment in many ways. The construction process itself can cause widespread disturbance, and although the building period may be only a few years, the effect on a fragile eco-system can be long-lasting. In the longer term, a large reservoir is bound to bring significant environmental changes. Whether these are seen as catastrophic, beneficial or neutral will depend on the situation – in the geographical and biological sense – and certainly on the points of view and interests of those concerned. A recent survey, for instance, reports that the '*primary purpose or benefit*' of 35% of dams in the USA is '*recreation*', with hydroelectricity accounting for only 2% (DOE, 2001).

Catastrophes

The most obviously harmful effect of a large dam is when it fails. The companion volume to this book, discussing the risks and other environmental impacts associated with energy systems, draws attention to some estimates:

> During the twentieth century, some 200 dam failures outside China are thought to have resulted in the deaths of more than ten thousand people. And within China, in one year alone, 1975, it is estimated that almost a quarter of a million people perished in a series of hydroelectric dam failures.
>
> Sullivan, 1995

The 10 000 deaths mentioned here could have been much greater. In 1971, an earthquake severely damaged the Lower San Fernando Dam north of Los Angeles, leaving only '*a shattered wall of dirt*' (USGS, 1995). Had the reservoir been at its maximum height, some 15 million tonnes of water could have been released on to the 80 000 inhabitants of the valley below.

Silt

The Aswan High Dam, built in the 1960s, supplies a large part of Egypt's electricity; as planned. But the land downstream no longer receives the soil and nutrients previously carried by the annual Nile floods. An agricultural system in existence for millennia has largely been destroyed, to be replaced by irrigation and the use of fertilizers. Meanwhile the silt is accumulating behind the dam, reducing its useful volume and the hydro potential of the site. This latter problem is not confined to Aswan. The Hoover Dam for instance (Figure 5.11), now nearly 70 years old, lost about one sixth of its useful storage volume in its first 30 years – although the loss rate fell when the Glen Canyon dam was built, 370 miles upstream.

BOX 5.12 Itaipú

The hydroelectric plant at Itaipú, on the Paraná River between Brazil and Paraguay, is currently the world's largest single power station, with a capacity of 12.6 GW. The first of eighteen 700 MW generators came on line in 1984 and the last in 1991. The effective head is 118.4 metres and the reservoir area 1350 km^2, with an average water flow of 9000 cubic metres per second, peaking at times to over 30 000 m^3 s^{-1}. The claimed efficiency of the Francis turbines is 93.8%, and of the generators 98.6%.

The plant supplies 95% of Paraguay's power and about a quarter of Brazil's, requiring a feature that must be unique in power stations: half the 18 generators produce 50 Hz AC power and the other half 60 Hz AC. Moreover, as Paraguay (50 Hz) uses only a fraction of its share, the rest is sold to Brazil, where it is rectified, transmitted as DC, and then re-converted to provide the 60 Hz supply.

The total output in 2000 was 93.4 TWh, achieving the contracted annual load factor of 85%. It was claimed that the sceptics who called Itaipú a white elephant, producing unwanted power in the wrong place, had been proved wrong. However, the output is essentially dependent on the rainfall in the enormous catchment area and on the demand for power in Brazil – both of which have shown severe fluctuations in recent years.

Nevertheless, two more 700 MW units are due to become operational in 2004, and it is possible that Itaipú will be the first power station to generate more than 100 billion kWh in a year.

Sources: Itaipú, 2003; Krauter, 2000

Figure 5.26 The Itaipú dam and power plant

Methane

Until recently, hydro was listed amongst the renewables that produce virtually no greenhouse gases. But a report by the *World Commission on Dams* (WCD, 2000; Anon, 2001) put this view into question. It drew attention to the fact that vegetable matter that normally decays in the air to produce carbon dioxide (CO_2) can decay anaerobically, producing methane (CH_4), when land is flooded for a hydro scheme. Methane is a much more potent greenhouse gas than CO_2, so this raises the question whether hydroelectricity should be included with the fossil fuels as a significant contributor to global warming.

According to the WCD report...

> All large dams and natural lakes in the boreal and tropical regions that have been measured emit greenhouse gases [...] some values for gross emissions are extremely low, and may be ten times less

> than the thermal option. Yet in some cases the gross emissions can be considerable, and possibly greater than the thermal alternatives.
>
> WCD, 2000

In other words, whilst some lakes and reservoirs generate relatively little methane, the net greenhouse gas emissions from others can be comparable to those from fossil fuel plant of the same capacity. It appears that the emissions are sometimes greater than can be accounted for by the main flooded area, and that contributions must come from the entire catchment area. The report observed that this is clearly a site-specific issue, requiring site by site assessment. It noted that...

> ...in boreal climates (like Canada and Scandinavia) available studies so far suggest that emissions from hydropower reservoirs are low. For Brazil, of ten dams studied, emissions vary from dam to dam with a 500-fold difference between lowest and highest. The lowest emissions are on similar levels to Canadian lakes and reservoirs, the highest annual gross emissions reach the ranges of thermal energy plants, although life cycle assessment, and determination of net emissions, is needed before definitive comparisons can be made.
>
> WCD, 2000

The data are as yet patchy, but this issue puts a further question mark against new large hydroelectric installations, in at least some locations.

Decommissioning

The account of the Galloway Hydros in Section 5.2 mentions the effect on the salmon passing up the rivers to spawn as a major environmental issue. France has many dams constructed during the early twentieth century on rivers previously used by Atlantic salmon and other fish, and in the 1990s, as concessions became due for renewal, stringent requirements were introduced for the construction of fish ladders or similar passages. One consequence was the *decommissioning* of dams deemed unsuitable for renewal, on environmental or economic grounds (ERN, 1999).

These relatively small French examples were designed in part as pilot schemes, in advance of the much larger requirement that could arise throughout Europe as more of the dams built for hydroelectricity during the early and mid-twentieth century become due for renewal or renovation (Section 5.14). Recent years have seen calls in the USA for the decommissioning of dams on environmental grounds. (See, for example, www.amrivers.org or www.friendsoftheriver.org – or for a different view, www.ussdams.org/ [all accessed 28 October 2003].)

Social effects

From a child's book on energy:

> They built a dam and made a lake in the place where Ahmed lived. Ahmed and his family had to leave the farm. His grandfather had lived there. Ahmed was born in that place. He was sad to go.

Just so. Cost-benefit analysis, it has been said, usually means that I pay the cost and you get the benefit.

The Aswan and Kariba dams involved the relocation of some 80 000 and 60 000 people respectively, whilst the rising water behind the Three Gorges Dam (Box 5.13) will submerge about 100 towns and displace over a million people. It is estimated that during the second half of the twentieth century, some 10 million people were displaced by reservoirs in China alone.

But even for the people immediately affected, the building of dams can have very different consequences. For those living in a valley which will become a reservoir it means the loss of your family home, possibly your livelihood, and often your entire community. In contrast, for people living on a river which periodically overflows its banks, the barrage and embankments of a hydroelectric scheme can bring freedom at last from devastating floods. And on the smaller scale, the changes that mean the loss of a beloved riverside walk to some may be welcomed by others as an opportunity for new leisure activities.

Small-scale systems

There is general consensus that small-scale hydro plants have fewer deleterious effects than large systems. In some respects this is evidently true – few people have been displaced from their homes by the installation of little 5 MW plants, and deaths from the collapse of dams across small streams seem rare. But not everyone agrees with the overall conclusion:

> Recent legislation has differentiated between projects with capacities above or below 10 MW, favouring smaller projects. There is no scientific or technical justification for this, and it may lead to greater environmental impacts.
>
> WEC, 2003b

The claim, mainly by proponents of large-scale hydro, is that a general world view – '*small is beautiful*' – has been allowed to override detailed analysis. It is true that the efficiencies and the load factors of SSH plants tend to be lower, and in some cases the 'reservoir area' per unit of output is greater, increasing evaporation, and perhaps methane emission. But all these factors vary greatly from site to site, and generalization is difficult.

Comparisons

It should not be forgotten that the choice may not be hydroelectricity or nothing, but hydroelectricity or some other form of power station. Despite the 'penalties' discussed above, hydroelectricity scores relatively well in comparison with many other options. Section 10.7 of Chapter 10 summarizes a review in the companion volume of the impacts of different electricity generating systems, and of attempts to quantify the **external costs** – the 'extras' that should be added to basic generating costs to take into account the deleterious effects of each system. Current issues for hydro power include the question of methane emissions, discussed above, and the costing of long-term compensation for the people displaced by major new hydroelectric installations. Nevertheless, on the criteria used in these studies, hydro appears amongst the least harmful sources of electricity.

5.13 Integration

Even if a potential source of electric power is acceptable environmentally and financially, other factors remain that affect its viability. Few large power stations operate in isolation, and the extent to which the proposed plant can form a useful part of a supply **system** is important.

Power stations as elements in a system

From the point of view of the operator of the system, the characteristics of the ideal power station would be:

1. constant availability
2. a reserve energy store to compensate for variations in input
3. no correlation in input variations between power stations
4. rapid response to changing demand
5. an input which matches annual variation in demand
6. no sudden and/or unpredictable changes in input
7. a location which does not require long transmission lines.

Few if any sources meet all these criteria, but each compromise with the ideal adds to the effective cost.

Almost all hydroelectric plants score well on item 4, and in regions with cold, dark, wet winters, on 5 as well – unless the water is locked up as ice. And *sudden* unplanned fluctuations in input (6) are rare, at least in large plants.

How well hydro performs on items 1–3 depends in part on the type of plant. A high-head installation with a large reservoir will normally have little difficulty in maintaining its output over a dry period, whereas the water held behind the low dam of a run-of-river plant may not be sufficient to compensate for periods of reduced flow. A serious drought can of course affect all hydro plant over a wide region, so it cannot be said fully to satisfy the third requirement.

The final criterion is the real hurdle. Hydro locations are determined by geography, and whilst run-of-river plant may sometimes be near major centres of population, this is rare for high-head systems.

Is the case different for *small-scale* hydro? Smaller plants are predominantly run-of-river, or perhaps served by relatively small reservoirs, in either case a less reliable supply. On the other hand, scattered sites with different rainfall patterns could result in increased reliability. Local small-scale plant can reduce the need for long-distance transmission, reducing energy losses and costs, although the generating cost per unit of output may be greater.

Overall, hydroelectricity ranks reasonably well in terms of the above criteria. And it may also offer a bonus. A hydro plant with a large reservoir not only maintains its own reserve of energy – it might provide a store for the surplus output of other power stations.

Pumped storage

As electricity became an ever more central feature of mid-twentieth century life, the need arose for 'backup' systems that could respond quickly to a sudden surge in demand or the failure of a base-load plant. Hydro plants with reservoirs, scoring well on Items 1, 2 and 4 above, were an obvious choice for this role.

And there was a further possibility. During periods of low demand, the surplus power from coal-fired plants could be used to increase the stored energy by pumping water up to the reservoir. This feature became even more relevant with the growth of nuclear power, as nuclear reactors cannot easily follow the normal hour-to-hour daily variations in demand.

The economic viability of pumped storage depends on two nice technological facts. The first is that a suitably designed generator can be run 'backwards' as an electric motor: the machine which converts mechanical energy into electrical energy can perform the reverse process. And a suitably designed turbine can also run in either direction, either extracting energy *from* the water as a turbine or delivering energy *to* the water as a pump. The complete reversal is thus **turbo-generator** to **electric pump** (Figure 5.27). The machines must of course be designed for this dual role, but the cost saving is obviously significant.

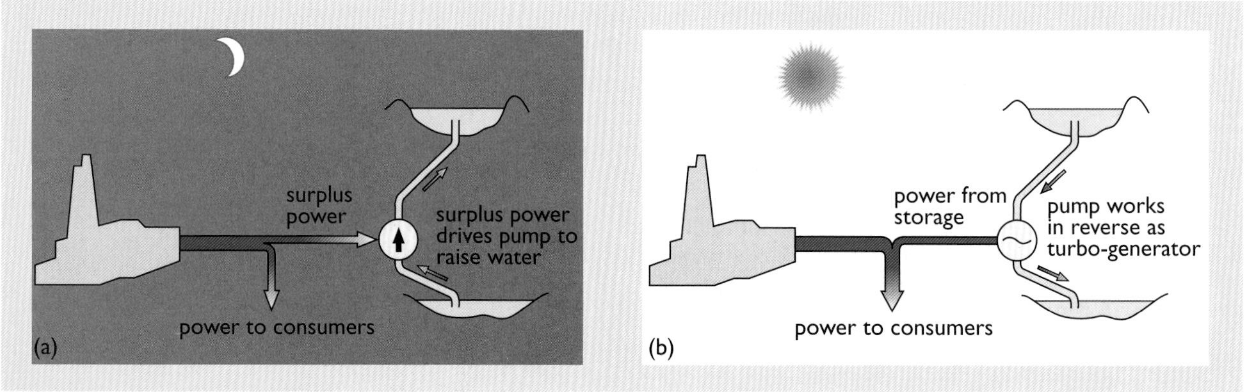

Figure 5.27 Pumped storage system: (a) at time of low demand; (b) at time of high demand

There will, as always, be losses associated with the conversion processes, but turbines and generators are very efficient, and nearly 80% of the input electrical energy can be retrieved. The value of the system is enhanced by its speed of response; any of the six 300 MW Francis turbines of the Dinorwig storage plant in Wales can be brought to full power in just 10 seconds if initially spinning in air, and even from complete standstill takes only one minute.

The location must of course be suitable. A low-level reservoir of at least the capacity of the upper one must be available – or must be constructed. Sites such as Cruachan (Figure 5.28), where the mountains rise from a large loch or lake, are obviously ideal. Pumped storage can be combined with 'normal' hydroelectric generation in locations where the potential exists. The upper reservoir will in any case have a local catchment area, so

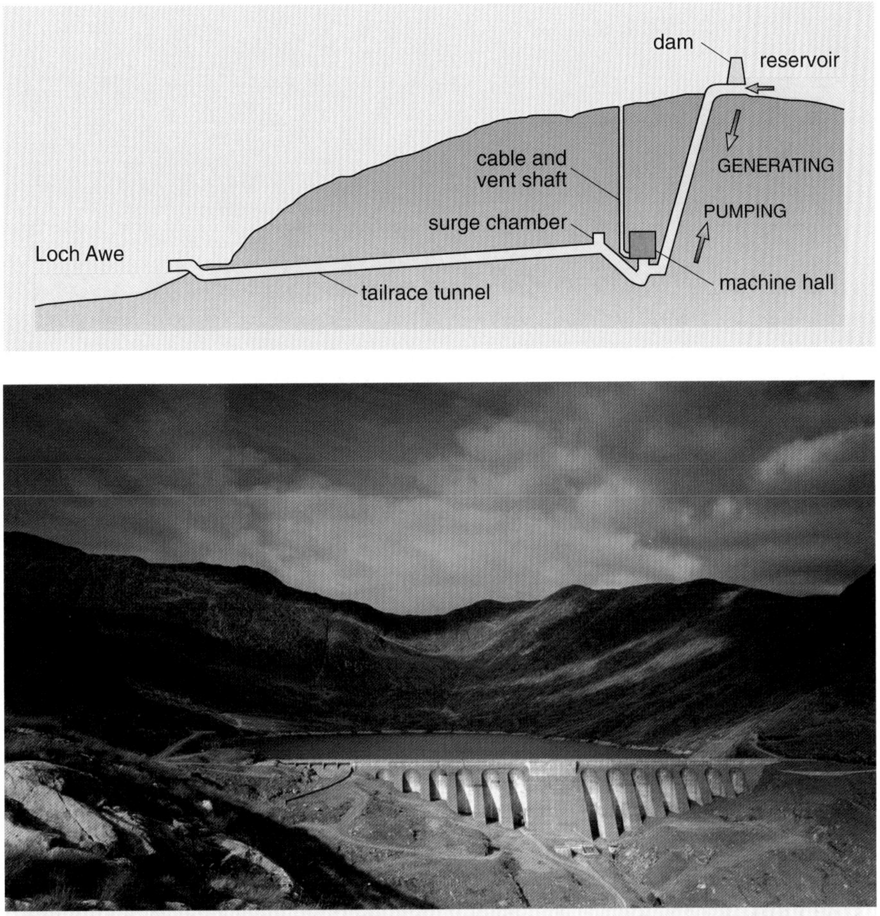

Figure 5.28 Cruachan pumped storage plant. The reservoir of this Scottish plant, commissioned in 1965, can store 10 million cubic metres of water at an operating head of about 370 m. Running the four 100 MW reversible machines for an hour at full capacity, as electric pumps or turbo-generators, raises or lowers reservoir level by about a metre. (top) the installation; (bottom) the dam

there may be a positive *net* output from the plant. Very high heads would have the advantage of needing smaller reservoirs for a given energy storage, but the Pelton wheels and Turgos most suited to high heads cannot be used as pumps. In contrast, if the head is sufficiently low for propeller-type turbines, the switch from pumping to generating could be achieved by reversing the pitch of the blades instead of changing the direction of rotation.

Pumped storage, at present the only practicable and economically viable way to store electrical energy in very large quantities, plays an increasing role in national – and even inter-national – power systems. In 2001, it contributed 86 GW of the 420 GW total hydro capacity in the thirty countries of the OECD (IEA, 2003). And its role as rapid backup could well expand if power from 'intermittent' renewables such as wind and solar power continues to increase. (It could, however, be superseded in the future by a completely different system, using hydrogen as the store and fuel cells to retrieve the energy - see Section 10.6 of Chapter 10)

5.14 Economics

No matter how elegant the technology, few will buy it if it is going to lose money. Potential investors need to know how much each kilowatt-hour of output will cost, taking all relevant factors into consideration. These factors include plant data, such as its initial capital cost, operation and maintenance costs, predicted lifetime and load factor, and also external factors such as the discount rate, or cost of borrowing money over a period of time. (For a brief introduction to the financial terms used in the following, see Appendix 2, or for a more detailed account, Chapter 12 of the companion volume.

Capital costs

Hydroelectricity is a very well-established technology, and much of the above information is easily available. The water-control systems, turbo-generators and output controls are standard items, covering a power range from a few hundred watts to hundreds of megawatts. The expected lifetime of the machinery is 25–50 years, and of the external structures, 50–100 years. Nevertheless, it is difficult to generalize meaningfully about 'the cost of hydroelectric power'.

The reason lies in the combination of heavy 'front-end loading' and extremely site-specific construction costs. In other words, the dominant factor in determining the total cost per unit of output is the initial capital cost, and a major part of this can be the civil engineering costs, which vary greatly from site to site.

Evidence on current costs of large-scale hydro schemes world-wide is conflicting and in some cases uncertain. Table 5.7 shows the quoted capital cost per kW of capacity for three major plants described in this chapter (see Boxes 5.11, 5.12 and 5.13).

Table 5.7 Capital costs for three major plants

plant	date	planned capacity	capital cost per kW
Itaipú	1984–91	12.6 GW	$1600
Gabcikovo-Nagymaros[1]	1977–	0.88 GW	$1200
Three Gorges[2]	1993–	18.2 GW	$1200

1 Original data based on the full scheme

2 Expected final cost, stated at completion of the dam in June 2003.

The huge plants are hardly typical, however. Data that are more representative of present costs in the industrialized countries come from a recent study of hydro potential in the USA (Hall *et al*, 2003). Total development costs, i.e. initial capital plus running costs, were assessed for over 2000 sites with potential capacities in the range 1–1300 MW. About 1000 of these, with a total potential of 17 GW, are green-field (blue-water?) sites, with no existing dams or hydro plants. For these, the estimated development cost, based on data for similar existing plants, falls mainly in the range \$2000–\$4000 per kW – very different from the figures in Table 5.7. It should, however, be noted that these are the remaining undeveloped sites in a country already using most of its accessible hydro potential.

A striking fact revealed in the study is the dominance of the initial costs. The civil engineering works typically account for 65–75% of the total, and meeting the environmental and other criteria necessary for a license can add a further 15–20%. In all, 85–95% of the development cost is 'site' cost, whilst the turbo-generator and control systems account for only 10% or so.

There are of course no fuel costs, and operation and maintenance costs are so low that in assessing the total development cost, their net present value (NPV, see Appendix 2) typically adds just 1–2% to the initial capital cost.

A UK study (ETSU, 1999) found very similar results for green-field sites, although the balance between 'site' and 'machinery' costs differed (probably a consequence of different definitions). The UK study revealed regional variations, with site averages of £1250 per kW for Wales, £1400 for Scotland and £1800 for England, where the available heads tend to be lower. But the overall range, £1000 to £2500 per kW, is remarkably similar to that for capital costs in the US study.

Both studies confirm the view that the largest plants and those with the highest heads tend to have the lowest unit costs. In either country, someone proposing a modest 5 MW plant with a relatively low head of 25 m would need an 'up-front' total of over 20 million dollars, or £13M. Given that the same sum would buy about *fifty* megawatts of combined-cycle gas turbine plant, it is easy to see why investors tend to 'forget' the fuel costs of the latter.

Refurbishing

Two other options have been considered – and implemented – in recent years by countries with little remaining accessible hydro capacity:

- the installation of hydro plants at existing dams constructed for other purposes
- the refurbishing or up-grading of existing hydro plants.

Both approaches offer the possibility of increased hydro capacity at lower cost and with fewer environmental consequences than green-field development. The two studies mentioned above are in agreement on the lower cost of these options (at least for the smaller plants they examine) with estimates of perhaps half the cost of green-field development when the dam already exists, and as little as a third, per 'new' kW, for refurbishment and up-grading of older plant.

In the UK, the Elan Valley scheme (Box 5.10) is one small-scale example that used both the above options. But larger plants, with output up to 20 MW, can qualify for support under the 2002 UK Renewables Obligation (see Section 10.8 of Chapter 10), and this has undoubtedly influenced producers in their decisions to invest in up-grading. Scottish and Southern Energy (SSE), one of two main owners of hydro plants in Scotland, have a £250 million program, within which contracts worth £4M were awarded in early 2003 for the refurbishment of five plants. All are rated at under 20 MW and will qualify for support when refurbished. Finlarig

(Figure 5.20) is one of the five, and is to have new runners, spears and nozzles (EEF, 2003, GE Power, 2003).

On the still larger scale, Scottish Power is refurbishing two of the 100 MW pump-turbines of the Cruachan pumped storage plant (Figure 5.28) at a cost of £16M. The project is expected to increase their power output by more than 30% (Power Technology, 2003).

Other countries are adopting similar policies. The USA and Canada have extensive refurbishment and up-grading programs for large-scale hydro, and in Europe, hydro capacity is expected to rise by over 20 GW during the present decade, largely through upgrading and modernization of existing schemes. This is to be achieved mainly by a 10% increase in output from the major producers: Austria, Italy, Norway, Spain, Sweden and Switzerland.

Investing in hydroelectricity

Hydroelectric plants often have long lives. As a result, according to the accepted methods of calculation (Appendix 2), and using historic cost data, the cost of each unit of electricity produced today by a plant built several decades ago becomes very small indeed: much less than the current price at which that unit can be sold. It seems, therefore, that hydroelectricity has been a profitable investment. If this is so, and hydro also has the desirable features listed in Section 5.13, why are producers in many countries preferring to build gas turbine plants?

Part of the answer lies in environmental constraints, and the fact that in the industrialized countries little accessible hydro potential remains. But even in the US, some 20 GW of additional capacity – potentially a 25% increase in the country's hydro output – is accessible at competitive costs (Hall *et al*, 2003). However, as we have seen, most of the costs of hydro are concentrated at the start, whilst those of favoured systems such as the gas turbine are spread more uniformly over their lives. The accepted methods for assessing the cost per unit of future output favour the more uniform distribution.

Suppose we compare a hydro plant and a CCGT plant which have the exactly same annual output, and the same total lifetime costs (construction, maintenance, operation, fuel). The hydro plant has a longer life and therefore a greater *lifetime* output, so it seems obvious that it produces the cheaper power. But future costs and future earnings are both subject to discounting. This reduces the present value of the *fuel costs* for the CCGT plant and of the *output earnings* for both – but with more effect on the more distant earnings in the later years of the hydro plant. The net result is that the hydro appears to have a higher lifetime cost, whilst the two plants appear to have very similar lifetime earnings. Annuitizing the costs, the alternative approach described in Appendix 2, leads to the same conclusion. On the basis of these methods, the gas turbine is therefore the better investment.

In the words of a report some years ago:

> It is paradoxical that investment in hydro schemes looks extremely favourable in retrospect … but extremely uncertain in prospect.
>
> Munasinghe, 1989

Or in one specific case:

> The Puueo plant [in Hawaii], first operated in 1918, is still generating electricity. ... Long ago, the plant paid off its initial capital costs and completed its economic life. Therefore, according to a cost-benefit economic study, the electricity which the plant is producing has no value. But the fact is that the plant continues to produce ... the same power that it did 60 years ago and there should be some way to place a value on it ...
>
> Miyabara, 1981

There is much truth in the observation that all power producers wish they had invested in hydroelectricity twenty years ago but unfortunately can't afford to do so now – and that they held this same view twenty years ago!

BOX 5.13 **Three Gorges**

China's Three Gorges project on the Yangzi River is the latest and largest of a series of hydro developments world-wide that have attracted major opposition on environmental and social grounds. Originally proposed in 1919 by Sun Yat Sen, the project has had a varied political history, culminating in its approval in 1992 against the unprecedented opposition of a third of the Congress delegates.

Opposition, locally and internationally, has centred on the displacement of the 1.13 million people who have lost their homes, farms and work-places. Other serious concerns have included the problem of silt that is predicted to block harbours upstream, increase flooding in some areas and ultimately reduce the plant output. And then there is the loss of one of China's most valued landscapes. Table 5.8 is an interesting summary of the main arguments.

The international outcry led the World Bank and some other major financial institutions to dissociate themselves from the project, but support was found

Table 5.8 Summary of the arguments in favour of and against the dam

Issue	Criticism	Defense
Cost	The dam will far exceed the official cost estimate, and the investment will be unrecoverable as cheaper power sources become available and lure away ratepayers.	The dam is within budget, and updating the transmission grid will increase demand for its electricity and allow the dam to pay for itself.
Resettlement	Relocated people are worse off than before and their human rights are being violated	15 million people downstream will be better off due to electricity and flood control
Environment	Water pollution and deforestation will increase, the coastline will be eroded and the altered ecosystem will further endanger many species	Hydroelectric power is cleaner than coal burning and safer than nuclear plants and steps will be taken to protect the environment
Local culture and natural beauty	The reservoir will flood many historical sites and ruin the legendary scenery of the gorges and the local tourism industry	Many historical relics are being moved, and the scenery will not change that much.
Navigation	Heavy siltation will clog ports within a few years and negate improvements to navigation	Shipping will become faster, cheaper and safer as the rapid waters are tamed and ship locks are installed.
Power generation	Technological advances have made hydrodams obsolete, and a decentralized energy market will allow ratepayers to switch to cheaper, cleaner power supplies	The alternatives are not viable yet and there is a huge potential demand for the relatively cheap hydroelectricity
Flood control	Siltation will decrease flood storage capacity, the dam will not prevent floods on tributaries, and more effective flood control solutions are available	The huge flood storage capacity will lessen the frequency of major floods. The risk that the dam will increase flooding is remote

Source: ChinaOnLine, 2000

Figure 5.29 The Three Gorges Dam shortly after completion in June 2003

elsewhere and work started in 1993. The dam, a mile long and 181 metres high, was completed in early 2003, and the sluices were finally closed at midnight on 1 June. The river was released from its five-year diversion and by early morning the water behind the dam was over 100 m deep. Two weeks later it was almost a metre *above* its intended final level.

The planned capacity is 18.2 GW, and at the time of writing (August 2003) the first two of the twenty-six 700 MW turbo-generators are undergoing tests. It is claimed that the project is on time for completion in 2009, and within budget (see Table 5.8 above).

Sources: China, 2003; ChinaOnLine, 2000; IRN, 2003; ThreeGorgesProbe, 2003

5.15 Future prospects

World total electricity production in 2002 was 16 000 TWh from a installed generating capacity of about 3600 GW. The corresponding figures for hydroelectricity are 2700 TWh from a capacity of 740 GW – roughly 16% of the world's output from 20% of the world's capacity. In the decade 1992–2002, hydro output rose at barely more than half the 2.7% annual increase in world electricity consumption, and its percentage contribution to the total has fallen continuously for some years.

The *technical potential* for hydroelectricity is estimated as 14 000–15 000 TWh a year – only a little less than the world's present *total* electricity consumption. Why then is its contribution falling behind those from other sources? The high initial capital cost is one reason; but some 14 GW of new hydro capacity is built each year, so it is clearly not a total impediment. It is however relevant, because, as we saw in Table 5.3, the wealthiest regions of the world have the least remaining hydro potential.

The critical factor for these wealthier countries is not therefore the technical potential but the *practicable* potential, i.e. the acceptable amount of development – the '*how much do we want to use*' potential. In the view of many people, countries such as the USA, Switzerland and other parts of Europe (including the UK) have already reached the practicable potential for large-scale hydro. No room at all for further development.

Not everyone agrees, of course. A spokesperson for the US hydro industry used a suitable moment during August 2003 to argue the case:

> During Thursday's blackout, affecting an estimated 50 million people from New York City to Michigan, hydropower facilities were the first to be put into service to initiate grid stability and restore power. Hydropower's unique operational characteristics allow it to generate power almost immediately while other sources can take hours to days to come back into service. ... Nationally, 19 626 MW

of potential hydropower, if developed, would raise predicted capacity margins by 16 percent...

Source: NHA, 2003

However, as we have seen, the costs of meeting ever more stringent regulations could make this extra 20 GW very expensive. Environmental costs play an increasing role in reducing the *economic potential* of hydro – the financially viable resource.

For the wealthier parts of the world, then, up-grading and refurbishment, discussed in Section 5.14, may be the principal remaining options.

In the poorer countries of Asia, Africa and South America, the situation is rather different. Very large-scale plants have been constructed in these regions in recent decades, and are undoubtedly supplying 'new' power to many people. However, it is by no means clear that investment in power plant is succeeding in one of its main aims, of encouraging industrial growth. Concern has been expressed that power station construction alone has been accounting for a quarter or more of government investment in many countries, with rates of return which have shown a steady fall over several decades.

China is of course the exception to all generalizations about hydroelectricity. Her small-scale hydro output is reportedly greater than the *total* hydro output of any but a handful of other countries; her total hydro output has risen by 86% in the last decade; and the world's largest-ever power station is only one of several current major developments. Yet China's *per capita* electricity consumption is still only a quarter of the UK value (and a tenth of the USA level). Her technical hydro potential is said to be nearly 2000 TWh a year, or about eight times the present output. According to the World Energy Council (WEC, 2003b), the economic potential is about 1200 TWh y^{-1}, some 200 TWh y^{-1} of which is already planned or under construction. Unless there is some major change, it seems probable that China will continue to account for a large part of any world growth in hydroelectricity, both large- and small-scale.

Small-scale hydro (SSH)

We have seen that is not easy to assess the world output from SSH, so it is not surprising to find that the world resource is even more uncertain. A decade ago, estimates suggested a world economic SSH potential in the range 200–300 TWh y^{-1} (IEA, 1993). The fact that in 2003 world SSH output is estimated to be in the region of 150 TWh y^{-1} (Section 5.11) does not of course mean that half the potential has been developed. The implication is rather that we should be cautious about *any* estimates of SSH output or potential.

As we have seen, many countries have SSH programs, but only in China does it make a significant contribution, and there is little indication that this situation is due for a major change – in the industrialized countries or in the rest of the world.

World-wide, there appear to have been two main factors acting against small-scale development on the Chinese scale. One is the predilection of most governments and major utilities companies everywhere for large centralized systems. The other, as we have seen, is financial. From the investors' point of view, the fact that the hydro plant will still be producing power in the year 2050, long after its capital is paid off, is of little immediate interest. (It may however interest their great grand-children, particularly when they realise that the alternatives have exhausted the world's gas reserves.)

To see large-scale growth in small-scale hydro, therefore, some change is necessary. The options appear to be:

- a reduction in the initial cost – perhaps as a result of technological improvements or by greater standardization of systems;
- a financial structure reflecting environmental and long-term benefits;
- in the wealthier countries, government support for small-scale systems, and encouragement of their development elsewhere through overseas aid programs.

Each of these options probably requires public expenditure though this need not be unduly onerous. In the UK during the 1990s, the NFFO subsidies for small-scale hydro added rather less than thirty pence per year to the average household electricity bill.

References

Anon (2001) 'Hydro Damned', *Renew*, no. 129, Jan/Feb, p. 9.

Boyle, G., Everett, B. and Ramage, J. (2003) *Energy Systems and Sustainability*, Oxford, Oxford University Press in association with the Open University.

BP (2003) 'Statistical Review of World Energy', BP. Available at www.bp.com/subsection.do?categoryId=95&contentId=2006480, BP (accessed 12 December 2003).

China (2003) *The Three-Gorges Project*, Embassy of the People's Republic of China in the USA Available at www.china-embassy.org/eng/c2718.html.

ChinaOnLine (2000) Available at www.chinaonline.com/refer/ministry-profiles/threegorgesdam.asp [accessed 24 August 2003].

CONAES (1979) *Energy in Transition 1985–2010*, Washington, D.C., National Research Council, National Academy of Sciences.

DOE (2001) *Hydropower Facts: Primary Purpose or Benefit of U.S. dams*, U.S. Department of Energy Hydropower Program. Available at inel.gov/hydropower/facts/facts.htm [accessed 25 August 2003].

Dorf, R. C. (1987) *Energy resources and policy*, Addison Wesley.

DTI (1999) *New & Renewable Energy:* Prospects *for the 21st Century*, London, Department of Trade and Industry. Available at www2.dti.gov.uk/renew/condoc [accessed 24 August 2003].

DT1 (2003) *Digest of UK Energy Statistics 2003*, London, Department of Trade and Industry. Chapter 7: Renewables. Available at www2.dti.gov.uk/energy/inform/dukes/dukes2003/07main.pdf or www.etsu.com/RESTATS/Publications/07main.pdf [both accessed 22 August 2003].

EEF (2003) *SSE to refurbish five hydro-electric power stations*, European Energy Focus Jan/Feb Available at www.europeanenergyfocus.com/pages/hydropower/janfeb03/058b.htm [accessed 25 August 2003].

ERN (1999) *European Rivers Network* Available at www.rivernet.org, select *Dam Decommissioning.*

ETSU (1989) *Small Scale Hydroelectric Generation Potential in the UK*, Department of Energy Report No. ETSU-SSH-4063, London.

ETSU (1994) An assessment of Renewable Energy for the UK, Energy Paper R82, London, HMSO.

ETSU (1998) Small Hydro – Five Go Live in Elan Valley, New Review, Issue 35, Feb.

ETSU (1999) *New and Renewable Energy: Prospects in the UK for the 21st Century: Supporting Analysis* (ETSU R-122) Available at www2.dti.gov.uk/renew/condoc/support.pdf [accessed 24 August 2003].

ETSU (2000) *Monitoring of successful renewables obligation small hydro projects* (ETSU/H/01/00049/00/REP) Available at www.dti.gov.uk/energy/renewables/publications/pdfs/rep49.pdf [accessed 25 August 2003].

ETSU (2003) *Renewable Generating Capacity from the Renewables Obligation*. Available at www.etsu.com/RESTATS renewables obligations.html [accessed 22 August 2003].

Fleischer, Tamás (1993) *Jaws on the Danube: Water management, regime change and the movement against the Middle Damube Hydroelectric Dam.* Available at www.vki.hu/~tfleisch/PDF/pdf93/CAPA93an.pdf [accessed 22 September 2003].

GE Power (2003) GE Power Systems, press release, 18 Feb. Available at www.gepower.com [accessed 25 August 2003].

Hall, D. G. *et al* (2003) *Estimation of Economic Parameters of U.S. Hydropower Resources*, Idaho National Engineering and Environmental Laboratory (INEEL/EXT-03-00662) Available at hydropower.inel.gov/pdfs/Project ReportFINAL with Disclaimer-3Jul03.pdf [accessed 25 August 2003].

Hill, G. (1984) *Tunnel and Dam: the Story of the Galloway Hydros*, South of Scotland Electricity Board (now Scottish Power)

IEA (1993) *Renewables in Power Generation: Towards a Better Environment, Appendix 3*, Paris, International Energy Agency. Now out of print, but Appendix available at www.iea.org/pubs/studies/files/benign/pubs/append3e.doc [accessed 19 August 2003].

IEA (2003) *Renewables Information 2003*, Paris, International Energy Agency. Available at www.iea.org/stats/files/renew2003.pdf [accessed 20 August 2003].

IRN (2003) *IRN's Three Gorges Campaign*, International Rivers Network. Available at www.irn.org/programs/threeg/index.shtml [accessed 28 August 2003].

Itaipú (2003) *Itaipú Binacional – A Maior Hidrelétrica do Mundo* Available at www.itaipu.gov.br/english/main.htm [accessed 24 August 2003].

Krauter, S. (2000) *Itaipú, Largest Power Plant on Earth*, Grupo Solar, Federal University of Rio de Janiero. Available at www.solar.coppe.ufrj.br/itaipu.html [accessed 25 August 2003].

Maurer, E. A. (1997) *New Market for Small-scale Hydro*, Caddet Renewable Energy Newsletter, Issue 4/97.

McGriff, B. (1999) *The Danube River* Available at user.intop.net/~jhollis/danube.htm [accessed 25 August 2003].

Miyabara, T. and Goodman, L. J. (1981) 'Hawaii, USA: hydroelectric development' in Goodman, L. J. *et al.*, *Small Hydroelectric Projects for Rural Development*, Pergamon, London.

Munasinghe, M. (1989) 'Power for development, electricity in the Third World', *IEE Review*, March.

NHA (2003) *Hydro Brought New York Out of the Dark*, National Hydropower Association press release, 19 August.

Power Technology (2003) Cruachan awarded to VA Tech Hydro consortium, Power Technology, 20 Feb Available at www.power-technology.com/contractors/renewable/vahydro/press1.html [accessed 25 August 2003].

Slovakia (2000) *Gabcikovo Dam Dispute*. Available at www.slovakia.org/history-gabcikovo.htm [accessed 31 March 2003].

Sullivan, L. (1995) The Three Gorges Project: damn if they do?, *Current History*, September.

ThreeGorgesProbe (2003) at www/ThreeGorgesProbe.org [accessed 25 August 2003].

Tong, Jiangdong (2002) *Small hydro on a large scale*, Global Forum for Sustainable Energy. Available at www.gfse.at/papers/TONG.pdf [accessed 22 August 2003].

Upadhyay, D. (1997) 'Pico hydro in Nepal – An overview', Pico hydro, Issue 1, October.

USGS (1995) 'The Los Angeles Dam Story', USGS. Available at quake.wr.usgs.gov/prepare/factsheets/LADamStory/ (accessed 12 December 2003).

WCD (2000) *Dams and Development: A New Framework for Decision-making*, World Commission on Dams. Available at www.dams.org (see also www.unep-dams.org) [both accessed 31 March 2003].

WEC (2003a)) *Survey Of Energy Resources, Overview*, Energy Information Centre, World Energy Council, London. Available at www.worldenergy.org/wec-geis/pulications/reports/ser/overview.asp [accessed 15 August 2003].

WEC (2003b) *Survey Of Energy Resources, Hydropower*, Energy Information Centre, World Energy Council, London. Available at www.worldenergy.org/wec-geis/pulications/reports/ser/hydro/hydro.asp [accessed 15 August 2003].

WWF (1997) *Danube River loses in the Hague,* WWF press release, 25 September.

WWF (1998) *New Danube dam could threaten Hungary's accession to EU,* WWF press release, 18 February.

Zinke, A. (2002) Gabcikovo: 10 Years After, *Danube Watch*, 2/2002. Available at www.icpdr.org/pls/danubis/docs/FOLDER/HOME/NEWSANDEVENTS/DWO2-2/DWO202p14.htm [accessed 31March 2003].

Chapter 6

Tidal Power

by David Elliott

6.1 Introduction

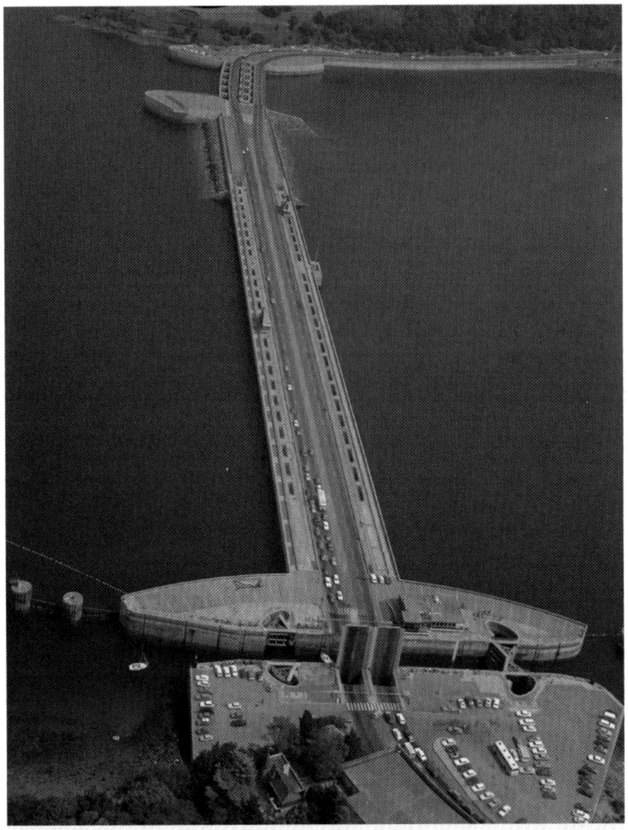

Figure 6.1 A view of La Rance tidal scheme

The rise and fall of the seas represents a vast, and as King Canute reputedly discovered, relentless natural phenomenon.

The use of tides to provide energy has a long history, with small tidal mills on rivers being used for grinding corn in Britain and France in the Middle Ages. More recently, the idea of using tidal energy on a much larger scale to generate electricity has emerged, with turbines mounted in large **barrages** – essentially low dams – built across suitable estuaries (see Box 6.1).

A medium-scale 240 MW scheme has been built at the Rance Estuary in France (see Figure 6.1) and a number of small schemes have been built around the world. There have also been proposals for barrages with several gigawatts of generating capacity. An artist's impression of the proposed 8.6 gigawatt (GW) Severn Barrage, stretching 16 kilometres across the Severn Estuary in the UK, is shown in Figure 6.2. If built, this would generate 17 terawatt-hours of electrical energy per year (TWh y^{-1}), the equivalent of nearly 5% of the electricity generated in the UK in 2002. As we shall see, there are also several other possible sites for large-scale projects around the world.

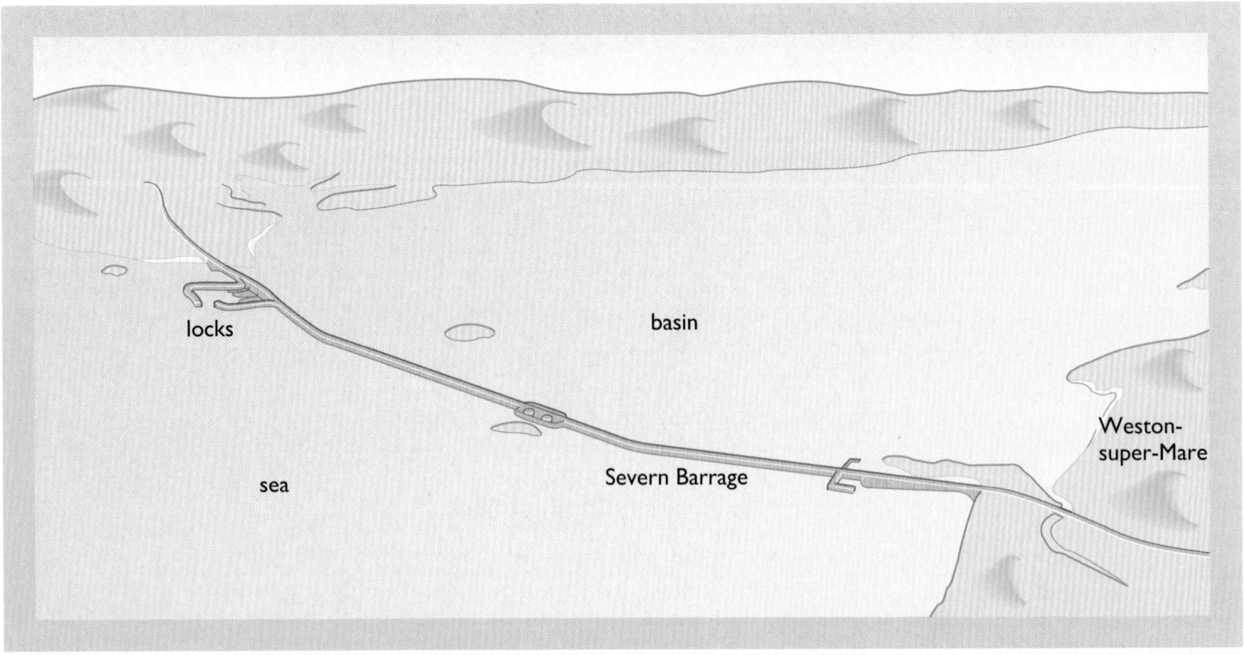

Figure 6.2 Artist's impression of the proposed Severn Barrage

BOX 6.1 A brief history of tidal power

As mentioned in the main text, small tidal mills, not unlike traditional watermills, were used quite widely on sections of rivers in the Middle Ages for grinding corn, but the idea of exploiting the full power of the tides in estuaries is relatively recent.

There have been a number of proposals for various types of crossing for the Severn Estuary, the UK's largest estuary, but so far none have been taken forward. For example, a barrage concept (albeit with no provision for power generation) attributed to Thomas Telford was put forward in 1849. The first serious proposal involving electricity production came in 1920 from the Ministry of Transport. This was followed by a major study by the Brabazon Commission, which was set up in 1925. Its 1933 report focused on a barrage crossing the estuary along the 'English Stones line', not far from the modern Severn Bridges. It was to have 72 turbines with a total installed capacity of 804 MW and to incorporate road and rail crossings. The scheme was not followed up. It was reassessed in 1944, but, again, not followed up.

During the 1960s and 1970s a number of schemes were proposed, each crossing along different lines. These proposals culminated in a new government-supported study by the Severn Barrage Committee, which was set up in 1978 under Professor Sir Hermann Bondi. The Committee reported in 1981 (Department of Energy, 1981), concluding that it was 'technically feasible to enclose the estuary by a barrage located in any position east of a line drawn from Porlock due North to the Welsh Coast'. Of all the possible crossing lines, three were favoured, the most ambitious being from Minehead to Aberthaw, which, it was estimated, could generate 20 TWh y^{-1} from 12 GW of installed capacity.

Subsequently, the less ambitious, but still very large, so-called 'inner barrage', on a line (first proposed by E. M. Wilson in 1966) from Weston-super-Mare to Lavernock Point, became the favourite, and was pursued by the Severn Tidal Power Group (STPG) industrial consortium. It was initially conceived of as generating approximately 13 TWh y^{-1} from 7 GW installed, although this was later upgraded to 17 TWh y^{-1} from 8.64 GW installed.

Enthusiasm for tidal schemes was fuelled in part by the success of the French scheme on the Rance Estuary in Brittany, near St Malo (Figure 6.1). This was constructed between 1961 and 1967 and the first output from its 240 MW turbine capacity was achieved in 1966. This structure includes a road crossing. Apart from a problem with the generator mountings in 1975, it has operated very successfully. Subsequently, a much larger 15 GW scheme was proposed, to enclose a vast area of sea from St Malo in the south to Cap de Carterel in the north, the so-called 'Isle de Chausey Project'. This has not been followed up.

Although large-scale schemes have been proposed for the Bay of Fundy in Canada and also in Russia, the only significant tidal plants to be built to date, other than La Rance, are an 18 MW single unit, using a 'rim generator' (see Figure 6.12), at Annapolis Royal in Nova Scotia, Canada, completed in 1984; a 400 kilowatt (kW) unit in the Bay of Kislaya, 100 km from Murmansk in Russia, completed in 1968; and a 500 kW unit at Jangxia Creek in the East China Sea.

Whilst a number of other schemes have been considered around the world, the main focus in recent years in the field of tidal power has been on the Severn Barrage 'inner barrage' concept and on the Mersey Barrage, both in the UK. However, as we shall see, barrages on the Humber and several smaller UK estuaries have also been considered.

The nature of the resource

It is important at the outset to distinguish *tidal energy* from *hydro power*. As we saw in Chapter 5, *hydro* power is derived from the hydrological climate cycle, powered by solar energy, which is usually harnessed via hydroelectric dams. In contrast, *tidal* energy is the result of the interaction of the gravitational pull of the moon and, to a lesser extent, the sun, on the seas. Schemes that use tidal energy rely on the twice-daily tides, and the resultant upstream flows and downstream ebbs in estuaries and the lower reaches of some rivers, as well as, in some cases, tidal movements out at sea.

Equally, we must distinguish between tidal energy and the *energy in waves*. Ordinary waves are caused by the action of wind over water, the wind in turn being the result of the differential solar heating of air over land and sea (see Chapter 8). If we consider wave energy, like hydroelectric energy, to be a form of solar power, tidal energy could be called 'lunar power'. Such distinctions are not helped, however, by the terminology which is often used – for example, the term 'tidal wave' is used to describe the occasionally dramatic surges of water (which are neither waves nor tides!) that can be produced by under-sea earthquakes. There also exist large climate-driven water flows in the oceans, which are the result, ultimately, of solar heating. The Gulf Stream is one such example.

The energy in these various movements of water can, in principle at least, be tapped. The rise and fall of the tides can be exploited without the use of dams across estuaries, as was done in the traditional **tidal mills** on the tidal sections of rivers, as mentioned earlier. A small pond or pool is simply topped up and closed off at high tide and then, at low tide, the trapped water is used to drive a water wheel, as with traditional watermills.

There is also the possibility of using turbines mounted independently in the rapidly flowing **tidal currents** created due to the effects of concentration in narrow channels, for example between islands or other constrictions. In addition, it may be possible to harness some of the energy in the larger scale ocean streams such as the Gulf Stream. Some recent developments in the tidal current and ocean stream areas are discussed later in Section 6.8.

For the moment, however, we will be focusing on extracting energy using tidal barrages across estuaries. In most of these systems, the water carried upstream by the **tidal flow** – usually called the **flood tide** – is trapped behind a barrage. The incoming tide is allowed to pass through sluices, which are then closed at high tide, trapping the water. As the tide ebbs, the water level on the downstream side of the barrage reduces and a **head** of water develops across the barrage. The basic technology for power extraction is then similar to that for low-head hydro: the head is used to drive the water through turbine generators. The main difference, apart from the salt-water environment, is that the power-generating turbines in tidal barrages have to deal with regularly varying heads of water.

Basic physics

The variation in tidal height is due primarily to gravitational interaction between the earth and the moon. This gravitational force, combined with the rotation of the earth, produces, at any particular point on the globe, a twice-daily rise and fall in sea level, this being modified in height by the gravitational pull of the sun and by the topography of land masses and ocean beds. (For the sake of completeness, we should note that the gravitational pull of the sun and the moon also causes tidal phenomena in the atmosphere and in the earth.) A detailed analysis of this interaction between earth, moon and sun is quite complex, but we will attempt to describe it in simple terms.

The first part of the explanation is relatively straightforward. Starting first with just the earth and the moon, the gravitational pull of the moon draws the seas on the side of the earth *nearest* to the moon into a bulge *towards* the moon. That gives us one tide per day at any one point, as the planet rotates through the bulge. But what about the second tide each day? This is more difficult to explain. Sometimes it is explained in simple terms by saying that the waters that make up the bulge facing the moon are drawn from the seas at each side of the earth, but the water at the far side is 'left behind', at its original level. However that does not really explain fact that the second tide is roughly the same height as the first. Neither does the fact that the water in the seas *furthest* from the moon experiences slightly less of the lunar pull, being further away.

The full explanation of the bulge that forms on the side of the planet furthest away from the moon depends on a more complicated analysis, based on understanding the effect of the relative movements of the earth and moon. This is described in Box 6.2 below.

BOX 6.2 The earth and the moon

A useful mathematical analysis of the generation of tides is given in *Renewable Energy Resources* (Twidell and Weir, 1986). This identifies *two* processes at work in relation to the earth and the moon: a centrifugal effect as well as a gravitational effect.

The first process, the centrifugal effect (that is, the tendency of any mass in motion to try to continue in a straight line rather than be constrained to move in a circle), is the result of the fact that the earth and the moon rotate around each other, somewhat like a 'dumb-bell' being twirled. However, this giant dumb-bell does not rotate around the half-way point between the earth and the moon. Since the earth is much larger than the moon, their common centre of rotation is close to the earth; in fact it is just below its surface (see Figure 6.3). The mutual rotation around this point produces a relatively large *outward* centrifugal force acting on the seas on the side of the earth *furthest* from the moon, bunching them up into a bulge. There is also a smaller centrifugal force, directed *towards* the moon, that acts on the seas *facing* the moon. (This force is smaller since here the distance from the earth's surface to the common rotation point, just below the surface, is smaller.)

The second process, the gravitational effect, relates to the gravitational pull of the moon, which draws the seas on the side of the earth *nearest* to the moon into a bulge *towards* the moon, whilst the seas *furthest* from the moon experience a reduced lunar pull.

There is thus, to summarize, a small centrifugal force and an increased lunar pull acting on the seas facing the moon, and a larger centrifugal force and a decreased lunar pull acting on the seas on the other side of the earth. The end result, on the basis of this analysis, is essentially a rough symmetry of forces, small and large, on each side of the earth, producing tidal bulges of roughly the same size on each side of the earth. In practice, the bulges may differ significantly, due, for example, to the tilt of the earth's axis in relation to the orbit of the moon and to local topographic effects.

As the earth rotates on its axis, the lunar pull will maintain the high tide patterns, as it were 'under' the moon. That is, the two high tide configurations will in effect be drawn around the globe as the earth rotates, giving, at any particular point, *two* tides per day (or, more accurately, two tides in every 24.8-hour period), occurring approximately 12.4 hours apart. Since the moon is also moving in orbit around the earth, the timing of these high tides at any particular point will vary, occurring approximately 50 minutes later each day.

For further discussion, see *Waves, Tides and Shallow Water Processes* (Open University, 1989).

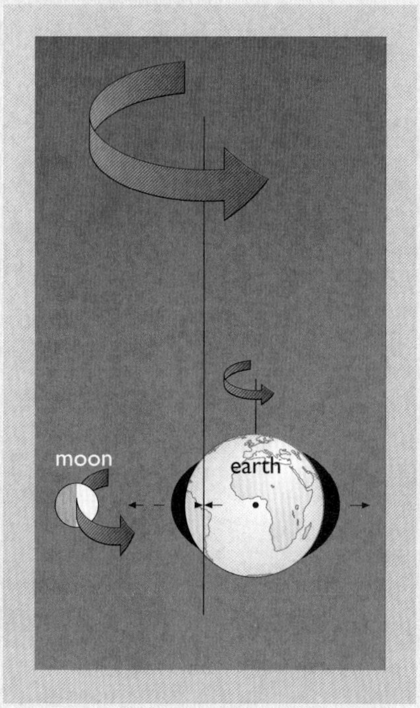

Figure 6.3 Relative rotation of the earth and the moon (not to scale)

The basic pattern described in Box 6.2 is also modified by the pull of the sun. Although the sun is much larger than the moon, its distance from the earth is much greater, and the moon's gravitational influence on the seas is therefore approximately twice that of the sun. The final impact depends on their relative orientation.

When the sun and the moon pull together (in line), whether both pulling on the same side of the earth or each on opposite sides, the result is the very high **spring tides**; when the sun and moon are at 90° to each other, the result is the lower **neap tides**. The period between neap and spring tides is approximately 14 days – that is, half the 29.5-day lunar cycle (Figure 6.4). The ratio between the height of the maximum spring and minimum neap tides can be more than two to one.

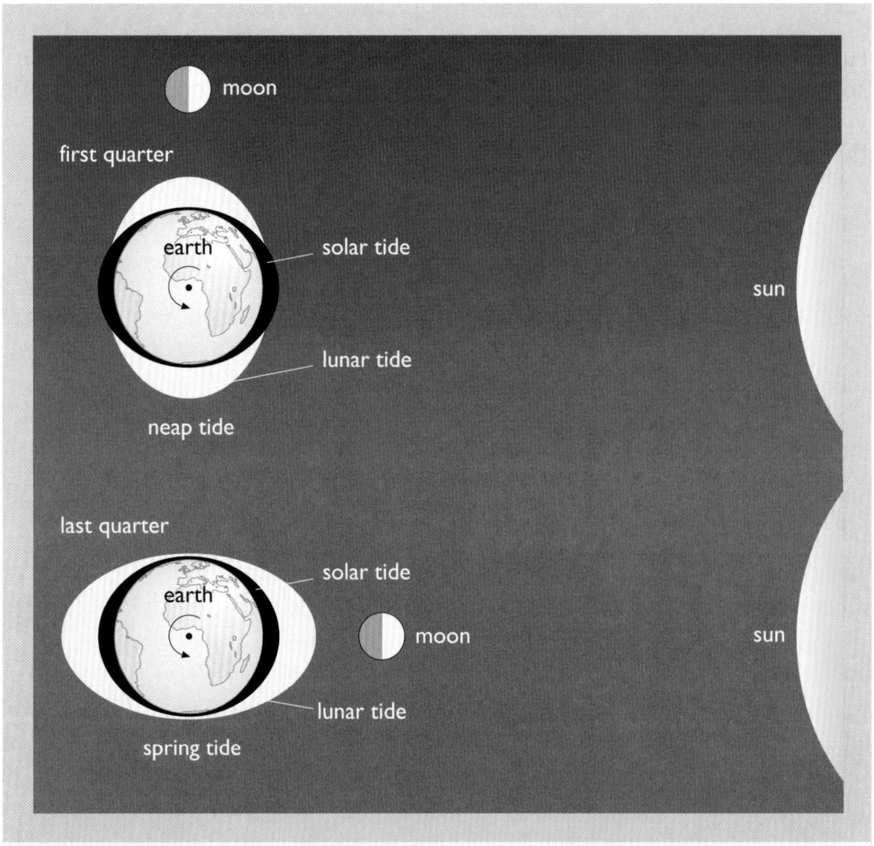

Figure 6.4 Influence of the sun and the moon on tidal range (not to scale)

The analysis of interactions presented above is simplified. In reality, there are other factors that complicate the tidal patterns and can alter them dramatically.

Some of these other factors do not have much direct relevance to tidal power generation. For example, although the tides are not seasonal in climate terms, the weather can play a significant (and sometimes destructive) role, with high winds and major storms occasionally combining with high tides to produce very high 'surge' tides. Barrages have to be designed to withstand these surges, but any energy gains from them would be marginal.

However, there are other factors that *are* more relevant to power generation, even if their significance is usually relatively small. For example, the basic twice-monthly moon–sun, spring–neap tide pattern outlined above is modified by the fact that the moon's orbit is not circular but *elliptical*. There are also longer-term orbital variations, for example a semi-annual cycle caused by the inclination of the moon's orbit in relation to that of the earth's orbit around the sun, which gives a variation of approximately 10% in the height of the tides. And finally, to complete the complex picture of the nature and cause of tides, the tides are modified in some locations by **Coriolis forces**. These are due to the spin of the earth, and deflect tidal currents from the paths that they would otherwise have taken.

Apart from relatively minor perturbations such as these, the overall effect of the basic sun–moon–earth interaction is that, *in mid-ocean*, the typical tidal variation or **tidal range** will be approximately 0.5 metres.

However, the tidal ranges experienced in practice at coastal sites are sometimes significantly modified and amplified by *local* topographic variations, for example in shallow coastal waters and in estuaries. As the tide approaches the shore, and the water depth decreases, the tidal flow is concentrated and can be increased to reach up to, typically, 3 metres. If the tide then enters a suitably shaped estuary, it can be funnelled and therefore heightened even further, up to 10–15 metres at some sites (such as the Severn Estuary), with complex **resonance effects** playing a major role. Such resonances are like the vibrations that can be set up in the sound boxes of some musical instruments, amplifying certain frequencies of the original sound. Whether any particular resonance can be set up depends on the shape of the 'cavity' in which it is established. The size of the cavity has to be matched to the wavelength of the sound, or some multiple of it. Given the variations in depth and width of estuaries, it is hardly surprising that in practice the resonance patterns that emerge, when and if tidal resonances occur, are often very complicated. This point is discussed further in Box 6.3, which also opens up the intriguing idea of resonance across entire oceans.

Even without a full understanding of resonances, it is fairly easy to see that, given the right 'funnel' configuration, the tidal range will increase as the tide moves upstream, and as the depth and width of the estuary decreases. However, there are also frictional effects: for example, energy is lost as the tidal flow moves over differing estuary bed materials. So in practice, the extent to which the tidal range is magnified at any point depends on the balance between the energy losses and the concentration of the tidal flow by the topography. The frictional losses will usually begin to outweigh the concentration gains at some point upstream when the funnel-like layout of an estuary gives way to a more parallel-sided, flat-bottomed river configuration. On the Severn, this 'natural' optimal point normally occurs around the site of the existing Severn Bridge, where the tidal range reaches 11 metres.

Occasionally, in some long estuaries, dramatic tidal effects can occur further upstream. For example, rather than producing a relatively slow rise, as normally happens in the main part of the Severn Estuary, the upstream tidal flow further up the Severn can be concentrated so abruptly that it rises into an almost vertical step or wave, the so-called Severn Bore. A similar effect occurs on other long estuaries, including the Humber in the UK and the Hoogly near Calcutta in India.

BOX 6.3 Resonances

Resonance effects, occurring both locally (for example in estuaries) and also across the width of oceans, can play a major role in increasing tidal range.

An estuary or a complete ocean basin can behave like a resonant cavity: an enclosure or box in which resonances occur when the dimensions of the box are equal to the wavelength of an impinging vibration or oscillating 'signal'. If the dimensions are right, the waves trapped in the cavity reflect off the walls and reinforce or *amplify* the original signal by combining their amplitudes with that of the original wave – this is **constructive interference**. (The opposite case, where waves wholly or partially cancel one another out, reducing the overall amplitude, is **destructive interference.**) Constructive interference will also occur where the waves have wavelengths that are multiples of the cavity length, as long as they are exact (integer) multiples. Similarly, resonances can occur with waves at exactly one half or one quarter of the original wavelength.

The cyclic rise and fall of the tides represents an oscillating signal or vibration that can enable amplified resonances to be created in an ocean basin or estuary of appropriate dimensions. In very crude terms, this is something like the 'sloshing' effect you can create by moving with the right rhythm when lying in a bath. The wave height you can obtain depends on the amount and phasing of energy you put in, but also on the shape of the bath.

As it happens, the distance between North America and Europe, approximately 4000 km, turns out to be just about right for creating a resonance, given the 12-hour tidal cycle, with a wavelength of twice the width of the ocean.

However, this is a simplified picture, where the cavity within which the oscillations take place, that is, the ocean basin, is assumed to have vertical sides which will reflect the standing waves perfectly. In practice, local effects at each coast must also be taken into account when analysing the tidal range that results from resonance effects – since the sea bed rises near the coast this will modify the resonance effect. In the case of the North Atlantic, the end result is that the initial open sea tidal range (of 0.5 metres) is enhanced to approximately 3 metres at each side of the Atlantic.

Modified resonances of this sort, albeit at different frequencies, should also occur in the Pacific, which is about four times the width of the Atlantic, the tidal cycle's wavelength in this case being *half* the width of the ocean. However, the width is irregular, and the result in practice is that resonant effects around the Pacific are often more complex and less dramatic. In some locations around the Pacific, the result is that only very small tides occur – that is, there is little resonatory enhancement. Interestingly, in some sites there is only one significant tide per day, since the resonances occur only on the basis of 24-hour periods – i.e the distance between coasts means that in a 12-hour cycle the waves at one coast are at their maximum amplitude, and at the other they are at their minimum.

In addition to the resonances created across *complete* ocean basins, *local* resonances in shallower coastal areas are possible, for example in the smaller basins defined by some land masses and in estuaries. These, combined with local topographical funnelling effects in some estuaries, can increase the tidal range from its typical coastal height of 3 metres to over 10 metres.

For example, the Irish Sea, Bristol Channel and Severn Estuary complex has a total length of approximately 600 km and a natural resonant period of approximately six hours, and resonances at a quarter wavelength (twice the tidal frequency) are therefore possible. There is also a half-cycle resonance in the area from Land's End to Dover along the English Channel (approximately 500 km at an average depth of approximately 70 metres).

The effects of resonance can be very much greater in estuaries, since they are more like closed cavities, but the resonances are often very complex, since the width and depth varies and changes in the nature of the estuary bottom introduce varying frictional losses. Local funnelling effects also play a part, as the tidal flow moves up narrowing channels (see Figure 6.5).

For more detailed discussion of resonance effects, see *Tidal Power* (Baker, 1991a).

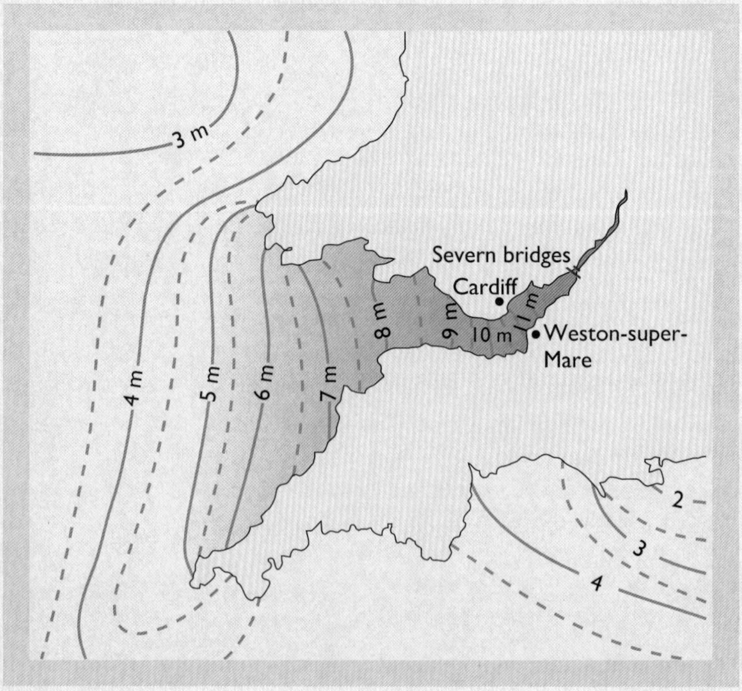

Figure 6.5 The effects of concentration of tidal flow in the Severn Estuary (tidal ranges in metres)

As can be seen, even leaving aside freak effects like the Severn Bore, there is a complex range of tidal phenomena. Fortunately for the designers of tidal barrages, the end results – that is, the tidal patterns in estuaries – although very much site-specific, are predictable and reliable. The tides will continue to ebb and flow, on schedule, indefinitely.

But is the energy in the tides *really* 'renewable'? We have seen that the primary mechanism in tide generation is the gravitational interaction between the earth and the moon. The rotation of the earth draws the resulting tidal bulges across the seas, or, more precisely, the water in the seas rises into a bulge as the water rotates with the planet. The rotation of the earth is being very gradually slowed by this process (by approximately one-fiftieth of a second every 1000 years); because of frictional effects it takes energy to drag the water along, especially through areas where there are topographical constrictions. However, the extra frictional effect that would be produced by even the widespread use of tidal barrages would be extremely small. The influence on the moon's orbital velocity (which is also being very slowly reduced by the tidal interaction) would be even smaller.

Power generation

The basic physics and engineering of tidal power generation are relatively straightforward.

Tidal barrages, built across suitable estuaries, are designed to extract energy from the rise and fall of the tides, using turbines located in water passages in the barrages. The **potential energy**, due to the difference in water levels across the barrage, is converted into **kinetic energy** in the form of fast-moving water passing through the turbines. This in turn is converted into **rotational kinetic energy** by the blades of the turbine, the spinning turbine then driving a generator to produce **electricity**.

The average power output from a tidal barrage is roughly proportional to the square of the tidal range. The mathematical derivation of this is fairly simple, as is demonstrated by the analysis in Box 6.4.

Clearly, even small differences in tidal range, however caused, can make a significant difference to the viability and economics of a barrage. A mean tidal range of at least 5 metres is usually considered to be the minimum for viable power generation, depending, of course, on the economic criteria used. As the analysis in Box 6.4 indicates, the energy output is also roughly proportional to the area of the water trapped behind the barrage, so the geography of the site is very important. All of this means that the siting of barrages is a crucial element in their viability.

Many studies have been carried out on tidal power in the UK, dating from the early 1900s onwards (see Box 6.1). This is hardly surprising, as the UK holds about half the total European potential for tidal energy, including one of the world's best potential sites, the Severn Estuary. There is also a range of possible medium- and small-scale sites, including locations on the Mersey, Wyre and Conwy. The total UK tidal potential is, in theory, around 53 TWh y^{-1} (ETSU, 1990), which is about 14% of UK electricity generation in 2002. The contribution to electricity consumption that could be achieved in the UK and elsewhere in practice would depend on a range of *technical*, *environmental*, *institutional* and *economic* factors. Although these factors interact, we can explore each in turn before attempting a synthesis.

BOX 6.4 Calculation of power output from a tidal barrage

Let us assume that we have a rectangular basin behind a barrage which has a constant surface area A, and a high-to-low tidal range R (see Figure 6.6). When the tide comes in, it is freely allowed to flow into the basin, but when the tide goes out, the water in the basin is held there, at the high tide level. When the sea has retreated to its low tide level, the surface of the water held behind the barrage will be at a height R above the sea.

Given a rectangular basin, the centre of gravity of the mass of water will be at a height $R/2$ above the low tide level. The total volume of water in the basin will be AR and, if the density of the water is ρ it will have a mass ρAR, i.e. ρ multiplied by the volume of water (A times R). This water could all now be allowed to flow out of the barrage through a turbine to the low tide level. The maximum potential energy E available per tide if all the water falls through a height of $R/2$ is therefore given by the mass of water (ρAR) times the height ($R/2$) times the acceleration due to gravity (g); that is, $E = \rho ARg(R/2)$. The basin could then be allowed to fill on the next incoming tide and the cycle repeated again and again. If the tidal period is T, then the average potential energy that could be extracted becomes E/T or $\rho AR^2g/2T$.

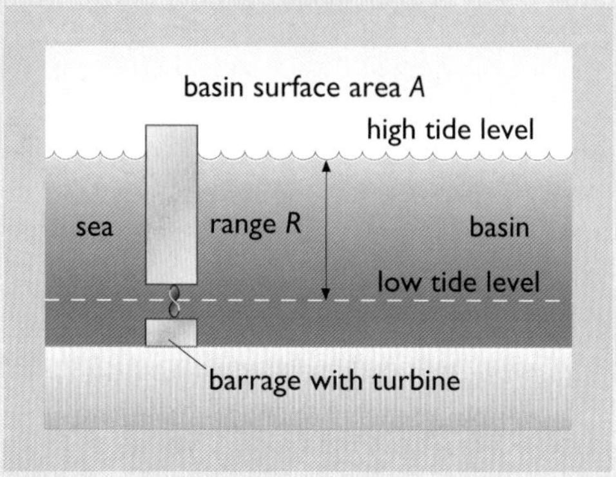

Figure 6.6 Power generation from tides

6.2 Technical factors

The input energy source for a barrage, the rise and fall of the tides, follows a roughly sinusoidal pattern (see the sea level curves in Figures 6.7–6.9). As we have seen, the tides have a 12.4-hour cycle, with the peak-to-trough height variation, the tidal range, varying from site to site as a result of complex resonance and funnelling effects.

Given the complexity of estuary configurations, the actual resonances and funnelling effects are very difficult to model accurately, with variations in depth, width and friction over differing estuary bed materials introducing many local variations.

However, it is well worth the effort required to analyse these effects when deciding on the precise siting and orientation of a barrage, since they will have a major effect on its output. Indeed, it may be possible to locate and/or operate a barrage so as to 'tune' the estuary to be more resonant, and thus to increase power output. Certainly, any disturbance that might reduce existing resonance effects should be avoided.

In addition to the basic issues of location and orientation, a second set of factors that influences the likely power output of a barrage relates to its *operational pattern.*

Power can be generated from a barrage in three main ways. The most commonly used method is **ebb generation**. Here the incoming tide is allowed to pass through the barrage sluice gates with the turbines idling (that is, without generating power). The water is trapped behind the barrage at high tide level by closing the sluices. The head of water is then passed back through the turbines on the *outgoing* ebb tide in order to generate power; see Figure 6.7. Alternatively, **flood generation** uses the *incoming*

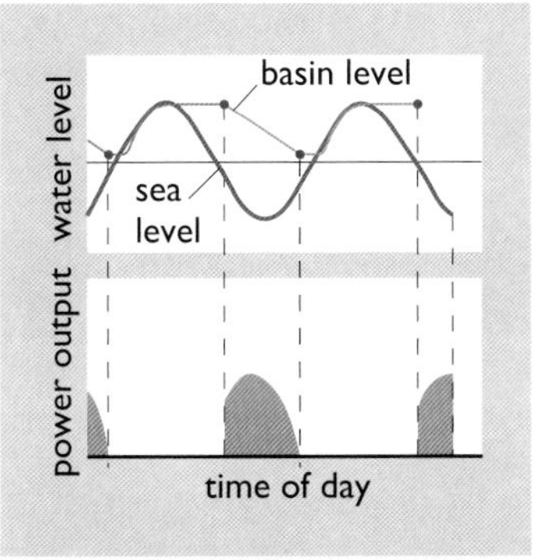

Figure 6.7 Schematic diagram of water levels and power outputs for an ebb generation scheme

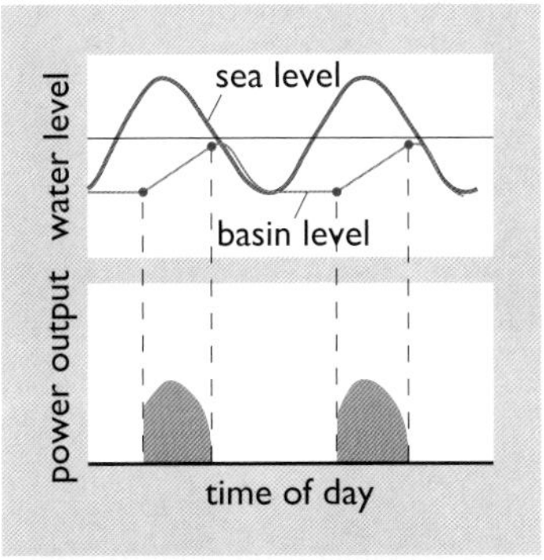

Figure 6.8 Schematic diagram of water levels and power outputs for a flood generation scheme

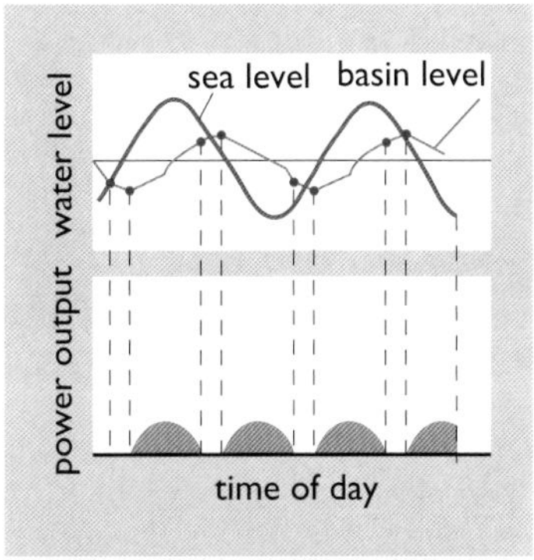

Figure 6.9 Schematic diagram of water levels and power outputs for a two-way generation scheme

tide to generate electricity as it passes through the turbines mounted in the barrage (Figure 6.8). In either case, two bursts of power are produced in every 24.8-hour period. **Two-way operation**, on both the ebb and the flood, is also possible (Figure 6.9).

The basic technology for power production is well developed, having much in common with conventional low-head hydro systems (see Chapter 5). Figure 6.10 is an artist's impression of the typical layout of a power generation scheme.

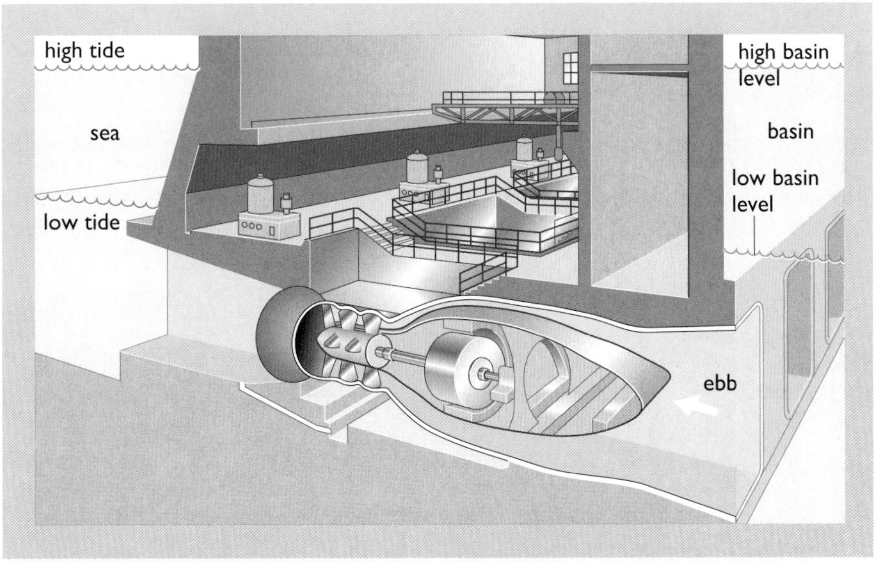

Figure 6.10 Artist's impression of the typical layout of a power generation scheme

A number of different configurations is possible. At La Rance, a so-called **bulb** system is used, with the turbine generator sealed in a bulb-shaped enclosure mounted in the flow (Figure 6.11). However, the water has to flow around the large bulb and access (for maintenance) to the generator involves cutting off the flow of water.

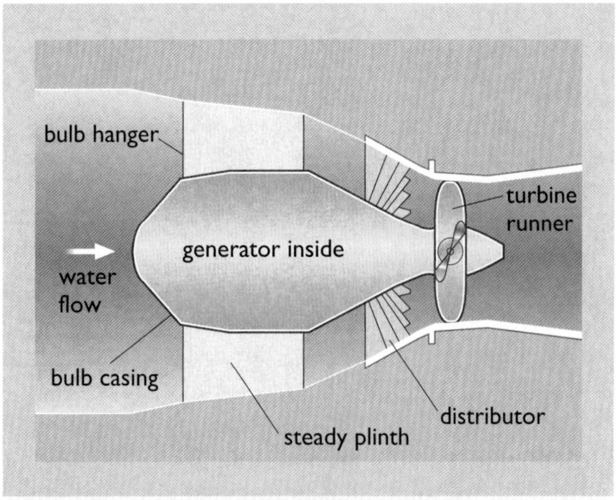

Figure 6.11 Bulb turbine as used at La Rance

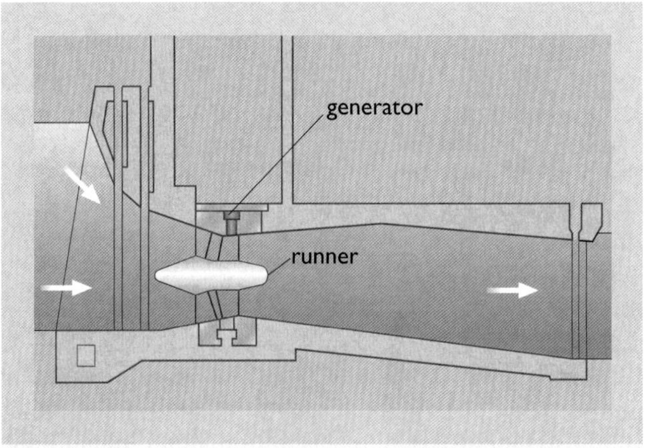

Figure 6.12 Straflo or rim generator turbine as used at Annapolis Royal

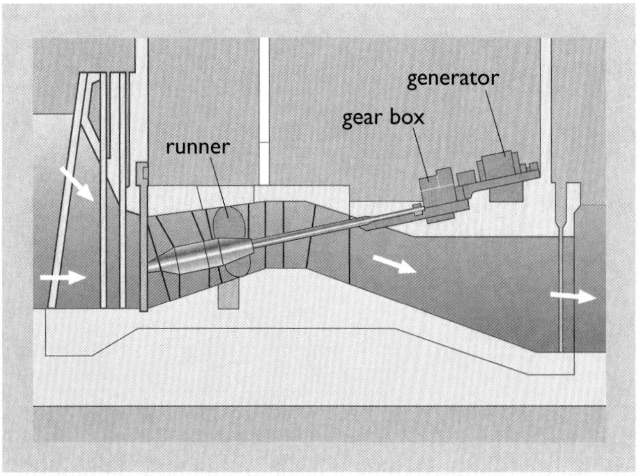

Figure 6.13 Tubular turbine

These problems are reduced in the **Straflo** or **rim generator** turbine (as used at Annapolis Royal in Canada), with the generator mounted radially around the rim and only the runner (that is, the turbine blades) in the flow (Figure 6.12). However, although it is more efficient than the bulb design because the water flow is not so constricted, this design introduces extra problems with the sealing between the runner blades and the radial generator.

Alternatively, there is the **tubular** turbine configuration, with the runner set at an angle so that a long (tubular) shaft can take rotational power out to an external generator (Figure 6.13). This design also avoids constricting the flow of the water, and, since the generator is not in a confined space, there is room for a gearbox, which can allow for efficient matching to the higher speed generators usually used with hydro plants. Several such units have been used in hydro power plants in the US, the largest being rated at 25 MW. However, there have been problems with vibration effects in the long drive shaft, and, so far, bulb turbines have proved to be the most popular with barrage designers.

As mentioned above, the rotational speeds of the turbines in tidal barrages tend to be lower than those for turbines in hydro plants (50–100 revolutions per minute as compared to about ten times this for typical hydro generators), and therefore wear is also reduced. But because large volumes of water have to pass through in a relatively short time, large numbers of turbines are required in a large-scale barrage. For example, the proposed ebb generation Severn Barrage would have 216 turbines, each rated at 40 MW, giving a total installed capacity of 8640 MW.

In simple ebb or flood generation, this large installed capacity is used only for a relatively short period (three to six hours at most) in each tidal cycle, producing a large but short burst of power, which may not match demand. Two-way operation using reversible-pitch turbines gives a more nearly continuous output, but these turbines are more complex and costly, and, although the output will be more evenly distributed in time, there will be a net decrease in power output for each phase compared with a simple ebb generation scheme. This is because, in order to be ready for the next cycle, neither the ebb nor the flow generation phases can be taken to completion: it is necessary to open the sluices and reduce water levels ready for the next flood cycle, and vice versa for ebb generation (see Figure 6.9). Furthermore, the turbine blade design cannot be optimized for flow in both directions, and efficiency must be compromised to achieve two-way operation.

Flood pumping is another option for power generation. Here, the turbine generators are run in reverse and act as motor-pump sets, powered by electricity from the grid. Additional water is thus pumped behind the barrage into the basin, to provide extra water for the subsequent ebb generation phase.

In addition, as will be discussed in more detail later, many different types of **double-basin** system have been proposed (see, for example, Figure 6.14), often using pumping between the basins. Excess power generated during periods of low demand by the turbines of the first basin can be used to pump water into the second basin, ready for the latter to use for generation when power is required.

Whatever the precise configuration chosen for a barrage, the basic components are the same: **turbines**, **sluice gates** and, usually, **ship locks**, to allow passage of ships, all linked to the shore by **embankments**.

The turbines are usually located in large concrete units. For the Severn Barrage, the use of large concrete (or steel) **caisson** structures to house the turbines has been proposed. These could be constructed on shore, in dry dock facilities, and then floated onto site and sunk into place

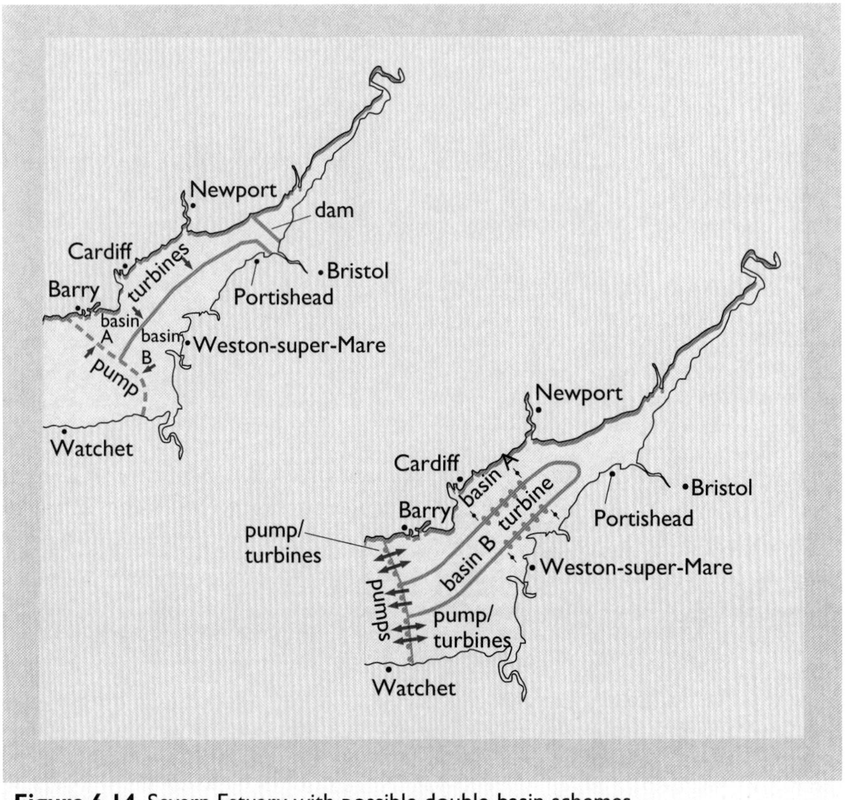

Figure 6.14 Severn Estuary with possible double-basin schemes

Sluice gates are another essential operational feature of a barrage, to allow the tide to flow through ready for ebb generation, or back out after flow generation. These can also be mounted in caissons.

The rest of a barrage is relatively straightforward to construct. La Rance, for example, has a rock-filled embankment, whilst the proposed Severn Barrage would use sand-filled embankments faced with suitable concrete or rock protection.

Details of the existing Rance Barrage and of the proposed Severn and Mersey barrages are given in Boxes 6.5, 6.6 and 6.7, respectively.

BOX 6.5 La Rance

The 740-metre long Rance Barrage was constructed between 1961 and 1967. It has a road crossing and a ship lock (see Figure 6.15) and was designed for maximum operational flexibility. It contains 24 reversible (that is, two-way) pump turbines (each of 10 MW capacity), operating in a tidal range of up to 12 metres, with a typical head of approximately 5 metres.

The operational pattern initially adopted at La Rance was to optimize the uniformity of the power output by using a combination of two-way generation (which meant running the turbines at less than the maximum possible head of water) and incorporating an element of pumped storage. For spring tides, two-way generation was favoured; for neap tides and some intermediate tides, direct pumping from sea to basin was sometimes carried out to supplement generation on the ebb.

Although some mechanical problems were encountered in 1975, which subsequently led to two-way operation mostly being avoided, overall the barrage has been very successful. Typically the plant has been functional and available for use more than 90% of the time, and net output has been approximately 480 GWh per year, with, in some years, significant availability gains from pumping.

The construction of the barrage involved building two temporary coffer-dams, with the water then being pumped out of the space in between to allow work to be carried out in dry conditions (see Figure 6.16). River water was allowed past via sluices, but the reduced ebb and flow resulted in effective stagnation of the estuary and the subsequent partial collapse of the ecosystem within it. Since construction, exchange of water between the open sea and the estuary has restored the estuarine ecosystem, but because there was no monitoring it is difficult to establish what changes the barrage caused to the original environment in the estuary.

Figure 6.16 La Rance coffer-dam during construction of the barrage

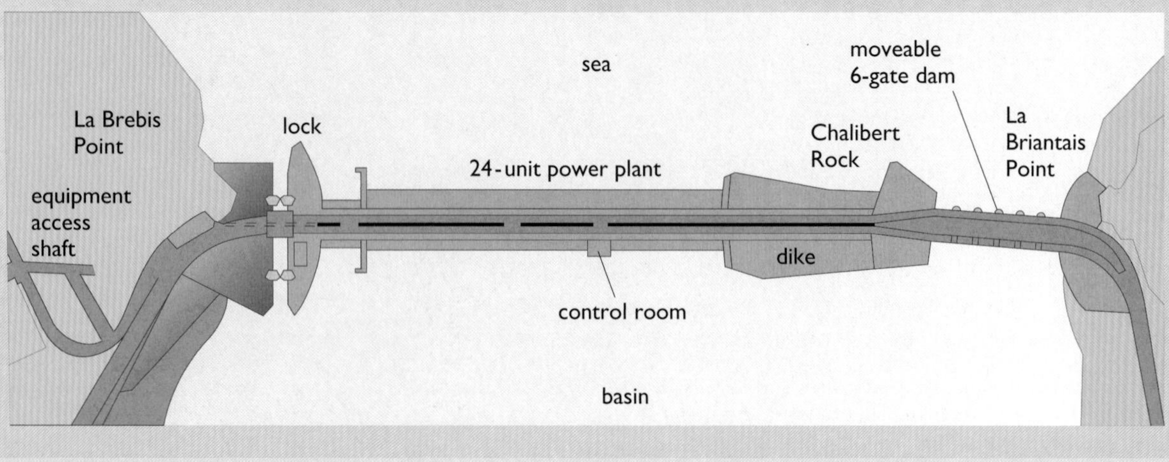

Figure 6.15 Layout of the Rance Barrage

BOX 6.6 The Severn Barrage 'reference project'

Basic data for the Severn Barrage 'reference project' are given below (Department of Energy, 1989).

Number of turbine generators:	216
Diameter of turbines:	9.0 metres
Operating speed of turbines:	50 rpm
Turbine generator rating:	40 MW
Installed capacity:	8640 MW
Number of sluices, various sizes:	166
Total clear area of sluice passages:	35 000 m^2
Average annual energy output:	17 TWh
Operational mode:	ebb generation with flood pumping
Length of barrage:	
total:	15.9 km
including: powerhouse caissons	4.3 km
sluice caissons	4.1 km
other caissons	3.9 km
embankments	3.6 km
Area of enclosed basin at mean sea level:	480 km^2
Construction cost, excluding public roads and grid strengthening (at April 1988 prices):	**£8280 million**

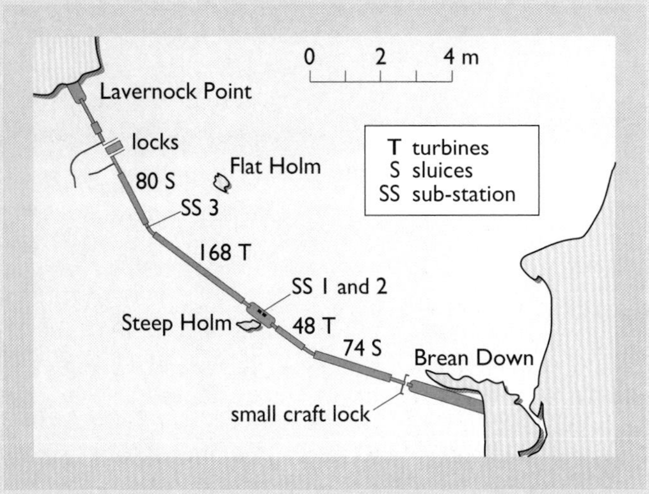

Figure 6.17 Layout of the proposed Severn Barrage

BOX 6.7 The proposed Mersey Barrage

As initially conceived by the Mersey Barrage Company, the station would have 700 MW installed capacity comprising twenty-eight 25 MW turbines and 20 sluices. The preferred method of construction would be *in situ* construction of the New Ferry Lock, and caisson construction for turbine and sluice housings, either *in situ* or off-site (in a dry dock) for the Dingle Lock.

Construction cost (at July 1989 prices) was estimated to be £880 million.

The construction period would be about five years following parliamentary approval and detailed design.

The barrage would generate 1.4 TWh of electricity per annum throughout its operational life of at least 120 years.

A revised configuration subsequently reduced the estimated cost to £847 million (at October 1992 prices) and increased the output to 1.45 TWh y^{-1}.

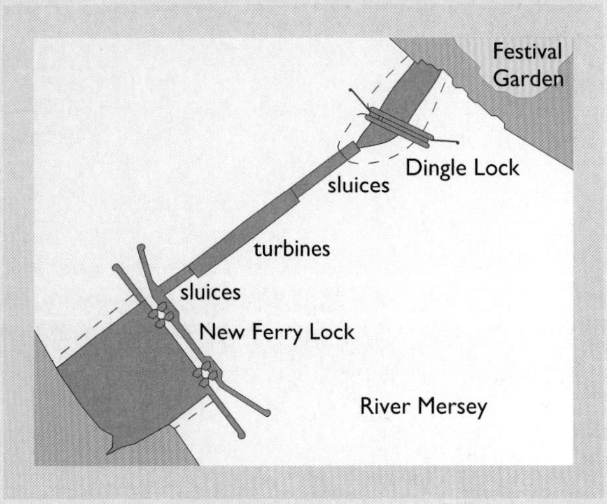

Figure 6.18 Layout of the proposed Mersey Barrage, in Liverpool, UK

6.3 Environmental factors

The construction of a large barrier across an estuary will clearly have a significant effect on the local ecosystem. Some of the effects will be negative, and some will be positive. Much research has gone into trying to ascertain the probable final outcome, focusing mainly on the proposed Severn Tidal Barrage.

The Department of Energy (1987) summarized the potential environmental effects of the Severn scheme as follows:

> The construction of a barrage would result in higher minimum water levels and slightly lower high water levels in the basin. Currents will be reduced and extreme wave conditions will, in many places, be less severe. The changes that will occur to the tides and currents during construction and then later during the operation of a barrage will cause changes in sediment characteristics and in the salinity and quality of the water. These factors have a major bearing on the estuary's environment and ecology.

The most obvious potential impact of any barrage would be on local wildlife, that is, fish and birds, many of the latter being migratory. The UK's estuaries play host to approximately 28% of European swans and ducks and to 47% of European geese. There are also large populations of fish: the Severn, for example, is well known for its salmon and eels (elvers). Many of these species rely on the estuaries for food, and access to that supply might be affected by a tidal barrage.

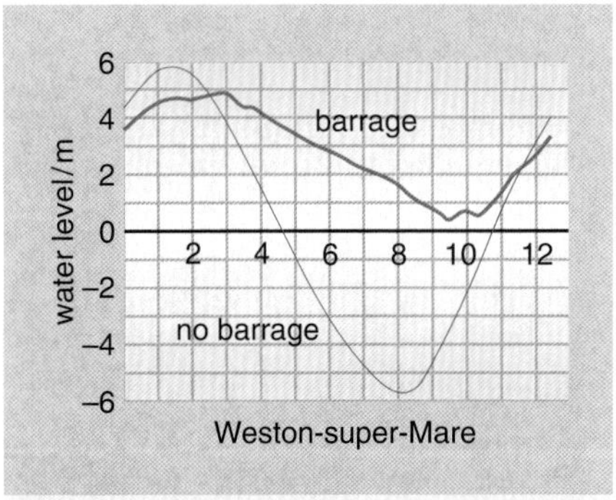

Figure 6.19 Tide level curves with and without the Severn Barrage (Source: Department of Energy, 1989)

The proposed Severn Barrage would decrease the currently large area (200 km^2 or more) of mud flats exposed each day, since the water level variations behind the barrage would be significantly reduced (see Figure 6.19). Some species (for example, mud-wading birds) feed on worms and other invertebrates from the exposed mud flats, and could be adversely affected.

However, the barrage could have a compensating impact on the level of silt and sediment suspended in the water. The waters in the Severn Estuary currently carry in suspension much silt churned up by the tides, making the water impenetrable to sunlight. With the barrage in place and the tidal ebbs and flows reduced, some of this silt would drop out, making the water clearer. Given this change in **turbidity**, sunlight would penetrate further down, increasing the biological productivity of the water, and therefore increasing the potential food supplies for fish and birds. The net impact for some species might thus be positive: some might not find a niche in the new ecological balance, whilst others previously excluded from the estuary might become established. Similar issues would apply for salt marshes that might be exposed daily by the tides at other potential barrage sites.

This rather simplified example illustrates the general point that there are complex interactions at work, which often make it difficult to predict the net outcome. For a more detailed analysis, see Box 6.8.

Similar interactions and trade-offs occur in relation to other ways in which barrages can impact on their surroundings. Clearly, the construction of a

BOX 6.8 Environmental impact

In a review of tidal power issues, Baker (1991b) identifies the most important environmental issues as probably being water quality and sediment movements, 'because these govern to a large extent the ecology of the estuary'. Baker continues:

> One unavoidable change in water quality behind a barrage is that, in the upper estuary away from the immediate effects of the barrage, salinity will reduce as a result of the reduction in the volume of sea water entering the estuary each tide. Thus freshwater species will extend their domain seawards, while the brackish water zone, which tends to be relatively impoverished, will move downstream.
>
> In the basin behind and clear of a barrage, tidal current velocities will be reduced, particularly during the ebb. This will have a major effect on the power of the currents to erode and transport sediments. The effect will be a general 'freezing' of sediments which are normally mobile, especially during spring tides. This will in turn reduce the turbidity of the water and provide a more stable regime for organisms which live in muddy deposits. One result could be an increase in invertebrate populations, which would benefit wading birds.

The Severn Barrage

The Severn project has so far been the main focus of study in the UK, although there have also been studies on the Mersey.

The Severn Barrage Project: General Report (Department of Energy, 1989) includes summaries of some of the key ecological and conservation problems. It indicates that whilst the levels of micro-organisms (phytoplankton, etc.) and even inter-tidal plants are relatively low as a result of the harsh hydro-dynamic and sedimentary regime in the area around the proposed barrage (with high turbidity being a major feature), there are significant populations of invertebrates, and therefore of fish and birds which feed off them. The area is a major site for waders and shelduck, there being four overwintering bird species of international importance and two species of national importance. Parts of the estuary are protected as 'Sites of Special Scientific Interest', and there are also a number of other areas with potential protected status.

The above report identifies a number of opportunities for 'ameliorative and creative conservation measures', ranging from the relatively straightforward provision of wildlife sanctuaries, coastal lagoons and protected intertidal areas, to the more complex provision of a quasi-tidal regime for those plant and animal species dependent on tidal rhythms. The report concludes 'The continuation of a formal dialogue between the development and conservation interests is recommended in order to achieve a nature conservation strategy'.

Comparisons with other schemes

Whilst it is, of course, important to remember that the impacts of any barrage will be site-specific, it is informative to examine the case of the La Rance barrage in France.

No detailed preliminary environmental studies were done, but as noted earlier (Box 6.5), the construction phase caused serious dislocation, and the operator (Electricité de France) reported that initially the barrage had a significant effect on the estuary's biological productivity. However, gradually most flora and fauna returned, and some new species appeared. The fishing is now said to be better and there are more migratory birds. The effects on sedimentation are said to be low.

If the turbines at La Rance are brought up to speed rapidly, strong undercurrents and waves along the shore can be generated, which can present hazards to leisure craft (for example). Otherwise, though, the operation of the barrage seems unproblematic. Leisure sailing has increased, and to aid navigation the barrage operator publishes details of operational plans three days in advance.

Damage to fish passing through the turbines or sluices is another potential problem (as it is for hydroelectric plants, see Chapter 5), and while there has been no evidence of damage at La Rance, there has been significant mortality to shad (a local fish) at Annapolis Royal in Canada. Consequently, a sonic generator which produces one-second 'hammer blow' pulses has been installed in order to warn the fish. It appears to be effective. A special fish pass has also been installed, as with some run-of-river hydro schemes, and as was proposed in the plans for the Severn Barrage.

Mersey Barrage

Initial studies of the proposed Mersey Barrage indicated that water quality should not deteriorate to affect flora or fauna, but that the barrage could affect the habitats of birds around the estuary, although such effects could be 'considerably offset as the design and operations regime of the barrage is developed'. Although siltation might increase, it was felt that the barrage should not significantly affect fish life in the river.

See *Tidal Power from the River Mersey: A Feasibility Study Stage 1* (ETSU, 1988) and subsequent Mersey Barrage Company (MBC)/Energy Technology Support Unit (ETSU) reports.

barrage across an estuary will impede any shipping, even though ship locks are likely to be included; however, the fact that the sea level behind the barrage would on average be higher could improve navigational access to ports, the net effect depending on tidal cycles and the precise location of the barrage and of any ports.

Visually, barrages present fewer problems than comparable hydro schemes. Even at low tide, the flank exposed would not be much higher than the maximum tidal range. From a distance, all that would be seen would be a line on the water.

Barrages could also play a useful role in providing protection against floods and storm damage, since they could be operated to control very high tidal surges and limit local wave generation. However, for some sites, due to the changed tidal patterns (with the tide upstream staying above mid-level for longer periods), there might be a need for improved land drainage upstream.

A barrage would, of course, have some effect on the local economy, both during the construction phase and subsequently, in terms of employment generation and local spending, tourism and, in particular, enhanced opportunities for water sports. Depending on the scale and the site, there could also be the option of providing a new road or rail crossing, as has happened with the Rance Barrage. The incorporation of public roads was part of the plan proposed for both the Severn and the Mersey barrages.

Whether these local infrastructural improvement options represent environmental benefits or costs depends, of course, on your views on industrial and commercial development (some conservation and wildlife groups, for example, baulk at the prospect of increased tourism), but many people would be likely to welcome local economic growth. Indeed, that was the message from local populations faced with barrage proposals. Whilst a minority of special interest groups, including conservationists and preservationists, have opposed barrages, local commercial and civic interests and the wider public have on the whole been supportive of such plans (see Box 6.9).

The environmental case against barrages is, however, a far from trivial one. For example, when the Severn Tidal Power Group (STPG) proposal was debated in the 1980s, it was opposed by the Royal Society for the Protection of Birds, which sees barrages as inherently damaging, reducing habitats for key species, particularly migrant birds. This problem could clearly be compounded if several barrages were to be built.

The debate became quite heated at times, with some national and local environmental pressure groups coming out strongly against barrages. Friends of the Earth, for example, argued that whilst each project should be assessed on its merits (see Box 6.9), in general large barrages are likely to have a negative net impact, and the organization opposed the proposed Severn and Mersey barrages.

Some of the opposition described above relates not so much to specific points about the potential environmental impact as to more general strategic questions relating to 'opportunity cost' aspects. For example, Friends of the Earth argues that any money spent on large barrages would be better spent on energy conservation.

BOX 6.9 Some reactions to UK barrage proposals

Reactions to hypothetical schemes, rather than actual projects, are, of course, somewhat unreliable, but the rather drawn-out and intermittent public debate over the Severn Barrage, dating from the 1981 Bondi Report (Department of Energy, 1981) onwards, has thrown up some interesting responses.

Although the Severn Tidal Power Group (STPG) carried out a consultation exercise (Department of Energy, 1991) which reviewed responses from some 300 individuals and organizations in 1990/91, in general the STPG industrial consortium has adopted a relatively low public profile, leaving the field relatively free for objectors to express their views. Some objections have been quite harsh, with some even expressing a fear of the creation of a 'stinking lake'.

For example, during a parliamentary debate on the Severn Tidal Barrage in October 1987, Michael Stern, MP for Bristol North West, said:

> The Severn Barrage has been compared to a large, inefficient, activated sewage treatment plant. It operates by aeration and agitation by wind and tidal coverage, and variation of that agitation from violent on stormy days and spring tides to relatively mild on neap tides and during calm weather. When the waters of the Severn Estuary are quiescent, large pools of liquid mud with a high affinity for pollutants settle in the estuary.
>
> One effect of the interposition of a barrage in the estuary will be greatly to increase the frequency and depth of those pools of liquid mud. That in itself may not be too much of a worry, but the pools have an affinity for pollutants, and if they remain anaerobic for 14 days or more because of the process of stratification – they will undoubtedly do so once the tidal range of the Severn is cut off – large amounts of toxic, evil-smelling gases such as hydrogen sulphide, methane and other reduced sulphur compounds will begin to be emitted. Imagine the extension of that process over a period of years, and having to live anywhere along the shores of the Severn Estuary with the build-up of these vast smells which are an inevitable risk when the tide is cut off in an area that has been subject to one of the greatest tidal ranges in the world.

Although these anxieties may be exaggerated, others feared that the quiet retirement town of Weston-super-Mare might be disrupted, not least as a consequence of the presence, albeit only temporary, of the tens of thousands of construction workers needed. The STPG has estimated the peak number of workers at 35 000, although not all of these would be in or near Weston.

On balance, however, on the English side most local opinion seemed to favour the project, often on the grounds of local economic and indirect employment benefits (tourism, etc.), but also as an alternative to the construction of more nuclear power plants. This view is reinforced, for some, by Weston's proximity to the nuclear site at Hinkley. On the Welsh side, there was also strong support on the basis of employment and economic gains, although some nationalists felt that the project was a scheme for the benefit of English power consumers only.

The main objections, however, came from environmental and wildlife groups, nationally and locally. The Royal Society for the Protection of Birds strongly opposed barrages on any estuaries. Friends of the Earth, striving to balance, on the one hand, its commitment to renewable energy and, on the other, its concern for environmental protection, suggested that the onus of proof of net benefit should be placed on intending developers. However, on balance it remained unconvinced that there was a case for large schemes like the Severn or Mersey barrages, though it left the door open for smaller schemes.

This opposition is partly motivated by strategic issues. Many environmentalists, for example, felt that, in general, smaller projects are preferable, not just for environmental reasons, but because they might provide an opportunity for a more decentralized form of energy production. Although barrages are portrayed as inevitably centralized, in reality some smaller barrages might be owned and managed by regional or even local bodies, including municipalities.

While, as noted above, the STPG adopted a fairly low-profile approach with regard to public consultation, the Mersey Barrage Company attempted to involve local environmental pressure groups in a negotiation process from the start. However, in spite of this some strong objections emerged, with wildlife protection groups concerned that the barrage would endanger large areas of salt marsh, which are of international importance to wintering waders and other birds. Shipping interests also expressed concern, as did some local industrialists concerned about maintaining access.

In the end, should barrage proposals proceed in the UK, what is likely to emerge (assuming outright opposition is avoided) are compromises over siting, design and operation. For example, as we shall see, there are operational patterns that could allow larger areas of mud flats to be exposed for longer periods, to the benefit of some species, albeit perhaps with some cost penalties. There are also other, non-barrage, costs. For example, as the quotation from the speech by Michael Stern MP in Box 6.9 highlights, it would be vital to invest in cleaning up sewage plant outflow and the reduction of emissions from industrial plants, since the tides could no longer be relied on to flush the estuaries so vigorously. Although there would still be significant flushing effects (for instance, the Severn Barrage would involve a water exchange of 2.08 cubic kilometres twice a day), most proposals so far have included plans for investing in new plants to clean up local pollution emissions, and the fact that current waste disposal policies are inadequate should not be seen as an argument against barrages. Rather, the advent of barrages could help to ensure that emissions into estuaries are cleaned up. Indeed, investment in cleaning up is occurring in any case, to meet rising European Union standards.

A final point on scale should be made. Barrages are inevitably fairly large structures, but some environmentalists argue that smaller barrages are preferable to large ones, in that they might have less environmental impact. That could clearly be true on a *pro-rata* basis for individual barrages, but the net effect of a lot of small units compared with, say, one large unit, is far from clear, with the site-specific nature of impacts complicating any general analysis. To take the Severn as an example, there has been some support for a smaller barrage on the English Stones line, near the existing Severn Bridge. This would be cheaper to build and could also provide a base for another river crossing (motorway or rail). However, initial studies indicated that at that site, further up the river, siltation could be a major problem, and the scale of the local environmental impact has yet to be ascertained.

Clearly, whether we are talking about small or large barrages, the environmental questions must be treated seriously. In the end, however, it would seem to come down to a matter of strategic choice. Assuming that there are significant negative local impacts, these would have to be weighed up against the role that barrages can play in resolving some of our global environmental problems, such as global warming caused by CO_2 emissions from the burning of fossil fuels.

The Severn Barrage Project: General Report summed up this point as follows:

> If renewable energy sources are to be utilized to increase diversification of electricity generation and to reduce pollution, the Severn Barrage remains the largest single project which could make a significant contribution on a reasonable time scale. Once completed, the project would represent an insurance against escalation of fuel prices.
>
> (Department of Energy, 1989)

6.4 Integration

The electricity produced by barrages must usually be integrated with the electricity produced by the other power plants that feed into the national grid power transmission network.

The key problem in feeding power from a tidal barrage into national grid networks is that with conventional ebb or flow generation schemes the tidal energy inputs come in relatively short bursts at approximately twelve-hour intervals. Typically, power can be produced for five to six hours during spring tides and three hours during neap tides, within a tidal cycle lasting 12.4 hours (Figure 6.20).

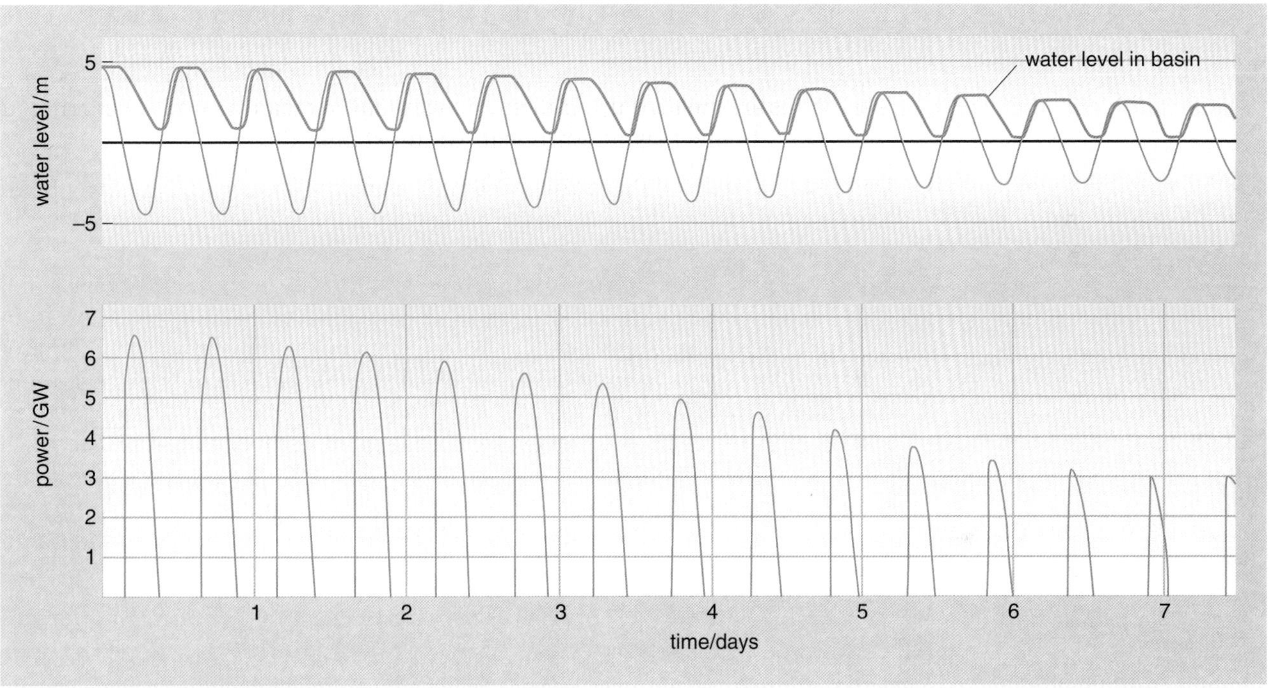

Figure 6.20 Water level and power output of the proposed Severn Barrage over a spring–neap tide cycle (Source: Watt Committee on Energy, 1990)

Clearly, such availability of power from a barrage will not always match the pattern of demand for power on the grid. With a large, well-developed grid, as in the UK, with a large load and many other types of power plant connected into it, this problem might not be too severe, depending, of course, on the size of the barrage output. The two daily bursts of tidal power could be used to reduce the load on older, less efficient and/or more expensive 'low merit-order' plants, for example older coal-fired plants. The barrage would thus be operating in a 'fuel-saving' mode, the predictability of the tides allowing for the process of substitution to be planned well in advance. Even so, with a barrage the size of that proposed for the Severn, absorbing all the power would clearly represent a significant task. At peak, it would be generating over 8 GW of power, which represents a sizeable proportion of the UK's total installed (in 2000) capacity of approximately 75 GW.

In some other countries, with smaller grids and relatively low demands, weaker interconnections, longer transmission distances and relatively few other types of generating plant, a large burst of tidal power input could represent a major integration problem. In this situation, the option of two-way generation, on the incoming flood as well as the ebb, appears more attractive, since that would give *four* bursts of power for each 24-hour cycle. However, as was noted earlier, although the phasing of the power is improved in two-way generation, thus potentially matching to loads more effectively, there is in fact a slight net loss of power as compared to one-way generation, since, in order to be ready for the reversed flow, neither the ebb nor the flood generation phase can be taken to completion. The installation of the more complex two-way turbines also involves extra cost and it is difficult to optimize the design of the turbines for two-way operation. The end result is that two-way generation could cost 10–15% more, and, in the case of the Severn Barrage, the overall benefit was seen as negligible.

There is also the option of using modified turbines run on grid power (e.g. during off-peak periods) to pump extra water behind a barrage during the flood phase. This could, to a limited extent, help to re-phase the power output, the extra head and extra volume of water helping to increase the subsequent ebb generation period. The barrage would thus be used as a short-term pumped storage reservoir. In addition, a bonus would be obtained by pumping at low head during low tide, since the incoming tide would in effect lift the extra water to high-tide level, ready for subsequently generating power on the ebb. Typically, between 5 and 15% extra output can be gained, with little additional capital cost and with no loss of generating efficiency. This was the favoured option for all proposed UK tidal energy schemes (Severn, Mersey, Conwy, Wyre) and, as we have seen, is carried out on the Rance Barrage. That said, the overall *economic* advantages of pumping may be fairly small (you have to pay for the electricity used), and depend crucially on tidal timing, which will not necessarily coincide with off-peak grid power availability.

Another option we noted earlier for providing more nearly continuous power is to construct two basins (see Figure 6.14). In one version of this concept, one basin is filled at high tide, whilst the other is kept low (by emptying it regularly at low tide level), so that the difference in levels between the two is maximized and the power output from turbines running from the high to the low basin can be more nearly continuous. Another possibility is to have two separate basins, each with their own set of turbines. The first operates on ebb generation, whilst the second acts essentially as a reservoir, filled at high tide and used for generation only when no ebb power from the first basin is available – so there are two reservoirs operating in sequence, the second one only being emptied after the first one has been emptied.

There are many other double-basin concepts, with most of them using (modified) turbines to pump water 'uphill'. For example, the first basin could operate in the normal manner, but any excess power, during low demand periods, could be used to pump water out of the second basin to keep it below tide level, so that power could be generated from the second basin when needed by filling it through its turbines. However, with or without pumping, these configurations all involve building what amounts to two barrages, which increases the overall cost significantly. To date, no double-barrage systems have been seriously considered, simple ebb generation being seen as the most economic option overall.

In addition to these basic design options, there is, of course, a variety of adjustments that can be made to detailed design (for example, optimizing the number, location and size of the turbines and sluices) to give maximum efficiency, as well as ways of optimizing operational performance. For example, ebb generation schemes usually start up when the difference in level between the basin and the sea is approximately half the tidal range, with generation continuing until the levels are equalized, when the next tide is allowed in. However, other patterns are possible, taking the local tidal patterns into account.

Clearly, as we have seen in Section 6.3, the system design and operation pattern will have environmental as well as integration and economic implications: for example, they will define the duration of the low sea level behind the barrage. From an environmental point of view, flood generation is seen as less attractive than ebb generation, partly because the estuary behind the barrage would be kept at a low level for longer periods (as power was absorbed from the incoming flow), leaving the mud flats or salt marshes uncovered for longer periods. In contrast, double-basin schemes would probably be left at near, or even above, high tide level overnight, since this is a low demand period. Either way, wildlife feeding patterns and food supplies would be affected.

With two-way generation, the water in the basins remains closer to mean sea level. Even so, where navigation must be maintained to ports upstream of a barrage (for example, on the Severn and the Mersey), two-way generation is not likely to be acceptable because there would still be a reduced high water level within the basin. This applies even more to the case of flood generation.

Whichever system is used, the overall economic viability of tidal power might be enhanced, in theory at least, if several barrages were in operation, given the fact that the tidal maxima occur at slightly different times around the coast. For example, the Solway Firth and Morecambe Bay are approximately five to six hours out of phase with the Severn, so that the output from these and other possible barrage sites could be fed into the grid to provide a more nearly continuous net contribution, although of course at neap tides this input would be low.

Finally, the linking of power from barrages to the national grid could present some practical problems. As with many other types of large, new, power plant, extra grid connections would have to be made, and in some circumstances existing local grid lines strengthened to carry the extra power. Fortunately, most potential barrage sites in the UK are reasonably near to existing power lines, so problems such as those of implementing deep-sea wave power (much of which would have to be transmitted from the north of Scotland; see Chapter 8) would be avoided. However, in the case of the proposed Severn Barrage new power lines would probably be required. These extra grid connections would add to the total system cost. For the Severn Barrage, these costs have been estimated at some £850 million, out of a total barrage cost of £8280 million at 1988 prices (see Table 6.1 in the next section), although the STPG has argued that some grid strengthening in the area is needed anyway.

As can be seen, the key issue in terms of integration is *cost*, whether this is in terms of basic grid linkages or in terms of systems for allowing power to be produced on a more nearly continuous basis.

The discussion above is based on the assumption that the UK's existing national power system remains basically unchanged, with single or multiple barrage inputs simply fed into the grid. However, if renewable energy systems of various types were to proliferate, with the problem of intermittent inputs therefore growing, it could be that the pumped storage capacity that barrages potentially represent would become more attractive economically. The pumped storage option has not been followed up fully because there are disagreements about the value of this option, a key issue being that 'carrying' an extra head of (stored) water beyond the next ebb phase would interfere with the normal cycle of operations. But as already mentioned, pumping to extend the ebb can make sense, particularly if cheap off-peak electricity can be used and the stored power employed to meet demand peaks. However, the tidal cycle will shift in and out of phase with the off-peak electricity supply and consumer demand cycle. Overall, it could be argued, that if longer-term storage is wanted, it is better to build a separate conventional, high-head pumped storage reservoir on land, like the 1800 MW Dinorwig scheme in Wales. It is worth noting, though, that even though barrages are low head, in principle it is possible to store a lot of water behind one, possibly much more cheaply than in excavated underground spaces like Dinowrig.

In addition it is possible that there might be other forms of synergy between tidal and other renewable systems, for example the installation of wind turbines along barrages, in the same way as some wind farms have been constructed on causeways in harbours. The wind plant might even be used at times for pumping water behind the barrage, although the energy contribution that could be made, even if wind turbines were located regularly along the entire length of a barrage, would be relatively small compared with the output of the barrage. For example, if, say, thirty 2-MW wind turbines were installed along parts of the 16-km long Severn Barrage, their total annual electricity output would be just over 1% of that from the barrage.

For a useful discussion of the integration issue in relation to tidal energy, see *Renewable Energy Sources* (Watt Committee on Energy, 1990) and *Tidal Power from the Severn Estuary, Volume I* (Department of Energy, 1981).

6.5 Economic factors

The overall economics of tidal barrages depends both on their operational performance and on their initial capital cost. Table 6.1 gives a breakdown of the STPG's 1988 estimates of the capital and running costs for the Severn Barrage (Department of Energy, 1989). Note that civil engineering works are the single largest element in the total cost, closely followed by turbine manufacture and installation. As noted at the foot of Table 6.1, in 2002 the total estimated capital cost of the Severn Barrage, which had been put at £8280 million in 1988, was updated to £10–15 billion. These estimates exclude interest on borrowed capital accrued during the course of construction, although interest *is* taken into account in the calculation of

Table 6.1 Severn Barrage 'reference project' capital and annual recurring costs

	Cost / £ million[1]
Pre-construction phase	
Feasibility and environmental studies, planning and parliamentary costs	60
Design and engineering	130
Barrage construction	
Civil engineering works[2]	4900
Power generation works	2400
On-barrage transmission and control	380
Management, engineering and supervision	300
Land and urban drainage sea defences	30
Effluent discharge, port works	80
Barrage capital cost total	**8280**
Off-barrage transmission and grid reinforcement	
With all transmission lines overhead	850
Extra cost for 10% of transmission lines underground	380
Annual costs	
Barrage operation and maintenance	40
Off-barrage costs	30
Annual costs total	**70**

1 The costs presented in Table 6.1, and in Box 6.6 earlier, are in 1988 money terms. In 1991 these estimates were updated, in line with inflation, with the total being put at over £10 billion. More recently an STPG report *The Severn Barrage – Definition Study for a New Appraisal of the Project*, commissioned by DTI to feed into the 2002 Energy Review (ETSU, 2002), concluded that the Barrage would cost between £10–15bn at current prices.

2 Excluding public road across the barrage (estimated to cost from £135 million to £207 million depending on links provided into road network).

Source: Department of Energy (1989)

generation costs given in Table 6.2. This raises a key issue for the economics of large capital-intensive projects like this which have long construction times. The timescales involved mean that some of the interest on the capital borrowed to pay for building has to be paid before any money is recouped in the operation.

In addition to the significant initial capital costs, the period during which power can be produced, at least for simple single-basin ebb or flow systems, is clearly less than for a conventional power plant. For example, because it would only operate during tidal cycles, the 8.6 GW turbine capacity of the Severn Barrage could only offer the same output, averaged out over year, as conventional plant with around 2 GW of generating capacity. In other words, the barrage requires a large investment in expensive capacity which is only used intermittently and can therefore only replace a limited amount of conventional plant output. The precise 'capacity credit' that can be attributed to barrages (in effect, their value as replacements for conventional plants) will depend in practice on the scale and timing of the outputs of

Table 6.2 Severn Barrage generation costs (pence per kilowatt-hour)[1]

Discount rate	2%	5%	8%	10%	12%	15%
Basic capital[2]	1.05	2.25	3.42	4.15	4.83	5.77
Time component[3]	0.11	0.63	1.64	2.60	3.82	6.18
Total capital cost	1.16	2.88	5.06	6.75	8.65	11.95
Annual cost	0.53	0.50	0.47	0.47	0.46	0.46
Total cost of energy at barrage boundary	1.70	3.37	5.53	7.22	9.12	12.42
Off-barrage transmission (all overhead lines)	0.12	0.28	0.49	0.64	0.80	1.08
Total cost of energy including overhead transmission lines	1.81	3.65	6.01	7.85	9.92	13.50
Extra cost for 10% of transmission lines being underground	0.05	0.11	0.18	0.23	0.28	0.36

1 In 1988 money terms. Figures rounded.

2 On the basis that all the capital cost is incurred instantaneously at the time when capital spending ceases.

3 Extra cost arising from the actual timing of the incidence of capital expenditure (equivalent to interest during construction).

Source: Department of Energy (1989)

the plants they can off-load, not all of which will be able to generate continuously either. The most convenient way to compare systems is therefore by using their **load factors** – in effect the percentage of time the plant can deliver power. The average annual load factor for the Severn Barrage is estimated at around 23%. By comparison, the average annual plant load factor for nuclear stations is currently around 77%, and for combined cycle gas turbines it is around 84%.

Thus, compared with most other types of power plant, tidal projects have a relatively high capital cost in relation to the usable output, with, consequently, long capital payback times and low rates of return on the capital invested, the precise figures depending on the price that can be charged for the electricity.

For the Severn Barrage, the STPG has estimated that, depending on the price at which the electricity could be sold, the internal rates of return on the large capital outlay would be around 6–8% (1988 figures). This is unlikely to be attractive to private sector investors, which will expect much higher rates of return. To put it another way, if commercial rates of return were expected, the price charged for electricity would have to be substantially higher than is currently considered acceptable (see Table 6.2 and Figure 6.21).

An estimate by the UK Government's Renewable Energy Advisory Group put the cost of electricity from the Severn Barrage at 5–7 pence per kilowatt-hour (p per kWh) at an 8% discount rate, or 10–14p per kWh at a 15% discount rate, both at 1991 prices (Renewable Energy Advisory Group, 1992).

Another perspective is given by Table 6.3, based on evidence given to the House of Commons Select Committee on Energy in 1991/2. This is a comparison of costs of electricity and capital repayment times for three UK schemes. The capital payback times in Table 6.3 are the number of years that it would take to pay back, from sales of electricity at the stated prices, the full capital cost, plus interest charges on the borrowed capital, including interest accrued during construction. As can be seen, payback times can be 20 years or more, with the smaller barrages, which generally have higher generation costs, having to charge higher electricity prices.

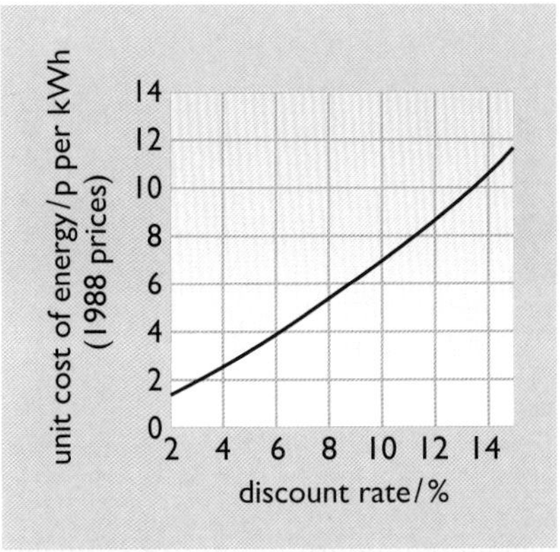

Figure 6.21 Severn Barrage electricity cost versus discount rate (Source: Department of Energy, 1989)

Of course, once the capital and interest costs had been paid off, a tidal barrage would be generating profit for the rest of its life (at least 100 years), apart from the relatively small operation and maintenance costs. Although, as Tables 6.1 and 6.2 indicate, there are running costs (approximately 1% of the total capital cost), tidal barrages, like all renewable energy systems that are based on natural flows, have no fuel cost. After initial construction, they can generate power for many years without major civil engineering effort, the low-speed turbines needing replacement perhaps only every 30 years.

Table 6.3 Tidal power electricity cost comparisons and capital repayment periods (November 1991 estimates)

	Capital cost/ £ million	Running costs per annum/ £ million	Electricity price/ pence per kWh
Severn (17 TWh y^{-1})	10 200	86	6p (16.5-year payback) 5p (20-year payback)
Mersey (1.4 TWh y^{-1})	966	17.6	6.75p (25-year payback)
Conwy	72.5	0.6	8.6p (15–20-year payback)

Source: House of Commons Select Committee on Energy (1992)

It is therefore sometimes argued that it is misleading to try to assess barrages just on a conventional 'discount rate' basis over a financial project period, which may be relatively short compared with the life of the barrage. For example, the House of Commons Select Committee on Energy, when reviewing tidal power as part of the review of renewable energy that it carried out in 1991/2 (House of Commons Select Committee on Energy, 1992), was keen for a 'total life cost' approach to be adopted, at least for use in comparisons. In this approach, the capital cost would be averaged out over a plant's lifetime.

On this sort of basis, capital projects with very long lives, like barrages or hydroelectric plants, appear to be very good long-term investments. The average cost of the Mersey Barrage if spread over its lifetime, for example, works out at only about 2p per kWh. The STPG has argued that, given reasonable maintenance and regular turbine replacements, the Severn Barrage could, in fact, last almost indefinitely.

However, given the 'short-termism' of many UK financial institutions, such arguments are unlikely to appeal to the private sector, which normally takes the view that the maximum loan period for large-scale industrial investments should not exceed 20 years. It is usually argued that instead the state should invest in long-term projects of this sort, for the long-term benefit of the nation as a whole.

The benefits that could flow from the construction of tidal barrages are more than just the provision of relatively cheap power: they include environmental benefits, for example reduction in carbon dioxide emissions, diversity, security and sustainability of supply, local and regional employment gains and, in some cases, possible new road or rail transport crossings.

However, to date, the UK Government has been unwilling to provide financial support for tidal barrage construction, preferring, as a matter of policy, to leave such developments to the private sector, although around £14 million was allocated for background research and feasibility studies during the 1980s and 1990s.

With the privatization of the UK electricity industry in 1990, and the liberalization of the electricity market, the emphasis, in terms of new construction projects, has shifted to smaller power plants, notably combined cycle gas turbines. In this context, smaller tidal schemes such as the Mersey Barrage (rated at 700 MW) or even the Wyre Barrage (64 MW) might be more likely to be viable, because of their lower capital cost (less than £1 billion at 1991 prices for the Mersey; £90 million for the Wyre), at least part of which could conceivably be raised from private finance.

The Mersey Barrage Company (MBC) did try to make use of the UK Government's Non-Fossil Fuel Obligation (NFFO) cross-subsidy scheme, which was launched in 1990. Under this scheme a levy on fossil-fuelled power generation was used to subsidize non-fossil based generation. In 1991, the MBC proposed that it should be paid a 6.75p per kWh unit cost for 25 years, of which 2p per kWh for 25 years at zero discount rate would be paid 'up front' as an advance during construction, through the NFFO cross-subsidy. This advance would be equivalent to over £700 million of the £1 billion construction cost. To meet part of the remaining £300 million cost, the MBC proposed an electricity price of 4.75p per kWh over 25 years. However, this would require a continuing NFFO payment, since the pool (wholesale) price for electricity at that point was around 3p per kWh (MBC, 1991). Subsequently MBC proposed some alternative approaches avoiding an advance payment, but inevitably this involved an even higher NFFO payment, e.g. 11p per kWh over 7.5 years (Haws and McCormick, 1993). The government at the time felt that overall the project was too costly and decided not to support further development work.

Subsequently, the NFFO scheme was phased out and from 2002 was replaced by the Renewables Obligation, which requires electricity supply companies to obtain a growing proportion of their power from renewable energy sources. However this has, in effect, a 5p per kWh price ceiling (approximately 3p per kWh greater than average wholesale electricity prices in 2002) and so far no barrage project has obtained support under this scheme.

Approaching the problem from a somewhat different angle, in 1993 the STPG submitted a funding proposal based on the idea of joint private–

public support, possibly using a 'Hybrid Bill' (Wardle *et al.*, 1993). This proposal was also turned down, and, according to the STPG, the option of 'going it alone' with just private finance was 'a trifle unlikely' (*Electrical Review*, 15–28 April 1994).

Since then, the STPG has continued to make regular attempts to gain support for the Severn Barrage. For example, the STPG has argued that the barrage could meet 60% of the UK Government's target of obtaining 10% of the UK's electricity from renewables by 2010. However, in 1999, the Government refused funding for a new appraisal, arguing that tidal power was not likely to become economically viable within the next 20 years and that there was no suitable buyer for the amount of electricity that would be generated (RCEP, 2000).

In the early 2000s, private–public funding initiatives became increasingly popular in government circles, and this might still offer one way forward in the future, but, so far, the STPG has not managed to find a way to fund the project.

As can be seen, the economics of tidal power is complex. Quite apart from the problems of funding and the vagaries of finance capital, interest rates, etc., on the 'supply' side, there are technical and environmental uncertainties, with trade-offs between operational efficiency and likely impact. To these uncertainties must be added the 'demand' side uncertainties, with the price that can be charged for tidal electricity having to be estimated over the very long-term and compared with conventional fuel prices.

Although it is likely that fossil fuel, and possibly nuclear, electricity prices will increase over time, so that tidal energy projects will become more attractive, under present conditions tidal barrages appear to be relatively unattractive commercial investment options. Interestingly, however, in its study of possible responses to climate change the Royal Commission on Environmental Pollution did recommend that 'in view of the large amount of energy that would be available', tidal barrages should 'be kept under consideration as an option for the longer term' (RCEP, 2000).

6.6 Tidal energy potential

United Kingdom

In general, the best potential tidal barrage sites in the UK are on the west coasts of England and Wales, where the highest tidal ranges are to be found. Despite its indented coastline, the tidal energy potential of Scotland is very small (1–2 TWh y^{-1}) due to its generally low tidal range.

As we have seen, the practical potential of tidal power in the UK depends crucially on economics, as well as on environmental factors. In theory, the exploitable potential, assuming that every practical UK scheme was developed, could rise to approximately 53 TWh y^{-1}, or around 14% of UK electricity generation in 2002. Approximately nine-tenths of this potential (48 TWh y^{-1}) lies in eight large sites, each offering between 1 TWh y^{-1} and 17 TWh y^{-1}, while one-tenth relates to 34 small sites, each providing somewhere in the range 20–150 GWh y^{-1} (ETSU, 1990).

The Severn Barrage, if built, would make the largest contribution, approximately 17 TWh y^{-1}, but initial estimates have suggested that the Wash, the Mersey, the Solway Firth, Morecambe Bay and possibly the Humber, amongst others, could also make significant contributions (Table 6.4 and Figure 6.22). In addition to these larger sites, there are many smaller estuaries and rivers that could be used. Feasibility studies have been carried out on the Loughor Estuary (8 MW) and Conwy Estuary (33 MW) in Wales, the Wyre (64 MW) in Lancashire, and the Duddon (100 MW) in Cumbria.

Table 6.4 An early assessment of some potential tidal barrage sites in the UK

	Range/ m	Length/ m	Capacity/ MW	Output/ GWh
Severn – Outer line	6.0	20 000	12 000	19 700
Severn – Inner line	7.0	17 000	7200	12 900
Solway Firth	5.5	30 000	5580	10 050
Morecambe Bay	6.3	16 600	3040	5400
Wash	4.45	19 600	2760	4690
Humber	4.1	8300	1200	2010
Thames	4.2	9000	1120	1370
Dee	5.95	9500	800	1250
Mersey	6.45	1750	620	1320
Milford Haven	4.5	1150	96	180
Cromarty Firth	2.75	1350	47	100
Loch Broom	3.15	500	29	42
Loch Etive	1.95	350	28	55
Padstow	4.75	550	28	55
Langstone Harbour	3.13	550	24	53
Dovey	2.90	1300	20	45
Hamford Water	3.0	3200	20	38

Source: Baker (1986)

As an example, the preliminary feasibility study for the Wyre Barrage estimated that electricity could be generated at a cost of 6.5p per kWh (at 1991 prices), assuming an 8% discount rate over a 20-year period (ETSU, 1991). In a subsequent ETSU report, this estimate was raised to 6.8p per kWh in 1992 cost terms, but this was still lower than the equivalent figures for the Severn and the Mersey Barrages, which it quoted as, respectively, 7.2p per kWh and 7.1p per kWh, in 1992 costs, all at an 8% discount rate (ETSU, 1994). However, Conwy came out as more expensive at 8.7p per kWh, and a study of Duddon estimated that its generation cost was also 8.7p per kWh.

Overall, the total UK potential for small tidal schemes (that is, schemes of up to 300 MW capacity) is estimated at 2% of electricity requirements, with the most economic of these providing 1.5% of requirements.

Figure 6.22 Some potential locations for barrages in the UK

World

Tidal power availability is clearly very site-specific. However, as Table 6.5 indicates, there are numerous potential sites for tidal barrages around the world, in Russia, Canada, the US, Argentina, Korea, Australia, France, China and India, with an estimated total potential of perhaps as much as 300 TWh y^{-1}. Figure 6.23 indicates the locations of some of these sites. Although the main potential for tidal energy is from a small number of large barrages, there are also many potentially suitable sites for small- to medium-scale units.

Currently, in terms of large schemes, the main sites of interest outside the UK are the Bay of Fundy in Canada, and Mezeh and Tugur in Russia. Smaller schemes of interest include Garolim Bay in Korea, the Gulf of Kachchh (Kutch) in India, Secure Bay in Australia and a project at São Luís in Brazil.

The pace of development of tidal energy may be relatively slow compared with that of some of the other renewables, but the potential is nevertheless substantial. The total energy dissipated from tides globally is approximately 3000 GW, of which approximately 1000 GW is dissipated in accessible shallow sea areas.

In practice, since there are geographical access and location constraints for siting barrages, and they also have to be reasonably near to major grid lines, the realistic available resource is much smaller. The realistically recoverable resource has been put at 100 GW (Jackson, 1992). Although that is only about 15% of the existing global hydroelectric capacity, it still represents a significant resource.

Table 6.5 Some world locations for potential tidal power projects

Country	Mean tidal range/m	Basin area/ km^2	Installed capacity/ MW	Approx. annual output/ TWh per year	Annual plant load factor/%
Argentina					
San José	5.8	778	5040	9.4	21
Golfo Nuevo	3.7	2376	6570	16.8	29
Rio Deseado	3.6	73	180	0.45	28
Santa Cruz	7.5	222	2420	6.1	29
Rio Gallegos	7.5	177	1900	4.8	29
Australia					
Secure Bay	7.0	140	1480	2.9	22
Walcott Inlet	7.0	260	2800	5.4	22
Canada					
Cobequid	12.4	240	5338	14.0	30
Cumberland	10.9	90	1400	3.4	28
Shepody	10.0	115	1800	4.8	30
India					
Gulf of Kachchh (Kutch)	5.0	170	900	1.6	22
Gulf of Cambay (Khambat)	7.0	1970	7000	15.0	24
Korea (Rep)					
Garolim	4.7	100	400	0.836	24
Cheonsu	4.5	–	–	1.2	–
Mexico					
Rio Colorado	6–7	–	–	54	–
USA					
Passamaquoddy	5.5	–	–	–	–
Knik Arm	7.5	–	2900	7.4	29
Turnagain Arm	7.5	–	6500	16.6	29
Russian Federation					
Mezeh	6.7	2640	15000	45	34
Tugur[1]	6.8	1080	7800	16.2	24
Penzhinsk	11.4	20530	87400	190	25

1 7000 MW variant also studied

Adapted from a table on the World Energy Council website, http://worldenergy.org [accessed 23 December 2003]

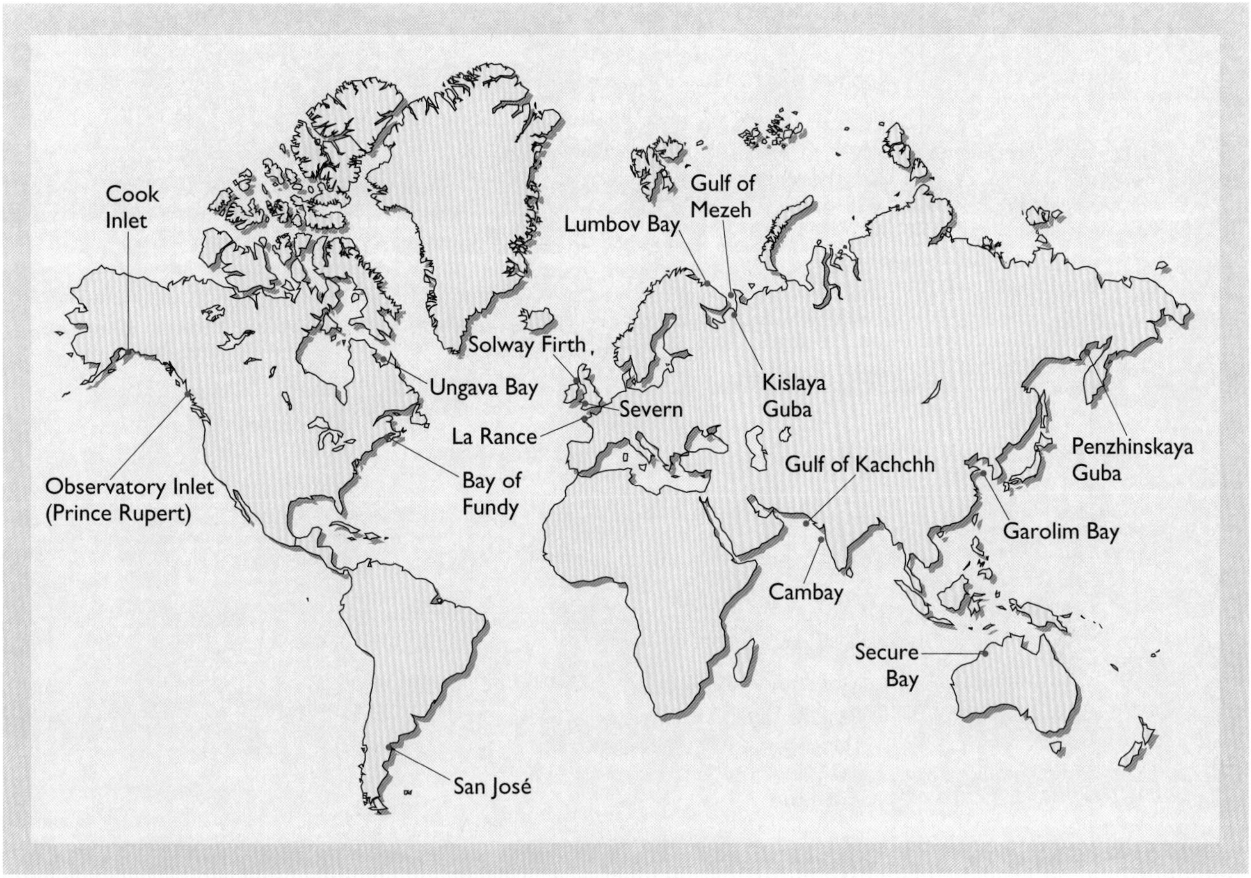

Figure 6.23 Some world locations for potential tidal power projects

Moreover, this figure does not take into account the possibility of larger and more ambitious (if very speculative) proposals for very large barrages, such as a barrage across the Wash and even huge barrages across the Bering Straits and the Irish Sea, though the latter proposals would be considered by many engineers to be technologically unrealistic.

Somewhat less ambitiously, but still speculatively, there have also been proposals by the US company Tidal Electric for offshore 'bounded reservoir' systems, consisting of circular dams trapping water at high tide, with the water then being used to drive turbines in the usual way – see Figure 6.24. The bounded 'low head' reservoirs would be constructed from rocks, as in some conventional breakwaters or causeways, and the developers claim that because this would be cheaper than building major 'high head' dams, offshore tidal generation could be competitive with conventional energy sources such as the burning of fossil fuels. These reservoirs have the added advantage that although, as with conventional barrages, they would involve the creation of a head of water, they would not involve blocking off an estuary and so should have a lower environmental impact.

Potential sites for bounded reservoirs are being investigated around the world, in Alaska, Africa, Mexico, China, and a 1000 MW project is being considered in India. Three sites are also under consideration in the UK, off the coast of Wales: see Box. 6.10.

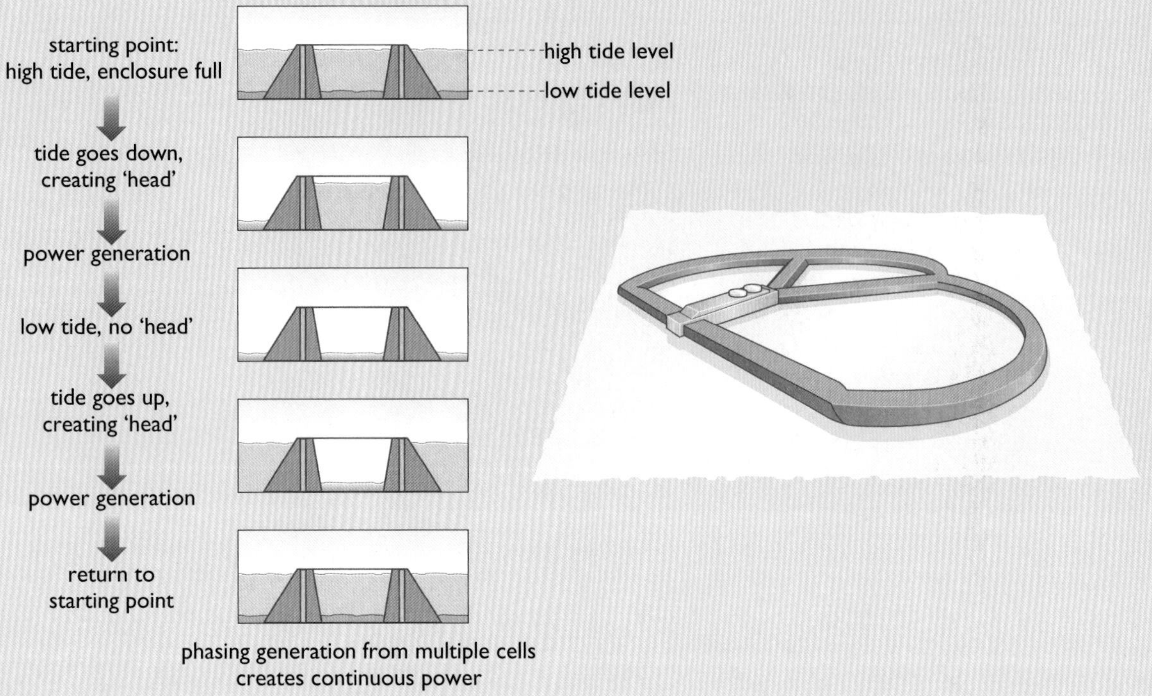

Figure 6.24 Tidal Electric 'bounded' reservoir

BOX 6.10 Offshore tidal lagoon proposals for Wales

The US company Tidal Electric has come up with proposals for three impounded offshore tidal reservoir projects or 'tidal lagoons' off the coast of Wales. The most advanced in terms of planning is a 30 MW tidal power project near Fifoots Point in the Bristol Channel, near the site of an existing 360 MW coal-fired power plant, with which, it is proposed, it would share management staff, facilities, and a 400 MW grid connection.

The other two projects are a 30 MW plant in Swansea Bay, being considered in cooperation with the Environment Trust, and a 423 MW project at Rhyl off the North Wales coast. The reservoir of the Rhyl scheme would be 14 km long and 3 km wide, and would be subdivided into segments, with each being filled and emptied in turn, so as to provide more nearly continuous output. This is clearly an ambitious project, which will have to obtain positive environmental assessment and the necessary financial backing, but if it does go ahead, it will represent the largest single renewable energy contribution in the UK.

For further details see http://www.tidalelectric.com [accessed on 29/7/2003].

6.7 Tidal barrages: conclusions

The technology of tidal power generation using barrages is well developed, and useful operating experience has been obtained from the Rance Barrage and other smaller projects.

Construction techniques have developed since La Rance was built and there is now much experience with large-scale civil engineering at sea in relation to oil rig and flood protection projects, for example in the use of large concrete caissons floated into place. The use of these new techniques, as envisaged by the STPG for the Severn Barrage, could cut civil engineering costs by up to 30% compared with La Rance, although there could be additional difficulties in placing caissons in strong currents.

As we have seen, the main uncertainties relating to any barrage system are environmental impact and investment provision. Some of the environmental impacts will, of course, be positive and of global significance – for example, the carbon dioxide emissions that result from power generated via fossil fuel plants could be avoided if this power could be provided through tidal barrages instead. The negative impacts are more localized and further research might show how these can be reduced. One of the key factors will probably be how people perceive barrages, in which case prior public consultation is vital. In the UK this has been done to a limited extent for both the Severn and the Mersey projects, but more work would need to be done to avoid unnecessary conflicts when and if and these projects went ahead, with design compromises perhaps emerging in response to public concerns.

By contrast, it is much more difficult to respond to the financial constraints. Despite all attempts to cut capital costs, they remain stubbornly high. No doubt design and technology improvements could improve plant economics, and, more importantly, new construction technology could reduce capital costs, but the potential for cost reduction is considered to be limited in the short- to medium-term. Certainly, this was the view of the UK Government in 1992, when it commented, in its response to the House of Commons Select Committee on Energy report on renewable energy, that it saw 'little scope for cost reductions via an R and D programme' (DTI, 1992).

So, in many ways, the issue is now out of the hands of the engineers, and depends much more on the attitudes of both the public and the private sectors towards investment in large-scale projects in the UK.

In general, tidal power, on a small or a large scale, would seem to require at least some state support if it is to be developed significantly in the UK. However, this looks unlikely in the foreseeable future. In March 1994 the Department of Trade and Industry announced that its tidal programme would be closed, with tidal power being relegated to 'watching brief' status (DTI, 1994). As was noted earlier (Section 6.5), subsequent attempts to obtain funding for a reappraisal of the Severn Barrage have not been successful, and, given their generally higher generation costs, there has been no support for small barrages. Although, as we have seen the Wyre barrage might be slightly cheaper than the Severn or Mersey, the 1994 ETSU review noted that 'ministers declined to support further development work on the project'. Over all, ETSU's conclusion was that, on the assumption of a 'zero commercial potential by current UK criteria', there was no justification for further research and development work, although the door was left open by the suggestion that it might still be worth investigating 'whether reverse economics of scale apply to tidal barrages' so as to 'see if small schemes are more financially attractive'.

However ETSU subsequently concluded in its 'supporting analysis' for DTI's 1999 review *New and Renewable Energy: Prospects for the 21st Century* that 'there are no significant economies of scale and so the high capital cost and construction periods associated with barrage construction act as a major barrier to exploitation' (ETSU, 1999).

But with concerns about the impacts of climate change growing, a shift in viewpoint might yet occur. In its 2000 study of Climate Change and UK energy policy, the Royal Commission on Environmental Pollution included a 2.2 GW average contribution from the Severn Barrage in three of its four scenarios for the year 2050, omitting it only from the scenario in which nuclear and/or coal fired generation expanded (RCEP, 2000). For the moment, however, the only barrage-type projects that seem to stand a chance of going ahead in the UK are those of the impounded reservoir type, as proposed for installation off the coast of Wales (see Box 6.10). But elsewhere things may be different. For example, in 2002, the Korean Water Resources Corporation announced that it was planning to build what it called 'the world's largest tidal-power plant' in Sihwa Ho, Gyeonggi province, bordering the West Sea. It would have an installed capacity of 252 MW and would generate power on the ebb from a tidal rise of 5.64 metres.

6.8 Tidal streams

Although the prospects for tidal barrages look relatively limited for the present, another approach to tidal power generation is being followed up with some enthusiasm – the use of tidal streams.

Instead of using costly and potentially invasive barrages located in estuaries to exploit the *vertical* rise and fall of tides, and the *potential* energy of heads of water trapped behind dams, it is possible to harness the *horizontal* flow of tidal currents – that is the *kinetic* energy of the tides. The energy in the fast, free-flowing tidal currents or tidal streams which exist at some sites can in principle be tapped using relatively simple, submerged, wind turbine-like rotors.

As was noted in Section 6.1, despite the common usage of the terms 'tidal current' and 'tidal stream', not all these natural energy flows in oceans are actually 'tidal' in the sense of being driven by the moon's gravitational pull. Some of these flows, especially the larger offshore ones such as the Gulf Stream, are the result of complex interactions between warm and cold layers of water in the oceans around the world, and the associated effects of varying salinity, and are, in effect, solar-driven. Strictly speaking, the correct term to use for these flows is **ocean currents** or **ocean streams**.

The idea of using these various types of ocean and estuarine energy flows is as yet relatively undeveloped, but it has been estimated that, for example, the power flowing through the north channel of the Irish Sea is equivalent to 3.6 GW, while the flow through Pentland Firth in Scotland is the equivalent of 6.1 GW. Tidal streams tend to be concentrated through narrow channels and around headlands, and conventional sailing charts can be a good source of information for identifying potential areas of interest for the deployment of tidal current devices.

Flows of this sort could be harnessed using arrays of large-diameter rotors, supported on pontoons tethered to the sea bed, or fixed to the bottom. One advantage of this concept would be that, unlike barrages, units could be constructed on a modular basis and installed incrementally. There would be problems of fouling (for example, by seaweed, lost fishing nets, etc.), tethering and power take-off to overcome, but on the other hand, expensive and environmentally intrusive barrages would not be required. There would be little visual impact, as most of the turbine would be under water, and the system would also be virtually silent.

The basic physics of tidal turbines is similar to that of wind turbines, as outlined in Chapter 7 – the power available is equal to the density of the water, times the area swept by the rotor, times the cube of the water velocity. Typical sea current velocities can be up to 3 ms^{-1}, compared with, say, the 7 ms^{-1} wind speeds for wind turbines, so that tidal turbines will have lower rotational speeds than wind turbines. However, given that the density of the working fluid is much higher, the power output from a tidal stream machine can be much larger than from a wind turbine of equivalent blade size – or alternatively, water turbines can be much smaller than wind turbines and still generate as much power. The power output will of course be even larger if sites with higher water speeds can be used.

That said, the output from a tidal current turbine would be likely to be lower than that from an equivalent-sized conventional turbine in a barrage, which would have the advantage of the funnelling effect of estuaries and the creation of an enhanced head of water. To compensate for this it should be possible to use larger rotors in free streams than is possible in barrage-mounted turbines, thus bringing the outputs of the two systems into line.

In 1993 ETSU estimated that the total UK tidal stream resource at 58 TWh y^{-1}, or around 19% of electricity requirements at that time, of which 46.5 TWh was attributed to sites where machines of over 100 kW size could be installed (ETSU, 1993). However, ETSU felt that the cost of this electricity could be high. The report's preliminary estimates were that bottom-mounted, fixed-orientation 'tide mill' devices of 100 kW or more rated power, located at the best sites (Pentland Firth and the Channel Islands) would produce electricity at some 10p per kWh at an 8% discount rate or 16p per kWh at a 15% discount rate. On this basis, work on tidal stream technology was initially given a low priority in the renewable energy development strategy adopted by the Department of Trade and Industry.

Subsequently, however, following a review by the Marine Foresight Panel set up by the Office of Science and Technology, support for tidal stream projects has grown, in part because the report argued that economies of scale achieved through series production of turbines could reduce these figures, albeit only after a substantial development programme (OST, 1999). In addition, as we shall see, a range of novel devices has emerged in the UK and elsewhere, and there is now considerable enthusiasm for this new area of development.

Practical projects in the UK

In 1994 a prototype two-bladed 10 kW tidal current turbine was tested in the Corran Narrows in Loch Linne, near Fort William in Scotland, by a consortium involving IT Power, the National Engineering Laboratory and Scottish Nuclear.

The £200 000-project was based on a rotor of 3.9 metres in diameter submerged in the sea at a depth of 10 metres. It operated at around 35–40 rpm in tidal currents with velocities up to 2.5 m s^{-1}. In a fully operational system the rotor unit would be supported by a cable attached to a floating buoy and also tethered to the sea bed. It could swivel around

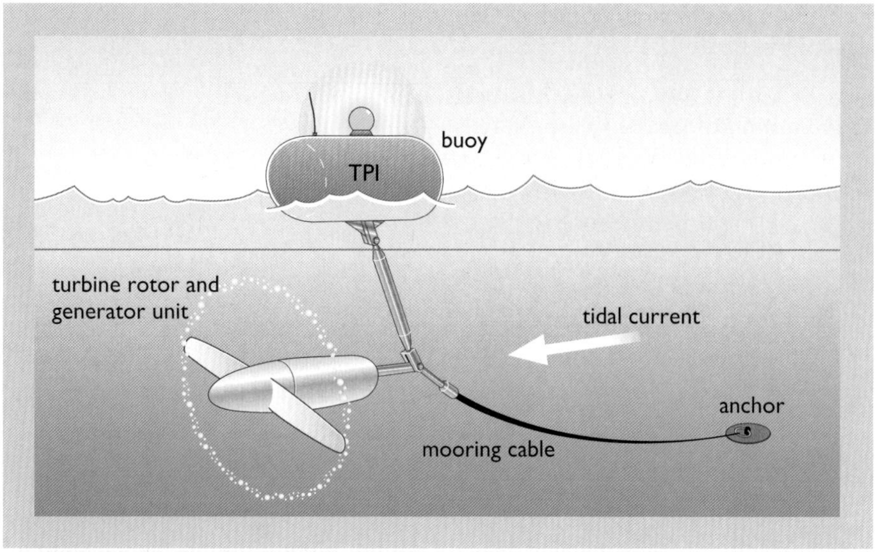

Figure 6.25 The IT Power tidal concept

Figure 6.26 The Marine Current Turbine

on the change of the tide to absorb power from tidal currents in either direction (see Figure 6.25). Full-scale systems could be rated at up to 1 MW each, with 20-metre diameter rotors.

This free-floating, mid-depth swivelling system concept, as developed by the IT Power consortium, might in some locations have cost advantages over fixed bottom-mounted devices, e.g. in places where there is sufficient room for the structure to swing around. However, the need for flexible marine cable power take-off arrangements could potentially add to the cost, and IT Power subsequently developed a system with a swivelling rotor mounted on a fixed steel pile driven into the sea bed, the **Marine Current Turbine**: see Figure 6.26.

Figure 6.27 The Marine Current Turbines Ltd 300-kW 'Seaflow' prototype on test off the north Devon Coast. The turbine can be raised out of the water, as shown, for easy access

IT Power has established a new company, Marine Current Turbines Ltd, which is developing this concept, starting with a 300 kW **Seaflow** prototype which was tested off the coast of north Devon near Lynmouth during 2003, with joint EU and DTI funding (about £1.3 million from DTI and about £0.6 million from the EC). Figure 6.27 shows the test rig in place in the summer of 2003, with the turbine raised out of the water.

Environmental impacts are likely to be minimal, but nevertheless consultation with local people has been taking place in order to try avoid any conflicts.

Regarding the costs, Marine Current Turbines Ltd is quite optimistic. Giving evidence to a House of Commons Select Committee in 2001, Peter Fraenkel, the company's technical director, commented that 'with our base line model, which is a 30 MW installation, we estimate about 3.8p per kilowatt hour when it goes commercial' and, although, he added, the consultants appointed to monitor the project, Binnie, Black and Veatch, had come out with 4.3p per kilowatt hour, 'we are not very far apart on that'. Fraenkel went on to say that 'For the very first machines it is very scale-dependent, and obviously with a single one-off 1 MW machine we are looking at something like 7p per kilowatt hour' (House of Commons Select Committee on Science and Technology, 2001).

The next stage is to develop a larger commercial model of 600 kW or so. Following that, the aim is to build a 10 MW 'current farm' array, using several machines. All being well, commercial manufacturing is then seen as starting with, by around 2010, around 300 turbines a year being produced, each worth £1 million.

As can be seen from Figures 6.26 and 6.27 the technology is based on the use of monopiles – single steel tubes which can be set in holes drilled in the sea bed, and then coupled with two- or three-bladed propeller units. Peak spring tidal currents of 4–5 knots (~2–2.5 m s^{-1}) and a depth of between 20 and 35 metres are seen as necessary for economic exploitation. Marine Current Turbines Ltd says it has identified many potentially suitable sites around the UK coast (see Figure 6.28), and elsewhere in the world. For more information see http://www.marineturbines.com [accessed on 29/7/2003].

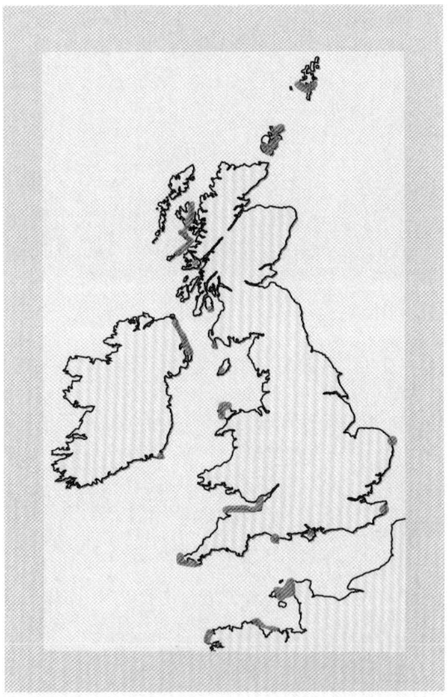

Figure 6.28 Possible sites in the UK for the Marine Current Turbine

Associated work on general resources and turbine optimization studies are being carried out by The Robert Gordon University in Scotland: see http://www2.rgu.ac.uk/subj/mes/cee/main/Research.htm [accessed on 29/7/2003].

The Engineering Business group, based in Northumberland in the UK, has also developed a novel design – the **Stingray**. This has a set of totally submerged hydroplanes which oscillate up and down in the tidal flows and drive a generator mounted on the seabed (see Figures 6.29 and 6.30). A 150 kW prototype was installed and tested in 2002 and again 2003, in Yell Sound in the Shetlands, supported by a £1.1 million DTI grant. Following further development work there are plans for 3- or 5 MW pre-commercial

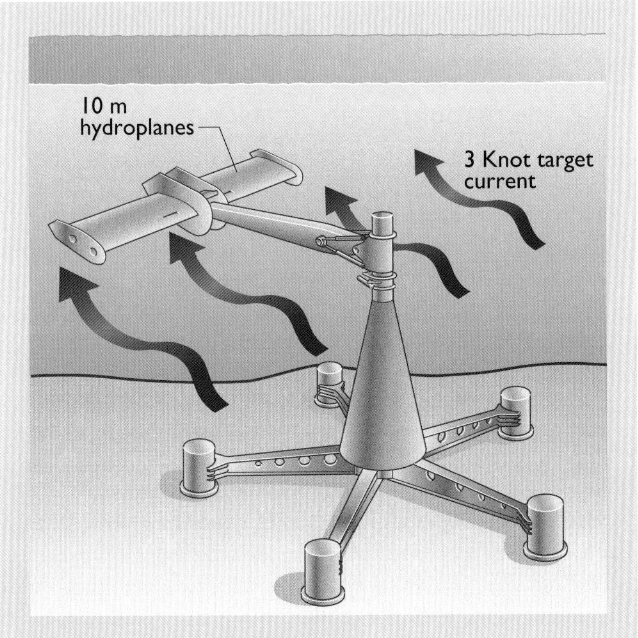

Figure 6.29 Stingray tidal generator

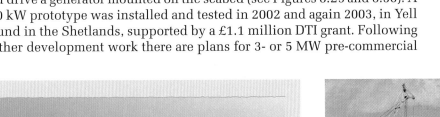

Figure 6.30 Stingray being prepared for test in 2002

versions, with, ultimately, several devices being linked in an array. For more details see http://www.engb.com [accessed on 29/7/2003].

Interestingly, given the long history of interest in developing a tidal barrage project in the Severn Estuary, in 2003 DTI allocated £1.6 million to a new company, Tidal Hydraulic Generators Ltd, to develop its tidal rotor design, with the plan being to install five rotors on the bed of the estuary, in between the two bridges. The design is based on conventional propeller type turbines, of 8 metres in diameter, mounted on a frame resting under its own weight on the estuary bed 40 metres down.

Tidal current projects and concepts around the world

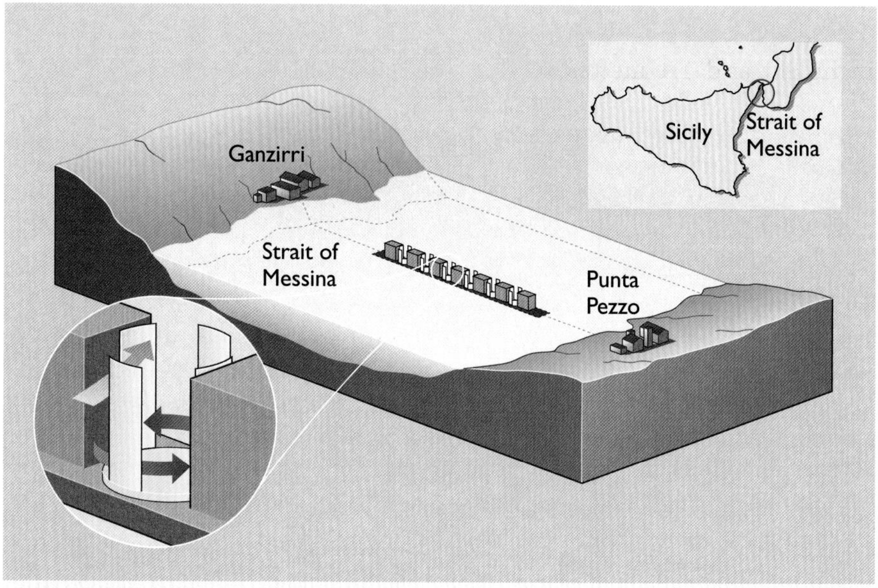

Figure 6.31 The Straits of Messina project

The idea of using tidal currents is also being explored around the world. A device similar to the undersea horizontal axis propeller design being developed by IT Power/Marine Current Turbines Ltd was tested in 2003 by a group involving Hammerfest Stroem, Statoil and ABB off the coast from Hammerfest in Norway. However, in most cases the emphasis has been on using vertical-axis turbines. For example, an EU-funded study has been made of the potential for tidal current generation in the Straits of Messina, between Sicily and mainland Italy. The proposal was to locate 100 vertical-axis turbines on the sea bed, at a depth of 100 metres; see Figure 6.31.

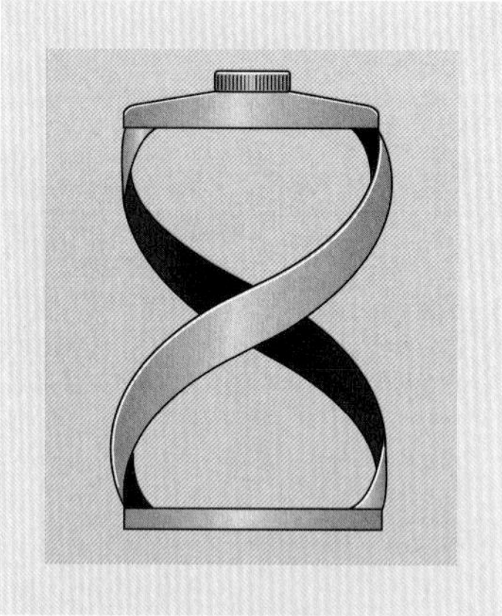

Figure 6.32 Gorlov's helical turbine

Even more ambitiously, Dr Alexander Gorlov of Northeastern University in Boston, USA, has been developing ideas for tidal current devices for use with ocean currents further out to sea. Dr Gorlov is the inventor and patent holder of the **Gorlov Helical Turbine** (Figure 6.32) – a variant of the Darrieus vertical-axis wind turbine design (see Chapter 7). Small prototypes have been tested and Gorlov is looking to large-scale applications in offshore locations. In particular, he is interested in the Gulf Stream, and notes that:

> The Gulf Stream transports approximately 80 000 000 cubic meters of water per second past Miami's front door. This is more than 50 times the total of all the rivers in the world. The surface velocity sometimes exceeds 8.2 feet per second (2.5 meters per second). Thus Nature has concentrated and transfigured part of the incoming solar radiation in the great current systems of the world, such as the American Gulf Stream, the Japanese Kuroshiwo Stream, the African Agulhas-Somali Stream and others. The total power of the kinetic energy of the Gulf Stream near Florida is the equivalent to approximately 65 000 MW.

Clearly, there exists a large energy resource and Gorlov has ambitious plans for exploiting it. His aim is to construct a 'farm' of one hundred power modules with 656 triple-helix Gorlov turbines mounted in columns in a lattice array. Each turbine would measure 0.83 metres in height and 1.1 metres in diameter. This power farm would generate around 136 MW in a current flow of 2.5 m s^{-1} or 30 MW in 1.5 m s^{-1}. The power would be used to electrolyse sea water to generate hydrogen gas which would then be piped, or tanked by ship, to the land. Whether this ambitions project will prove to be technologically and environmentally feasible remains to be seen.

Equally ambitious, but further advanced, a Canadian company called Blue Energy has developed a 'tidal fence' concept, in which H-shaped vertical-axis turbines are mounted in a modular framework structure (Figure 6.33). This concept would have much less environmental impact than the equivalent-scale barrage, as it would have no effect upon tide levels and therefore potentially less direct effect on the feeding and breeding habitats of birds and animals.

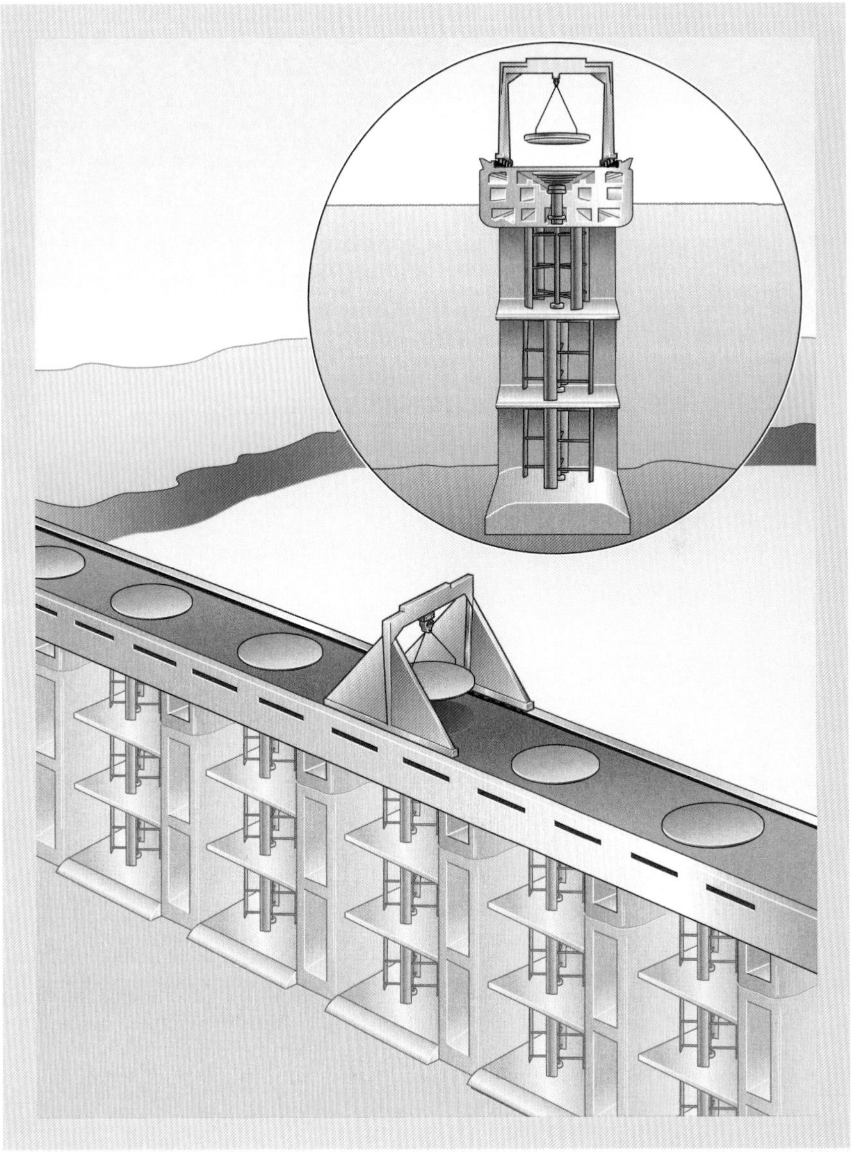

Figure 6.33 A tidal fence

Blue Energy is involved in two contracts which were signed with the Philippines Department of Energy in

November 1998. The first is for a demonstration tidal power plant, expected to generate 50 MW during peak tidal flow periods and 30 MW on average. The second is a 1000-MW peak capacity plant which should deliver on average 600 MW. The plants will consist of arrays of vertical-axis turbines mounted in a permeable causeway, or tidal fence, between two islands, extracting power from tidal current flows.

Blue Energy is also exploring ideas for an even larger project, a 4-km long tidal fence between the islands of Samar and Dalupiri in the San Bernardino Strait (see Phase 1 in Figure 6.34), where tidal currents range up to 8 knots (4–5 knots, or about 2–2.5 m s^{-1}, is usually considered to be sufficient for economically viable tidal current devices). The power plant would have a total estimated generating capacity of 2200 MW at peak tidal flow and 1100 MW on average. The estimated cost for the Dalupiri passage project (Phase 1 of four potential phases) is $US2.8 billion and Blue Energy says that 'the project could help the Philippines to become a net exporter of electrical power'. It adds that 'the modular nature of the Blue Energy Power System allows for power to be generated in the fourth year of the project, with the installation of the first module in the chain, which gradually increases to full capacity by project completion in year six. If built, this project would be one of the largest renewable energy developments in the world.'

Blue Energy has some even more ambitious proposals. Completing Phases 2, 3 and 4 (see Figure 6.34), at a cost of $US38 billion, would, it says, lead to the availability of around 25 000 MW of generating capacity. This could

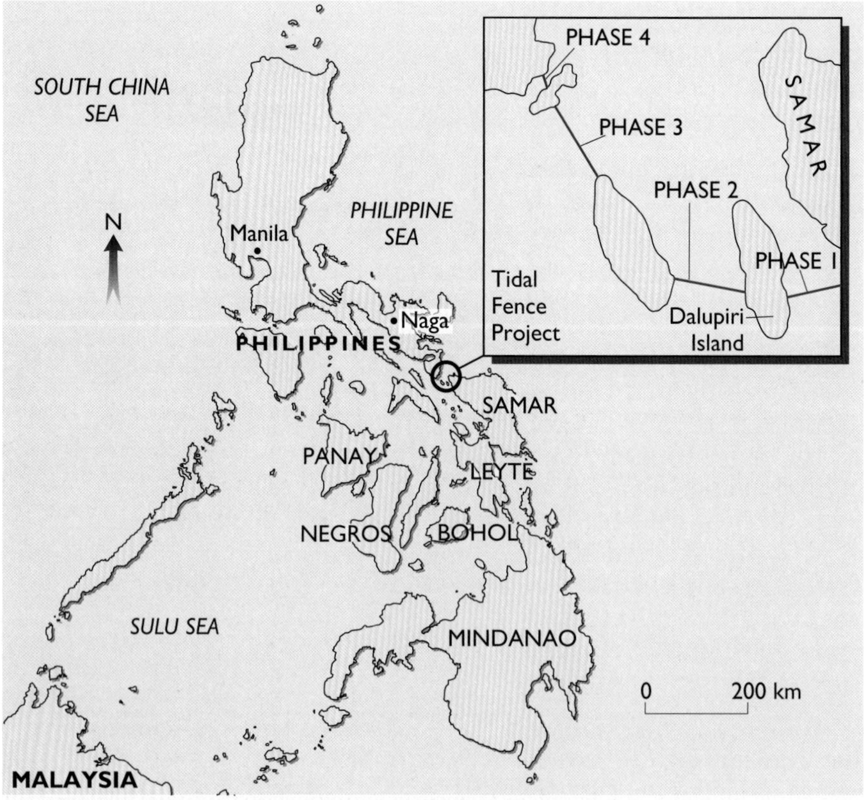

Figure 6.34 Site for Blue Energy's proposed tidal fence

provide a key element in a pan-Asian power grid, interlinking the various power systems in the region, and helping to balance local peak demands and local variations in availability of renewable energy inputs around the region. Blue Energy is also looking at potential sites elsewhere in the world. For details of current projects see http://www.bluenergy.com/ [accessed on 29/7/2003].

6.9 Tidal current turbines: the next stage

As the proposals for such major schemes suggest, the tidal current option has a lot of potential, although so far no full-scale system exists, and therefore costs are still speculative. Its advocates are convinced that the technology can be competitive with conventional energy systems, but there is still some way to go before this can be confirmed. Given that this is a new area of development it may be that, rather than aiming for giant projects, it would make sense to focus initially on smaller-scale, incrementally developed projects, which can allow more room for innovative development.

One novel idea is the **Rochester Venturi** (RV), invented by Geoff Rochester at Imperial College, London, in 1997. The RV device works by exploiting the pressure drop that occurs at a constriction when water flowing in a pipe accelerates – this is known as the venturi effect. This can provide a suction action to pump a relatively small quantity of water around a secondary water circuit, which drives a turbine mounted on the surface of the water, so there are no moving parts underwater. This secondary flow, the developers say, 'can be maintained for a good fraction of the tidal cycle, so the output is much less sharply peaked than it is from a barrage. This means that, for an equivalent electrical energy output, the generating capacity can be less and the duty cycle greater, thereby increasing the economic viability of the project. Tests on a scale model have shown that it should be possible, even on conservative estimate, to convert 20% of the total water power dissipated in the RV into electricity'.

The RV Power Company, which has been set up to exploit this concept, aims to produce a 2 MW version, following initial tests on a small prototype. It claims that the fully developed device should be able to produce electrical power 'at between 2 and 10 pence per kWh, with the most viable sites clustering around 2 pence or 3 pence per kWh. The greatest uncertainties in our costings come from site-specific issues, marine civil engineering estimates and planning permission' (House of Commons Select Committee on Science and Technology, 2001).

The concept allows for some innovative applications. The suction effect from several venturis could be used in parallel to drive one large turbine, so the concept is very flexible. For example, venturis could be incorporated into offshore wind turbine bases. The developers have also a proposal for several small units mounted on the Thames Barrier.

In 2003, the San Francisco Board of Supervisors agreed to test an RV device Device, under the name 'HydroVenturi', in San Francisco Bay, near the Golden Gate Bridge. The prototype project was reported as being likely to cost \$2 million and its supporters claimed that if fully developed the system could ultimately produce up to 2000 MW – twice the power needed by the city on a peak day. The early estimate for a 1000 m-wide partial barrier of Venturi devices 50 m below the surface in San Francisco Bay was that it could generate 452 GWh per year, assuming only 10% energy conversion efficiency. However, the developers claim that well-tuned test systems have attained an efficiency of 24%, although they say that, given that there is no shortage of energy in the water flow, efficiency is not the key issue. The total energy extracted would be kept smaller than the natural fluctuations in the energy flux in the water, so the environmental impact would be small. See http://www.rvcogen.com [accessed on 29/7/2003] for more details.

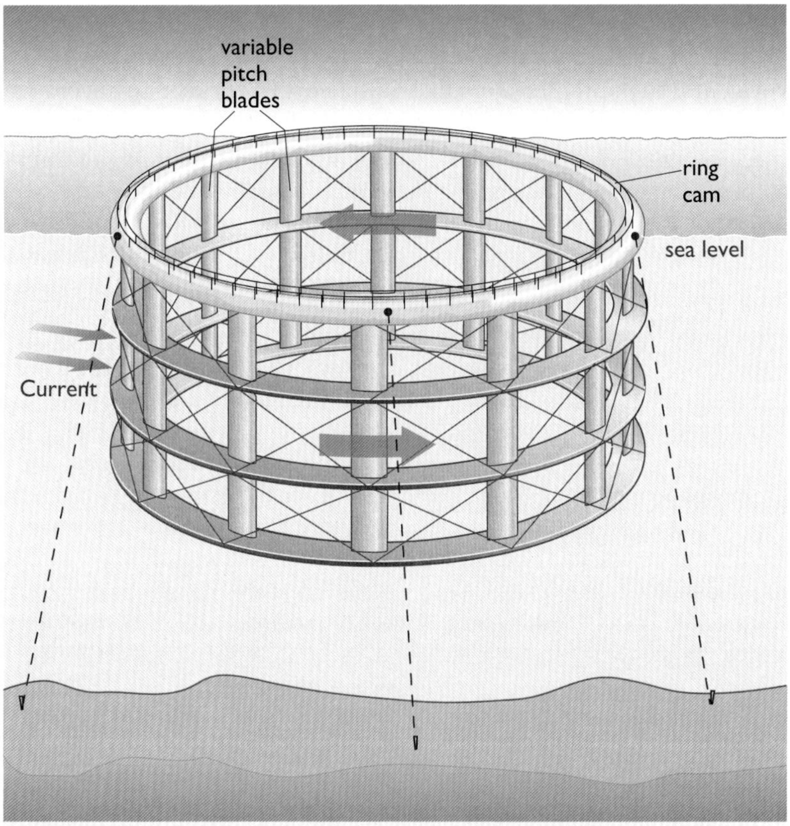

Figure 6.35 Salter's proposed 'Polo' turbine. The cylindrical rotor assembly rotates on roller bearings in a ring cam system which is anchored to the sea bed

Clearly, novel ideas are emerging in this field and some could be scaled up to become significant energy sources. In this context, it is interesting to note that the Edinburgh University wave energy pioneer Stephen Salter (see Chapter 8) has also developed an idea for a tidal current device. Nicknamed 'the Polo', the device consists of a series of vertical-axis water turbine blades mounted in a cylinder-shaped structure, which can rotate on bearings in a tethered ring which floats on the surface (Figure 6.35). At full scale the ring would be 50 metres in diameter and the blades would be 20 metres deep. The complete 600-tonne design could have a generation capacity of more than 12 MW.

6.10 Tidal current assessment

Although tidal current technology is in its infancy, the prospects do look promising. Stephen Salter has suggested that the development of tidal current systems could probably overtake that of wave technology, despite wave technology's head start. He told a House of Commons Select Committee that:

> The uncertainties about tidal streams are lower, I think they can take a lot of technology from wind, and I think they are a more predictable environment; so I would expect that would reach commercial viability sooner than wave energy. The problem is that it is not such a large resource, and we can use all of it and still want more, whereas wave energy is such a big resource that it is worth going for, even if it looks hard to start with.

However, although practitioners like Salter are clearly keen to develop novel devices, the official view of the prospects for tidal current technology initially remained cautious. In its review of UK renewable energy options, carried out between 1997 and 2000, the Department of Trade and Industry concluded that at most sites it would be expensive, so that the contribution that it might be expected to make was relatively small. Thus, although a back-up report for the DTI review produced by the Energy Technology Support Unit (ETSU, 1999) put the potential UK tidal stream resource at 36 TWh y^{-1}, it suggested that the UK might only obtain 0.7 TWh from tidal streams by 2010, from 322 MW of installed capacity. The ETSU report also included a cost resource curve for vertical-axis tidal stream devices such as Salter's Polo, which suggests that in principle some devices might generate at around 1.5p per kWh. The scale of the resource at that cost is limited, but Salter seems to have focused on it as a niche market for establishing tidal stream systems.

The overall conclusion was that, like tidal barrages, tidal current technology did not merit much effort for the time being. In the consultation paper produced in 1999, as part of the DTI review of renewables, tidal barrages appear in DTI's ranking of technological options in the 'very long-term' category – defined as 'technologies that are unlikely to be worth pursuing extensively at this time except at the fundamental research level'. However, tidal stream devices did not even seem to merit that level of commitment – this option did not appear on the ranking chart. The DTI report commented 'assessment of inadequately understood technologies will be considered periodically which could lead to reclassification and further support. A technology falling into this group could be tidal stream' (DTI, 1999).

But not everyone was quite so pessimistic about the prospects for tidal stream technology. For example, although it agreed that tidal stream technology was still relatively underdeveloped, the UK Marine Foresight Panel, which was set up to feed ideas about future possibilities into the UK's Technology Foresight Programme, commented that 'since so little work has been completed in this area, the learning curve is still steep and valuable results should be obtained from relatively small further investment in R and D' (OST, 1999). In addition, in its report on UK energy policy and climate change published in 2000, the Royal Commission on Environmental Pollution suggested that tidal stream devices might play a role in helping the UK to reduce its carbon dioxide emissions by up to 60%. Its future energy scenarios for 2050 included five hundred 1-MW tidal current devices installed in fast moving tidal flows around the UK, with a total average annual power rating of around 0.25 GW (RCEP, 2000).

Subsequently, the House of Commons Select Committee on Science and Technology looked at wave and tidal current technology again, and concluded that both merited significantly more attention. Indeed, in its report published in May 2001, the Committee was quite forceful on this issue arguing that 'given the UK's abundant natural wave and tidal resource, it is extremely regrettable and surprising that the development of wave and tidal energy technologies has received so little support from the Government'.

In fact, the UK Government had by that time expanded research and development support for tidal and wave energy to around £3 million, but most of this was for generic development work rather than new devices and very little went to tidal projects. However, perhaps as a result of the Select Committee's admonitions, the situation then changed. In June 2001, DTI decided to match the EC's provisional allocation of €1 million for IT Power's 300 MW Seaflow project off the Devon coast, and provided approximately £1 million to allow the project to go ahead. Subsequently, in November 2001, as part of a £100 million increase in funding for new renewables, the UK Cabinet Office Performance and Innovation Unit recommended that £5 million be allocated to wave and tidal power work (PIU, 2001). Following up on this, in January 2002, DTI announced that it was providing £1.1 million towards the development of a full-scale prototype of the 'Stingray' seabed mounted hydroplane mentioned earlier. However, these allocations, and other subsequent allocations, such as the £1.6 million awarded to Tidal Hydraulic Generators Ltd, were still only focused on prototype 'demonstration' projects. The strategic 'routemap' developed by DTI suggested that tidal current technology would not be deployed on a significant scale for power generation until after 2010; see http://www.dti.gov.uk/renewable/routemap.htm [accessed on 29/7/2003].

So, although tidal current power systems are being allocated some funding by DTI, and a wave and tidal energy test centre has been set up in the Orkneys, the tidal current option is still evidently seen, at least by DTI, as only likely to be deployed on a significant scale in the relatively long-term future.

References

Baker, A.C. (1986) 'The development of functions relating cost and performance of tidal power schemes and their application to small scale sites', *ICE Symposium on Tidal Power*, Thomas Telford, pp.331–44.

Baker, A.C. (1991a) *Tidal Power*, Peter Peregrinus.

Baker, A.C. (1991b) 'Tidal power', *Energy Policy*, vol. 19, no. 8, pp.792–7.

Department of Energy (1981) *Tidal Power from the Severn Estuary, Volume I*, Energy Paper 46 (The Bondi Report), HMSO.

Department of Energy (1987) *Information Bulletin on the Severn Barrage*, Issue 2, HMSO.

Department of Energy (1989) *The Severn Barrage Project: General Report*, Energy Paper 57, HMSO.

Department of Energy (1991) *The Severn Barrage Project Report on Responses and Consultations*, TID 4090 p1, Severn Tidal Power Group/ Energy Technology Support Unit.

DTI (1992) *Government Response to the Energy Select Committee Report on Renewable Energy*, 16 July, HMSO.

DTI (1994) *New and Renewable Energy: Future Prospects for the UK*, Energy Paper 62, HMSO.

DTI (1999) *New and Renewable Energy: Prospects for the 21st Century.*

ETSU (1988) *Tidal Power from the River Mersey: A Feasibility Study Stage 1*, Mersey Barrage Company Limited/Energy Technology Support Unit.

ETSU (1990) *Renewable Energy Research and Development Programme*, Report R56, Energy Technology Support Unit.

ETSU (1991) *River Wyre – Preliminary Feasibility Study*, TID 4100, Energy Technology Support Unit.

ETSU (1993) *Tidal Stream Energy Review*, Report 05/00155, Energy Technology Support Unit.

ETSU (1994) *An Assessment of Renewable Energy for the UK*, Report R82, Energy Technology Support Unit.

ETSU (1999) *New and Renewable Energy: Prospects in the UK for the 21st Century*, Energy Technology Support Unit Report, R122.

ETSU (2002) *The Severn Barrage – Definition Study for a New Appraisal of the Project*, Report T/09/00212/REP, STPG report commissioned by DTI.

Haws, T. and McCormick, J. (1993) 'Economic and financial development of the Mersey Barrage', *IEE Conference Proceedings 'Clean Power 2001'*, Conference Paper 385, November 17–19, pp.72–7.

House of Commons Select Committee on Energy (1992) *Renewable Energy*, Fourth Report Session 1991–1992, HMSO.

House of Commons Select Committee on Science and Technology (2001) *Wave and Tidal Energy*, Seventh Report Session 2000–2001, HC291, HMSO.

Jackson, T. (1992) 'Renewable energy: summary paper for the Renewable Energy Series', *Energy Policy*, vol. 20 no. 9, pp 861–83.

MBC (1991) Memorandum submitted by MBC to the House of Commons Select Committee on Energy, 20 November, reprinted in Energy Committee Fourth Report *Renewable Energy*, vol. III, Minutes of Evidence and Appendices, 43-III, 11 March 1992, HMSO.

Open University (1989) S330 *Oceanography, Book 4: Waves, Tides and Shallow Water Processes*, The Open University in association with Pergamon Press.

OST (1999) *Energies From the Sea: Towards 2020*, Marine Foresight Panel publication, Office of Science and Technology, Department of Trade and Industry.

PIU (2001) *Building the Future of the Environment*, Performance and Innovation Unit, Cabinet Office.

RCEP (2000) *Energy – The Changing Climate*, Royal Commission on Environment Pollution, ISO.

Renewable Energy Advisory Group (1992) *Report to the President of the Board of Trade*, Energy Paper 60, HMSO.

Twidell, J. W. and Weir, A. J. (1986) *Renewable Energy Resources*, E and F Spon.

Wardle, D.G., Gibson, J.P. and McGlynn, R.F. (1993) 'The present status of the Severn Barrage project studies', *IEEE Conference Proceedings 'Clean Power 2001'*, Conference Paper 385, November 17–19, pp.78–83.

Watt Committee on Energy (1990) *Renewable Energy Sources*, Report 22, Elsevier Applied Science.

Chapter 7

Wind Energy

by Derek Taylor

7.1 Introduction

Wind energy has been used for thousands of years for milling grain, pumping water and other mechanical power applications. Today, there are several hundred thousand windmills in operation around the world, many of which are used for water pumping. But it is the use of wind energy as a pollution-free means of generating *electricity* on a significant scale that is attracting most current interest in the subject. Strictly speaking, a wind*mill* is used for milling grain, so modern 'windmills' tend to be called **wind turbines,** partly because of their functional similarity to the steam and gas turbines that are used to generate electricity, and partly to distinguish them from their traditional forbears. They are also sometimes referred to as **wind energy conversion systems (WECS)** and those used to generate electricity are sometimes described as **wind generators** or **aerogenerators**.

Figure 7.1 The 70 watt rated Marlec 'Rutland Wind Charger'. Many thousands of small wind turbines like these are in use world-wide

Attempts to generate electricity from wind energy have been made (with various degrees of success) since the end of the nineteenth century. Small wind machines for charging batteries have been manufactured since the 1930s (see Figure 7.1). It is, however, only since the 1980s that the technology has become sufficiently mature to enable a viable large-scale industry to evolve, centred around the manufacture of large turbines for electricity production. The cost of wind turbines fell steadily between the early 1980s and the early 2000s. Wind is now one of the most cost-effective methods of electricity generation available, in spite of the relatively low current cost of fossil fuels. The technology is continually being improved to make it both cheaper and more reliable, so it can be expected that wind energy will become even more economically competitive over the coming decades.

Wind power is also one of the fastest growing renewable energy technologies world wide. A total of 31 000 MW of wind generating capacity had been installed by the end of 2002. This is about four times the capacity that had been installed by the end of 1997, implying an average growth rate of around 40% per annum.

As we shall see, an understanding of the machines and systems that extract energy from the wind involves many fields of knowledge, including meteorology, aerodynamics, electricity and planning, as well as structural, civil and mechanical engineering.

This chapter begins with a description of the atmospheric processes that give rise to wind energy. Wind turbines and their aerodynamics are then described, together with various ways of calculating their power and energy production. This is followed by discussions of the environmental impact and economics of wind energy, together with an examination of recent commercial developments in the field and a discussion of its future potential. The final section looks at offshore wind power, which seems likely to be one of the most challenging areas of wind energy development in coming decades.

7.2 The wind

The earth's wind systems are due to the movement of atmospheric air masses as a result of variations in atmospheric pressure, which in turn are the result of differences in the solar heating of different parts of the earth's surface.

As described in Chapter 2, one square metre of the earth's surface on or near the equator receives more solar radiation per year than one square metre at higher latitudes. The curvature of the earth means that its surface becomes more oblique to the sun's rays with increasing latitude. In addition, the sun's rays have further to travel through the atmosphere as latitude increases, so more of the sun's energy is absorbed *en route* before it reaches the surface. As a result of these effects, the tropics are considerably warmer than the high latitude regions. A simplified explanation of the way in which this differential warming creates the earth's wind systems is given in Box 7.1.

Atmospheric pressure is the pressure resulting from the weight of the column of air that is above a specified surface area, and it is measured by means of a barometer (Figure 7.3). The unit of atmospheric pressure is known as the bar (see also Chapter 5 *Hydroelectricity* and Appendix A2). One bar is approximately normal atmospheric pressure at sea level.

BOX 7.1 The earth's wind systems

Like all gases, air expands when heated, and contracts when cooled. In the atmosphere, warm air is lighter and less dense than cold air and will rise to high altitudes when strongly heated by solar radiation.

A low pressure belt (with cloudy and rainy weather patterns) is created at the equator due to warm humid air rising in the atmosphere until it reaches the **tropopause** (the top of the **troposphere**). At the surface the equatorial region is called the 'doldrums' (from an old English word meaning dull) by early sailors who were fearful about becoming becalmed.

At the tropopause in the Northern Hemisphere the air moves northwards and in the Southern Hemisphere, it moves southwards. This air gradually cools until it reaches latitudes of about 30 degrees, where it sinks back to the surface, creating a belt of high pressure at these latitudes (with dry clear weather patterns). These 30-degree latitudes are known as the 'horse latitudes' – a name that dates back to when the Western Hemisphere was being colonized. The crews of sailing ships, becalmed for long periods in the light and variable winds, were forced to throw horses overboard to preserve their drinking water supply. The majority of the world's deserts are found in these high-pressure regions.

Some of the air that reaches the surface at these latitudes is forced back towards the low-pressure zone at the equator. These air movements are known as the '**trade winds**'. On reaching the equator these air movements complete the circulation of what is known as the **Hadley cell** – named after the scientist (George Hadley) who first described them in 1753.

However, not all of the air that sinks at the 30 degree latitudes moves toward the equator. Some of it moves poleward until it reaches the 60 degree latitudes, where it meets cold air coming from the poles at what are known as the 'polar fronts'. The interaction of the two bodies of air causes the warmer air to rise and most of this air cycles back to the 30-degree latitude regions where it sinks to the surface, contributing to high-pressure belt. This completes the circulation of what is known as the **Ferrel cell** named after William Ferrel who first identified it in 1856.

The remaining air that rises at the polar fronts moves poleward and sinks to the surface at the poles as it cools. It then returns to the 60-degree latitude region completing the circulation of what is known as the **polar Hadley cell** or **polar cell**.

As the Earth is rotating, the winds are subjected to a phenomenon known as the **Coriolis Effect**, which was first identified by Gustave-Gaspard de Coriolis in 1835. 'North bound' winds are caused to veer 'right' ('westerlies') in the Northern Hemisphere and to the 'left' ('trade winds') in the Southern Hemisphere. Likewise, 'south bound' winds veer 'right' ('trade winds') in the Northern Hemisphere and 'left' ('westerlies') in the Southerly Hemisphere. Figure 7.2 shows the overall pattern of global wind circulation.

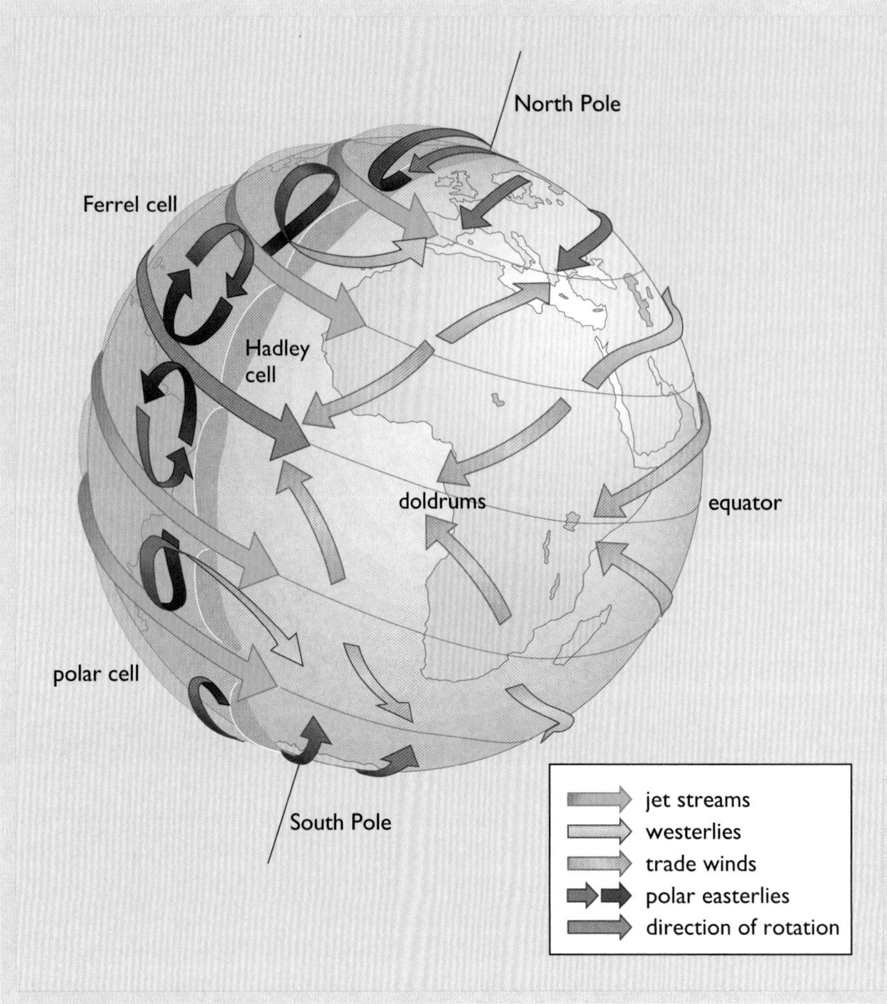

Figure 7.2 The global wind circulation (source: Burroughs *et al.*, 1996)

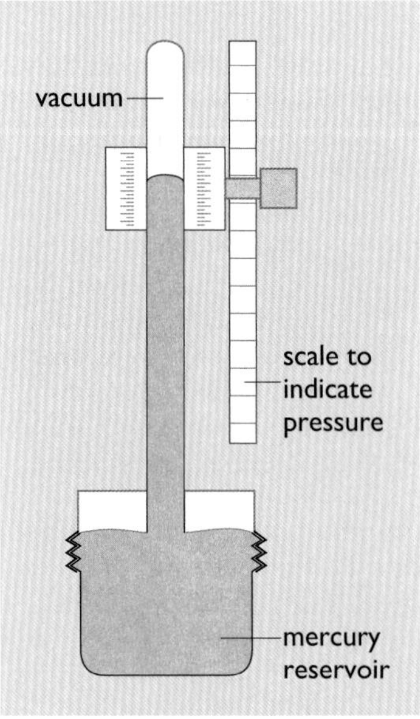

Figure 7.3 Fortin barometer, an example of a barometer used to measure atmospheric pressure. Variations in atmospheric pressure acting on the mercury in the reservoir cause the mercury in the column to rise or fall

Barometers are usually calibrated in millibars (mbar), that is, thousandths of a bar. The average atmospheric pressure at sea level is about 1013.2 mbar. Pressure is also measured in pascals (Pa): one pascal is defined as one newton per square metre (see Appendix A2)

On the weather maps featured in weather forecasting television programmes or in newspapers, you will notice that there are regions marked ‘high’ and ‘low’, surrounded by contours (Figure 7.4). The regions marked ‘high’ and ‘low’ relate to the atmospheric pressure and the contours represent lines of equal pressure called **isobars**. The high pressure regions tend to indicate fine weather with little wind, whereas the low pressure regions tend to indicate changeable windy weather and precipitation.

In addition to the main global wind systems shown in Box 7.1 there are also local wind patterns, such as sea breezes (Figure 7.5) and mountain-valley winds (Figure 7.6).

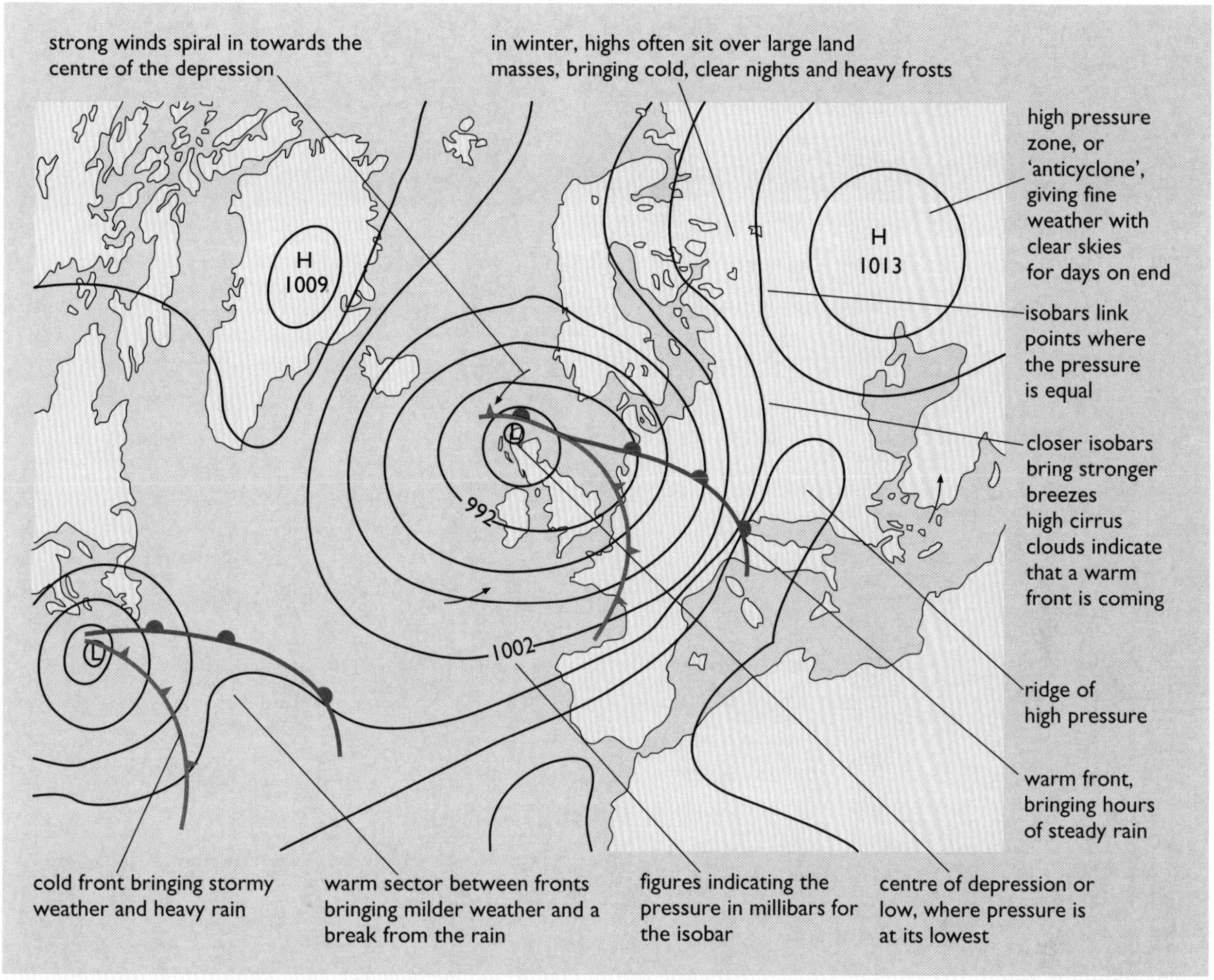
strong winds spiral in towards the centre of the depression
in winter, highs often sit over large land masses, bringing cold, clear nights and heavy frosts
high pressure zone, or 'anticyclone', giving fine weather with clear skies for days on end
H
1009
H
1013
isobars link points where the pressure is equal
closer isobars bring stronger breezes
high cirrus clouds indicate that a warm front is coming
992
L
L
1002
ridge of high pressure
warm front, bringing hours of steady rain
cold front bringing stormy weather and heavy rain
warm sector between fronts bringing milder weather and a break from the rain
figures indicating the pressure in millibars for the isobar
centre of depression or low, where pressure is at its lowest

Figure 7.4 Typical weather map showing regions of high (H) and low (L) pressure

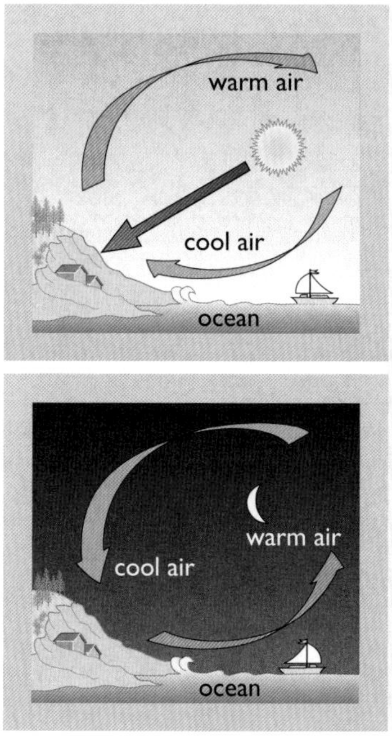
warm air
cool air
ocean
cool air
warm air
ocean

Figure 7.5 Sea breezes are generated in coastal areas as a result of the different heat capacities of sea and land, which give rise to different rates of heating and cooling. The land has a lower heat capacity than the sea and heats up quickly during the day, but at night it cools more quickly than the sea. During the day, the sea is therefore cooler than the land and this causes the cooler air to flow shorewards to replace the rising warm air on the land. During the night the direction of air flow is reversed

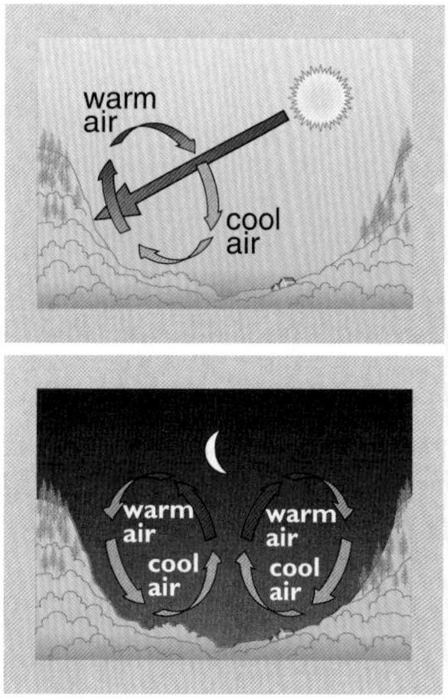
warm air
cool air
warm air
cool air
warm air
cool air

Figure 7.6 Mountain-valley winds are created when cool mountain air warms up in the morning and, as it becomes lighter, begins to rise: cool air from the valley below then moves up the slope to replace it. During the night the flow reverses, with cool mountain air sinking into the valley

Energy and power in the wind

The energy contained in the wind is its kinetic energy, and as we saw in Chapter 1 the kinetic energy of any particular mass of moving air is equal to half the mass, *m*, of the air times the square of its velocity, *V*:

kinetic energy = half mass × velocity squared

i.e. kinetic energy = $0.5mV^2$ (1)

where *m* is in kilograms and *V* is in metres per second ($m\ s^{-1}$).

We can calculate the kinetic energy in the wind if, first, we imagine air passing through a circular ring or hoop enclosing a circular area *A* (say 100 m^2) at a velocity *V* (say 10 $m\ s^{-1}$) (see Figure 7.7). As the air is moving at a velocity of 10 $m\ s^{-1}$, a cylinder of air with a length of 10 m will flow through the ring each second. Therefore, a volume of air equal to 100 × 10 = 1000 cubic metres (m^3) will pass through the ring each second. By multiplying this volume by the density of air, ρ (which at sea level is 1.2256 $kg\ m^{-3}$), we obtain the mass of the air flowing through the ring per second (in $kg\ s^{-1}$). In other words:

mass (*m*) of air per second = air density × volume of air flowing per second

= air density × area × length of cylinder of air flowing per second

= air density × area × velocity

i.e. $m = \rho AV$

Substituting for *m* in (1) above:

kinetic energy *per second* = $0.5\ \rho\ AV^3$ (joules per second)

where ρ is in kilograms per cubic metre ($kg\ m^{-3}$), *A* is in square metres (m^2) and *V* is in metres per second ($m\ s^{-1}$).

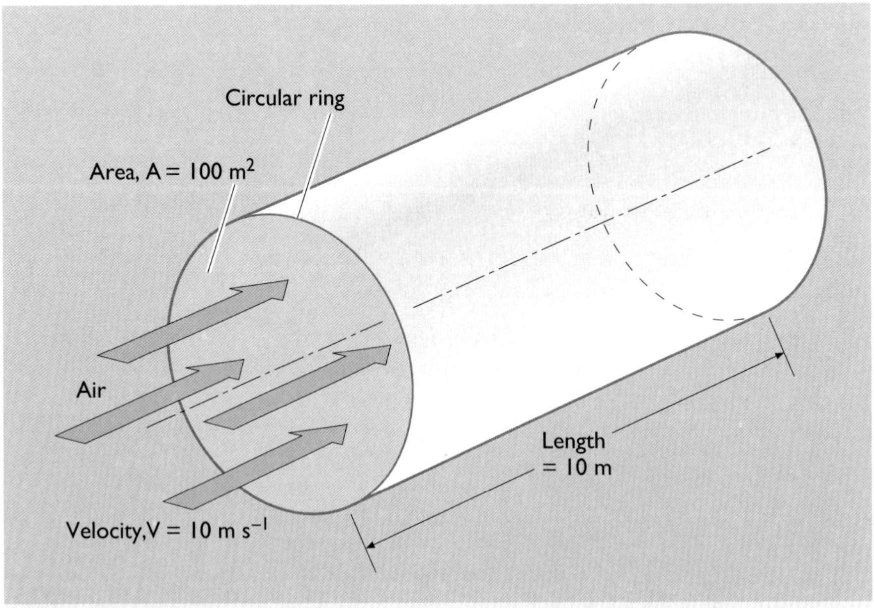

Figure 7.7 Cylindrical volume of air passing at velocity V (10 $m\ s^{-1}$) through a ring enclosing an area, A, each second

If we recall that energy per unit of time is equal to power, then the power in the wind, P (watts) = kinetic energy in the wind *per second* (joules per second)

i.e. $P = 0.5\,\rho\,AV^3$

The main relationships that are apparent from the above calculations are that the power in the wind is proportional to:

- The density of the air. The density is lower at higher elevations in mountainous regions; but average densities in cold climates may be up to 10% higher than in tropical regions.
- The area through which the wind is passing; and
- The cube of the wind velocity. Wind velocity therefore has a strong influence on power output. For example, if wind velocity is 4 ms^{-1} instead of 3 ms^{-1}, power increases by a factor of more than two.

Note that the power contained in the wind is *not* in practice the amount of power that can be extracted by a wind turbine. This is because losses are incurred in the energy conversion process (see Section 7.4 on aerodynamics). In addition, some of the air is 'pushed aside' by the rotor and bypasses it without generating power.

7.3 Wind turbines

A brief history of wind energy

Wind energy was one of the first non-animal sources of energy to be exploited by early civilisations. It is thought that wind was first used to propel sailing boats, but the static exploitation of wind energy by means of windmills is believed to have been taking place for about 4000 years (Golding, 1955).

Windmills have traditionally been used for milling grain, grinding spices, dyes and paint stuffs, making paper and sawing wood. Traditional wind pumps were used for pumping water in Holland and in East Anglia in the UK, and, because they often used identical forms of sails and support structures, they were (and are) often also referred to as windmills.

Many early windmills were of the *vertical axis* type. Some examples are shown in Figure 7.8 and include the following:

Screened windmills. These employ screens or walls around the windmill as shown in Figure 7.8, which are positioned to screen the windmill sails from the wind during the 'backward' part of the cycle.

'Clapper' windmills. These are so called because the moveable sails 'clap' against stops as the rotor turns with the wind (forwards), maximising their air resistance, but align themselves with the wind (like a weather vane) when on the part of their cycle in which they are moving into the wind (backwards), so reducing their air resistance (Figure 7.8).

Cyclically pivoting sail windmills. These are similar to the 'Clapper' windmills, but use a more complex mechanism to achieve progressive changes in sail orientation. The pitch angle of each sail is cyclically adjusted

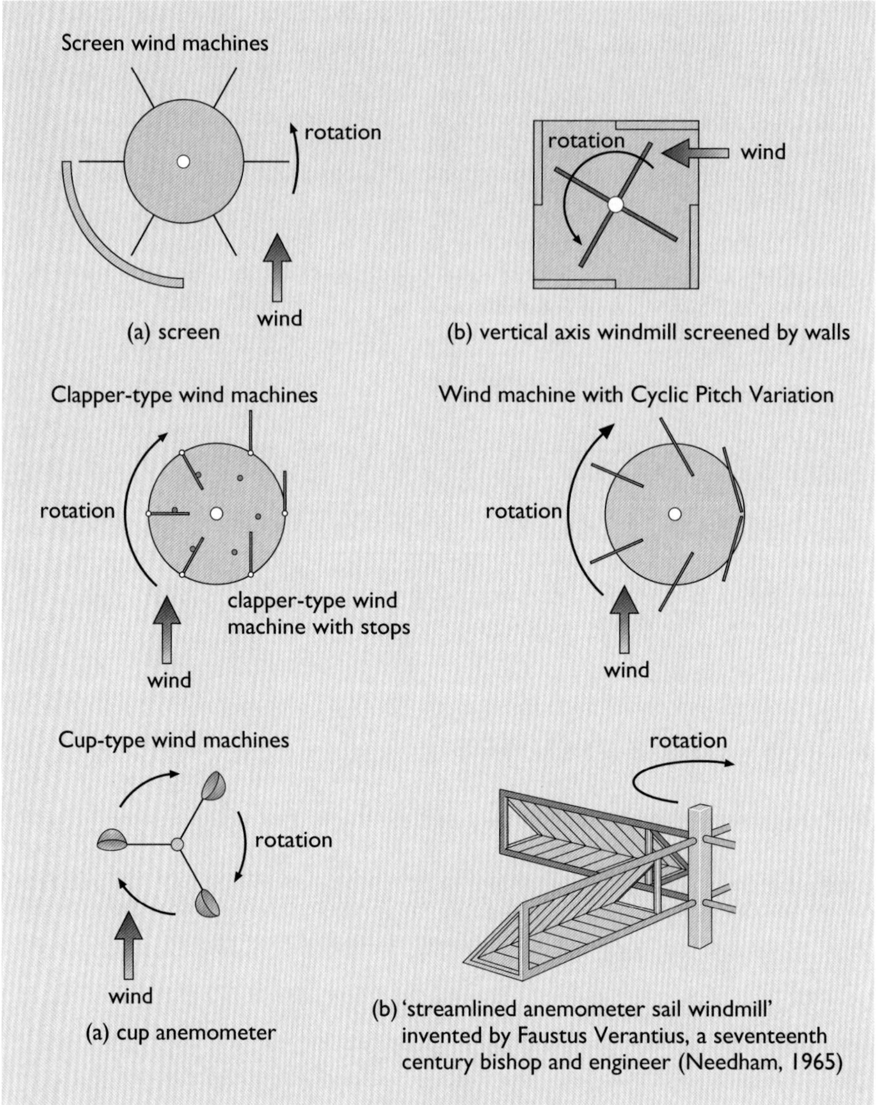

Figure 7.8 Some examples of traditional vertical axis windmills

according to its position during its rotation cycle and to the direction of the wind. This gives a difference in resistance either side of the windmill's rotation axis, causing it to rotate when exposed to a wind stream (Figure 7.8).

Differential resistance or cup type windmills. In these windmills, the blades are shaped to offer greater resistance to the wind on one surface compared with the other (Figures 7.8a and b). This results in a difference in wind resistance on either side of the windmill axis, so allowing the windmill to turn. A modern example of this type of wind-driven device is the cup anemometer, an instrument used for measuring wind speed. The simple 'S' type windmill (Figure 7.9) is also an example of this type – as is the 'Savonius Rotor' (a 'split-S' shaped rotor as shown in Figure 7.9). Savonius Rotors are used for powering fans in trucks and vans and have been used for simple 'do-it-yourself' windmills. They are produced as 'micro' wind generators, including variants with helically twisted semi-cylindrical 'cups'.

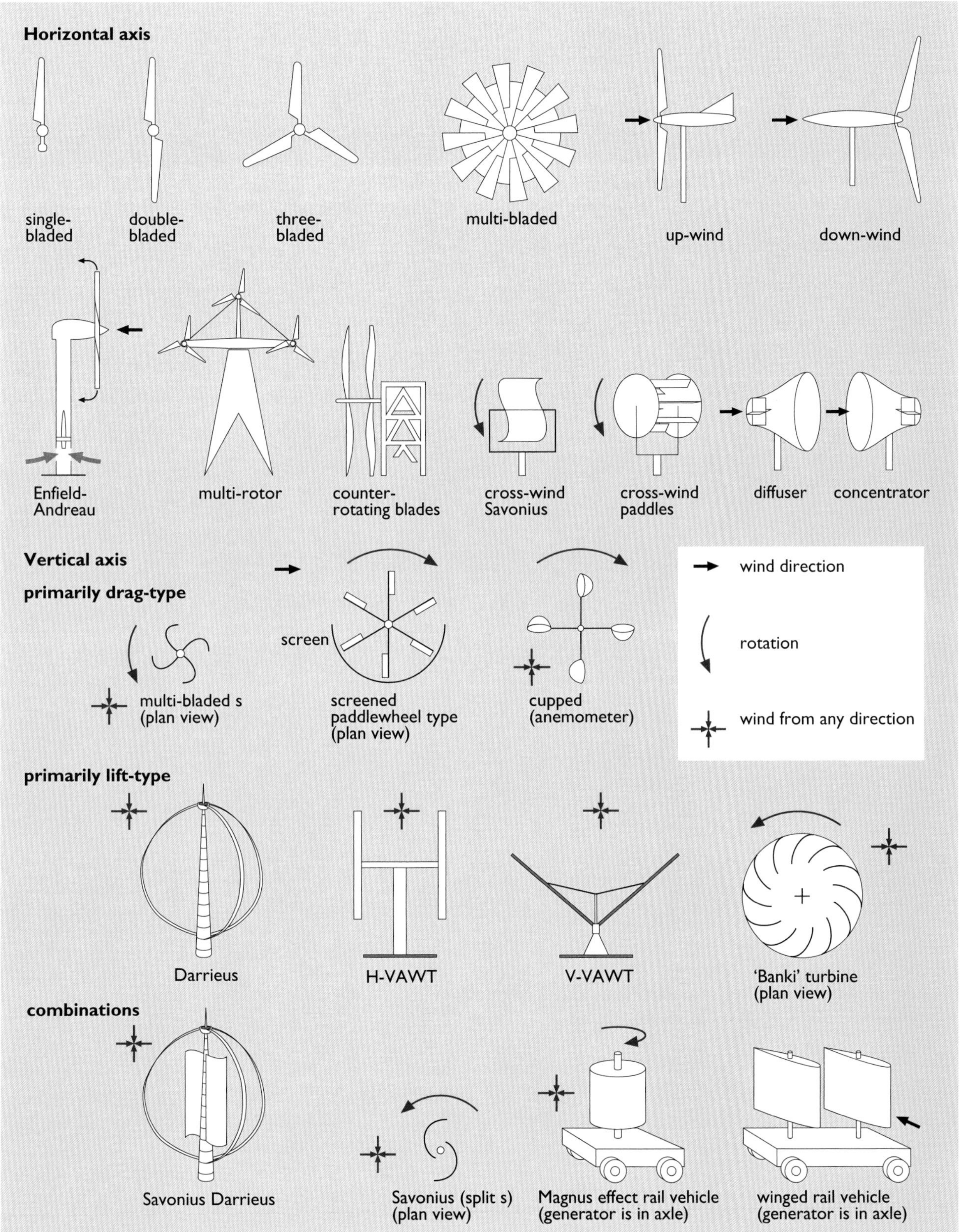

Figure 7.9 Some examples of the machines that have been proposed for wind energy conversion (source: adapted from Eldridge, 1975. For further information on these machines see Eldridge, 1975 and Golding, 1955)

The more familiar *horizontal-axis* windmills are thought to have appeared in Europe in the twelfth century. The first recorded reference in England dates from 1185 AD, which mentions a windmill in the village of Weedley in Yorkshire (Reynolds, 1970). These traditional machines consisted of radial arms supporting sails that rotated about a horizontal axis, in a plane that faced into the direction from which the wind was blowing. The sails or blades themselves were set at a small oblique angle to the wind and moved in a plane at right angles to the wind direction. Another characteristic of these windmills is that their rotation axes were usually aligned with the wind direction.

Figure 7.10 Traditional north European tower windmill

In the Mediterranean regions of Europe, the traditional windmills took the form of triangular canvas sails attached to radial arms. In northern Europe, such windmills were characterized by long, rectangular sails (usually four in number), consisting of either canvas sheets on lattice frameworks, so-called 'common sails', or 'shutter-type sails', which resembled venetian blinds. The latter were closed when the windmill started to operate, progressively opened or closed to regulate the speed of the mill in different wind strengths and opened fully when the windmill was 'parked'.

In northern Europe there were two main forms of windmill. One was the less common 'post mill', in which the whole body of the windmill was moved about a large upright post when the wind direction changed; the other was the much more common 'tower mill' (Figure 7.10), in which the rotor and cap were supported by a relatively tall tower, usually of masonry – although one variant of the tower mill known as the 'smock mill' was constructed from timber. In the tower mill, only the cap (in combination with the rotor and its shaft) were moved in response to changes in wind direction. The sails turned fairly slowly and provided mechanical power. There are still a few enterprises using windmills to mill grain – either independently as flour producers or in combination with bakers selling cakes and bread made from windmill-stoneground wholemeal flour.

At their zenith, before the Industrial Revolution, it is estimated that there were some 10 000 of these windmills in Britain (Golding, 1955) and they formed a familiar feature of the countryside.

Wind turbine types

The variety of machines that has been devised or proposed to harness wind energy is considerable and includes many unusual devices. Figure 7.9 shows a small selection of the various types of machines that have been proposed over the years.

Modern wind turbines (apart from a few innovative designs) come in two basic configurations: horizontal axis and vertical axis. Horizontal axis turbines are predominantly of the 'axial flow' type, whereas vertical axis turbines are generally of the 'cross flow' type. They range in size from very small machines that produce a few tens or hundreds of watts to very large turbines producing as much as 5 megawatts of power.

Horizontal axis wind turbines

Horizontal axis wind turbines (HAWTs) predominantly have either two or three blades, or a large number of blades. The latter have what appears to be virtually a solid disc covered by many solid blades (usually of slightly cambered sheet metal construction) and are described as **high-solidity** devices. They include the multi-blade wind turbines (Figure 7.11) used for water pumping on farms.

In contrast, the swept area of wind turbines with two or three blades is largely void: only a very small fraction of this area appears to be 'solid'. These are referred to as **low-solidity** devices (see Box 7.2).

Modern *low-solidity* HAWTs evolved from traditional windmills: their rotors resemble aircraft propellers and they are by far the most common form of wind turbines manufactured today. They have a clean streamlined appearance, due to wind turbine designers' improved understanding of aerodynamics, derived largely from developments in aircraft wing and propeller design. Their rotors generally have two or three wing-like blades (Figures 7.12 and 7.13). They are almost universally employed to generate electricity. Some experimental single-bladed HAWTs have also been produced (Figure 7.14).

Figure 7.11 Multi-bladed wind pump

Figure 7.12 A two-bladed horizontal-axis wind turbine (WEG 400 kW)

Figure 7.13 A three-bladed horizontal-axis wind turbine (Howden 330 kW)

Figure 7.14 A single-bladed horizontal-axis wind turbine (MBB 600 kW)

BOX 7.2 Effect of the number of blades

The speed of rotation of a wind turbine is usually measured in either revolutions per minute (rpm) or radians per second (rad s^{-1}). The **rotation speed** in revolutions per minute (rpm) is usually symbolized by N and the **angular velocity** in radians per second is usually symbolized by Ω (and sometimes by ω). The relationship between the two is given by:

$$1\,\text{rpm} = \frac{2\pi}{60}\,\text{rad s}^{-1} = 0.10472\,\text{rad s}^{-1}$$

Another measure of a wind turbine's speed is its **tip speed**, U, which is the **tangential velocity** of the rotor at the tip of the blades, measured in metres per second. It is the product of the angular velocity, Ω, of the rotor and the **tip radius**, R (in metres),

i.e. $U = \Omega R$

Alternatively, U can be defined as:

$$U = \frac{2\pi RN}{60}$$

By dividing the tip speed, U, by the **undisturbed wind velocity**, V_0, upstream of the rotor, we obtain a non-dimensional ratio known as the **tip speed ratio**, which is usually symbolized by (λ). This ratio provides us with a useful measure with which to compare wind turbines of different characteristics.

$$\lambda = \frac{U}{V} = \frac{\Omega R}{V}$$

A wind turbine of a particular design can operate over a range of tip speed ratios, but will usually operate with its best efficiency at a particular tip speed ratio, i.e., when the velocity of its blade tips is a particular multiple of the wind velocity.

The optimum tip speed ratio for a given wind turbine rotor will depend upon both the number of blades and the width of each blade. As we have seen, the term '**solidity**' describes the fraction of the swept area that is solid. Wind turbines with large numbers of blades have highly solid swept areas and are referred to as **high-solidity** wind turbines; wind turbines with small numbers of narrow blades are referred to as **low-solidity** wind turbines. Multi-blade wind pumps have **high-solidity rotors** and modern electricity-generating wind turbines (with one, two or three blades) have **low-solidity rotors**.

In order to extract energy as efficiently as possible, the blades have to interact with as much as possible of the wind passing through the rotor's **swept area**. The blades of a high-solidity, multi-blade wind turbine interact with all the wind at very low tip speed ratios, whereas the blades of a low-solidity turbine have to travel much faster to 'virtually fill up' the swept area, in order to interact with all the wind passing through. If the tip speed ratio is too low, some of the wind travels through the rotor swept area without interacting with the blades; whereas if the tip speed ratio is too high, the turbine offers too much resistance to the wind, so that some of the wind goes around it. A two-bladed wind turbine rotor with each blade the same width as those of a three-bladed rotor will have an optimum tip speed ratio *one-third higher* than that of a three-bladed rotor. A one-bladed rotor with a blade width the same as that of a two-bladed rotor will have *twice the optimum tip speed ratio* of the two-bladed rotor. Optimum tip speed ratios for modern low-solidity wind turbines range between about 6 and 20.

In theory, the more blades a wind turbine rotor has, the more efficient it is. However, large numbers of blades can interfere with each other, so high-solidity wind turbines tend to be less efficient overall than low-solidity turbines. Of low-solidity machines, three-bladed rotors tend to be the most energy efficient; two-bladed rotors are slightly less efficient and one-bladed rotors slightly less efficient still. Wind turbines with more blades can be generally expected to generate less aerodynamic noise (see Section 7.6) than wind turbines with fewer blades.

The mechanical power that a wind turbine extracts from the wind is the product of its angular velocity and the torque imparted by the wind. **Torque** is the moment about the centre of rotation due to the driving force imparted by the wind to the rotor blades. Torque is usually measured in newton metres (Nm). For a given amount of power, the *lower* the angular velocity the *higher* the torque; and conversely, the *higher* the angular velocity the *lower* the torque.

The pumps that are used with water pumping wind turbines require a high starting torque to function. Multi-bladed turbines are therefore generally used here because of their low tip speed ratios and resulting high torque characteristics.

Conventional electrical generators run at speeds many times greater than most wind turbine rotors so they generally require some form of gearing when used with wind turbines. Low-solidity wind turbines are better suited to electricity generation because they operate at high tip speed ratios and therefore do not require as high a gear ratio to match the speed of the rotor to that of the generator.

BOX 7.3 Swaffham, a 'wind-powered town'

In August 1999, the largest and tallest (at the time of its installation) land-based wind turbine in England was installed at the Ecotech environmental education centre on the outskirts of Swaffham, in Norfolk.

The German-manufactured 66 m diameter, 65 m hub height Enercon E66 wind turbine (Figure 7.15), is rated at 1.5 megawatts and uses a special variable-speed generator directly coupled to the wind turbine rotor, thus avoiding the need for a gear box. This improves the overall power transmission efficiency and also results in a very quiet turbine. The cut-in wind speed is 2.5 m s^{-1}, the rated wind speed is 13 m s^{-1} and the rotational speed ranges between 10 and 20.3 rpm. (**Cut-in wind speed**, **rated wind speed** and **shut-down windspeed** are illustrated in Figure 7.29.)

The turbine has viewing platform located at the top of the tower just underneath the nacelle. Many visitors have climbed the 300-step spiral staircase to experience close contact with a wind turbine and to see the view over the surrounding landscape.

The turbine's annual electricity production, over 3 GWh y^{-1}, is estimated to be equivalent to about a third of the electricity consumed in the homes in the nearby town of Swaffham. It is estimated to have produced energy equivalent to that used in its manufacture during the first five months of operation.

The turbine has been so popular locally that many of the inhabitants of Swaffham asked for a further wind turbine and this (**Swaffham II**) was installed in July 2003. Swaffham II is another Enercon turbine, this time rated at 1.8 MW, with a rotor diameter of 70 metres and a hub height of 85 metres. Its annual electricity production is estimated to be over 4 GWh y^{-1}, enough electricity for over 1000 homes.

The combined output of the two turbines is estimated to supply the equivalent of around 70% of Swaffham's total household electricity requirements, amply demonstrating the potential for town-scale wind energy. The project also shows how quickly wind turbines are evolving, with the second wind turbine estimated to be generating one third more electricity than the first.

Figure 7.15 View of the Enercon 66 wind turbine near Swaffham in Norfolk, UK

Vertical axis wind turbines

Vertical axis wind turbines (VAWTs), unlike their horizontal axis counterparts, can harness winds from any direction without the need to reposition the rotor when the wind direction changes. A description of how VAWTs operate is given in **Section 7.4**.

The modern VAWT evolved from the ideas of the French engineer, Georges Darrieus, whose name is used to describe one of the vertical-axis turbines that he invented in 1925. This device, which resembles a large eggbeater, has curved blades (each with a symmetrical aerofoil cross-section) the ends of which are attached to the top and bottom of a vertical shaft (see Figure 7.16). The Darrieus VAWT is the most advanced of modern VAWTs. Several hundred were manufactured in the USA and installed in wind farms in California in the 1980s. A small number have also been produced in Canada.

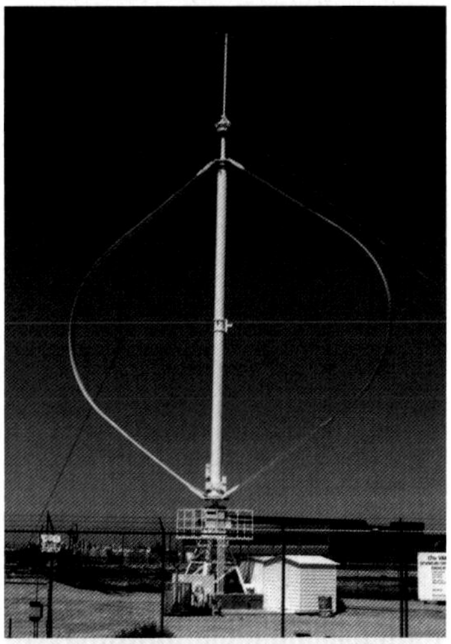

Figure 7.16 Seventeen metre diameter Darrieus-type VAWT at Sandia National Laboratories, New Mexico, US

The blades of a Darrieus VAWT take the form of a 'troposkien' (the curved, arch-like shape taken by a spinning skipping rope). This shape is a structurally efficient one, well suited to coping with the relatively high centrifugal forces acting on VAWT blades. However, the unusually-shaped blades are difficult to manufacture, transport and install. In order to

Figure 7.17 500 kW 'H'-type VAWT at Carmarthen Bay, Wales

Figure 7.18 'V'-type VAWT prototype at the Open University test site, Milton Keynes

overcome these problems, straight-bladed VAWTs have been developed: these include the 'H'-type vertical axis wind turbine (H-VAWT) and the 'V'-type vertical axis wind turbine (V-VAWT).

The H-VAWT (Figure 7.17) consists of a tower (usually housing a vertical shaft), capped by a hub to which is attached two horizontal cross arms that support the straight, upright, aerofoil blades. In the UK, this type of turbine was developed by VAWT Ltd, which built 125 kW and 500 kW prototypes at Carmarthen Bay and a 100 kW turbine on the Isles of Scilly in the 1980s. These were subsequently dismantled and only limited development of H-VAWTs has taken place since, although there is some continuing interest in variants of the H-VAWT for building-integrated wind energy applications.

The V-VAWT consists of straight aerofoil blades attached at one end to a hub on a vertical shaft and inclined in the form of a letter 'V'. Its main features include short tower, ground-mounted generator and ground level blade installation. An experimental prototype has been tested at the Open University (see Figure 7.18). A single bladed derivative of the V-VAWT (the 'Sycamore Rotor' invented and patented by the author and David Sharpe) is at an early stage of development and is being researched by the author and colleagues at the Open University and Altechnica.

VAWTs have a greater solidity than HAWTs, which usually results in a heavier and more expensive rotor. VAWTs are not at present economically competitive with HAWTs, but they continue to attract research as they should in principle offer significant advantages over HAWTs in terms of blade loading and fatigue, if built in very large sizes. The weight of a HAWT blade imposes reversing loads on the blade as it rotates upwards and downwards during each revolution. These reversing gravitational loads become significant for large scale turbines and may be a limiting factor on the maximum size of a HAWT. VAWT blades, on the other hand, do not experience reversing gravitational loads, so their maximum size is not limited by such loads.

7.4 Aerodynamics of wind turbines

Aerodynamic forces

When a force is transferred by a moving solid object to another solid object the second object will move in the same direction as the direction of motion of the first object unless subjected to another force. However, the method by which forces are transferred from a *fluid* to a solid object is very different and is similar to the way we experience forces whilst swimming.

Wind turbines are operating in an unconstrained fluid, in this case air. To understand how they work, two terms from the field of aerodynamics will be introduced. These are 'drag' and 'lift'.

An object in an air stream experiences a force that is imparted from the air stream to that object (see Figure 7.19). We can consider this force to be equivalent to two component forces acting in perpendicular directions, known as the *drag* force and the *lift* force. The magnitude of these drag and lift forces depends on the shape of the object, its orientation to the direction of the air stream, and the velocity of the air stream.

The **Drag force** is the component that is in line with the direction of the air stream. A flat plate in an air stream, for example, experiences maximum drag forces when the direction of the air flow is perpendicular (that is, at right angles) to the flat side of the plate; when the direction of the air stream is in line with the flat side of the plate, the drag forces are at a minimum. Traditional vertical axis wind mills and undershot water wheels are driven largely by drag forces.

Objects designed to minimize the drag forces experienced in an air stream are described as *streamlined*, because the lines of flow around them follow smooth, stream-like lines. Examples of streamlined shapes are teardrops, the shapes of fish such as sharks and trout, and aeroplane wing sections (aerofoils) (see Figure 7.20).

The **Lift force** is the component that is at right angles to the direction of the air stream. It is termed 'lift' force because it is the force that enables aeroplanes to '*lift*' off the ground and fly, though in other applications it may induce a *sideward* (as in a sailboat) or *downward* force (as in the spoiler aerofoil of a racing car). Lift forces acting on a flat plate are smallest when the direction of the air stream is at a zero angle to the flat surface of the plate. At small angles relative to the direction of the air stream (that is, when the so-called *angle of attack* is small), a low pressure region is created on the 'downstream' or 'leeward' side of the plate as a result of an increase in the air velocity on that side (Figures 7.21 and 7.22 show this effect on

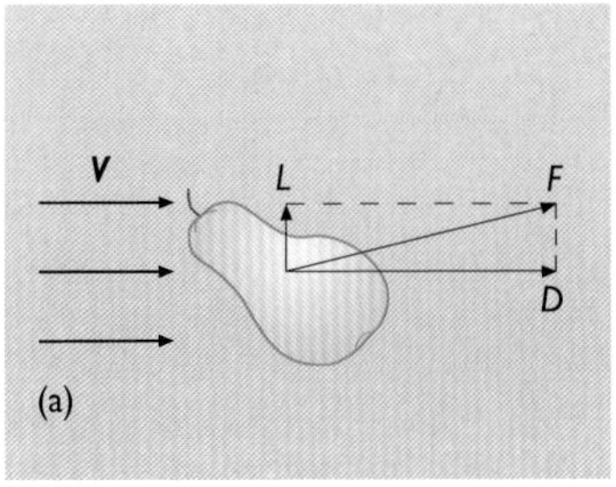

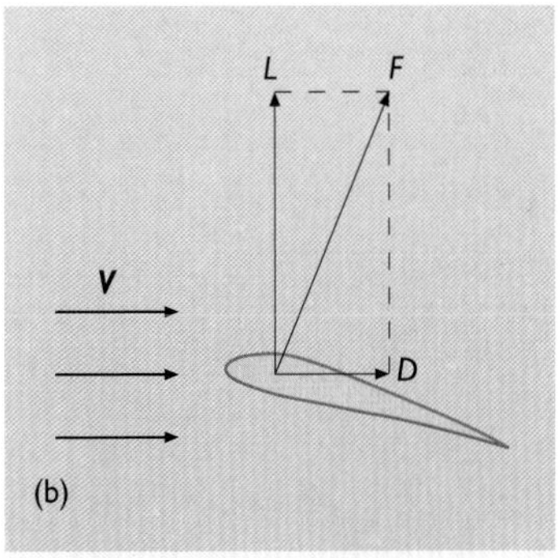

Figure 7.19 (a) and (b) An object in an air stream is subjected to a force, F, from the air stream made of two component forces: a drag force, D, acting in line with the direction of air flow; and a lift force, L, acting at 90° to the direction of air flow

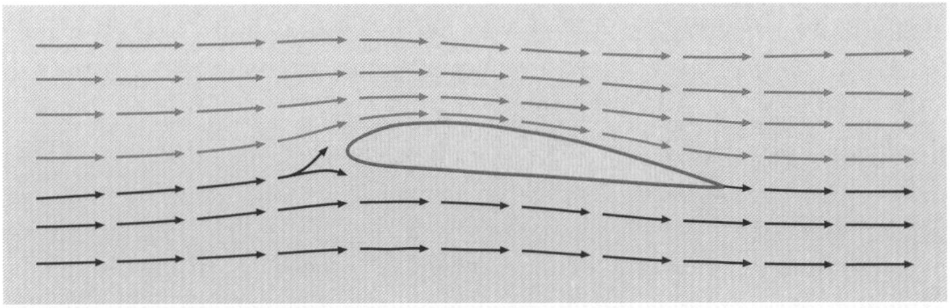

Figure 7.21 Streamlined flow around an aerofoil section

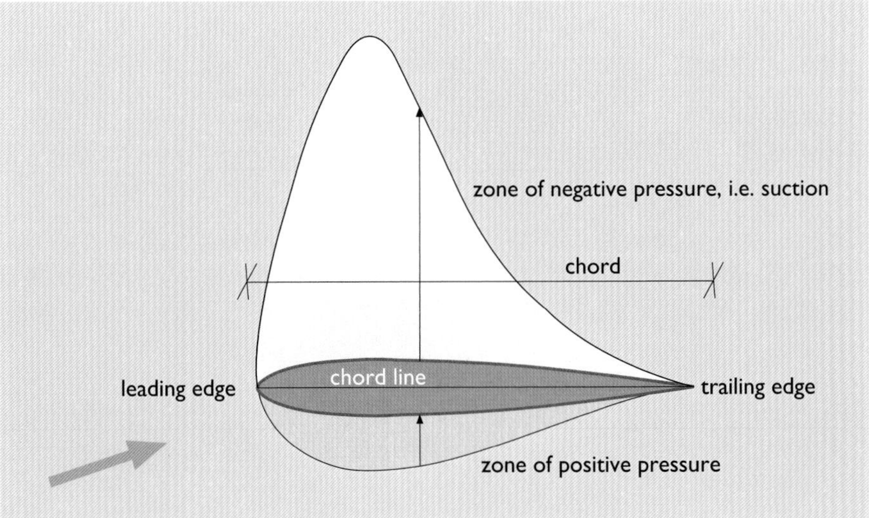

Figure 7.22 Zones of low and high pressure around an aerofoil section in an air stream

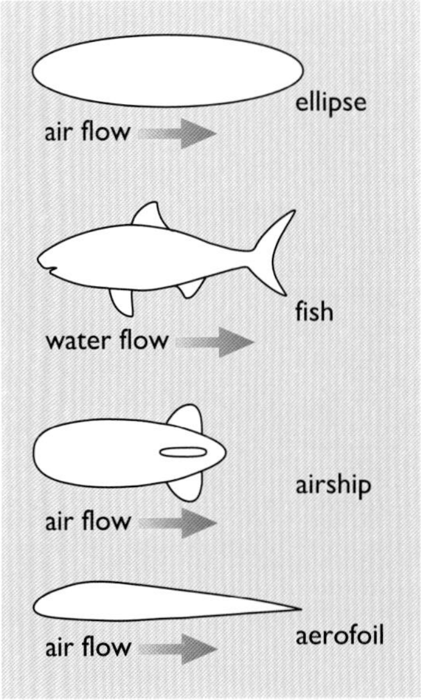

Figure 7.20 Some examples of streamlined shapes

aerofoil sections). In this situation, there is a direct relationship between air speed and pressure: the faster the airflow, the lower the pressure (i.e. the greater the 'suction effect'). This phenomenon is known as the **Bernoulli effect** after Daniel Bernoulli, the Swiss mathematician who first explained it. The lift force thus acts as a 'suction' or 'pulling' force on the object, in a direction at right angles to the airflow.

As well as enabling birds, aeroplanes and gliders to fly, it is lift forces that propel modern sailing yachts, and support and propel helicopters. They are also the forces that enable a modern wind turbine to produce power.

Aerofoils

Arching or cambering a flat plate will cause it to induce higher lift forces for a given angle of attack, and blades with a cambered plate profile work well under the conditions experienced by high-solidity, multi-bladed wind turbines. For low-solidity turbines, however, the use of so-called **aerofoil sections** is even more effective. The angle which an object makes with the direction of an airflow, measured against a reference line in the object, is called the **angle of attack** *α (alpha)*. The reference line from which measurements are made on an aerofoil section is usually referred to as the **chord line** (Figure 7.22). When employed as the profile of a wing, aerofoil sections accelerate the airflow over the more convex 'upper' surface. The high air speed thus induced results in a large reduction in pressure over the upper surface relative to the lower surface. This results in a 'suction' effect which 'lifts' the aerofoil-shaped wing. The strength of the lift forces induced by aerofoil sections is perhaps demonstrated most dramatically by their ability to support jumbo jets in the air.

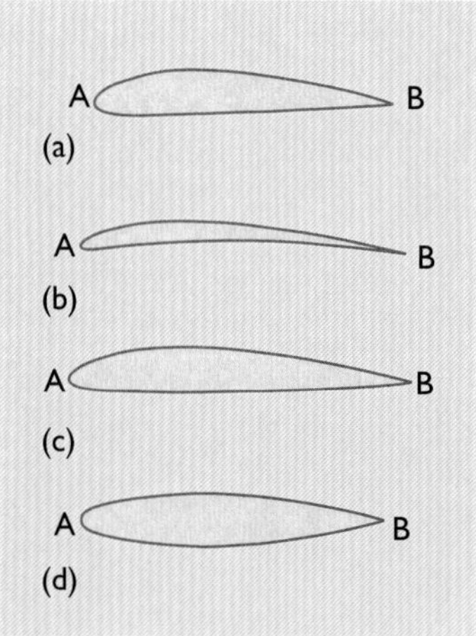

Figure 7.23 Types of aerofoil section: (a), (b) and (c) are various forms of asymmetrical aerofoil section and (d) is a symmetrical aerofoil section

There are two main types of aerofoil section: asymmetrical and symmetrical, as shown in Figure 7.23. Both have a markedly convex upper surface, a rounded end called the 'leading edge' (which faces the direction *from* which the air stream is coming), and a pointed or sharp end called the 'trailing edge'. It is the 'undersurface' of the sections that distinguishes the two types. The asymmetrical aerofoils are optimized to produce most lift when the underside of the aerofoil is closest to the direction from which the air is flowing. Symmetrical aerofoils are able to induce lift equally well (although in opposite directions) when the air flow is coming from either side of them. (Aerobatic display aeroplanes have wings with symmetrical aerofoil sections, which allow them to fly equally well when upside down.) When airflow is directed towards the underside of the aerofoil, the angle of attack is usually referred to as positive. (In the case of 'spoiler' aerofoils used on racing cars the normal 'underside' of the aerofoil is uppermost as the 'lift' force generated by the spoiler is acting downwards.)

The lift and drag characteristics of many different aerofoil shapes, for a range of angles of attack, have been determined by measurements taken in wind tunnel tests, and catalogued (e.g. in Abbott and von Doenhoff, 1958). The lift and drag characteristics measured at each angle of attack can be described using non-dimensional **lift** and **drag coefficients** (C_L and C_D) or as **lift to drag ratios** (C_L/C_D). These are defined in Box 7.3. Aerofoils can also now be designed with the aid of specially developed software, and new aerofoils are being designed and optimized to be more efficient in the aerodynamic conditions experienced by wind turbines. Figure 7.24 shows typical lift and drag coefficients, and lift to drag ratios, for one aerofoil section. Knowledge of these coefficients is essential when selecting appropriate aerofoil sections for wind turbine blade design. Lift and drag forces are both proportional to the energy in the wind.

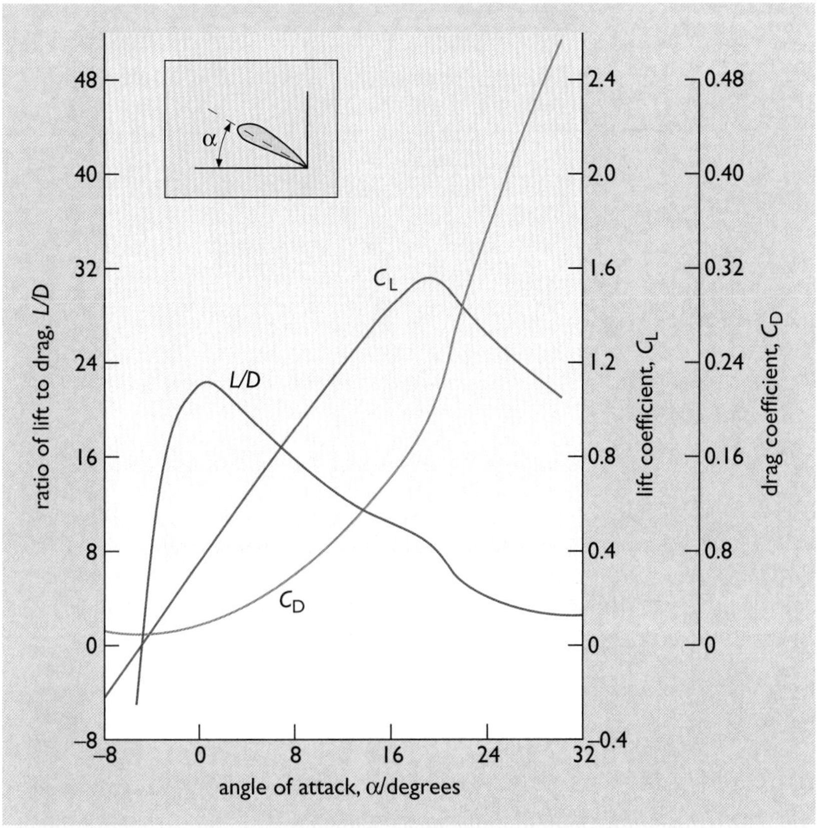

Figure 7.24 Lift coefficient, C_L, drag coefficient, C_D, and lift to drag ratio (L/D) versus angle of attack, α, for a Clark Y aerofoil section. The region just to the right of the peak in the C_L curve corresponds to the angle of attack at which stall occurs

Relative wind velocity

When a wind turbine is stationary, the direction of the wind as 'seen' from a wind turbine blade is the same as the undisturbed wind direction. However, once the blade is moving the direction from which it 'sees' the wind approaching effectively changes in proportion to the blade's velocity. (In the case of a moving vertical axis wind turbine blade, the direction from which the blade 'sees' the wind is also affected by its position during its rotation cycle – see Figure 7.28). Two-dimensional **vectors** are used to represent this effect graphically. A two-dimensional vector is a quantity that has both magnitude and direction. A velocity vector can be represented graphically in the form of an arrow, the length of which is proportional to speed, and the angular position of which indicates the direction of flow.

BOX 7.4 Aerofoil sections and lift and drag coefficients

The 'length' (from the tip of its leading edge to the tip of its trailing edge) of an aerofoil section is known as the **chord** (see Figure 7.24). The chord is also the same as the *width* of the blade at a given position along the blade.

Drag coefficient (C_D)

The drag coefficient of an aerofoil is given by the following expression:

where:

$$C_D = \frac{D}{0.5\rho V^2 A_b}$$

D is the drag force in newtons (N)

ρ is the air density in kilograms per cubic metre (kg m^{-3})

V is the velocity of the air approaching the aerofoil in metres per second (m s^{-1})

A_b is the blade area (i.e. chord × length) in square metres (m^2). In the case of a blade element, the area is equal to the mean chord × length of the blade element.

Lift coefficient (C_L)

The lift coefficient of an aerofoil is given by the following expression:

$$C_L = \frac{L}{0.5\rho V^2 A_b}$$

where L is the lift force in newtons.

The lift and drag coefficients of an aerofoil can be measured in a wind tunnel at different angles of attack and wind velocities. The results of such measurements can be presented in either tabular or graphical form as in Figure 7.24.

Each aerofoil has an angle of attack at which the lift to drag ratio (C_L/C_D) is at a maximum. This angle of attack results in the maximum resultant force and is thus the most efficient setting of the blades of a HAWT.

Another important characteristic relationship of an aerofoil is its angle of stall. This is the angle of attack at which the aerofoil exhibits *stall* behaviour. Stall occurs when the flow suddenly leaves the suction side of the aerofoil (when the angle of attack becomes too large), resulting in a dramatic loss in lift and an increase in drag (Figures 7.24 and 7.25). When this happens during the flight of an aeroplane, it can be extremely dangerous unless the pilot can make the plane recover. One of the methods used by wind turbines to limit the power extracted by the rotor in high winds takes advantage of this phenomenon.

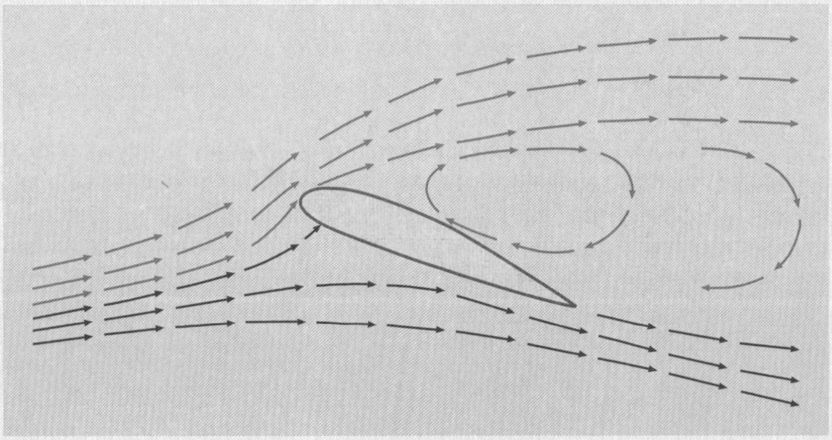

Figure 7.25 Aerofoil section in stall

The wind as seen from a point on a moving blade is known as the **relative wind** and its velocity is known as the **relative wind velocity** (usually symbolized by W). This is a vector which is the *resultant* (i.e. the vector sum) of the undisturbed wind velocity vector, V_0, and the tangential velocity vector of the blade at that point on the blade, u. (Note: the tangential velocity, measured in metres per second (ms^{-1}), is distinct from the angular velocity, which is measured in radians per second or in revolutions per minute (r.p.m.) – see Box 7.2). The angle from which the point on the moving blade sees the relative wind is known as **the relative wind angle** (usually symbolized by ϕ) and is measured from the tangential velocity vector, u.

Harnessing aerodynamic forces

Modern horizontal and vertical axis wind turbines make use of the aerodynamic forces generated by aerofoils in order to extract power from the wind, but each harnesses these forces in a different way.

In the case of a HAWT with fixed-pitch blades (with its rotor axis assumed to be in constant alignment with the undisturbed wind direction), for a given wind speed and constant rotation speed, the angle of attack at a given position on the rotor blade *stays constant throughout its rotation cycle.*

By contrast in a VAWT with fixed pitch blades, under the same conditions, the angle of attack at a given position on the rotor blade is *constantly varying throughout its rotation cycle.*

During the normal operation of a *horizontal* axis rotor, the direction from which the aerofoil 'sees' the wind is such that the angle of attack remains positive throughout.

In the case of a *vertical* axis rotor, however, the angle of attack changes from positive to negative and back again over each rotation cycle. This means that the 'suction' side reverses during each cycle, so a symmetrical aerofoil has to be employed to ensure that power can be produced irrespective of whether the angle of attack is positive or negative.

Horizontal axis wind turbines

Most horizontal axis wind turbines operate with their rotation axes in line with the wind direction and are so-called *axial flow* devices. The rotation axis is maintained in line with wind direction by a *yawing* mechanism, which constantly realigns the wind turbine rotor in response to changes in wind direction.

In addition to its swept area and rotor diameter, the performance (power output, torque and rotation speed) of an axial flow horizontal axis wind turbine rotor is dependent on numerous other factors. These include the number and shape of the blades and the choice of aerofoil section, the length of the blade chord, the tip speed ratio, the blade pitch angle, the relative wind angle and angle of attack at positions along the blade, and the amount of twist between the hub and tip.

Box 7.5 explains how the relative wind velocity, W, and relative wind angle, ϕ, both vary along the blade, together with their influence on the optimum blade pitch angle.

A few examples of cross-flow horizontal axis windmills have also been proposed historically (see Figure 7.9) and cross-flow horizontal axis turbines are under development for use in building integrated wind energy systems (see Section 7.8 below and Taylor, 1998). In this context cross-flow horizontal axis turbines function in an identical manner to the vertical axis wind turbines described below.

BOX 7.5 HAWT rotor blades wind forces and velocities

Figure 7.26 shows a section through a moving rotor blade of a HAWT. Also shown is a vector diagram of the forces and velocities at a position along the blade at an instant in time.

Because the blade is in motion, the direction from which the blade 'sees' the relative wind velocity, W, is the resultant of the tangential velocity, u, of the blade at that position and the wind velocity, V_1, at the rotor.

The tangential velocity, u, (in metres per second) at a point along the blade is the product of the angular velocity, Ω (in radians per second) of the rotor and the local radius, r, (in metres), at that point, that is:

$$u = \Omega r$$

The **wind velocity at the rotor**, V_1, is the undisturbed wind velocity upstream of the rotor, V_0, reduced by a factor that takes account of the wind being slowed down as a result of power extraction. This factor is often referred to as the **axial interference factor**, and is represented by a.

Albert Betz showed in 1928 that the *maximum* fraction of the power in the wind that can theoretically be extracted is 16/27 (59.3%). This occurs when the undisturbed wind velocity is reduced by *one-third*, in other words, when the axial interference factor, a, is equal to one-third. The value of 59.3% is often referred to as the **Betz limit** and it is believed to apply to both horizontal and vertical axis wind turbines.

The relative wind angle, ϕ, is the angle that the relative wind makes with the blade (at a particular point with **local radius** r along the blade) and is measured from the **plane of rotation**. [Note: If it were not for the fact

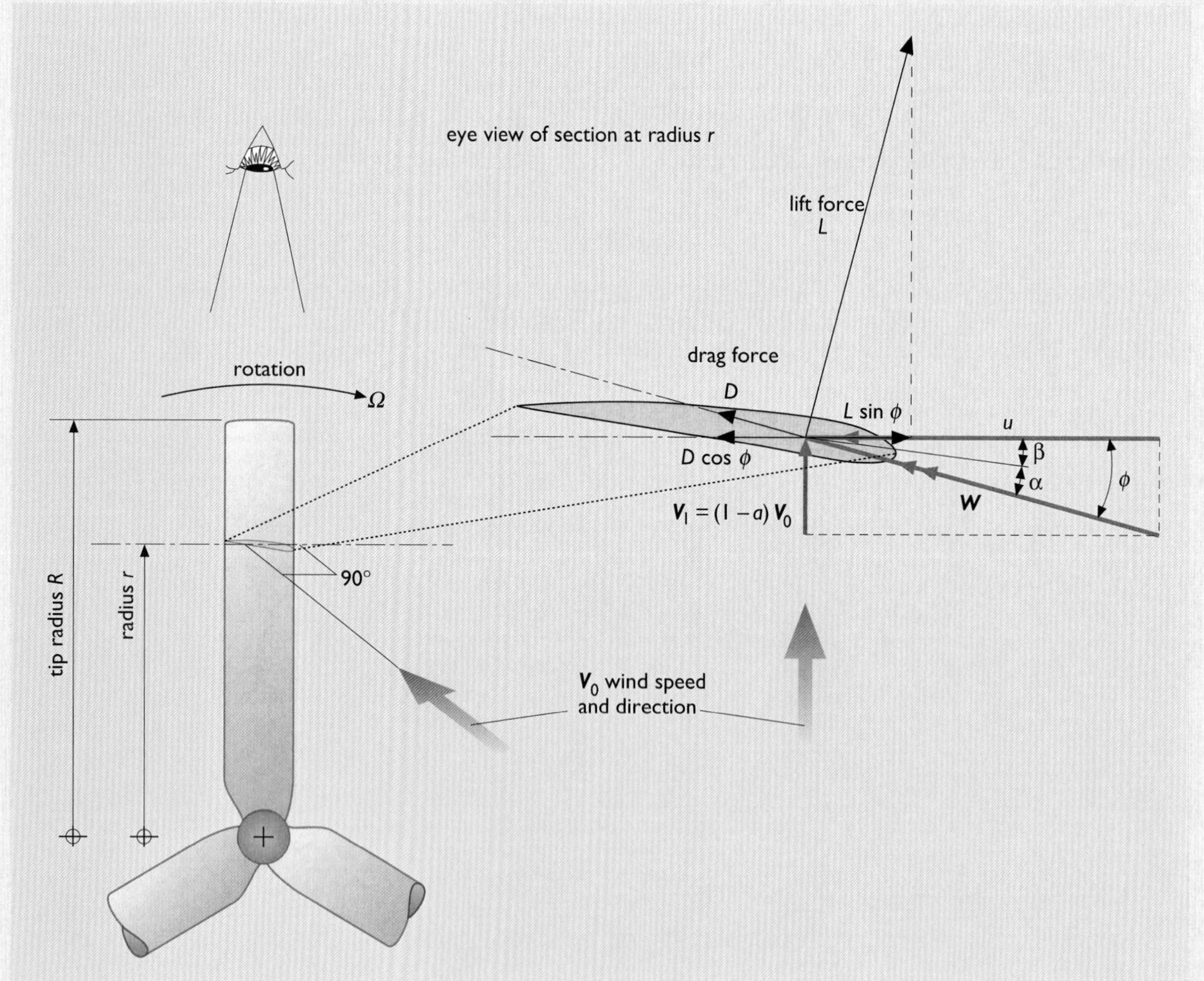

Figure 7.26 Vector diagram showing a section through a moving HAWT rotor blade. Notice that the drag force, D, at the point shown is acting in line with the direction of the relative wind, W, and the lift force, L, is acting at 90° to it

that the wind is slowed down as a result of the wind turbine extracting energy, the local tip speed ratio would be equal to the reciprocal of the tangent of the relative wind angle at the blade.] The angle of attack, α, at this point on the blade can be measured against the relative wind angle, ϕ. **The blade pitch angle** (usually represented by β) is then equal to the relative wind angle minus the angle of attack.

Because the rotor is constrained to rotate in a plane at right angles to the undisturbed wind, the **driving force** at a given point on the blade is that component of the aerofoil lift force that acts in the plane of rotation. This is given by the product of the lift force, L, and the sine of the relative wind angle, ϕ (that is, $L \sin \phi$). The **component of the drag force** in the rotor plane at this point is the product of the drag force, D, and the cosine of the relative wind angle, ϕ (that is, $D \cos \phi$).

The torque, $\boldsymbol{q}$ (that is, the **moment about the centre of rotation** of the rotor in the plane of the rotor), in newton metres (N m), at this point on the blade is equal to the product of the **net** driving force in the plane of rotation (that is, the component of lift force in the plane of rotation minus the component of drag force in the rotor plane) and the local radius, r. The **total torque,** Q, acting on the rotor can be calculated by summing the torque at all points along the length of the blade and multiplying by the number of blades. The power from the rotor is the product of the total torque, Q and the rotor's angular velocity, Ω.

The magnitude and direction of the relative wind angle, ϕ, varies along the length of the blade according to the local radius, r. This is because the **local tangential speed**, u, of a given blade element is equal to the rotor's angular velocity (Ω) times the local radius, r, of the blade element. As the tangential speed decreases towards the hub, the relative wind angle, ϕ, progressively increases. If a blade is designed to have a constant angle of attack along its length, it will have to have a built-in twist, the amount of which will vary (as the relative wind angle varies) progressively from tip to root. Figure 7.27 demonstrates the progressive twist of a HAWT rotor blade. Most manufacturers of HAWT blades use twisted blades, although it is possible to

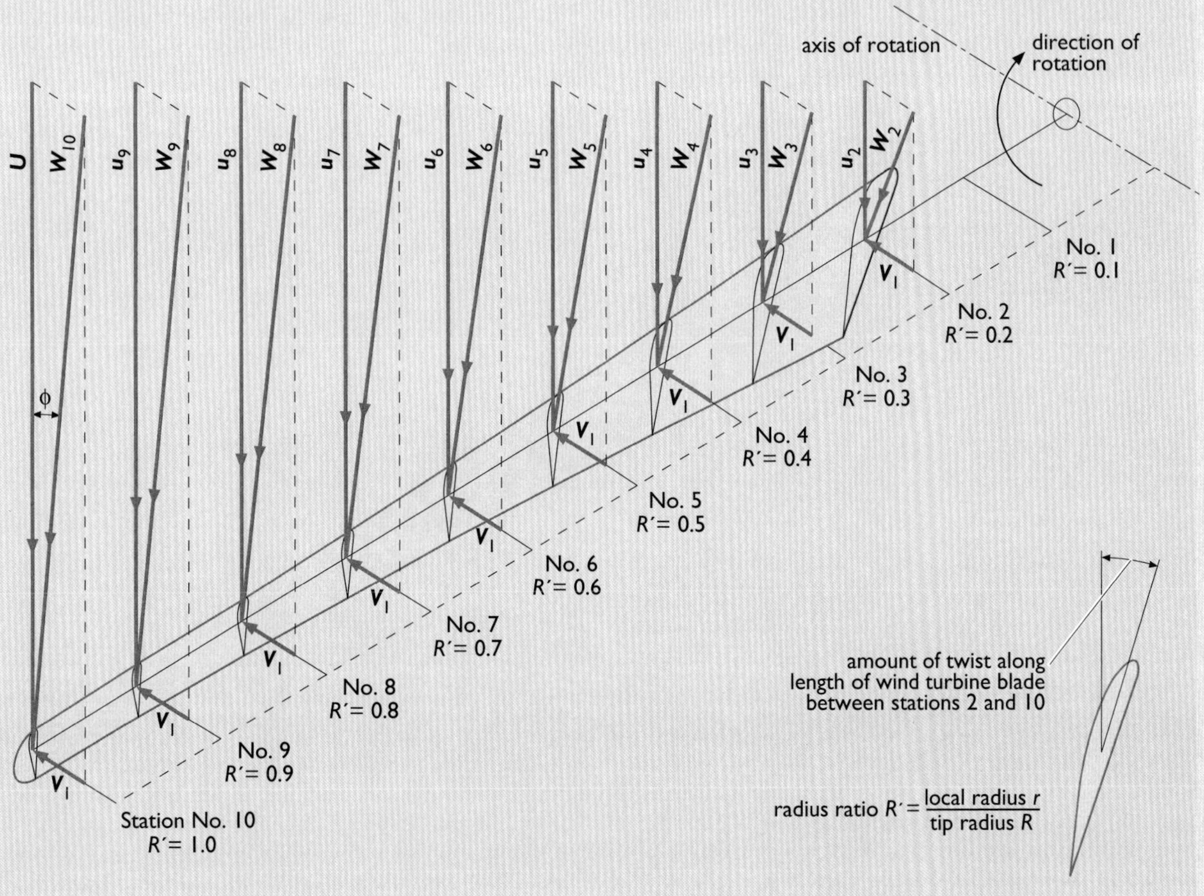

Figure 7.27 Three-dimensional view of an HAWT rotor blade design showing how the relative wind angle, ϕ, changes along the blade span

build functional HAWT rotor blades that are not twisted. These are cheaper, but less efficient and how well they function depends in part on both the aerofoil characteristics and the overall blade pitch angle.

Stall control

Let us assume that a wind turbine is rotating at a constant rotation speed, regardless of wind speed, and that the blade pitch angle is fixed. As the wind speed increases the tip speed ratio decreases. At the same time, the relative wind angle increases, causing an increase in the angle of attack.

It is possible to take advantage of this characteristic to control a turbine in high winds, if the rotor blades are designed so that above the rated wind speed they become less efficient because the angle of attack is approaching the **stall angle**. This results in a loss of lift, and thus torque, on the regions of the blade that are in 'stall'.

This method of so-called '**stall-control**' has been employed successfully on numerous fixed-pitch HAWT rotors and is also employed on most modern lift-driven VAWT rotors.

Vertical axis wind turbines

Modern VAWTs, unlike HAWTs, are 'cross-flow devices'. This means that the direction from which the undisturbed wind flow comes is at right angles to the axis of rotation; that is, the wind flows across the axis. As the rotor blades turn, they sweep a three-dimensional surface (a cylindrical surface in the case of an H-VAWT or a conical surface in the case of a V-VAWT), as distinct from the single circular plane swept by a HAWT's rotor blades.

In contrast to traditional vertical-axis windmills, the blades of modern vertical axis wind turbines extract most of the power from the wind as they pass across the front and rear – as distinct from the sides (relative to the undisturbed wind direction) – of the swept surface.

A vertical axis wind turbine will function with the wind blowing from any direction, but let us assume initially that it is blowing from one particular

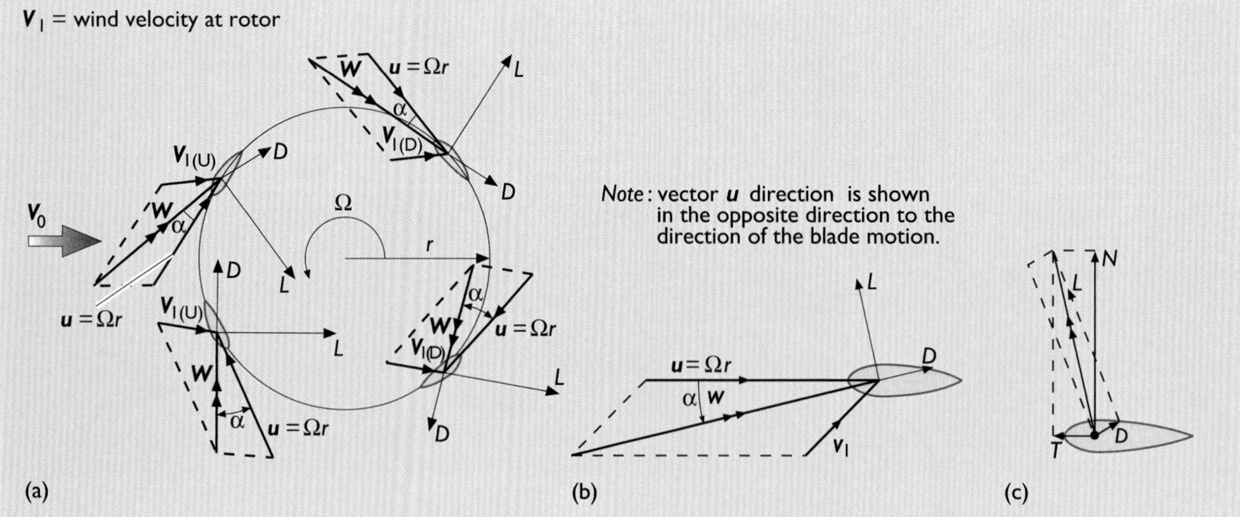

Figure 7.28 The lift and drag forces acting on VAWT rotor forces acting on rotor blades can be resolved into two components: 'normal', N, (that is, in line with the radius) and 'tangential', T, (that is, perpendicular to the radius). The magnitude of both components varies as the angle of attack varies. (a) Blade forces and relative velocities for a VAWT, showing angles of attack at different positions; (b) Detail: aerodynamic forces on a blade element of a VAWT rotor blade; (c) Normal (radial) and tangential (chord-wise) components of force on a VAWT blade. Note: $V_{I(U)}$ is wind velocity at rotor on upwind side; $V_{I(D)}$ is wind velocity at rotor on downwind side

direction and also that the setting angle of the blade is such that its chord is in line with a tangent to the circular path of rotation (that is, it has 'zero set pitch'). Clearly, the angle of the blade to the direction of the undisturbed wind changes from zero to 360 degrees over each cycle of rotation. It might appear that the angle of attack of the wind to the blade would vary by the same amount, and so it might seem impossible for a VAWT to operate at all. However, we have to take into account the fact that when the blade is *moving*, the relative wind angle 'seen' by the blade is the *resultant* (W) of the wind velocity V_1 at the rotor and the blade velocity u (see Box 7.5). Provided that the blade is moving sufficiently fast relative to the wind velocity (in practice, this means at a tip speed ratio (i.e. tip speed/wind speed) of three or more), the angle of attack that the blade makes with the relative wind velocity W will only vary within a small range (see Figure 7.28).

7.5 **Power and energy from wind turbines**

How much power does a wind turbine produce?

The power output of a wind turbine varies with wind speed and every turbine has a characteristic **wind speed-power curve**. The shape of a wind speed-power curve is influenced by the rotor swept area; the choice of aerofoil; the number of blades; the blade shape; the optimum tip speed ratio; the speed of rotation; the cut-in, rated and shut down wind speeds; the aerodynamic efficiency, gearing efficiency and generator efficiency. An example of such a curve is shown in Figure 7.29.

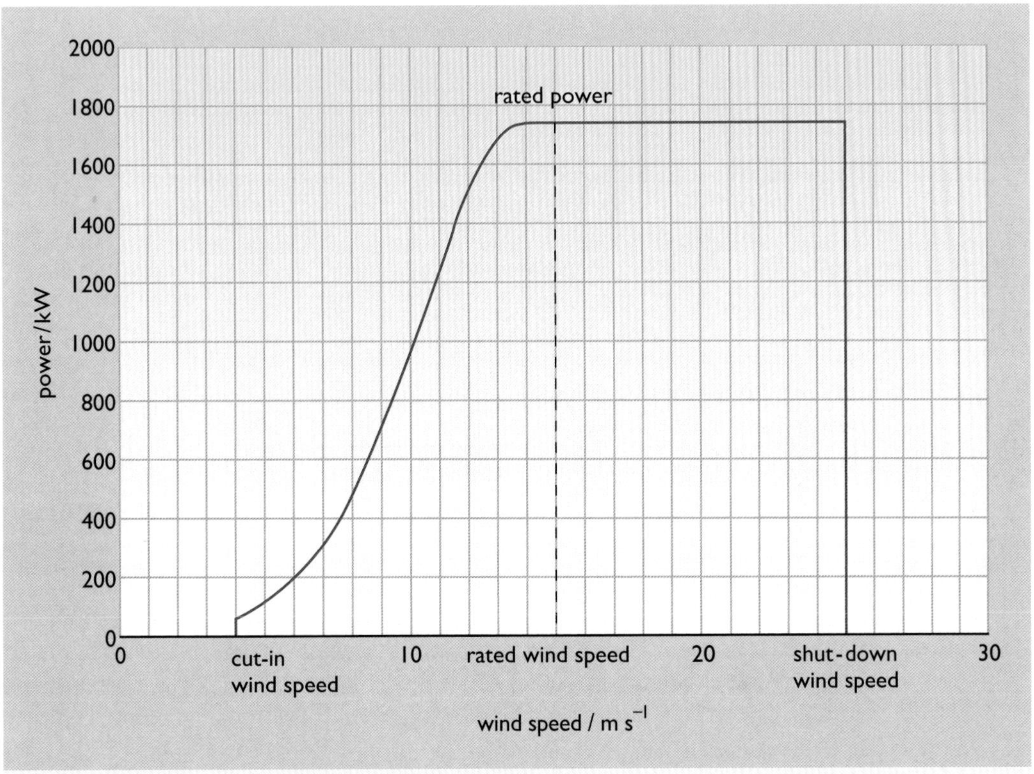

Figure 7.29 Typical wind turbine wind speed-power curve

How much energy will wind turbines produce?

The energy that a wind turbine will produce depends on both its wind speed-power curve and the **wind speed frequency distribution** at the site. The latter is essentially a graph or histogram showing the number of hours for which the wind blows at different wind speeds during a given period of time. Figure 7.30 shows a typical wind speed frequency distribution.

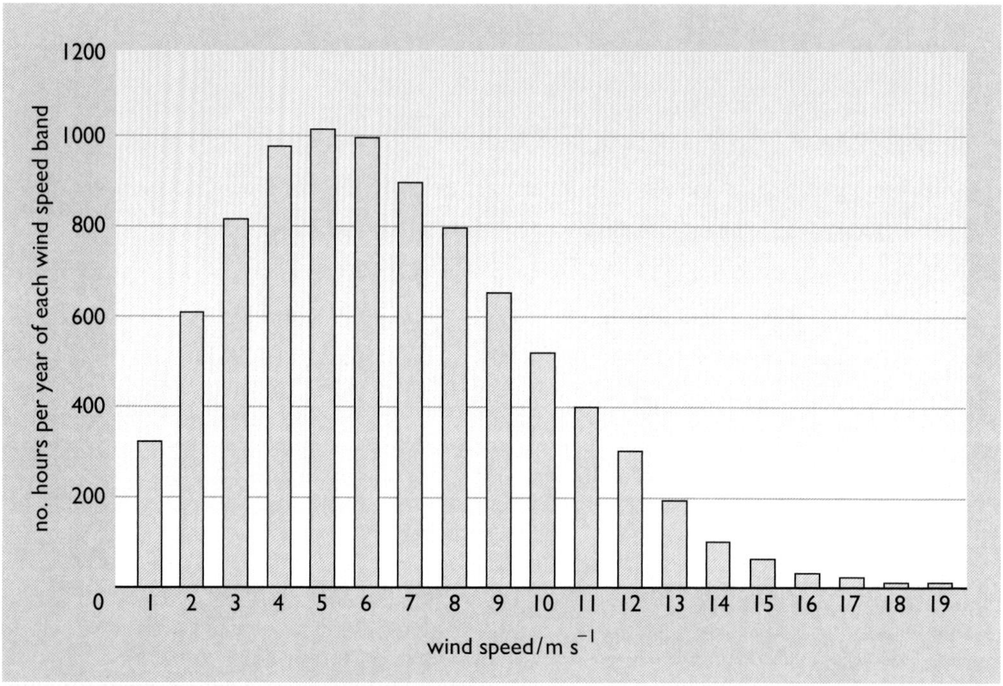

Figure 7.30 A typical wind speed frequency distribution

For each incremental wind speed within the operating range of the turbine (that is, between the cut-in wind speed and the shut-down wind speed), the energy produced at that incremental wind speed can be obtained by multiplying the number of hours of its duration by the corresponding turbine power at this wind speed (given by the turbine's wind speed-power curve). These data can then be used to plot a **wind energy distribution** such as that shown in Figure 7.31. The total energy produced is then calculated by summing the energy produced at all the wind speeds within the operating range of the turbine.

The best way to determine the wind speed distribution at a site is to carry out wind speed measurements with equipment that records the number of hours for which the wind speed lies within each given 1 m/s wide speed band, e.g., 0–1 m sec^{-1}, 1–2 m sec^{-1}, 2–3 m sec^{-1}, etc.

The longer the period over which measurements are taken, the more accurate is the estimate of the wind speed frequency distribution. Because of the V^3 law (i.e. the power in the wind is proportional to the cube of the wind velocity), a small error in estimating the wind speeds can produce a large error in the estimate of the energy yield.

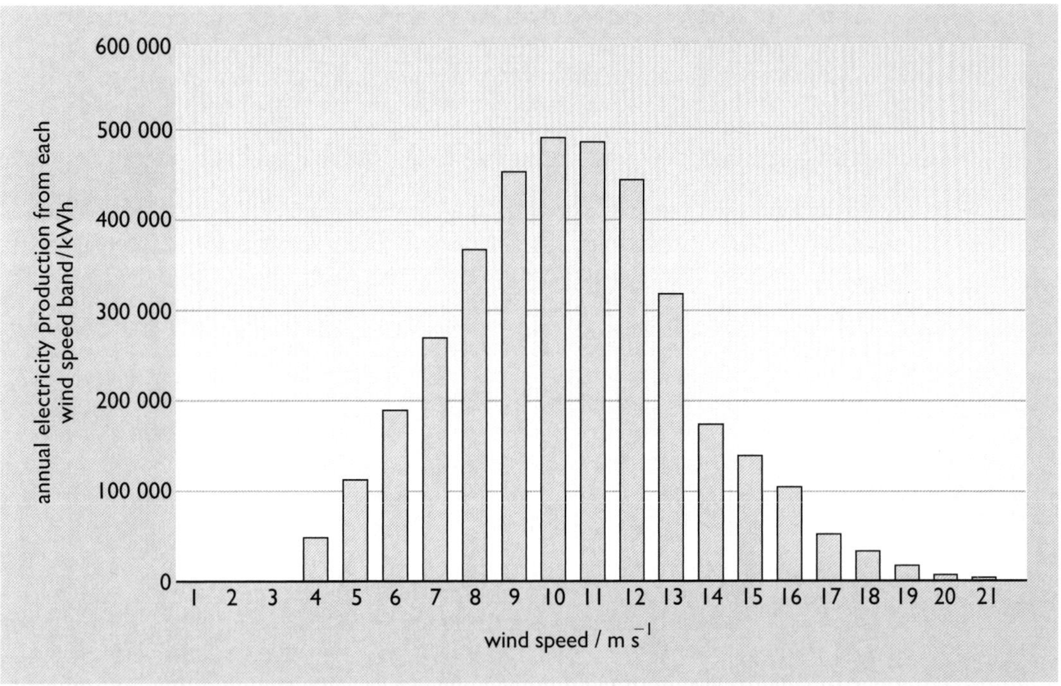

Figure 7.31 Wind energy distribution for the same site as in Figure 7.30, showing energy produced at this site by a wind turbine with the wind speed-power curve shown in Figure 7.29

Additional factors that affect the total energy generated include transmission losses and the **availability** of the turbine. Availability is an indication of the reliability of the turbine installation and is the fraction or percentage of a given period of time for which a wind turbine is available to generate, when the wind is blowing within the turbine's operating range. Current commercial wind turbines typically have annual availabilities in excess of 90%, many have operated at over 95% and some are achieving 98%.

If the mean annual wind speed at a site is known or can be estimated, the following formula (EWEA, 1991; Anderson, 1992, Beurskens and Jensen, 2001) can be used to make a *rough initial estimate* of the electricity production (in kilowatt-hours per year) from a number of wind turbines at that site:

$$\text{Annual electricity production} = K\, V_m^{\,3}\, A_t\, T$$

where:

$K = 3.2$ and is a factor based on typical turbine performance characteristics and an approximate relationship between mean wind speed and wind speed frequency distribution (see below)

V_m is the site annual mean wind speed in metres per second

A_t is the swept area of the turbine in square metres

T is the number of turbines.

This formula should be used with caution, however, because it is based on an average of the characteristics of wind turbines currently available and assumes an approximate relationship between annual mean wind speed and the frequency distribution of wind speeds that may not be accurate for an individual site. It does not allow for the different power curves of wind turbines that have been optimized either for low or high wind speed sites. The K factor of 3.2 given above assumes a well designed turbine suited to its site (Beurskens and Jensen, 2001).

Estimating wind speed characteristics of a site

It is expensive to carry out detailed measurements at a site, but there are a number of techniques that can be employed to give an approximate estimate of its wind speed characteristics.

Using wind speed measurements from a nearby location

This involves making use of existing wind speed measurements from one or more locations nearby and deriving the data for the proposed site by interpolation or extrapolation, taking into account differences between the proposed site and the sites for which measurements are available.

Using wind speed maps and atlases

Maps are available that give estimates of the mean wind speeds over the UK and many other countries. However, most of these maps were made using data from meteorological stations, which tend to be located in places that are often not appropriate for wind energy.

A *European Wind Atlas* (Troen and Petersen, 1989) has been produced by the Riso Laboratory in Denmark for the European Commission. This extensive document (over 650 pages) includes maps of various areas within the European Union (EU) (for example, Figure 7.32), which show the annual mean wind speed at 50 m above ground level for five different topographic conditions: sheltered terrain, open plain, sea coast, open sea, hills and ridges. The atlas includes a whole series of procedures for taking account of site characteristics to estimate the wind energy likely to be available. These procedures work quite well on sites with gentle topography but are not so good for very hilly terrain.

The Energy Technology Support Unit (ETSU) has also prepared a UK wind atlas. Using wind speed data from meteorological stations, a digital terrain model of the UK and a wind speed prediction computer model, ETSU has estimated an annual mean wind speed value for each 1 km Ordnance Survey grid square in the UK (Burch and Ravenscroft, 1992; further information at: www.bwea.com/noabl/download.htm).

Wind flow simulation computer models

A number of computer models have been developed that attempt to predict the effects of topography on wind speed. Data from the nearest wind speed measurement station, together with a description of its site, are required and local effects are taken into account to arrive at estimated wind data for the proposed wind turbine site. Two of the most popular models include NOABL and WASP. NOABL was used in the development of the UK wind atlas and WASP was used in the development of the European Wind Atlas (Figure 7.32). WASP also forms the basis of at least two proprietary wind speed assessment computer software models. Used with care, such models can be useful for carrying out initial assessments.

Wind resources at 50 m above ground level for five different topographic conditions										
	Sheltered terrain		Open plain		At a sea coast		Open sea		Hills and ridges	
	$m\ s^{-1}$	$W\ m^{-2}$	$m\ s^{-1}$	$W\ m^{-2}$	$m\ s^{-1}$	$W\ m^{-2}$	$m\ s^{-1}$	$W\ m^{-2}$	$m\ s^{-1}$	$W\ m^{-2}$
	>6.0	>250	>7.5	>500	>8.5	>700	>9.0	>800	>11.5	>1800
	5.0–6.0	150–250	6.5–7.5	300–500	7.0–8.5	400–700	8.0–9.0	600–800	10.0–11.5	1200–1800
	4.5–5.0	100–150	5.5–6.5	200–300	6.0–7.0	250–400	7.0–8.0	400–600	8.5–10.0	700–1200
	3.5–4.5	50–100	4.5–5.5	100–200	5.0–6.0	150–250	5.5–7.0	200–400	7.0–8.5	400–700
	<3.5	<50	<4.5	<100	<5.0	<150	<5.5	<200	<7.0	<400

Figure 7.32 Annual mean wind speeds and wind energy resources over Europe (EU Countries) (source: Troen and Petersen, 1989)

7.6 Environmental impact

Wind energy development has both positive and negative environmental impacts. The scale of its future implementation will rely on successfully maximising the positive impacts whilst keeping the negative impacts to the minimum.

Environmental benefits of electricity generation by wind energy

The generation of electricity by wind turbines does not involve the release of carbon dioxide or pollutants that cause acid rain, smog, or radioactivity or contaminate land, sea or water courses. Large-scale implementation of wind energy within the UK would probably be one of the most economic and rapid means of reducing carbon dioxide emissions. Over its working lifetime, a wind turbine is able to generate some 80 times the energy required to produce it (see the companion text, Chapter 13).

In addition, wind turbines do not require the consumption of water supplies, unlike many conventional (and some renewable) energy sources. This benefit could be of growing importance if water shortages occur with increasing frequency in the future.

Environmental impacts of wind turbines

Possible environmental impacts of wind turbines are noise, electromagnetic interference and visual impact, possibly including 'flicker' caused by sunlight interacting with rotating blades on sunny days. The physical components (e.g. visibility, sound levels and flicker zones) of these impacts can be quantified, but whether or not they are considered negative is partly a subjective matter.

A number of additional potential environmental factors are associated with offshore wind energy development. Concerns about possible impacts on fish, crustaceans, marine mammals, marine birds and migratory birds are currently the subject of ongoing research in order to confirm whether or not they are likely to be justified.

Wind turbine noise

Whilst wind turbines are often described as noisy by opponents of wind energy, in general they are not especially noisy compared with other machines of similar power rating (see for example Table 7.1 and Figure 7.33). However, there have been incidents where wind turbine noise has been cited as a nuisance. Currently available modern wind turbines are generally much quieter than their predecessors and conform to noise emission level requirements (see below).

There are two main sources of wind turbine noise. One is that produced by mechanical or electrical equipment, such as the gearbox and the generator, known as **mechanical noise**; the other is due to the interaction of the air flow with the blade, referred to as **aerodynamic noise**.

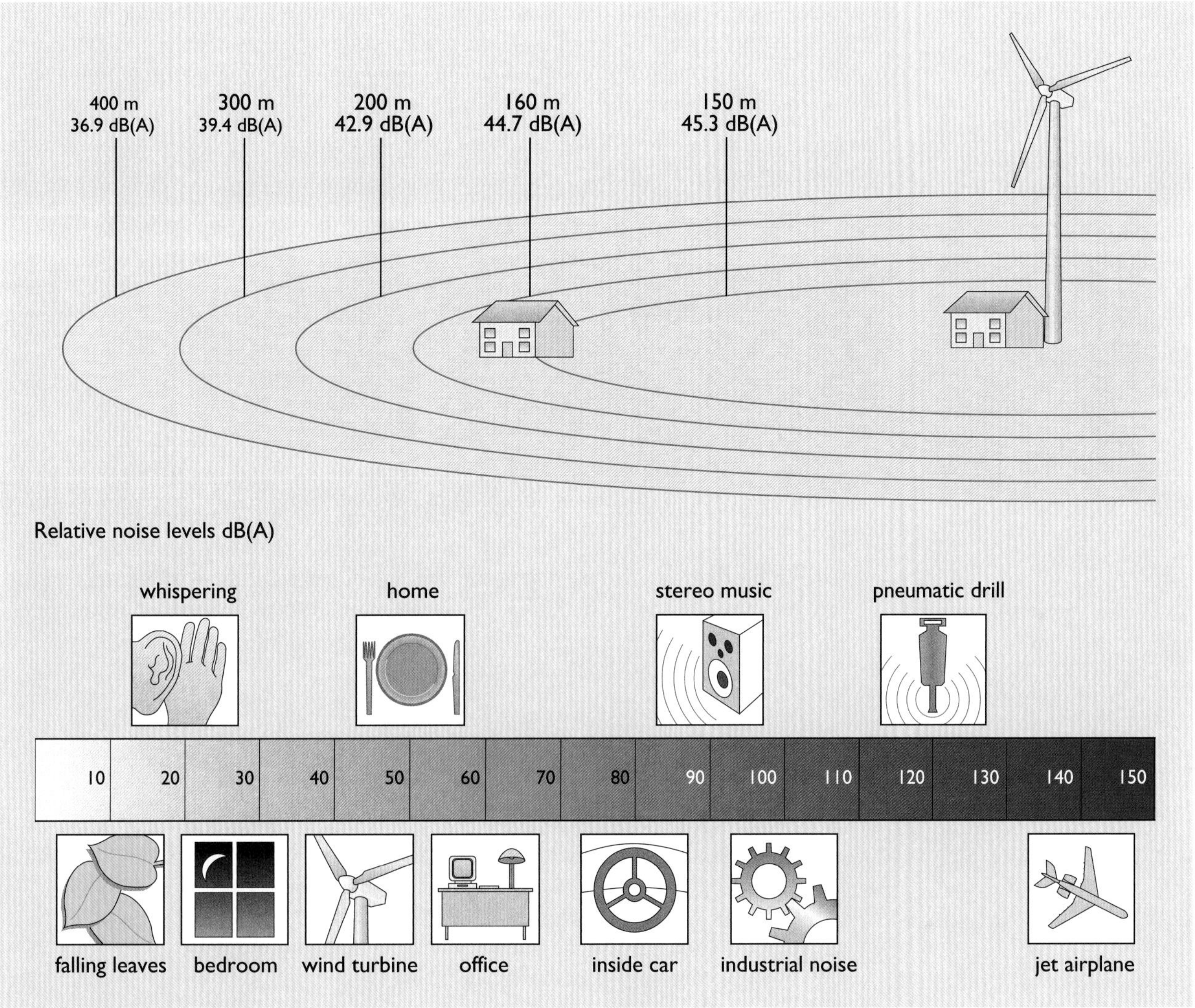

Figure 7.33 Noise pattern from a typical wind turbine (source: EWEA, 1991)

Table 7.1 Noise of different activities compared with wind turbines

Source/activity	**Noise level in dB(A)***
Threshold of pain	140
Jet aircraft at 250 m	105
Pneumatic drill at 7 m	95
Truck at 48 km h^{-1} (30 mph) at 100 m	65
Busy general office	60
Car at 64 km h^{-1} (40 mph)	55
Wind farm at 350 m	35–45
Quiet bedroom	20
Rural night-time background	20–40
Threshold of hearing	0

* dB(A): decibels (acoustically weighted)
(Source: Department of the Environment, 1993)

Mechanical noise is usually the main problem, but it can be remedied fairly easily by the use of special quieter gears, mounting equipment on resilient mounts, and using acoustic enclosures or eliminating the gearbox by opting for a direct-drive, low speed generator.

Aerodynamic noise

The aerodynamic noise produced by wind turbines can perhaps best be described as a 'swishing' sound. It is affected by: the shape of the blades; the interaction of the airflow with the blades and the tower; the shape of the blade trailing edge; the tip shape; whether or not the blade is operating in stall conditions; and turbulent wind conditions, which can cause unsteady forces on the blades, causing them to radiate noise.

Aerodynamic noise will tend to increase with the speed of rotation. For this reason, some turbines are designed to be operated at lower rotation speeds during periods of low wind speed. Noise nuisance is usually more of a potential risk in light winds; at higher wind speeds, background wind noise tends to mask wind turbine noise. Operating at a lower rotation speed will help to minimize aerodynamic noise emissions in low wind conditions. Increasing the numbers of blades is also likely to reduce aerodynamic noise.

Noise regulations, controls and reduction

Most commercial wind turbines undergo noise measurement tests according to either the recommended procedure developed by the International Energy Agency (Ljunggren, 1994 and 1997) or a procedure conforming to the Danish noise regulations. The measured noise level values from such tests provide information that enables the turbines to be sited at a sufficient distance from habitations to avoid noise nuisance. This standard procedure also allows manufacturers to identify any noise problem and take remedial action before the commercial launch of the machine. Figure 7.33 shows an indicative wind turbine noise pattern.

In Denmark, in order to control the effects of noise from wind turbines, there is a standard that specifies that the maximum wind turbine noise level permitted at the nearest dwelling in open countryside should be 45 dB(A). At habitations in residential areas a noise level of only 40 dB(A) is permitted. This noise limit has been demonstrated to be achievable with commercially-available turbines. In the UK, the noise limit at buildings near to roads is 68 dB(A), a value that must not be exceeded for more than 10% of the time over an 18-hour period. At present there are no standard maximum permitted noise levels specifically for wind turbines in the UK. However guidance is given in Planning Policy Guidance note 22 (PPG22) (Department of the Environment, 1993), referred to later in the section on planning, and more recently its successor: Planning Policy Statement 22 (PPS22) (ODPM, 2003).

Noise is a sensitive issue. Unless it is given careful consideration both at the wind turbine design and the project planning stages, taking into account the concerns of people who may be affected, opposition to wind energy development is likely.

By eliminating the gear box, the level of mechanical turbine noise can be considerably reduced. At least one manufacturer has developed a wind turbine with a low speed generator that enables it to be directly coupled to the rotor without the need for a gearbox. These turbines are very quiet and can be more comfortably sited close to buildings, as in the case of the Swaffham turbine described in Box 7.3.

There is also much ongoing research to reduce wind turbine noise. A useful overview of this work is included in Legerton (1992) and refinements in methods of measurement and noise propagation modelling are described in Kragh *et al*, 1999.

Electromagnetic interference

When a wind turbine is positioned between a radio, television or microwave transmitter and receiver (Figure 7.34) it can sometimes reflect some of the electromagnetic radiation in such a way that the reflected wave interferes with the original signal as it arrives at the receiver. This can cause the received signal to be distorted significantly.

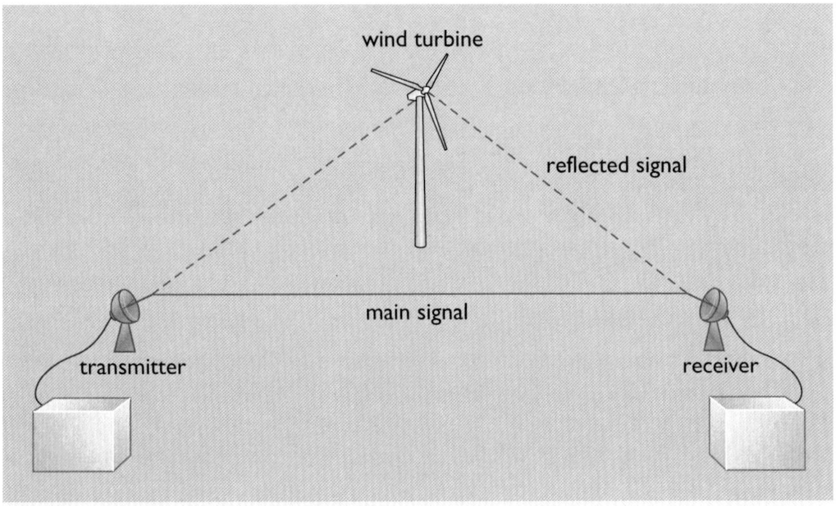

Figure 7.34 Scattering of radio signals by a wind turbine

The extent of electromagnetic interference caused by a wind turbine depends mainly on the materials used to make the blades and on the surface shape of the tower. If the turbine has metal blades (or glass-reinforced plastic (GRP) blades containing metal components), electromagnetic interference may occur if it is located close to a radio communications service. The laminated timber blades used in some turbines absorb rather than reflect radio waves so do not generally present a problem. Faceted towers reflect more than smooth rounded towers, due to their flat surfaces.

The most likely form of electromagnetic interference is to television reception. This is relatively easily dealt with by the installation of relay transmitters or by connecting cable television services to the affected viewers.

Microwave links, VHF Omni-directional Ranging (VOR) and Instrument Landing Systems (ILS) can also be affected by wind turbines. A method of determining an acceptable exclusion zone around radio transmission links has been developed which takes account of the characteristics of antennae (see Bacon, 2002). For a simplified approach to avoiding electromagnetic interference see Chignell, 1987.

Wind turbines and military aviation

The UK Ministry of Defence (MOD) has recently voiced concern about interference with military radar that could be caused by wind turbines. In addition the MOD is concerned that wind turbines (particularly those with large diameters and tall towers), when located in certain areas, will penetrate the lower portion of the low flying zones used by military aircraft, particularly in so called Tactical Training Areas. Whether these concerns are justified is still to be confirmed but the MOD's intervention has already impeded the development of a major wind farm in the north of England.

A working group has produced some interim guidelines (Wind Energy, Defence and Civil Aviation Interests Working Group, 2002) aimed at minimising risks to aircraft and air defence radar systems. It has also commissioned a number of studies (QineteQ, 2003; AMS, 2003) to attempt to clarify the precise nature of the effects of wind turbines on radar.

A review of wind turbines and radar operational experience, and of mitigating measures in the UK and overseas, was commissioned by Natural Power on behalf of a consortium of wind energy companies (Spaven Consulting, 2001). This review indicated that there are several examples of satisfactory coexistence of wind turbines and radar. Case studies from Denmark, the Netherlands and the USA indicate that technical solutions enabling radar stations to process-out the effects of wind turbines have been successfully applied. It seems that the software involved could be retrofitted to radar sites considered to be at risk. However the UK Civil Aviation Authority (CAA) has indicated that such 'radar track processing' is not acceptable in the UK.

A review of European experience and practice was commissioned by the UK Department of Trade and Industry (Jago and Taylor, 2002). This study focussed on experience in Denmark, Germany, Netherlands and Sweden. It concluded that in general that the effect of wind energy on military aviation was not a major problem.

> In Denmark the two seem to coexist most easily. In the Netherlands, also, aviation interests do not appear to impinge on wind energy developments. In Germany, frictions have appeared in the past between the two interests and may well increase in the future, but this has not prevented the rapid growth of wind development. In Sweden there has been a similar amount of interest in the issue as the UK, particularly with reference to the effects of turbines on technical systems. However, the tightest restrictions in Sweden come not from aviation specifically, but from other military activities unique to Sweden (the 'listening posts').
>
> Jago and Taylor, 2002

Visual impact

The visual perception of a wind turbine or a wind farm is determined by a variety of factors. These will include physical parameters such as turbine size, turbine design, number of blades, colour, the number of turbines in a wind farm, the layout of the wind farm and the extent to which moving rotor blades attract attention. Figure 7.35 compares wind turbines and other structures in the landscape.

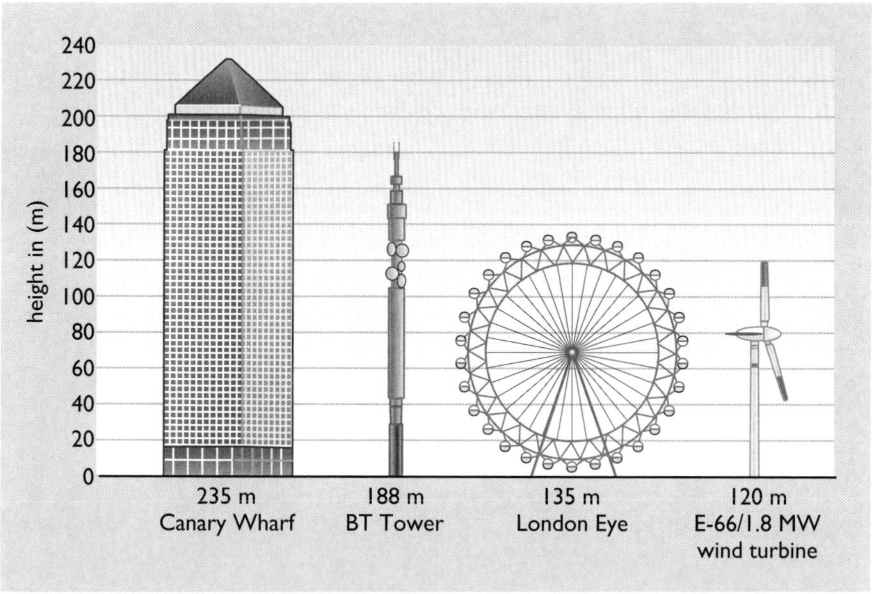

Figure 7.35 Comparison of wind turbines and other structures in the landscape

An individual's perception of a wind energy project will also depend on a variety of less easily defined psychological and sociological parameters. These may include the individual's level of understanding of the technology, opinions on what sources of energy are desirable and his or her level of involvement with the project. Newspaper and television reports are usually the only source of information most people have access to about wind energy and may well influence their opinions on the subject.

Much of the controversy about wind energy development has been due to opposition to changes to the *visual* appearance of the landscape. However, whether this is due to a visual dislike of wind turbines specifically, or simply to a general dislike to changes in the appearance of the landscape (as can happen when a new building is planned) is often unclear. Resistance to the visual appearance of new structures or unusual buildings is not a new phenomenon and opinions often change once structures become familiar.

When high visibility is an issue, wind farm developers are increasingly using more muted colours for their turbines: for instance 'battleship grey' has been chosen in one of the Danish off-shore wind farms.

Public attitudes to wind power

It sometimes appears from the media that people in the UK are generally opposed to wind power development and it seems likely that some of the negative publicity about wind power has influenced politicians and planning authorities. For some years, obtaining planning permission for wind farm projects was very difficult, though the success rate of new applications has recently improved (see below). Much of the pressure against wind power projects has been marshalled by a few vocal anti-wind farm groups, including a some national amenity groups, who believe that the visual appearance of wind turbines is not appropriate to the British countryside.

In the late 1980s and 1990s some twelve independent survey projects were carried out in order to attempt to establish public attitudes to wind power projects in the UK. These surveys canvassed the opinion of over 3000 people, the majority of whom were living close to existing or proposed wind farm sites.

Each of these surveys showed that a substantial majority of the residents who lived in areas containing the wind farms were in favour of wind power, both intellectually by supporting the idea of developing renewable energy sources, but also directly by supporting the construction of wind turbines within their locality. 80% of those surveyed supported their local wind farm; those who lived near an operating wind farm were more positive than those who did not.

When two-stage surveys were conducted, the first at the time of construction and the second some time after the wind farm had been installed, those interviewed were generally more positive at the time of the second survey. This is thought to be due to the fact that they had been able to see for themselves what an operating wind farm was like.

More recently a survey of over 2600 UK household bill payers was carried out by Ipsos in June 2003 (Ipsos, 2003) on behalf of the British Wind Energy Association (BWEA). The results of this survey indicated 74% were supportive of wind energy with 6% against and the majority of remaining bill payers neither supporting or opposing wind energy development. The BWEA has also summarized the conclusions of 42 surveys carried out between 1990 and 2002. These indicate an average 77% of the public in favour of wind energy, with an average 9% against (BWEA, 2002).

The most recent investigation was commissioned by the Central Office of Information (TNS, 2003) to assess attitudes towards renewable energy among the general public. This found that less than 20% of those surveyed would be resistant to a wind farm being developed in their area. Almost two thirds of the General Public agreed that they would be happy to have a 'clean' renewable energy generating station built in their area.

Wind turbines and birds

The main potential hazard to birds presented by wind turbines is that of fatal collisions caused by flying into the rotating blades. The incidence of bird fatalities from collisions is generally relatively low, with avian mortality running at 1 to 2 birds per turbine per year at worst, with most sites having much less than this or zero.

So far the worst location for bird strikes has been the Altamont Pass in California, where raptor species have been killed. However this raptor mortality does not seem to have been duplicated elsewhere so it may be due to special circumstances. These could include the fact that the early Californian wind farms consisted of large numbers of relatively small-diameter turbines (compared to current models) which rotated faster than later models. Also, many of the early turbines had space where birds could perch whereas more recent turbines are not so amenable.

Building on experience from Altamont Pass and other US wind energy projects, the US National Wind Co-ordinating Committee has produced a useful guide for studying the interactions between wind turbines and birds (NWCC, 1999).

To put the wind turbine bird strikes into context, however, in a single oil accident – the Exxon Valdez oil spill in Alaska's Prince William Sound in 1989 – more than 500 000 birds perished, or about 1000 times the estimated annual total in California's wind power plants (AWEA, 2002). Moreover, as the American Wind Energy Association (AWEA) points out:

> It is estimated that each year in the USA, 57 million birds die in collisions with vehicles, 1.25 million in collisions with tall structures (towers, stacks, buildings); and more than 97.5 million in collisions with plate glass (Kenetech Windpower, 1994)... An estimated 4 million to 10 million night-migrating songbirds collide with telecommunications towers each year...The ordinary pet household cat is another cause of bird mortality – household cats in the U.S. are estimated to kill 100 million birds each year.
>
> AWEA, 2002

Regarding the risk to birds from offshore wind turbines, one of the most thorough studies to date was carried out by Denmark's National Environmental Research Institute on the offshore wind farm at Tunø Knob. The location of this wind farm was deliberately chosen because of the large marine bird population, in order to monitor the interaction of birds and wind turbines. About 90% of the birds are eiders, and about 40% of the North Atlantic population of eiders winter in the Danish part of the Kattegat Sea. The Institute's conclusions were that the eiders keep a safe distance from the turbines but are not scared away from their foraging areas. 'The study showed that the offshore wind turbines have no significant impact on water birds.' (NERI, 1998).

Additional environmental factors

Additional environmental factors that should be considered in assessing the impact of a wind turbine installation include safety, shadow flicker and the possible impact on flora and fauna (see Department of the Environment, 1993).

A collaboration between English Nature, RSPB, WWF-UK and BWEA has yielded a guidance document (English Nature *et al.*, 2001) for nature conservation organisations and developers when consulting over wind farm proposals in England. This covers nature conservation, environmental impact assessments, the planning process and a checklist of possible impacts of relevance to nature conservation. It also touches on factors that should be considered for off shore wind turbines. (See also Section 7.9 below on offshore wind energy).

Planning and wind energy

Planning controls have a major influence on the deployment of wind turbines and some local authorities have developed policy guidelines on the planning aspects of wind energy. However, these aspects have been treated differently in different locations.

As already mentioned, the UK government's planning guidelines for renewable energy projects (Department of the Environment, 1993) encourage planners to look upon wind energy projects favourably, in part because of the importance of the environmental benefits of wind energy in terms of national commitments to targets for reducing CO_2 emissions. The guidelines include comments about noise, electromagnetic interference, visual aspects, shadow flicker and ecology. The Government has reviewed its planning guidance for wind energy and in 2003 published for consultation a Planning Policy Statement (ODPM, 2003) in order to assist the UK in meeting its national and international emissions targets. The statement proposes regional renewable energy targets, buffer zones and criteria-based policies for use in regional planning guidance.

Guidelines and checklists for developers and planners have also been prepared by Friends of the Earth (FOE, 1994) the British Wind Energy Association (BWEA, 1994) and Taylor and Rand (1991).

As mentioned above, the proportion of wind power projects that received planning approval compared to those that were refused has been increasing in recent years. In 2002 the proportion approved to refused (based on power rating) was 5 to 1.

Year	Refused (MW)	Approved (MW)
2000	60.675	78.68
2001	89.29	157.4
2002	122.65	621.12
2003 1st Quarter	33.9	567.25

Source BWEA, 2003

As a means of encouraging a more constructive approach to wind energy (and all renewable energy sources) the UK government, through its various regional offices, has sponsored regional resource assessment and planning studies. Each region has been required to come up with regional targets (see DTLR, 2000) for renewable energy technologies as part of its contribution to meeting the Government's aims of supplying 10% of UK electricity supplies from renewables and reducing CO_2 emissions by 20% – both by 2010. Another factor that will impact on planning and wind energy (and other renewables) is that the Department of Transport, London and the Regions (DTLR) is planning radical changes to the UK planning system. It published a consultative Green Paper (DTLR, 2001) on the subject in December 2001.

7.7 Economics

Calculating the costs of wind energy

The economic appraisal of wind energy involves a number of specific factors. These include:

- the annual energy production from the wind turbine installation;
- the capital cost of the installation;
- the annual capital charge rate, which is calculated by converting the capital cost plus any interest payable into an equivalent annual cost, using the concept of 'annuitization' (see Appendix A1);

BOX 7.6 A cost calculation procedure for wind energy

The cost per unit of electricity generated, g, by a wind farm can be estimated using the following formula:

$$g = C\,R/E + M$$

where:

C is the capital cost of the wind farm

R is the capital recovery factor or the annual capital charge rate (expressed as a fraction)

E is the wind farm annual energy output

M is the cost of operating and maintaining the wind farm annual output

The capital recovery factor, R, is defined as:

$$R = \frac{x}{1-(1+x)^{-n}}$$

where

x is the required annual rate of return net of inflation (expressed as a fraction)

n is the number of years over which the investment in the wind farm is to be recovered.

An estimate of the energy, E (in kilowatt-hours), can be made using the following formula:

$$E = (hP_rF)T$$

where

h is the number of hours in a year (8760)

P_r is the rated power of each wind turbine in kilowatts

F is the net annual capacity factor of the turbines at the site

T is the number of wind turbines

The operating and maintenance cost, M, per unit of output is defined by:

$$M = K\,C\,/\,E$$

where K is a factor representing the annual operating costs of a wind farm as a fraction of the total capital cost. The European Wind Energy Association (EWEA) has estimated this to be 0.025, that is 2.5% of capital cost. (See Figure 7.36). (Source: EWEA, 1991). While this percentage for annual operating costs can be lower, 2.5% is a good 'ball-park' figure for budgetary purposes if the actual value is unknown.

- the length of the contract with the purchaser of the electricity produced;
- the number of years over which the investment in the project is to be recovered (or any loan repaid), which may be the same as the length of the contract;
- the operation and maintenance costs, including maintenance of the wind turbines, insurance, land leasing, etc.

A detailed procedure for calculating the cost of wind energy is shown in Box 7.6.

As we have seen, the annual energy produced by a wind turbine installation depends principally on the wind speed-power curve of the turbine, the wind speed frequency distribution at the site, and the availability of the turbine.

As already noted in earlier Chapters **capacity factor** is widely used to describe the productivity of a power plant over a given period of time.

If a wind turbine were able to operate at full rated power throughout the year, it would have a capacity factor of one (i.e., 100%). However, in reality, the wind does not blow constantly at the full rated wind speed throughout the year, so in practice a wind turbine will have a much lower capacity factor. On moderate wind speed sites in the UK, with annual mean wind speeds equivalent to half the rated wind speed, a turbine capacity factor of 0.25 (i.e., 25%) is typical. However, on better wind sites, such as Carmarthen Bay in Wales, St Austell in Cornwall or the Orkney Islands, capacity factors of 0.35–0.40 or more are achievable.

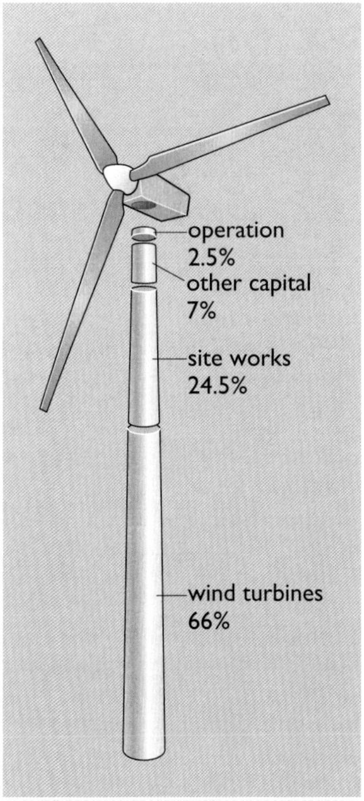

Figure 7.36 Breakdown of costs per unit of energy (kWh) from wind plant (source: adapted from EWEA, 1991)

The capital cost of wind turbines currently ranges from approximately £600 to £1000 per kilowatt of output, or £280 to £420 per square metre of rotor swept area. A breakdown of the components of the cost of energy from a typical wind turbine is shown in Figure 7.36. With 15 to 20 year contracts and on sufficiently windy sites, wind energy may be competitive with conventional forms of electricity generation, if the costs of the latter are calculated on a comparable basis. The cost of wind-generated electricity is very dependent on the way the plant is financed and this can strongly affect the price of the electricity produced.

As the cost of wind energy does not include the cost of fuel, it is relatively straightforward to determine, compared with the cost of energy from fuel-consuming power plants, which are dependent on estimates of future fuel costs. High or escalating fuel prices tend to favour zero (or low) fuel cost systems such as wind energy, but steady or falling fuel prices are less favourable to them.

Wind turbines are very quick to install, so they can be generating before they incur significant levels of interest on the capital expended during construction – in contrast to many other highly capital-intensive electricity generating plant (e.g. large hydro stations, tidal barrages and nuclear power stations).

The UK Government introduced in 2002 a Renewables Obligation (DTI, 2001) to succeed the Non-Fossil Fuel Obligation (NFFO). This, in effect, gives a subsidy to renewable electricity sources, including wind, and makes wind power projects more attractive to investors. The subject of costing and investing in energy projects is covered briefly in Appendix A1 and in more detail in Chapter 12 of the companion text.

7.8 Commercial development and wind energy potential

Wind energy developments world-wide

The present relatively healthy state of the wind energy industry is due largely to developments in Denmark and California in the 1970s and 1980s, and Germany in the 1990s.

In Denmark, unlike most other European countries that historically employed traditional windmills, the use of wind energy never ceased completely, largely because of the country's lack of fossil fuel reserves and because windmills for electricity generation were researched and manufactured from the nineteenth century until the late 1960s. Interest in wind energy took on a new impetus in the 1970s, as a result of the 1973 'oil crisis'. Small Danish agricultural engineering companies then undertook the development of a new generation of wind turbines for farm-scale operation.

It was California, however, that gave wind energy the push needed to take it from a small, relatively insignificant industry to one with the potential for generating significant amounts of electricity. A rapid flowering of wind energy development took place there in the mid-1980s, when wind farms began to be installed in large numbers. As a result of generous tax credits, an environment was created in which it was possible for companies to earn revenue both from the sale of wind-generated electricity to Californian utilities, and from the manufacture of wind turbines. The new Californian market gave Danish manufacturers an opportunity to develop a successful export industry, taking advantage of the experience acquired within their home market.

Since the 1980s, Europe has taken the lead in wind energy, with 23 000 MW of wind generating capacity (nearly three quarters of the world total) installed by the end of 2002. Germany in particular has been in the vanguard of deployment and by the end of 2002 had installed over 12000 MW.

In the UK, by contrast, progress has been much more modest. Some 1030 grid connected wind turbines had been installed by mid-2003, representing a combined capacity of some 586 MW. Figure 7.37 shows the location of wind energy projects in the UK in 2003.

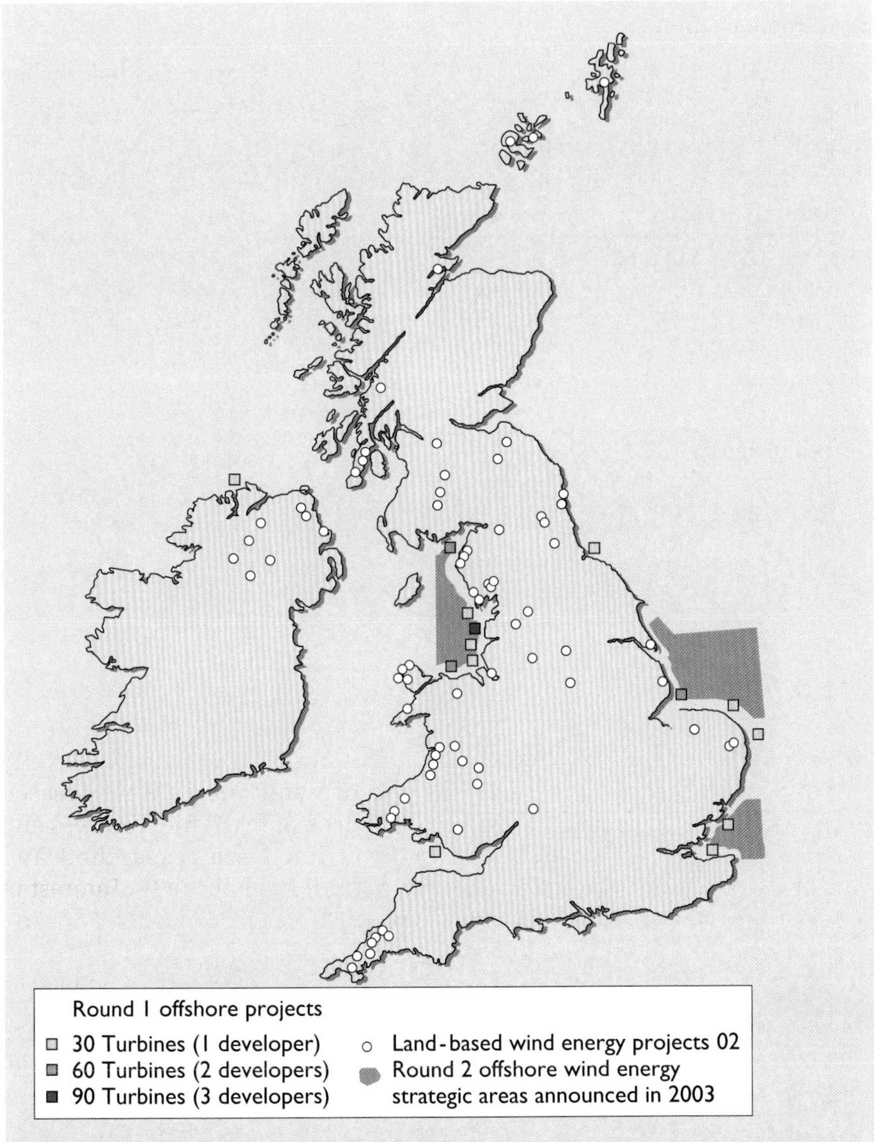

Figure 7.37 Location of UK land based wind energy projects and offshore Round One wind energy projects with Round Two offshore wind energy strategic areas. (Source BWEA, 2003; Crown Estates, 2003; DTI, 2003)

Table 7.2 Breakdown of Global Wind Power Capacity in 2002

	Megawatts	
	Start 2002	**End 2002**
Europe (EU)		
Germany	8754	12 001
Spain	3337	4830
Denmark	2489	2880
Italy	682	785
Netherlands	486	688
UK	474	552
Sweden	293	328
Greece	272	276
Portugal	131	194
France	93	145
Austria	94	139
Ireland	124	137
Belgium	32	44
Finland	39	41
Luxembourg	15	16
Total	**17 315**	**23 056**
Europe (non-EU)		
Norway	17	97
Ukraine	41	44
Poland	22	27
Latvia	2	24
Turkey	19	19
Czech Republic	6.8	7
Russia	7	7
Switzerland	5	5
Hungary	1	2
Estonia	1	2
Romania	1	1
Total	**9307**	**12 972**
North America		
USA	4275	4685
Canada	198	238
Total	**4473**	**4923**
Asia, Pacific + South and Central America		
India	1507	1702
Japan	275	415
China	400	468
Australia	72	104
Egypt, Morocco, Costa Rica, Brazil, Argentina, others	225 (est.)	225 (est.)
Total	**2479**	**2914**
World Total	**24 390**	**31 128**

Source: AWEA, 2003

Details of levels of wind generating capacity installed in various countries and continents in 2002 are given in Table 7.2.

Small-scale wind turbines

From the early days of wind generated electricity, small-scale wind turbines have been manufactured to provide electricity for remote houses, farms and remote communities, and for charging batteries on boats, caravans and holiday cabins. More recently they have been used to provide electricity for cellular telephone masts and remote telephone boxes.

Figure 7.38 Small scale wind turbine (source: Gazelle Wind Turbines Ltd)

Small-scale wind turbines are more expensive per kilowatt than medium-scale wind turbines. In most cases, the cost of the power they produce is not competitive with mains electricity (except in remote areas) and, because so much of the UK is connected to the National Grid, their potential has been limited. The need for batteries also tends to greatly increase the cost of such systems. However with the growing interest in 'net metering' (where a consumer who both exports and imports electricity pays only for the net number of units used), stimulated by building-integrated photovoltaic systems (see Chapter 3), in combination with availability of government capital grants for small wind turbines, there may be fresh opportunities for small wind systems that are grid-linked, avoiding the need for batteries.

The steady demand from people interested in obtaining electricity from pollution-free sources, or who are in locations where conventional supplies are not available, already provides enough support to sustain a significant number of manufacturers of small-scale wind turbines throughout the world (see Figures 7.1 and 7.40).

Local community and co-operatively-owned wind turbines

Local community windpower developments have been gaining support in recent years. These can take a variety of forms but in Denmark it usually involves a group of people from a local community buying a wind turbine or group of turbines. The local community benefits from the sale of the electricity produced, or makes use of it for its own purposes (Taylor, 1993).

More than 38% of the wind turbines in Denmark are owned and operated in this way and there is an active Danish Windmill Owners' Association, which provides support for the many people involved. Such wind energy co-operatives are also active in the Netherlands, Sweden and Germany.

This approach can encourage a positive attitude towards wind energy in communities that might be opposed to commercial wind energy developers from outside the area. A number of organisations have attempted to develop such projects in the UK, but so far there is only one community owned wind farm (Baywind), operating in Cumbria. In Wales, a community wind turbine has been installed by the Centre for Alternative Technology near Machynlleth and a community led scheme is being planned in Awel Aman Tawe. In 2003 the UK Government launched the 'Clear Skies Programme', a capital grant scheme to support community renewable energy schemes, including wind energy projects.

It is also being increasingly recognized that local people are likely to be more supportive of community wind turbines and interest in the concept of the 'village wind turbine' or 'town wind farm' appears to be increasing.

Since the further liberalisation of the UK electricity industry, it has become more feasible to propose local wind turbines to supply to local people. In addition, the Climate Change Levy is likely to encourage local companies and enterprises to participate, as wind-generated electricity is exempt from the levy.

Numerous single medium-scale and large-scale wind turbines are currently operating in the UK, either as community wind turbines, or wind turbines for factories, hospitals, supermarkets or housing projects. These range in size from about 60 kW to 1.5 MW, with a number of single 2 MW turbines being planned.

Wind energy and buildings

There may be opportunities for using buildings as a means of extracting wind energy by designing the building form to accelerate wind velocities. In doing so it is possible to reduce the size of wind turbine required for a given power output and to offset the topographical roughness effects which slow down local winds in urban and suburban environments.

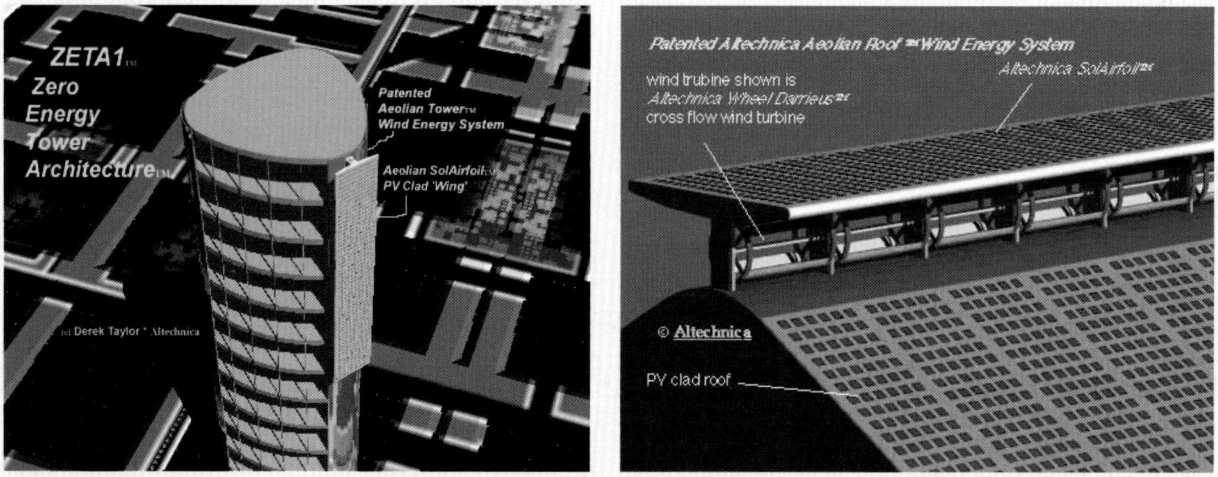

Figure 7.39 Aeolian Roof™ and Aeolian Tower™ building integrated wind energy systems. The concept is being researched by Altechnica in collaboration with the Open University (source: Derek Taylor)

The author has patented a concept using one or more wing-like 'planar concentrators'. These can be used individually, in tandem, in tri-plane configurations or in combination with a building roof (as in the 'Aeolian Roof', Figure 7.39, left) or in combination with the corners of tall buildings (as in the 'Aeolian Tower' Figure 7.39, right). Axial flow (or preferably cross-flow) wind turbines are located between the planes, or between the planes and the building roof or building corner, or between wings projecting out from the building. If successful, these devices could generate a high proportion of the electricity requirements of appropriately oriented energy-efficient buildings.

Wind energy potential

The World's total land based wind energy resource is estimated at over 53 000 TWh y^{-1} (Grubb and Meyer, 1993) as shown in Figure 7.40 (see Box 10.1 on resource terminology in Chapter 10: *Integration*). A report, Wind Force 12, published in 2002 by the European Wind Energy Association (EWEA), Forum for Energy and Development (FED) and Greenpeace (EWEA *et al*, 2002) used this estimate as the basis of a wind energy target of 12% of global electricity production by 2020.

To meet the target, some 1260 GW of capacity would need to be installed which would generate around 3000 TWh y^{-1}. This is equivalent to current electricity use in the European Union. Table 7.3 shows the suggested distribution of this capacity around the world and Table 7.4 shows the distribution throughout the EU. By 2040, the study further estimated that wind power could be supplying more than 20% of the world's electricity.

For the UK, the future potential for wind energy is also very large. In addition to the figures for on-shore wind shown in Table 7.4, Table 10.5 and Figure 10.5 in Chapter 10 give other estimates of the

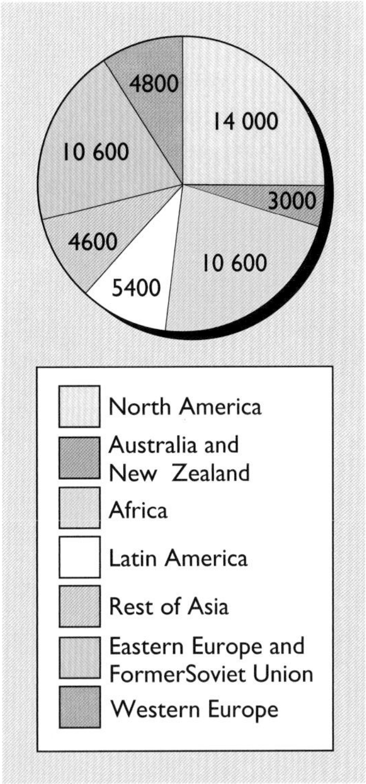

Figure 7.40 The world's land-based wind energy resources in TWh per year, as estimated in the *Wind Force 12* report. Medium-scale wind turbines are assumed, so the use of larger turbines would result in larger resource estimates (source: EWEA, 2002; based on Grubb and Meyer, 1993)

Table 7.3 Available *'land-based'* world wind resources (but also including OECD European offshore resources) and future electricity demand

Region of the world	Electricity demand by 2020 (TWh y^{-1})	20% of 2020 demand GW	Wind resource (TWh y^{-1})
OECD – Europe (land wind)	4515	903	630
OECD – Europe (offshore)	–	–	313
OECD – N. America	5729	1146	14000
OECD – Pacific	1745	349	3600
Latin America	2041	408	5400
East Asia	2081	416	–
South Asia	1695	339	–
China	3689	739	–
Middle East	907	181	–
Transition Economies	2615	523	–
Africa	864	173	4600
World Total	**25881**	**5177**	–

Source EWEA, 2002

Table 7.4 Technical potential for *land based* wind energy in the EU + Norway

Country	Total electricity consumption* (TWh y^{-1})	Technical wind potential capacity GW	Technical wind potential production (TWh y^{-1})	Up to 20% of consumption from wind (TWh y^{-1})	Surplus wind over 20% consumption (TWh y^{-1})
Austria	60	1.5	3	3	–
Belgium	82	2.5	5	5	–
Denmark	31	4.5	10	6.2	3.8
Finland	66	3.5	7	7	–
France	491	42.5	85	85	–
Germany	534	12	24	24	-
Great Britain	379	57	114	75.8	38.2
Greece	41	22	44	8.2	–**
Ireland	17	22	44	3.4	40.6
Italy	207	34.5	69	41.4	27.6
Luxembourg	1	0	–	–	–
Holland	89	3.5	7	7	–
Portugal	32	7.5	15	6.4	8.6
Spain	178	43	86	35.6	50.4
Sweden	176	20.5	41	35.2	22.8
Norway	116	38	76	23.2	
Total	**2500**	**315**	**630**	**366.4**	**244.8**

* Electricity consumption is based on OECD/IEA figures for 1989, extended by 3% per annum to 1995. The IEA *World Energy Outlook* (1998) records a total consumption for OECD-Europe in 1995 of 2678 TWh per year.

** Greece has an excess potential, but with resources scattered over many islands is unlikely to be an exporter for some time.

Source: EWEA, 2002

technical potential, practicable potential and costs of various renewables, including wind, by 2020–25. As can be seen, offshore and onshore wind are considered to have the greatest potential and lowest costs of all the renewables.

7.9 Offshore wind energy

The capital costs of energy from offshore wind farms are generally higher than those of onshore installations because of the extra costs of civil engineering for substructure, higher electrical connection costs and the higher specification materials needed to resist the corrosive marine environment. However, offshore wind speeds are generally higher than on land (apart from certain mountain and hill tops) and this, together with likely reductions in offshore costs as experience is gained in this environment, is expected to make offshore wind energy costs competitive in the medium term. Offshore, it is more feasible to utilize very large-scale wind turbines than on land, and this is also likely to improve the economic viability as more energy can be captured from one platform.

A demonstration offshore wind farm of eleven 450 kW wind turbines was installed in the seas around Vindeby in Denmark (Figure 7.41) and has been generating electricity since 1991. A further Danish offshore wind farm was constructed at Tunø Knob in 1995. A single 220 kW wind turbine was also installed in 1991 off the coast of Nogersund in the south of Sweden. An offshore wind farm consisting of four 500 kW wind turbines was constructed in 1994 in the Ijssel Lake in Dutch coastal waters.

Figure 7.41 Offshore wind farm located off the coast of Vindeby in Denmark, consisting of eleven 450 kW Bonus wind turbines

Offshore deployment of wind power is gaining further momentum with several European countries – led by Denmark – announcing ambitious plans for substantial offshore projects. If realised, these will mean that wind power will become a major provider of electricity in those countries.

A map of European offshore wind speeds shows that some of the areas with the highest mean wind speeds are located around the British Isles. See Figure 7.32. A resource study (Matthies *et al*, 1995) of the offshore wind energy potential from European seas estimated that within UK waters the potential annual electricity production from offshore wind is over 980 TWh per year (almost three times the current annual UK electricity consumption).

Experience gained from the operational offshore wind farms at Vindeby and Tunø Knob in Danish coastal waters has yielded considerable data about the various factors that influence the development and operation of wind farms at sea. The Vindeby wind farm costs were 85% higher than the cost of a land-based wind farm at that time, but electricity production for this project was 20% higher. There has been an increase in fishing yields in the locality – which was attributed to the foundations acting as an artificial reef and encouraging the growth of mussels – and in the period since the wind farm was built the variety of flora and fauna is said to have generally improved (Krohn, 1997).

Test results from Tunø Knob indicate that actual offshore wind energy output is 20–30% higher than estimated from wind speed prediction models. It also appears that wind speeds further offshore are higher than wind speed models predicted. Availability has been higher than expected with an average of 98% being achieved.

In February 1998, the Danish Government concluded an agreement with state owned utilities to develop 750 MW of offshore wind power by 2008, with a further target of 4000 MW planned for 2030. It is planned that the initial 750 MW will be made up of five 150 MW wind farms. However political changes in Denmark have resulted in modifications to this programme and the later projects are to be opened up for private development and international tender.

In addition to these projects, another offshore wind energy project has been constructed in shallow water near Copenhagen (Figure 7.42) by a collaboration between the Middelgrunden Wind Turbine Co-operative and Copenhagen Energy. Ten Bonus 2 MW wind turbines were installed in 2001, arranged along a slightly curved line 3.4 kilometres in length. Each turbine has a hub height of 64 metres and a rotor diameter of 76 metres. These are estimated to generate some 89 million kWh per year, equivalent to approximately 3% of the electricity consumed in Copenhagen. The largest offshore wind farm to date was completed in 2002 at Horns Rev in Danish

Figure 7.42 A view of the Middelgrunnden wind farm near Copenhagen. The first offshore cooperative wind farm (source: Danish Wind Turbine Manufacturers Association, 2001)

waters and consists of 80 turbines each rated at 2 MW. It generates some 600 GWh/year, enough electricity to supply some 150 000 Danish single family houses (Elsam, 2003).

The Government of the Netherlands is also intending to develop offshore wind energy, initially with one hundred 1 MW turbines, in 5 to 20 metre water depths. Another 100 MW wind farm is planned for 'semi-offshore' sites in water behind dikes. The Government sees these projects as a first step towards the development of larger schemes in the North Sea and it has plans to install 3000 MW of wind power capacity by 2020, half of it located offshore. The cost of electricity from these projects is expected to be between 2.3 and 4.6 p/kWh.

In addition the German Government plans substantial offshore wind energy developments. These are intended to supply 75 to 85 terawatt-hours of electricity per year by 2030 from two areas of the North Sea that have been allocated for the construction of some 4000 turbines.

Offshore wind energy in the UK

Offshore wind energy has significant potential in the UK. It is estimated (Border Wind, 1998) that the area of sea bed required to supply 10% of UK electricity from offshore wind would be of the order of 1200 km^2, and that to supply 40% of UK electricity would require an area of 4800 km^2. These are respectively some 0.007% and 0.03% of the total UK-controlled sea bed.

To demonstrate the importance of offshore wind energy to the UK, AMEC Border Wind carried out a study (Border Wind, 1998) which showed that it would be possible to produce 40% of the UK's annual electricity consumption by 2030 from an offshore wind energy deployment programme culminating in an installed capacity of 48 GW (at an assumed capacity factor of 33%). To achieve the target of 48 GW would involve the installation of 2500 MW each year, requiring 20 dedicated crane barges.

BOX 7.7 Blyth, the UK's first off-shore wind turbines

In 1992, Border Wind successfully constructed and operated a 'semi-offshore' wind farm consisting of nine 300 kW wind turbines built into the walls of Blyth Harbour on the Northumbrian coast in north east England. As it was necessary to construct the turbines adjacent to the sea, the project provided practical experience in some of the installation techniques that would be required for the installation of an offshore wind farm. (Border Wind was subsequently acquired by the major engineering company AMEC, becoming AMEC Border Wind.)

Building on this experience, the company obtained funding from the EU towards the construction cost of a demonstration off-shore wind power installation off the coast of Blyth Harbour. A joint venture company, Blyth Offshore Wind Limited (BOWL) was formed, consisting of AMEC Border Wind, Powergen Renewables, Shell and the Dutch utility Nuon. The installation was successfully completed and officially opened in December 2000 (Figure 7.43). It was the first offshore wind farm to be constructed in UK coastal waters and consists of two turbines each rated at 2 MW, manufactured by the Danish company Vestas. The turbines have been installed at a low tide depth of six metres on a submerged rocky outcrop, approximately one kilometre from the coast. This was the first offshore wind project in Europe to be located in the tough North Sea environment rather than the sheltered sea locations of the Danish offshore wind farms.

Rather than using the relatively expensive gravity foundations employed on the first Danish offshore wind farms, a more cost-effective foundation design based on a tubular steel pile known as a 'monopile' was employed. From a barge-based drilling rig, two hole sockets (one for each turbine tower) were drilled into the rocky outcrop and the monopiles were inserted into these sockets. Once a monopile was fixed in place, the tower was installed in two sections followed by the nacelle and a complete three bladed rotor.

Figure 7.43 View of the UK's first offshore wind turbines at Blyth, Northumberland (source: AMEC Border Wind)

To solve the problem of disposal of the drilling spoil without contaminating the seabed, the spoil was pumped along a 200 m long pipeline, depositing the material outside an exclusion zone to ensure that the existing lobster holes on the exposed rock head did not become blocked.

The two Vestas V66-2.0 turbines are specifically designed for offshore operation. Each has a rotor diameter of 66 m, rotational speed of 21 rpm, a cut-in wind speed of 4 m s^{-1}, rated wind speed of 17 m s^{-1} and a shut down wind speed of 25 m s^{-1}. The hub height is almost 68 m above the seabed.

An armoured submarine cable with an embedded fibre optic cable transfers power at 11 kV to the shore and permits communications between the turbines and the control building at the Port of Blyth, where the electricity is metered and transferred to Northern Electric, the local District Network Operator (DNO). Remote interrogation of the turbines' sensors via the fibre optic link allows the operators to minimize the need for visits to the turbines offshore.

Adverse weather and storm conditions were experienced during the installation period. These resulted in some delays, since lifting operations could only take place when the wind speed was 8 ms^{-1} or less and the swell no greater than 0.5 m. This experience highlighted the extra difficulties likely to be experienced by offshore wind energy installations. Future projects will require flexible and responsive planning and installation procedures to minimize the impact of adverse weather conditions on construction/installation schedules.

A problem encountered after installation was that one of the turbine blades was struck by lightning in 2001. This required both turbines to be shut down for a period, though the blade was not lost and remained attached to the hub. The blade was subsequently replaced and the turbines resumed operation in 2002.

The study also concluded that manufacturing enough turbines to achieve 48 GW by 2030 could technically be achieved by UK firms. It assumed that the offshore-specific or marine technology elements would provide the biggest opportunities for UK manufacturing firms, though it also concluded that there were possibilities for the local manufacture of turbines close to potential offshore wind energy sites. Substantial business opportunities would be created, with a total potential market by 2030 of £48 billion, a market for component supply of £24.9 billion and a market for installing offshore wind turbines of over £10 billion. The study also estimated that the number of jobs directly or indirectly related to such an offshore wind energy programme would be over 36 000 (reached by 2010). It concluded that such a programme would enable the UK to cut CO_2 emissions by some 21%.

In order to achieve these targets, strong support from Government would be needed to stimulate the level of activity required. Since the study was completed, there have been further advances in turbine size which increase the productivity from each platform, making it feasible to achieve the targets sooner than estimated.

The UK Government is supporting offshore wind farms with capital grants (£39 million) and, through the Crown Estates Office (which is responsible for the area of sea bed that surrounds the UK), has allocated licences for 13 sites (18 developments) around the UK (see Figure 7.39) for an initial Round 1 of offshore wind farm development licenses, based on installations of 30, 60 or 90 turbines. This would represent some 1.4 GW of capacity. The £39 million allocation should be sufficient to subsidize some 300 MW of capacity. In 2001, the Performance Innovation Unit of the UK Cabinet Office recommended an additional allocation of £25 million in capital grants to support a further 150 MW of offshore wind energy capacity, bringing the supported capacity up to 450 MW of the licensed capacity (PIU, 2001). At the time of writing (Autumn 2003) a number of the developments, including one at Scroby Sands near Great Yarmouth, are under construction.

Figure 7.44 North Hoyle Offshore Wind Farm located 4 to 5 miles off the North Wales coast near Rhyl was officially switched on in November 2003

The first of these projects the North Hoyle Offshore Wind Farm – located 4 to 5 miles off the North Wales coast near Rhyl – was officially switched on in November 2003 and estimated to be able to generate enough electricity for around 50 000 homes in the UK (see Figure 7.44). The project was completed three months after the first turbine was installed. It is expected to offset about 160 000 tonnes of carbon dioxide per year from 30 turbines each rated at 2 MW (National Wind Power, 2003).

In late 2002, the UK Government published for consultation a strategic framework for offshore wind energy development in the UK (Department of Trade and Industry, 2002). This outlined the strategic factors involved in expanding offshore wind energy in the UK beyond the Round 1 projects, including legal frameworks, allocation of rights, impacts, environmental assessment, consents, regulation and a proposed timetable for UK offshore wind energy expansion. The framework document identifies three primary strategic areas for the Second Round of UK offshore wind energy development (North West or 'Liverpool Bay', the Greater Wash and the Thames Estuary), plus other sites. It assumes that no wind turbines are installed within 5 km of the coast in order to minimize visibility from onshore. Table 7.5 shows the very large potential for offshore wind energy in these areas.

Table 7.5 Potential offshore wind generation resource in proposed strategic regions

Water depths	**5 to 30 metres**				**30 to 50 metres**			
Region All waters	**Area /km²**	%	**MW**	**TWh/yr**	**Area**	%	**MW**	**TWh/yr**
North West	3345	12	40 140	140	2067	4	24 804	87
Greater Wash	7391	27	88 692	310	946	2	11 352	40
Thames Estuary	2099	8	25 188	88	848	2	10 176	35
Other	14 431	53	173 172	606	45 441	92	545 292	1907
Total	**27 266**	**100**	**327 192**	**1144**	**49 302**	**100**	**591 624**	**2069**

Note: 'All Waters' denotes areas both within and outside UK territorial waters
Source: DTI, 2002

These estimates are based on 3 MW wind turbines installed at a separation spacing of 500 metres. The total figure of 919 GW, producing 3213 TWh per year, indicates the vast potential. This is almost 10 times UK electricity consumption in 2002, and nearly 100 times the Government's 2010 renewables target of 10% of electricity supplied (33.6 TWh per year). However these totals are estimates of the *potential*: they do not take account of any reductions necessary for conflicts of interest, environmental impact or availability of onshore electrical grid connections (such reductions are discussed in Chapter 10).

As part of the pre-qualification procedures for Round 1, the Crown Estates established a Trust Fund which is commissioning generic environmental studies of relevance to offshore wind energy development. These include impacts on birds and marine ecology and possible visual impact, plus the possible impacts on other marine activities. In July 2003, the Department of Trade and Industry published Phase One of a Strategic Environmental Assessment (SEA) of the identified strategic offshore development areas (BMT Cordah, 2003). It concluded that the potential constraints on achieving the draft offshore wind energy programme are not sufficient to prejudice its success, though longer term studies would need to confirm uncertainties about impacts on physical processes, birds, elasmobranches (electricity sensing species of fish) and cetaceans.

In July 2003, the UK Secretary of State for Trade and Industry announced Round 2, a major expansion in offshore wind energy (Department of Trade and Industry, 2003) in the three strategic areas mentioned above and shown on the map in Figure 7.37. Unlike the Round 1 projects, which were limited to a maximum of 30 turbines per project, much larger projects are possible, up to 1 GW in capacity and comprising hundreds of turbines. The DTI has said that expressions of interest suggest that Round 2 could deliver 'between 4 GW and 6 GW' total installed capacity. This is sufficient to power more than 3.5 million households, i.e. some 9 million people, more than the population of Greater London. It would also create over 20 000 jobs in manufacturing, installation and maintenance.

The DTI also stated that 'There is clear evidence that the biggest new contributor to our renewables target is going to be offshore wind and Government has a strong interest in encouraging it to develop quickly and successfully.' The Crown Estate has been requested by the DTI to invite developers to tender for sites in all three Strategic Areas in locations at least 8 km from the coastline.

This has the potential to be a step change in serious support for wind energy development in the UK. If and when these projects are successfully delivered, the UK would then have started to become a major generator of electricity from wind energy.

Wind energy continues to be one of the fastest growing energy technologies and it looks set to become a major generator of electricity throughout the world. Particularly in Europe, the offshore exploitation of wind energy is likely to become one of the most important means of reducing carbon dioxide emissions from the electricity sector.

References

Abbot, I. H. and von Doenhoff, A. E. (1958) *Theory of Wing Sections*, New York, Dover Publications inc.

AMS (2003) *Feasibility of Mitigating the Effects of Windfarms on Primary Radar*, ETSU W/14/00623/REP. Available from BWEA website at: www.bwea.com/aviation/W1400623%20part%201%20of%203.pdf [accessed 10 December 2003].

AWEA (2002) *Wind Energy Fact Sheet: Facts about Wind Energy and Birds*, American Wind Energy Association.

AWEA (2003) *Global Wind Energy Market Report*, American Wind Energy Association.

Anderson, M. (1992) *Current Status of Wind Farms in the UK*, Renewable Energy Systems.

Bacon, D. F. (2002) *Fixed-link wind-turbine exclusion zone method*, Radio Communications Agency.

BMT Cordah Ltd (2003) *Offshore Wind Energy Generation: Phase 1 Proposals and Environmental Report* (Offshore Wind SEA), Department of Trade and Industry, July.

Border Wind (1998) *Offshore Wind Energy: Building a New Industry for Britain*, Greenpeace.

Boyle, G., Everett, R. and Ramage, J. (eds) (2003) *Energy Systems and Sustainability,* Oxford University Press in association with the Open University, 620 pp.

Beurskens, J and Jensen, P. H. (2001) 'Economics of wind energy – Prospects and directions', *Renewable Energy World*, July-Aug.

Burch, S. F. and Ravenscroft, F. (1992) *Computer Modelling of the UK Wind Energy Resource: Final Overview Report,* ETSU WN7055, ETSU.

Burroughs, W. J., Crowder, B., Robertson, E., Vallier-Talbot, E. and Whitaker, R. (1996) *Weather – The Ultimate Guide to the Elements*, Collins.

BWEA (1994) *Best Practice Guidelines for Wind Energy Development*, British Wind Energy Association.

BWEA (2002) *Summary of 42 surveys carried out between 1990 and 2002 on public attitudes to wind energy*, British Wind Energy Association.

BWEA (2003) *Planning Progress*, available at BWEA Web site: www.britishwindenergy.co.uk [accessed 8 December 2003].

Chignell, R. J. (1987) *Electromagnetic Interference from Wind Turbines – A Simplified Guide to Avoiding Problems*, National Wind Turbine Centre, National Engineering Laboratory, East Kilbride.

Chris Blandford Associates (1994) *Wind Power Station Construction Monitoring Study*, Countryside Council for Wales.

Department of the Environment (1993) *Planning Policy Guidance Note, (PPG 22) Renewable Energy, Annex on Wind Energy,* HMSO. Update 2003 from Office of the Deputy Prime Minister.

Department of Trade and Industry (1994) *New and Renewable Energy: Future Prospects for the UK,* Energy Paper 62, HMSO.

Department of Trade and Industry (2001) *New and Renewable Energy*, The Renewables Obligation Statutory Consultation.

Department of Trade and Industry (2002) *Future Offshore: A Strategic Framework for the Offshore Wind Industry*, November, DTI.

Department of Trade and Industry (2003) *Offshore wind farms Round 2: Designed to provide a framework for rapid and successful expansion*, July, DTI.

DTLR (2000) *Planning Policy Guidance Note 11: Regional Planning*, Department of Transport, London and the Regions.

DTLR (2001) Planning Green Paper – Planning: Delivering a Fundamental Change, Department of Transport, London and the Regions.

Danish Wind Turbine Manufacturers Association (2000) *Wind Power Note*, DWTMA. Available at www.windpower.org/en/publ/annu9900.pdf [accessed 18 December 2003].

Eldridge, F. R. (1975) *Wind Machines*, Mitre Corporation.

Elsam (2003) *Horns Rev Offshore Wind Farm – a 160 MW wind power station*, Elsam Brochure, Denmark.

English Nature, RSPB, WWF-UK, BWEA (2001) *Wind farm development and nature conservation*, WWF-UK.

Esslemont, E. (1994) *Cemmaes Wind Farm Sociological Impact Study*, Harwell, Energy Technology Support Unit.

EWEA (1991) *Time for Action: Wind Energy in Europe,* European Wind Energy Association.

EWEA (2002) *Wind Force 12*, European Wind Energy Association and Greenpeace.

Farndon, J. (ed) (1992) *How the Earth Works,* Dorling Kindersley, London.

FOE (1994) *Planning for Wind Power: Guidelines for Project Developers and Local Planners*, Friends of the Earth.

Golding, E. W. (1955) *Generation of Electricity by Wind Power*, London, E. & F. N. Spon.

Grubb, M. and Meyer, N. (1993) 'Wind energy resources, systems and regional strategies' in Johansson, T. B. *et al.* (eds) *Renewable Energy – Sources for Fuels and Electricity*, Earthscan, pp. 157– 212.

Ipsos (2003) *Opinion Poll Survey of Householders' Opinions on Wind Energy Development*, Survey commissioned by British Wind Energy Association.

Jago, P. and Taylor, N. (2002) *Wind turbines and aviation interests – European Experience and Practice*, STAYSIS Ltd for the DTI, ETSU W/14/00624/REP (DTI PUB URN No 03/5151).

Kenetech Windpower (1994) *Kenetech Windpower Avian Research Program Update*, Kenetech Windpower, Washington D.C.

Kragh, J. *et al.* (1999) *Noise emission from wind turbines*, National Engineering Laboratory for the Energy Technology Support Unit (ETSU), ETSU W/13/00503/REP, February.

Krohn, Soren (1997) *Offshore Wind Energy: Full Steam Ahead*, Danish Wind Turbine Manufacturers Association Web page available at www.windpower.dk [accessed 8 December 2003].

Legerton, M. (ed.) (1992) *Wind turbine noise workshop Proceedings*, ETSU-N-123. Department of Trade and Industry/British Wind Energy Association.

Ljunggren, S. (ed.) (1994) *Recommended practices for Wind Turbine Testing 4. Acoustics Measurement of Noise Emission from Wind Turbines*, 3rd Edition. Submitted to the Executive Committee of the International Energy Agency Programme for Research and Development on Wind Energy Conversion Systems.

Ljunggren, S. (ed.) (1997) *Recommended practices for Wind Turbine Testing. 10. Measurement of Noise Emission from Wind Turbines*, 1st Edition. Submitted to the Executive Committee of the International Energy Agency Programme for Research and Development on Wind Energy Conversion Systems.

Matthies, H. G. *et al.* (1995) *Study of Offshore Wind in the EC*, JOUR 0072, Verlag Naturliche Energie.

Needham, J. (1965) *Science and civilisation of China*, Cambridge, Cambridge University Press.

NERI (1998) *Impact Assessment of an Off-shore Wind Park on Sea Ducks*, National Environmental Research Institute (NERI) Technical Report No. 227, Denmark.

NWCC (1999) *Studying wind energy/bird interactions: A Guidance Document*, National Wind Co-ordinating Committee, USA, December.

Office of the Deputy Prime Minister (ODPM) (2003) *Consultation on Draft New Planning Policy Statement 22 (PPS22): Renewable Energy*, available at www.odpm.gov.uk [accessed 8 December 2003].

Pepper, L. (2001) *Monitoring and Evaluations of Blyth Offshore Wind Farm: Installation and Commissioning*, AMEC Border Wind, ETSU W/35/00563/REP/1.

PIU (2001) *Renewable Energy in the UK – Building for the Future of the Environment* Resource Productivity and Renewable Energy Project, Cabinet Office.

QineteQ (2003) *Report on the development of a modelling tool for predicting the impact of wind turbines on radar systems*, ETSU W/14/00614/REP.

Spaven Consulting (2001) *Wind Turbines and Radar: Operational Experience and Mitigating Measures*. A report to a consortium of wind energy companies, downloadable from BWEA web site.

Reynolds, J. (1970) *Windmills and Watermills*, Hugh Evelyn Ltd Publishers, London.

Taylor, D. A. and Rand, M. (1991) *Planning for Wind Energy in Dyfed*, Milton Keynes, Altechnica.

Taylor, D. A. (ed.) (1993) *One-day Workshop on Local Community Wind Energy Projects*, Milton Keynes, Altechnica.

Taylor, D. A. (1998) 'Using buildings to harvest wind energy' in *Building Research and Information*, E & FN Spon.

TNS (2003) *Attitudes and Knowledge of Renewable Energy amongst the General Public – Report of Findings*, JN9419 and JN9385, Central Office of Information.

Troen, I. and Petersen, E. L. (1989) *European Wind Atlas*, Risø Laboratory, Denmark published for the Commission of the European Communities.

Wind Energy, Defence and Civil Aviation Interests Working Group (2002) *Wind Energy and Aviation Interests – Interim Guidelines*, ETSU W/14/00626/REP.

Further Reading

Burton, T., Sharpe, D., Jenkins, N. and Bossanyi, E. (2001) *Wind Energy Handbook*, Wiley. A large 600 page textbook covering technical topics related to wind energy.

Department of the Environment (1993) *Planning Policy Guidance Note (PPG 22)* Renewable Energy, Annex on Wind Energy, HMSO. This is an important document for those interested in government policy on the planning aspects of wind energy. It includes a number of recommendations on planning procedures. A consultation document (PPS 22) which reviewed PPG22 was released in 2003 by the Office of the Deputy Prime Minister – see References above.

Freris, L. L. (ed.) (1990) *Wind Energy Conversion Systems*, Prentice-Hall. A very comprehensive overview of topics related to wind energy with high technical level, authored by a number of contributors.

Golding, E. W. (1955) *Generation of Electricity from Wind Power*, London, E. & F. N. Spon. Whilst this book is somewhat out of date, it provides a good historical overview of early attempts to utilize wind energy for electricity generation.

Lancashire, S., Kenna, J. and Fraenkel, P. (1987) *Windpumping Handbook*, IT Publications. An excellent book on the practicalities of pumping water by means of wind power.

Legerton, M. (ed.) (1992) *Wind Turbine Noise Workshop Proceedings*, ETSU-N-123, Department of Trade and Industry / British Wind Energy Association. The proceedings of this workshop provide a good overview of work on wind turbine noise in the UK.

Manwell, J. F., McGowan, J. and Rogers, A. (2002) *Wind Energy Explained*, Wiley. A 570 page textbook which covers additional technical aspects of wind energy.

Internet Sources

Useful internet sites include the following:

UK Department of Trade and Industry at www.dti.gov.uk [accessed 9 December 2003].

Danish Wind Turbine Manufacturers at www.windpower.org [accessed 9 December 2003].

British Wind Energy Association at www.britishwindenergy.co.uk [accessed 9 December 2003].

American Wind Energy Association at www.awea.org [accessed 9 December 2003].

All of the above sites have links to other organisations involved in wind energy and most have a range of technical and non-technical publications and fact sheets about wind energy which can be downloaded. They are a useful source of the latest wind energy information.

Chapter 8

Wave Energy

by Les Duckers

8.1 Introduction

The possibility of extracting energy from ocean waves has intrigued people for centuries. However, although there are concepts over 200 years old, it was only in the 1970s that viable schemes began to emerge. In general, these modern wave energy conversion schemes have few environmental drawbacks, and the prospects that some of them may make a significant energy contribution in the longer term are promising. In fact, in areas of the world where the wave climate is energetic and where conventional energy sources are expensive, such as remote islands, some of these schemes may already be competitive.

The World Energy Council has estimated the worldwide wave power resource to be 2 TW, equivalent to an annual available energy resource of 17 500 TWh (Thorpe, 1999). Therefore for some countries – the UK is one example – wave energy offers a very large potential resource to be tapped.

Technological developments could enable wave energy to fulfil this promise. A number of shore-mounted and near-shore prototypes are already planned or in operation. Refinements of these prototype designs, and the development of tethered and floating offshore structures which could be deployed in large numbers in open water, could open up the possibility of harvesting vast quantities of energy from the oceans.

The 1999 Scottish Renewables Order led to three contracts being awarded to wave energy plants, and this is indicative of the level of commercial wave activity in the UK and EU. There are also commercial pilot plant developments in other parts of the world, including Australia and the US.

Recent history

The UK 'energy crisis' of 1973 prompted an increased interest in renewable energy, and especially in wave energy, as a potential source of electricity for the UK National Grid. Because of the enormous wave energy resource potentially available to the UK (see below and Section 8.4), a large number of device concepts were invented, mathematically modelled and experimentally tested, with support from commercial sponsors and the former Department of Energy (DEn), now part of the Department of Trade and Industry (DTI). Unfortunately, insufficient time and money was allocated to bring the various concepts and the associated technologies to maturity. Then in 1982, acting on advice from the Advisory Committee on Research and Development for Fuel and Power (ACORD), DEn scaled down the UK wave energy programme. See Ross (1995) for an interesting discourse on this.

Some of the research teams involved were however able to sustain a minimal amount of effort on wave energy projects. In 1989 a 75 kW prototype Oscillating Water Column (OWC) wave energy converter was installed on Islay in Scotland. This was fully funded by DEn following ACORD's recommendation that small-scale devices should be investigated as a source of energy for islands and remote communities where diesel normally provides the main energy source (ETSU, 1985).

Meanwhile, during this period of reduced funding in the UK, a number of other countries, notably Norway and Japan, increased their research and development programmes. With hydroelectric schemes supplying virtually all of its electricity Norway had little immediate domestic need for wave energy, but it was keen to develop an export market for wave energy technology. In contrast, Japan did require more energy sources, but its wave climate is very modest.

In the 1990s there was a revival of awareness amongst politicians and others in a number of countries of the potential of wave energy. In particular, a European Union initiative was launched (Garratti *et al.*, 1993), which provided funding for a small number of projects and led to the formation of a European Wave Energy Thematic Network. In the UK, a review was conducted by the Energy Technology Support Unit Chief Scientists' Group (ETSU) (Thorpe, 1992). This review estimated the generating costs for five representative 'main devices' to range from 6 to 16p per kWh, and concluded that the technical potential of the UK offshore wave energy resource is large (7–10 GW annual average, equivalent to 61–87 TWh per year), but that the practical potential would be much smaller because of operational and economic constraints. Further, it said that the main devices were unlikely to generate electricity competitively in the short- to medium-term, but that there might be some scope for reduction in generating costs (with changes to the design of devices in most cases).

Another review, commissioned by DTI (Thorpe, 1998), further considered the prominent devices, and was very positive about progress and prospects, particularly endorsing smaller structures, and reported potential costs of 5p per kWh or less. A later resource-cost analysis (Thorpe, 2001) gave an annual practical wave resource of about 30 TWh at a cost of 4p per kWh.

In 2001 a House of Commons Select Committee report recommended that the UK Government commit a large portion of its £100 million renewables budget to wave energy. In November 2001 the Cabinet Office Performance and Innovation Unit (PIU) proposed an allocation of £5 million to the development of wave and tidal energy schemes (PIU, 2001).

8.2 Introductory case studies

The case studies presented in Boxes 8.1 and 8.2 overleaf will assist the reader to understand the nature of waves and of schemes designed to harness wave energy. The TAPCHAN device is particularly valuable in that it has some inherent storage, whereas the oscillating water column (OWC) concept has been deployed in a number of countries and represents the most common form of wave energy converter, probably because of its simplicity and robustness.

Shoreline prototypes are generally of the oscillating water column type as discussed in Box 8.2. Offshore devices are more difficult to construct and maintain than shoreline devices, but they can harvest greater amounts of energy, as the waves in deep water have a greater energy content than those in the shallower water near to land. By using an area of ocean a few kilometres offshore as a 'wave farm', it would be possible to deploy an array of wave energy converters and hence capture large quantities of energy, which could then be transmitted back to shore via sub-sea electrical cables. These and many other aspects of wave energy conversion will be discussed later in this chapter, strating with the physical principles of wave energy in Section 8.3.

BOX 8.1 TAPCHAN

In 1985 a 350 kW prototype **TAPCHAN** wave energy converter, built by the company Norwave, commenced operation on a small Norwegian island some 40 km north-west of Bergen.

The name 'TAPCHAN' comes from the 'TAPered CHANnel' design of the scheme. The mouth of the channel is a 40 metre-wide horn-shaped collector. Waves entering the collector are fed into the wide end of the tapered channel, where they then propagate towards the narrow end with increasing wave height. The channel walls on the prototype are 10 m high (from 7 m below sea level to 3 m above) and 170 m long. Because the waves are forced into an ever-narrowing channel, their height is amplified until the crests spill over the walls into the reservoir at a level of 3 m above the mean sea level. The kinetic energy in the waves is thus converted into potential energy, and this is subsequently converted into electricity by allowing the water in the reservoir to return to the sea via a low-head Kaplan turbine system (see Chapter 5 on hydroelectricity for details of the Kaplan turbine). This powers a 350 kW generator that delivers electricity into the Norwegian grid.

The TAPCHAN concept is simple. With very few moving parts, its maintenance costs are low and its reliability high. The storage reservoir also helps to smooth the electrical output. We shall see in Section 8.3 that ocean waves have a random nature and so most wave energy converters produce a fluctuating power output. In contrast, TAPCHAN 'collects' waves in the reservoir, and so the output from the Kaplan turbine is dependent on the relatively steady difference in water levels between the reservoir and the sea. TAPCHAN therefore has an integral storage capacity which is generally not found in other wave energy converters. TAPCHAN systems are now being planned for other locations, including Indonesia.

In the 1990s Norwave considered methods for reducing the cost of construction of future TAPCHANs. Among those methods is a scheme for wave prediction, to allow the Kaplan turbine to run at a greater output for some short time before the arrival of a number of large waves. This reduces the level of water in the reservoir and so makes room for those large waves. This technique may permit the designers to build schemes with smaller reservoirs and hence reduce the construction costs. A second cost-reduction method that has been proposed is to fabricate a shorter channel, and this has been tried out on the existing prototype at Bergen by reducing the length of the existing channel. There were some technical difficulties with the dynamiting of the concrete channel, and the ensuing commercial problems have meant that this prototype is no longer in operation (Petroncini and Yemm, 2000).

TAPCHAN cannot be exploited economically everywhere in the world, simply because to make such a system effective, there are a number of features

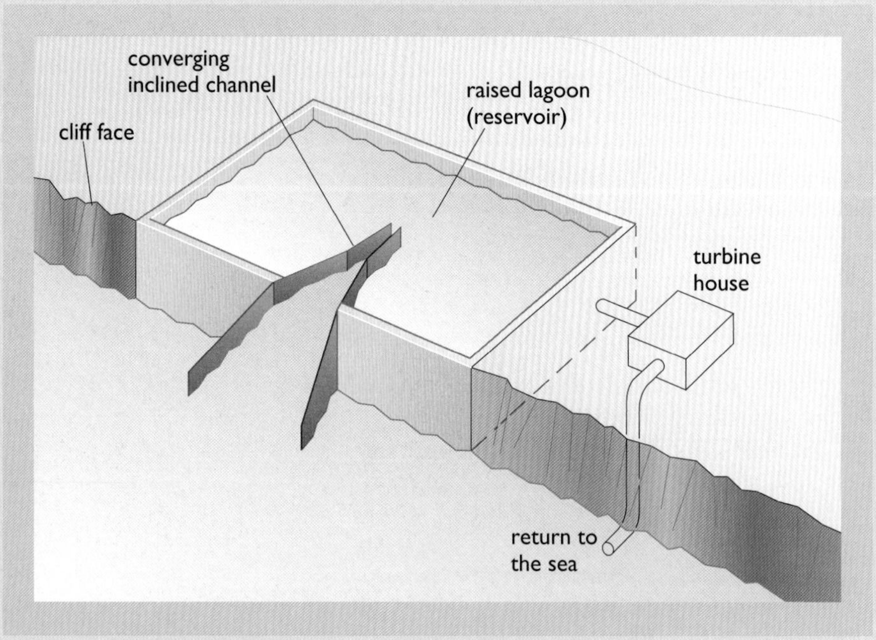

Figure 8.1 (a) The tapered channel wave energy conversion device: TAPCHAN

(b) Aerial photograph of the Norwegian TAPCHAN

which must coexist at any intended location. The requirements are:

- a good wave climate, i.e. high average wave energy, with persistent waves;
- deep water close to shore;
- a small tidal range (less than 1.0 m), otherwise the low-head hydro system cannot function properly for 24 hours a day (this therefore excludes most of the UK south of the Shetland Isles);
- a convenient and cheap means of constructing the reservoir, usually requiring a natural feature of the coastline.

BOX 8.2 The Islay shoreline gully oscillating water columns

Queen's University, Belfast, (QUB) has been working on the development of a shoreline gully **oscillating water column** (OWC) device for the Scottish Isles since 1985. After surveying several sites, the island of Islay was chosen for the first gully OWC scheme, which was installed in 1989 and decommissioned in 1999. It supplied the local grid with electricity on an intermittent basis from 1991 until 1999.

The approach was to develop a device which could be built cheaply on islands using established technology and techniques. It consisted of a wedge-shaped chamber made of reinforced concrete, open at the bottom, into which sea water was free to flow.

With experience gained from this natural gully project the team from QUB then collaborated with Wavegen of Inverness to develop what is referred to as a 'designer gully' OWC to overcome some of the limitations of the first Islay project. The principle modifications applied to the second Islay OWC were in the construction method and the shape of the oscillating column. The designer gully was excavated behind a natural rock wall, which was only removed at the end of the installation (see Figure 8.2). As with the first scheme, the sea water is contained as a water column which rises and falls with the waves. This acts as a piston, drawing air into and out of the top chamber with the rise and fall of the water column, through a cylindrical tube. The moving air drives a Wells turbine (see Box 8.4), which is directly coupled to an electrical generator. The original OWC had a horizontal floor at right angles to the back wall, causing turbulence and consequent loss of energy, so in order to improve the flow of water into and out of the oscillating chamber the channel of the designer gully OWC was built with a sloping floor to efficiently change the water motion from horizontal to vertical and vice versa (see Figure 8.3).

Construction of LIMPET, as the 'designer gully' OWC is known, was completed in September 2000, and Figure 8.4 shows the finished structure, consisting of a rectangular sloping chamber which ducts the airflow through two contra-rotating Wells turbines. Each turbine is coupled to a 250 kW induction generator, giving the device a 500 kW maximum power output. The scheme is one of three contracted under the Scottish Renewables Order (1999). Details of the first year of operation of LIMPET can be found in Boake *et al.* (2002).

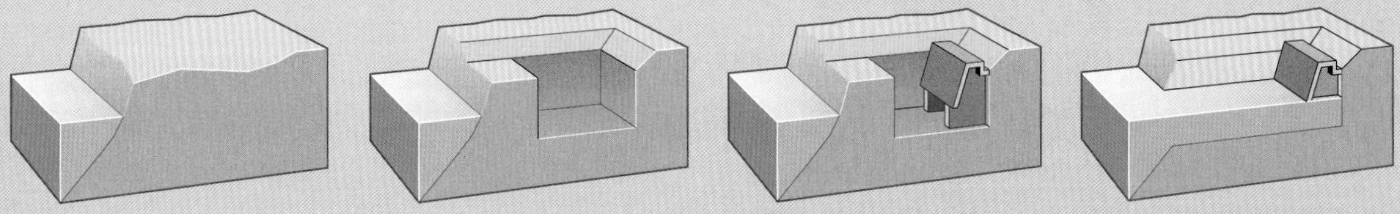

Figure 8.2 Construction sequence used by Queen's University and Wavegen for the 'designer gully' LIMPET OWC

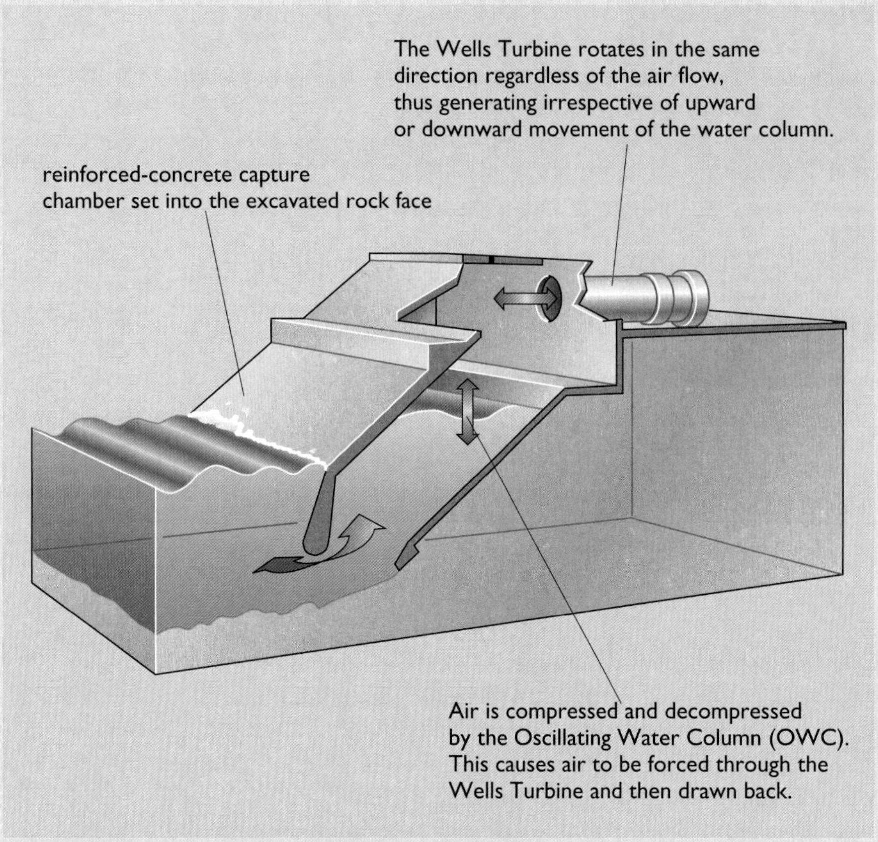

The Wells Turbine rotates in the same direction regardless of the air flow, thus generating irrespective of upward or downward movement of the water column.
reinforced-concrete capture chamber set into the excavated rock face
Air is compressed and decompressed by the Oscillating Water Column (OWC). This causes air to be forced through the Wells Turbine and then drawn back.

Figure 8.3 Outline of the LIMPET device on Islay

Figure 8.4 Photograph of the LIMPET OWC device

8.3 Physical principles of wave energy

Ocean waves are generated by wind passing over stretches of water. The precise mechanisms involved in the interaction between the wind and the surface of the sea are complex and not yet completely understood. Three main processes appear to be involved.

1. Initially, air flowing over the sea exerts a tangential stress on the water surface, resulting in the formation and growth of waves.
2. Turbulent air flow close to the water surface creates rapidly varying shear stresses and pressure fluctuations. Where these oscillations are in phase with existing waves, further wave development occurs.
3. Finally, when waves have reached a certain size, the wind can exert a stronger force on the up-wind face of the wave, causing additional wave growth.

Because the wind is originally derived from solar energy we may consider the energy in ocean waves to be a stored, moderately high-density form of solar energy. Solar power levels, which are typically of the order of 100 W m^{-2} (mean value), can be eventually transformed into waves with power levels of over 100 kW per metre of crest length.

Waves located within or close to the areas where they are generated are called **storm waves**. They form a complex, irregular sea. However, waves can travel out of these areas with minimal loss of energy to produce **swell waves** at great distances from the point of origin. The size of the waves generated by any wind field depends upon three factors: the wind speed; its duration; and the **fetch**, or distance over which wind energy is transferred into the ocean to form waves.

Although the continental shelf results in a large loss of power, the UK is well situated to make use of wave energy. In addition to being surrounded by very stormy waters, it lies at the end of a long fetch (the Atlantic Ocean) with the predominant wind direction being towards the UK. It therefore benefits from both storm and swell waves.

Waves are characterized by their wavelength, λ; height, H; and period, T (see Box 8.3). Greater amplitude waves contain more energy per metre of crest length than small waves. It is usual to quantify the *power* of waves rather than their energy content.

Typical sea state

A typical sea state is actually composed of many individual components, each of which is like the idealized wave described above. Each wave has its own properties, i.e. its own period, height and direction. It is the combination of these waves that we observe when we view the surface of the sea, and the total power in each metre of wave front of this irregular sea is of course the sum of the powers of all the components. It is obviously impossible to measure all the heights and periods independently, so an averaging process is used to estimate the total power, as follows.

1. By deploying a wave-rider buoy it is possible to record the variation in surface level during some chosen period of time. The average water height will always be zero, since the average value also defines the

Box 8.3 Wave characteristics and wave power

The shape of a typical wave is described as **sinusoidal** (that is, it has the form of a mathematical *sine* function). The difference in height between peaks and troughs is known as the **height**, *H*, and the distance between successive peaks (or troughs) of the wave is known as the **wavelength**, λ.

Suppose that the peaks and troughs of the wave move across the surface of the sea with a velocity, v. The time in seconds taken for successive peaks (or troughs) to pass a given fixed point is known as the **period**, *T*. The **frequency**, ν, of the wave describes the number of peak-to-peak (or trough-to-trough) oscillations of the wave surface per second, as seen by a fixed observer, and is the reciprocal of the period. That is, $\nu = 1 / T$.

If a wave is travelling at velocity v past a given fixed point, it will travel a distance equal to its wavelength λ in a time equal to the wave period *T*. So the velocity v is equal to the wavelength λ divided by the period *T*, i.e.:

$$v = \lambda / T$$

The power, *P*, (in kilowatts per metre) of an idealized ocean wave is approximately equal to the square of the height, *H* (metres), multiplied by the wave period, *T* (seconds). The exact expression is the following:

$$P = \frac{\rho g^2 H^2 T}{32\pi}$$

where *P* is in units of watts per metre, ρ is the density of water and *g* is the acceleration due to gravity (9.81 m s^{-2}).

Deep water waves

If the depth of water is greater than about half of the wavelength λ, the velocity of a long ocean wave can be shown to be proportional to the period as follows:

$$v = \frac{gT}{2\pi}$$

This leads to the useful approximation that the velocity in metres per second is about 1.5 times the wave period in seconds.

An interesting consequence of this result is that in the deep ocean the long waves travel faster than the shorter waves.

If both the above relationships hold, we can find the deep water wavelength, λ, for any given wave period:

$$\lambda = \frac{gT^2}{2\pi}$$

Intermediate depth waves

As the water becomes shallower, the properties of the waves become increasingly dominated by water depth. When waves reach shallow water, their properties are completely governed by the water depth, but in intermediate depths (i.e. between $d = \lambda / 2\pi$ and $d = \lambda / 4\pi$) the properties of the waves will be influenced by both water depth *d* and wave period *T*.

Shallow water waves

As waves approach the shore, the seabed starts to have an effect on their speed, and it can be shown that if the water depth *d* is less than a quarter of the wavelength, the velocity is given by:

$$v = \sqrt{g\,d}$$

In other words, the velocity under these conditions is equal to roughly three times the square root of the water depth *d* – it no longer depends on the wave period.

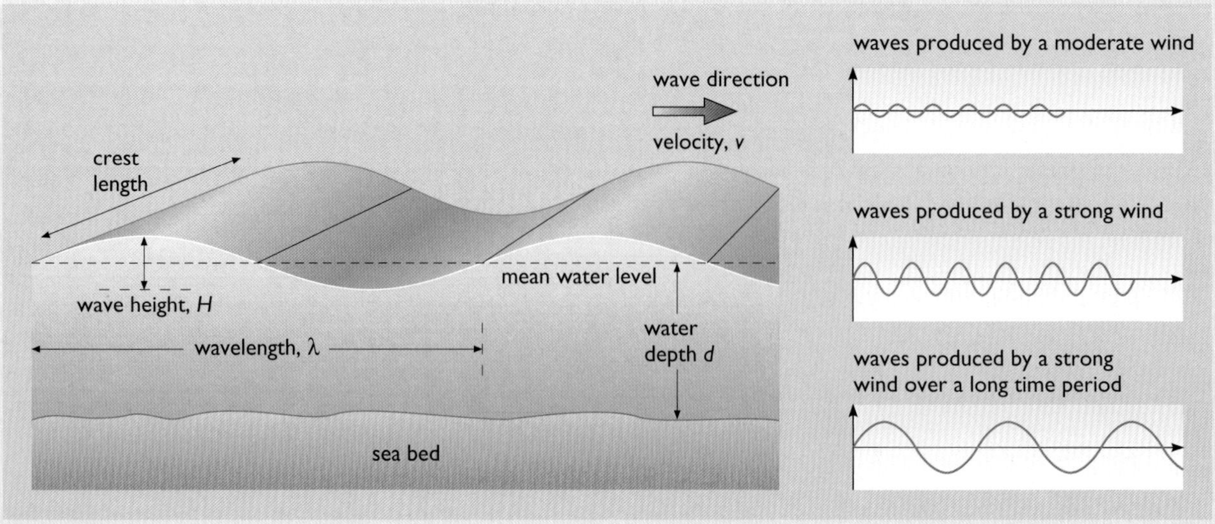

Figure 8.5 Characteristics of an idealized wave

Table 8.1 North Atlantic offshore wave conditions

	period / s	amplitude / m	power density / kW m^{-1}	velocity / m s^{-1}	wavelength / m
storm	14	14	1700	23	320
average	9	3.5	60	15	150
calm	5.5	0.5	1	9	50

zero value, but we can obtain a meaningful figure by calculating the **significant wave height**, H_s. This is defined as 4 × the *root mean square* of the water elevation – i.e. the instantaneous elevations are first squared, making all of the values positive, then the mean over a number of waves is calculated, then the significant wave height is calculated as four times the square root of the mean. The significant wave height is approximately equal to the average of the highest one-third of the waves (which generally corresponds to the estimation of amplitude made by eye, since the smaller waves tend not to be noticed).

2 The **zero-up-crossing period** T_e is defined as the average time – counted over ten crossings or more to get a reasonable average – between upward movements of the surface through the mean level. (Note that including the downward movements would give the half-period.)

3 For a typical irregular sea, it can then be shown that the average total power in one metre of wave crest is given by:

$$P = \frac{H_s^2 T_e}{2}$$

where P is in units of kW per metre length of wave crest.

Figure 8.6(a) illustrates a typical wave record and shows the significant wave height and zero-crossing period. Figure 8.6(b) overleaf shows two further wave records for the same location, recorded on different days.

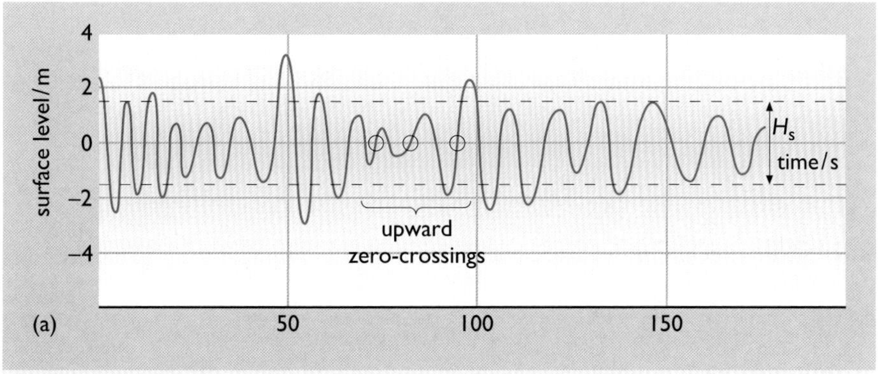

Figure 8.6 (a) A typical wave record. In this example the significant wave height H_s = 3 m. The successive upward movements of the surface are indicated with small circles. In this case there are 15 crossings in 150 seconds, so T_e = 10 seconds. From this, P (kW m^{-1}) = (3^2 × 10) / 2 (kW m^{-1}) = 45 kW m^{-1}

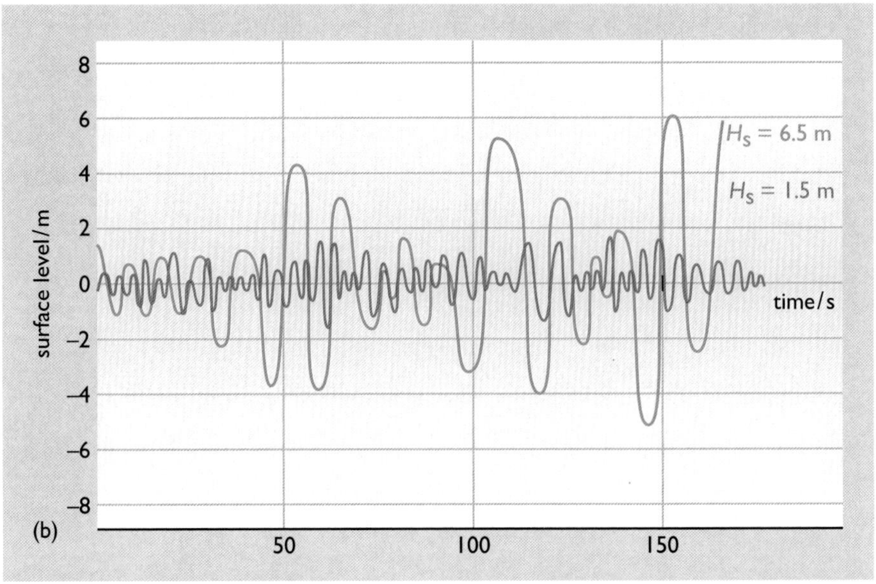

(b) Two wave records are shown here for the same location but represent recordings taken on different days

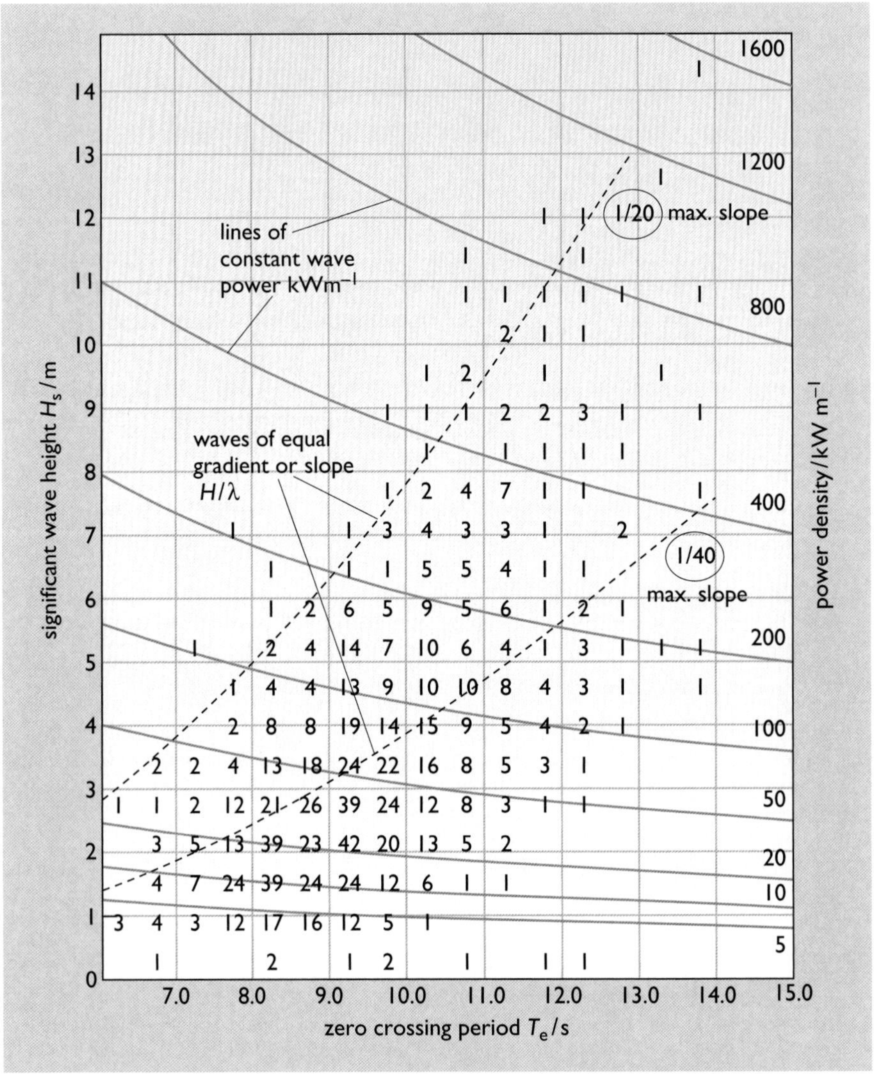

Figure 8.7 Scatter diagram of significant wave height (H_s) against zero crossing period (T_e) for 58°N 19°W in the north Atlantic. The numbers on the graph denote the average number of occurrences of each H_s and T_e in each 1000 measurements made over one year. The most frequent occurrences are at $H_s \sim 2$ m, $T_e \sim 9$ s

Variations in the wave power at any location

Sea level recordings made at different times or dates will of course differ, leading to different values of H_s and T_e. Suppose that each recording represents a time period of one-thousandth of a year or 8.76 hours. If we record the sea states at our chosen location over a whole year, characterizing each of them by their values of H_s and T_e, we can build up a statistical picture of the distribution of wave conditions at our chosen location. This picture, or scatter diagram, gives the relative occurrences in parts per 1000 of the contributions of H_s and T_e. The example of a scatter diagram opposite (Figure 8.7) is for the north Atlantic and shows that the waves at this location have a high average power density. In water 100 m deep at South Uist (Hebrides, Scotland), for example, the annual average is around 70 kW m^{-1} (or 613 000 kWh m^{-1} per year), whereas closer to shore where the depth is 40 m, the average power density is about 50 kW m^{-1} (or 438 000 kWh m^{-1} per year). These figures indicate that the north Atlantic is indeed a valuable wave energy resource.

Figure 8.8 shows estimates of the wave power density at various locations around the world. The areas of the world which are subjected to regular wind fluxes are those with the largest wave energy resource. South westerly winds are common in the Atlantic Ocean, and often travel substantial distances, transferring energy into the water to form the large waves which arrive off the European coastline.

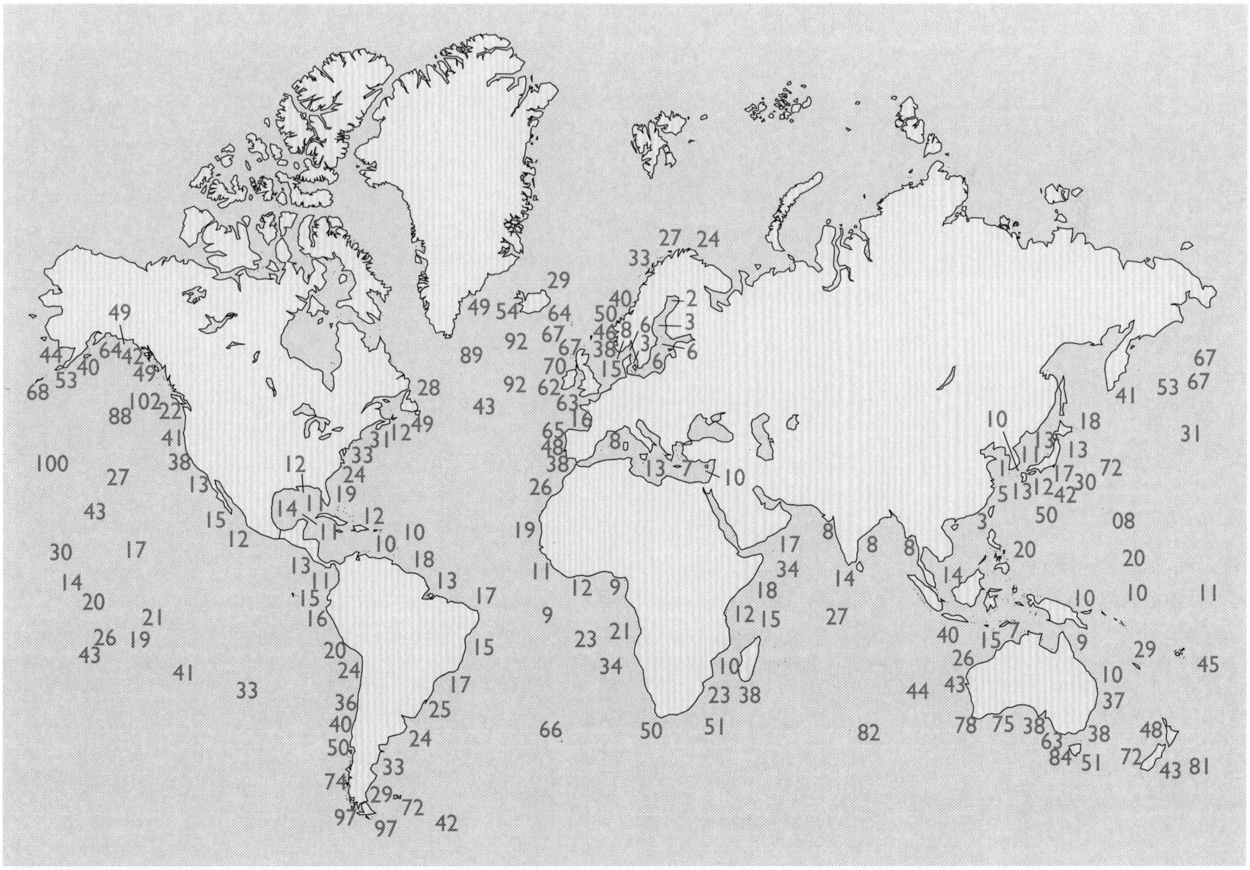

Figure 8.8 Annual average wave power in kilowatts per metre (kW m^{-1}) of crest length, for various locations around the world

Wave direction

The direction of waves travelling in deep water is obviously dictated by the direction of the wind generating them. Waves can travel vast distances across open water without much loss of energy. At any given location we can therefore expect to observe waves arriving from different sources, and hence different directions. For example, in the UK we might see waves approaching us from the south-west which were produced by the wind crossing the Atlantic, but at the same time find that some waves have been generated by a storm to the north of our position. It is easy to imagine that the resulting wave pattern will be complex, and indeed such patterns are commonly observed. A representation of the average power as a function of direction at a given location can be given by a 'directional rose' (Figure 8.9).

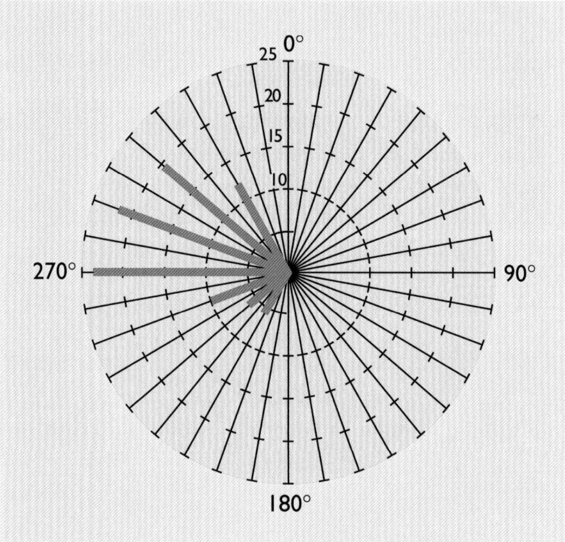

Figure 8.9 A directional rose for waves. The length of the line in each sector represents the average annual power in that sector. In this case most of the waves are coming from the west (Source: Thorpe, 1999)

What happens beneath the surface?

The surface profile of the ocean is the obvious evidence for the existence of waves, but we also need to understand the sub-surface nature of waves if we are to design schemes to capture energy from them (Figure 8.10).

Waves are composed of orbiting particles of water. Near to the surface, these orbits are the same size as the wave height, but the orbits decrease in size as we go deeper below the surface. The size of orbits decreases exponentially with depth.

To capture the maximum energy from a wave we could construct a device to intercept all of the orbiting parts of that wave. But this would be impractical and uneconomic, since the lowest orbits actually contain very little energy. In deciding how deep a structure to extract wave energy should be, it is useful to know that 95% of the wave energy is contained in the layer between the surface and a depth h equal to a quarter of the wavelength λ (i.e. $h = \lambda / 4$).

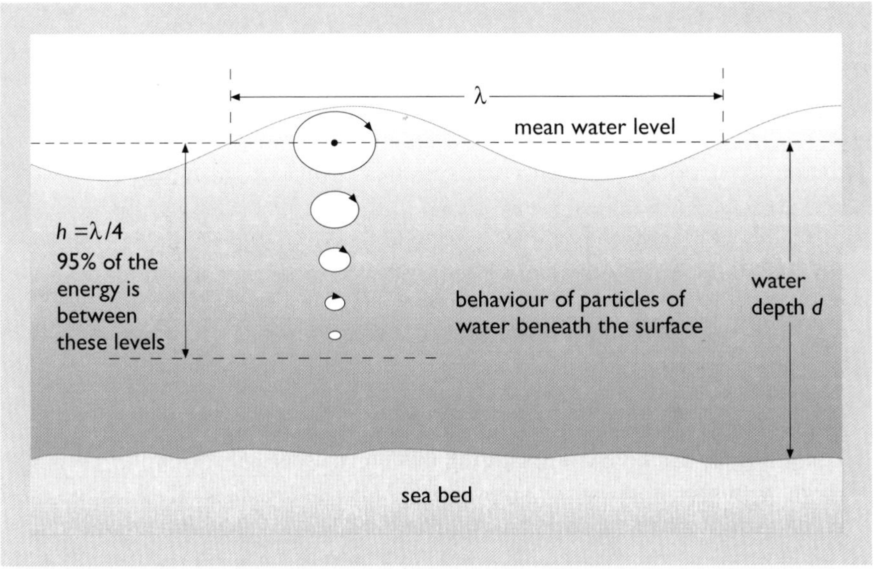

Figure 8.10 Behaviour of particles of water beneath the surface

Moving into shallow water

There are a few areas in the world where the shoreline is formed by a steep cliff which drops into reasonably deep water. These are the areas most suitable for shore-mounted wave energy converters because the incident waves have a high energy content. However, for most of the coastlines around the world the near-shore water is quite shallow. Due to the frictional coupling between the water particles at the greatest depths with the seabed, deep water waves gradually give up their energy as they move into shallower water and eventually run up the shore to the beach (see Box 8.3). The frictional effect becomes significant when the water depth is less than quarter of a wavelength, and the power loss can be several watts of metre per crest length for every metre travelled in-shore.

This loss of energy is very important, because it obviously reduces the wave energy resource. Typically, waves with a power density of 50 kW m^{-1} in deep water might contain 20 kW m^{-1} or less when they are closer to shore in shallow water, depending on the distance travelled in shallow water and the roughness of the seabed. That said, however, storm waves are also attenuated and are therefore less likely to destroy shoreline devices.

There is a further mechanism for energy loss as waves run up the beach: they form breaking waves, which are turbulent and energy-dissipating. We are usually happy to use breaking waves ('breakers') for leisure activities such as swimming and surfing, but breaking waves can be very damaging to structures such as wave energy converters, and so should be avoided when choosing suitable locations.

Structures must be carefully designed not only to perform their energy conversion tasks at an economic cost, but also to be able to withstand the worst wave loadings, as damage can be very expensive to repair.

Refraction

As ocean waves approach the shore they will usually be entering shallower water and, as we saw in Box 8.3, the velocity of the waves then becomes governed by the water depth. Shallower water means lower wave velocity. This in turn leads to **refraction** or change of direction. Imagine that a wave crest approaches shallow water at an angle, so that one end of the crest reaches shallow water first. This part then moves more slowly than the rest, changing its direction. The remainder of the crest progressively adopts this new direction as its velocity is also reduced on entering the shallow water. The effect of refraction, caused by the reducing depth and hence velocity, is gradually to change the direction of the crest to be roughly parallel with the shore (Figure 8.11).

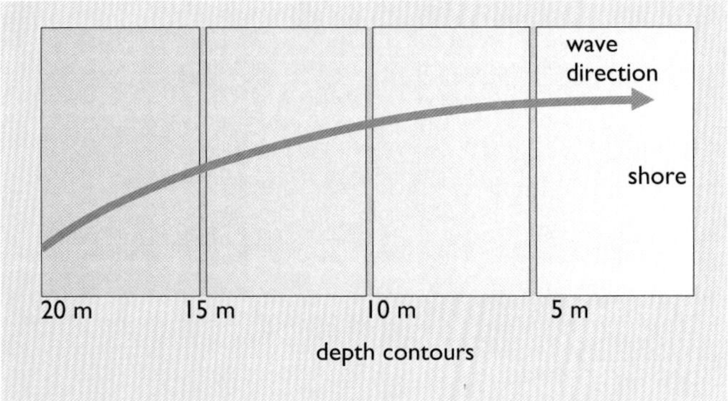

Figure 8.11 As waves travel from deep water into shallow water the effect on their velocity and the resulting refraction generally causes waves to approach a beach at right-angles to the shore

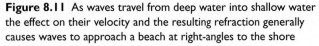

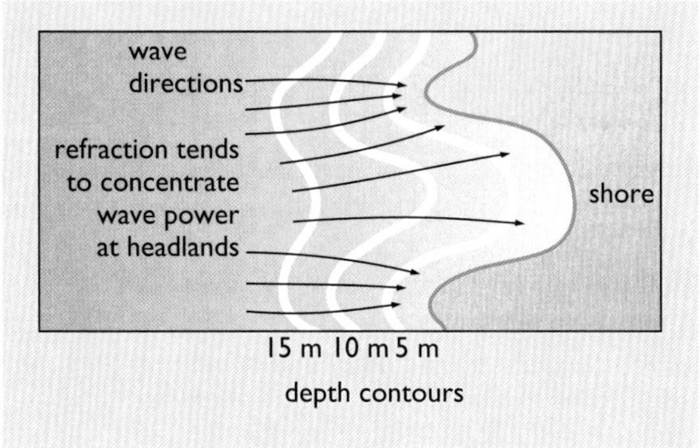

Figure 8.12 Concentration effects of refraction around a shoreline with headlands

Consider a shoreline with headlands (Figure 8.12). Notice how the varying water depth, as shown by the contours (white lines), causes refraction to occur. This concentrates the waves onto the headlands, and leaves the other areas with reduced wave density. Knowledge of the depth contours allows us to carry out a 'ray tracing' procedure, and with it we can identify those areas where the waves will be concentrated. Clearly such sites would be the most cost-effective for wave energy developments.

8.4 Wave energy resources

Wave power resources (the annual average kW per metre of crest length) around the world were shown in Figure 8.8. As already mentioned, the World Energy Council has estimated a total worldwide resource of 2 TW or 17 500 TWh per year (Thorpe, 1999), although this estimate might in fact prove to be rather conservative.

Regarding the UK, Thorpe (ETSU report, 1992) estimated that the *total* annual average wave power along the north and west side of the United Kingdom (i.e. from the south-west approaches to Shetland) ranges from less than 30 GW (equivalent to 260 TWh per year) at the shoreline to about 80 GW (equivalent to 700 TWh per year) in deep water. (UK electricity demand in 2002 was approximately 350 TWh.)

Figure 8.13 shows the areas of British waters and coastline included in the ETSU 1992 assessment of the UK wave energy resource. The proportion of this resource that could actually be harnessed to produce electrical power depends, of course, on various practical and technical constraints.

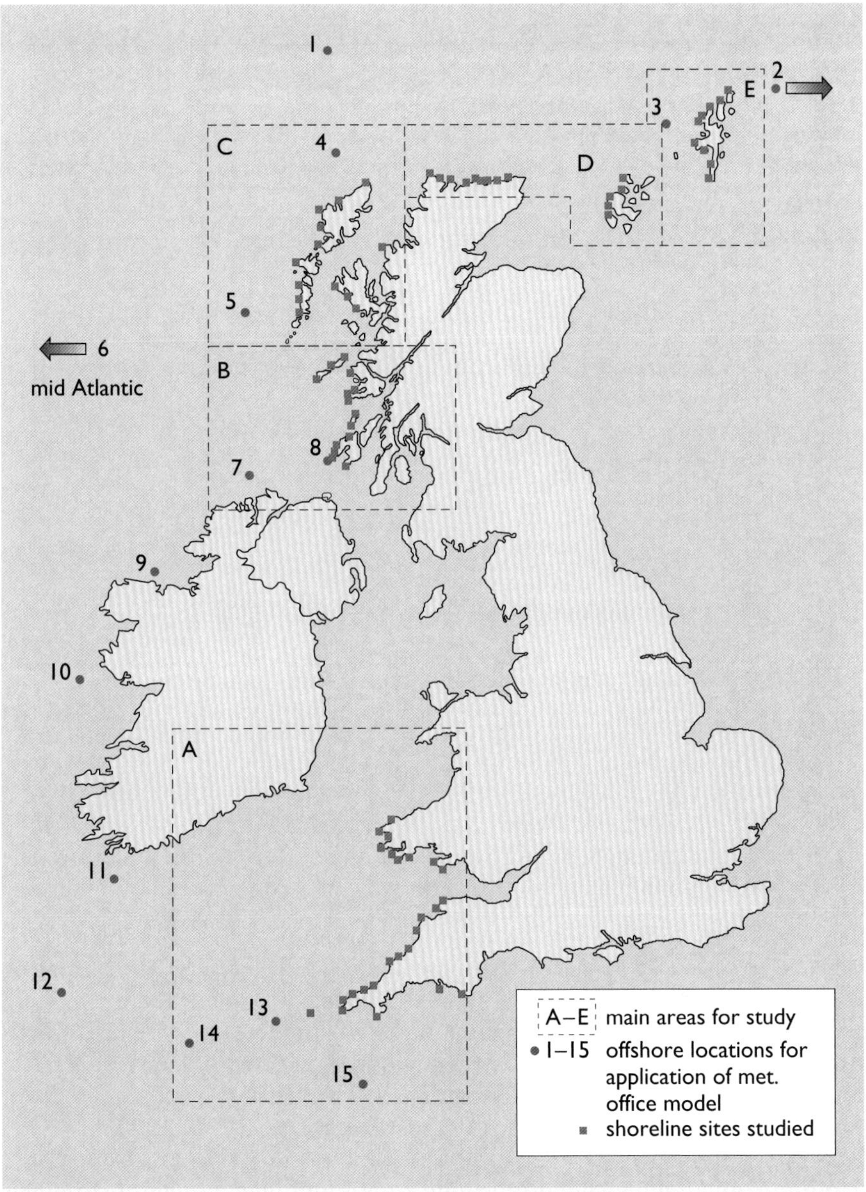

Figure 8.13 Wave power sites studied in the ETSU review of wave energy around the UK

Thorpe (1992) estimated that the *technical resource* – i.e. the resource technically available regardless of cost – is between 7 GW and 10 GW annual average power (equivalent to 61 to 87 TWh per year), depending on the water depth. However, the proportion of the technical resource that could be harnessed *economically* depends on the cost of wave energy relative to other energy sources (see Sections 8.6 and 8.8 below). Table 8.2 gives a breakdown of the resource at different water depths.

A later ETSU report (Thorpe, 1999) concluded that the annual available resources were 100–140 TWh for near-shore/on-shore devices and 600–700 TWh for offshore systems. From this, Thorpe (2001) suggests modest practical energy delivery of up to 2.5 and 50 TWh per year respectively.

Table 8.2 The natural and technical wave energy resource for the north and west side of the UK

water depth / m	average natural resource		average technical resource		average practical resource[2]	
	GW	TWh per year	GW	TWh per year	GW	TWh per year
100	80	700	10	87	5.7	50
40	45	394	10	8	0.24	2.1
20	36	315	7	61		
Shoreline	30	262	0.2[1]	1.75	0.05	0.4

1 The technical shoreline resource is very dependent on details of the local shoreline structure, for example the nature and shape of the rock formations and of gullies and beaches.

2 The practical potential represents what could reasonably be achieved, but the economic potential is lower, at 33 TWh per year, if the maximum cost of electricity is set at 4.0p per kWh.

Source: Thorpe 1992, 2001

8.5 Wave energy technology

In order to capture energy from sea waves it is necessary to intercept the waves with a structure that will react in an appropriate manner to the forces applied to it by the waves.

In order to convert the wave energy into useful mechanical energy (which can then be used to generate electrical energy) the key point is that there must be a central, stable structure, with some active part which moves relative to it under the force of the waves.

The main structure may be anchored to the seabed or seashore, but some part will be allowed to move in response to the force of the waves. Floating structures can also be employed, but still a stable frame of reference must be established so that the 'active' part of the device moves relative to the main structure. This can be achieved by taking advantage of inertia, or by making the main structure so large that it spans several wave crests and hence remains reasonably stable in most sea states.

The physical size of the structure of a wave energy converter is a critical factor in determining its performance. The appropriate size can be estimated by considering the volume of water involved in the particle orbits in a wave. In most circumstances a wave energy converter will have to have a swept volume which is similar to this volume of water in order to capture all of the energy contained in the wave. A variety of wave energy concepts is discussed below. The precise physical size and shape of each device will be governed by its mode of operation, but as a rough guide the swept volume must be of the order of several tens of cubic metres per metre of device width. A device with a swept volume much smaller than this would have a limitation on the total energy that it could capture from each typical wave cycle: although it might still be capable of capturing most of the energy from small waves it would be restricted in its response to larger waves, thereby reducing the overall efficiency.

There are many different configurations of wave energy converter, and a number of ways of classifying them have been proposed. One schematic representation of the various types is shown in Figure 8.14.

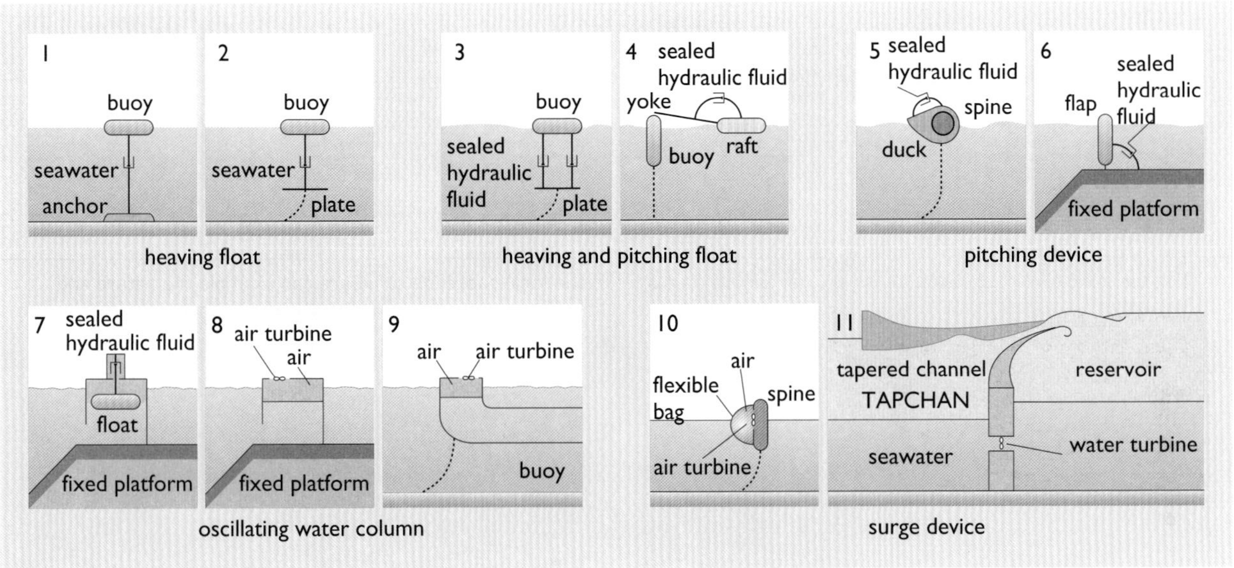

Figure 8.14 Schematic representations of various types of wave energy converter (Source: based on Falnes and Løvseth, 1991)

Wave energy converters can also be classified in terms of their location:

- fixed to the seabed, generally in shallow water,
- floating offshore in deep water, or
- tethered in intermediate depths;

and in terms of their geometry and orientation, as either:

- terminators,
- attenuators, or
- point absorbers.

Terminator devices have their principal axis parallel to the incident wave front, and they physically intercept the waves, whereas **attenuators** have their principal axis perpendicular to the wave front, so that wave energy is gradually drawn towards the device as the wave moves past it. **Point absorbers** also work by drawing wave energy from the water beyond their physical dimensions, but they have small dimensions relative to the incident wavelength. In principle, they could be extremely slim vertical cylinders which execute large vertical excursions in response to incident waves, but in practice the hardware involved tends to mean that they are a few metres in diameter and absorb energy from perhaps twice their own width. Tethered buoy systems, for example, act as point absorbers.

Because full-scale devices are generally very large as a result of needing to be on a similar scale to the waves that they capture, most developments of wave energy technology are therefore carried out at model scale. For this indoor tanks, lakes and reservoirs have been used by research teams. In 1976 a team at Edinburgh University in Scotland constructed a wide wave tank in which it was possible to create repeatable wave conditions, and so test the effects of varying specific design parameters of wave energy converters. The Edinburgh tank was rebuilt in 2002 and is shown in Figure 8.15 with its array of wave-makers.

Figure 8.15 The Edinburgh wave tank

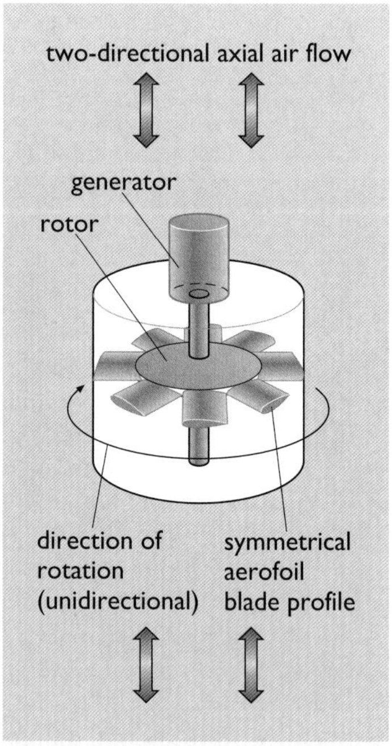

Figure 8.16 The Wells turbine

Fixed devices

Fixed seabed and shore-mounted devices are usually terminators, and these have been the most common types of wave energy converter to have been tested as prototypes at sea. Having a fixed frame of reference and with good access for maintenance purposes, they have obvious advantages over the floating devices, but have the disadvantage that they generally operate in shallow water and hence at reduced wave power levels. In addition, there appears to be a limited number of sites for future deployment.

The majority of devices tested and planned are of the **oscillating water column** (OWC) type, as described in Box 8.2. In these devices, an air chamber pierces the surface of the water and the contained air is forced out of and then into the chamber by the approaching wave crests and troughs. On its passage from and to the chamber, the air passes through an air turbine generator and so produces electricity. The **Wells turbine** is used in and proposed for many OWCs. This novel axial-flow air turbine rotates in one direction irrespective of whether the airflow is into or out of the chamber, and has aerodynamic characteristics particularly suitable for wave applications. Box 8.4 explains the action of the Wells turbine.

Box 8.4 The Wells turbine

The Wells air turbine, invented by Professor Alan Wells, is **self-rectifying** – that is to say, it can accept airflow in either axial direction. To achieve this, the aerofoil-shaped blade profile must be symmetric about the plane of rotation, untwisted and with zero pitch, i.e. the chord line must be in line with the plane of rotation. The vector diagrams in Figure 8.17 show how this occurs. Note that the diagrams are very much like those used in Chapter 7 to explain the properties of wind aerofoils. As the blade moves forward the angle of attack, which is the angle between the relative airflow velocity and the blade velocity, is small and this produces a large lift force (F_L). The forward component of F_L provides the thrust which drives the blade forwards.

The Wells turbine operates in much the same way as would a horizontal-axis wind turbine with symmetrical, untwisted blades and with zero pitch angle. Consider the nearest blade with air flowing in an upward direction (Figure 8.17a). If we now work in the frame of reference of the blade – we do this by making the blade appear to be stationary to us (even though it is moving) by considering the blade velocity vector to be in the opposite direction to the blade's actual direction of movement – we get Figure 8.17(b). Note that because the blade chord is in line with the plane of rotation the angle of attack α is the same as the relative wind angle ϕ referred to in Section 7.4 of Chapter 7. If we now resolve these vectors we get Figure 8.17(c). From this diagram we can see that there will be a net forward force on the blade acting in the plane of rotation if ($F_L \sin \alpha$) - ($F_D \cos \alpha$) is greater than zero. The reaction components are of little interest but the rotor bearings must be capable of carrying these forces. If the net forward thrust is greater than zero, then the blade will be driven forwards and can usefully extract energy from the airflow. The shape of the blade is extremely important here since it will dictate the values of the lift and drag coefficients C_L and C_D and hence the magnitude of the forward thrust.

There is a linear relationship between airflow and pressure drop for the Wells turbine rotating at a constant speed. This means that the Wells turbine has a constant impedance to airflow. Careful choice of the design parameters ensures that this impedance matches the requirements of the wave climate at the chosen location. Impedance matching like this maximizes the transfer of energy from the wave to the generator. The Wells is ideally suited to wave energy applications because of its constant impedance, whereas the impedance of a conventional air turbine varies with airflow. A conventional air turbine may have a superior peak efficiency, but a Wells will perform well over a range of air flows giving it a better wave cycle efficiency. A further beneficial characteristic of the Wells turbine is that, at the sizes typically employed, it can rotate at high speed (1500–3000 rpm) and so the electrical generator can be attached directly to the shaft of the turbine, obviating the need for a gear box to raise the generator speed.

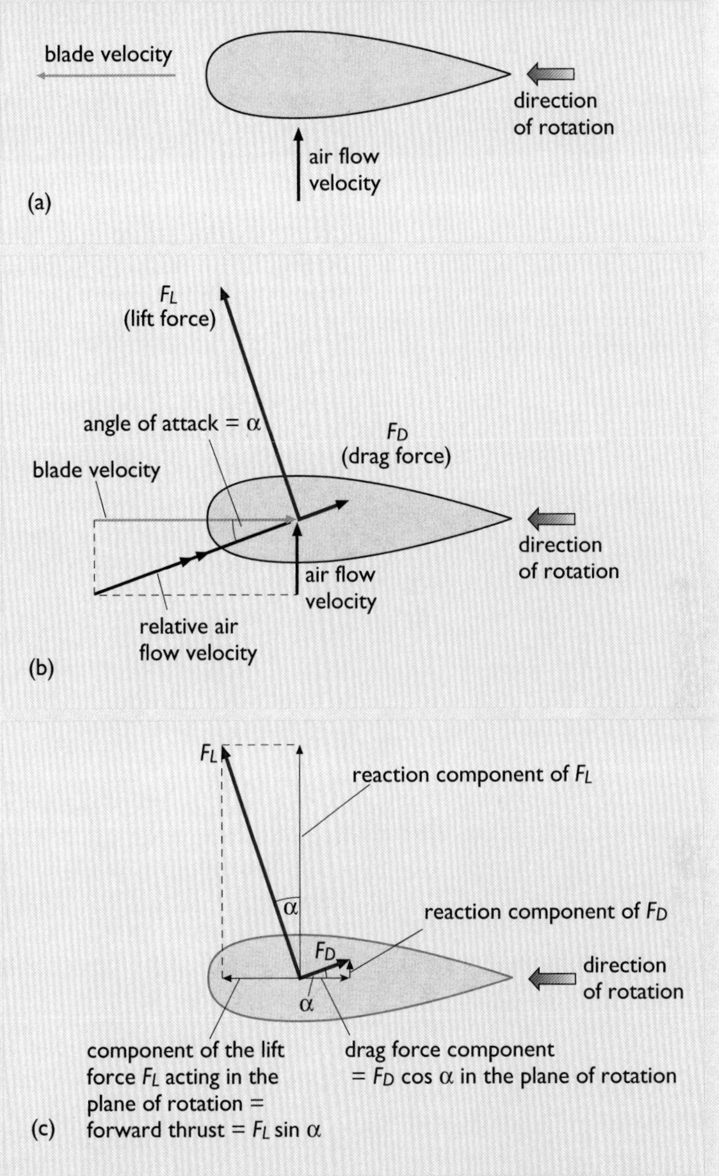

Figure 8.17 The Wells turbine: (a) airflow and blade velocity; (b) relative air velocity and lift and drag forces; (c) forces in the plane of rotation

Fixed oscillating water columns have been built in Norway, Japan, India, Azores and, as described in Box 8.2, Scotland (see Miyazaki, 1991; Ravindran *et al.*, 1995; Whittaker *et al.*, 1997; Duckers, 1998; Falcao, 2000; Heath *et al.*, 2000; Boake *et al.*, 2002).

The Kaimei, a converted barge fitted with a number of floating OWC devices, was first tested in Japan in 1977. In 1989 an OWC was installed at Sakata, on the north-west coast of Japan. Here a wave energy converter has been constructed in a 20-metre section of an extension to the harbour wall, incorporating a Wells turbine rated at 60 kW. The Japanese developers believe that, by functioning as both a breakwater and an energy generator, the cost-effectiveness of the system is increased.

The Norwegian OWC was a multi-resonant oscillating water column (MOWC) designed and manufactured in 1985 by Kvaerner Brug, one of Norway's leading engineering companies. It was located on the same island as the TAPCHAN prototype. The chamber was set back into a cliff face, which falls vertically to a water depth of 60 metres. This created two harbour walls at the entrance to the device, which had the effect of allowing the system to absorb energy over a wider range of wave periods. The oscillating airflow was fed through a 2-metre diameter Wells turbine rotating within the speed range 1000–1500 rpm. The turbine was directly coupled to a 600 kW generator, and the output passed through a frequency converter before being fed to the grid. The performance exceeded predictions and provided electricity at relatively low cost.

Unfortunately, two severe storms in December 1988 tore the column from the cliff, and to date the system has not been replaced. Future designs would therefore have to be much more robust, and would probably involve setting the column into the body of the cliff, as was been done with the small UK OWC on the island of Islay in Scotland, as already described in Section 8.2 of this chapter. (The first Islay OWC was set into a natural narrowing gully in the rocks and so was able to benefit from the tapered channel effect. The later 'LIMPET' OWC on Islay is installed in a 'designer gully' – see Box 8.2.)

Other, non-OWC fixed device prototypes have also been tested in Japan. A number have used a mechanical linkage between a moving component, such as a hinged flap, and the fixed part of the device. An example of this approach is the **Pendulor** (see Figure 8.18).

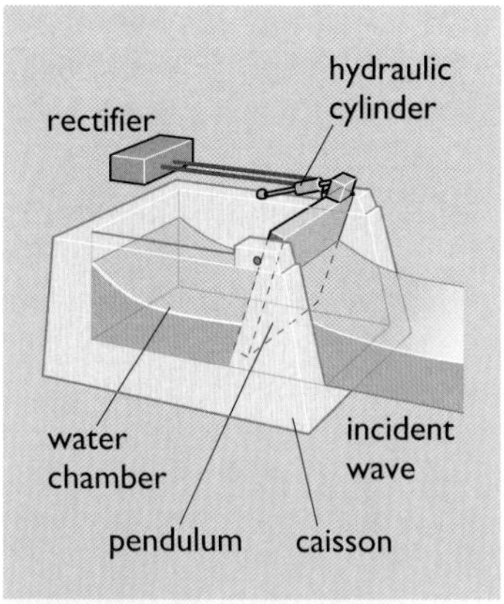

Figure 8.18 The Japanese Pendulor device

The Pendulor is a gate, hinged at the top, which is fitted one-quarter of a wavelength from the back wall of a caisson. This is at the first antinode (i.e. a point of maximum amplitude of a series of waves) and so the gate is subjected to maximum movements resulting from the wave. Note that the gate can be located at the antinode for only one particular wavelength. In the regions of Japan where Pendulor devices have been tested, the seas generally have wavelengths close to the design wavelength for much of the year.

A push–pull hydraulic system converts the mechanical energy from the movement of the Pendulor gate into electrical energy. Two prototypes with nominal power outputs of 5 kW have been operational on Hokkaido, Japan since the early 1980s. Further installations are planned, both in Japan and Sri-Lanka, where a four-chamber, 250 kW scheme is planned. Costs are estimated at 7p per kWh (Watabe *et al.*, 1999).

As we have seen, shore-mounted devices often have to be located in shallow water and so generally have a lower wave power density incident upon them than floating devices. They do, however, have the advantage of being closer to a grid and more easily maintainable. In addition, the seabed attenuates storm waves which could otherwise destroy the turbine.

The only other major drawback of shore-mounted devices is that of geographical location: to optimize output, they need to be positioned in an area of small tidal range, otherwise their performance may be adversely affected. It is also worth noting that mass production techniques are unlikely to be totally applicable to shore-mounted schemes, as site-specific requirements will demand a tailored design for each device, adding to the production costs.

Floating devices

Floating wave energy conversion devices include the **Duck**, **Clam** and **Pelamis** from the UK; floating OWCs such as the **Whale** and the **Backward Bent Duct Buoy (BBDB)** from Japan; and floating tapered channels (described as **Floating Wave Power Vessels, FWPV**) from Sweden. Denmark has versions of both the BBDB and FWPV, called **Swan DK3** and **Wave Dragon**, respectively (see Meyer and Nielsen, 2000). These devices should be able to harvest more energy than fixed, on-shore devices, since the wave power density is greater offshore than in shallow water and there is little restriction to the deployment of large arrays of such devices.

OWCs: The Whale and the Backward Bent Duct Buoy

The Whale and BBDB, both terminator devices, are illustrated in Figures 8.19 and 8.20.

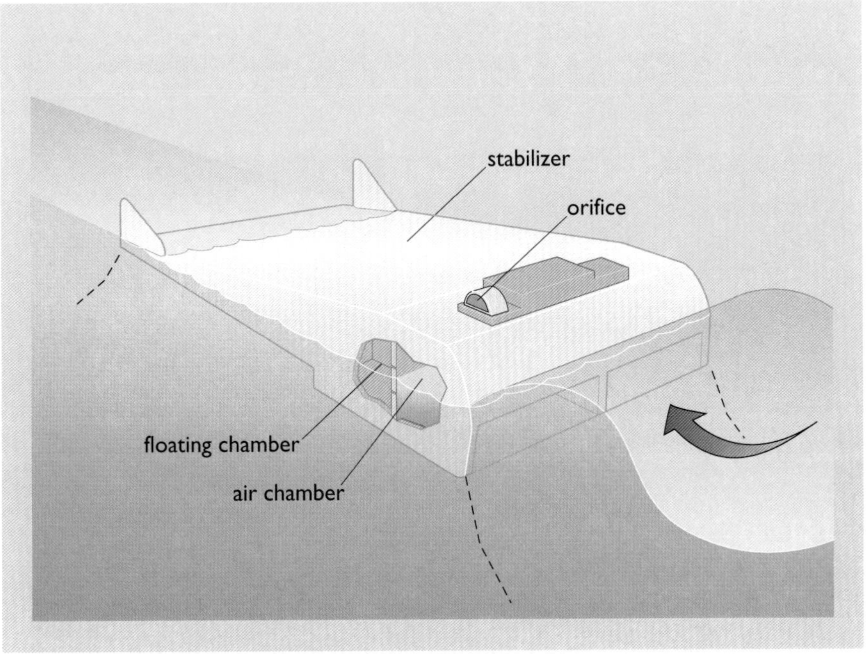

Figure 8.19 The Whale floating wave energy converter

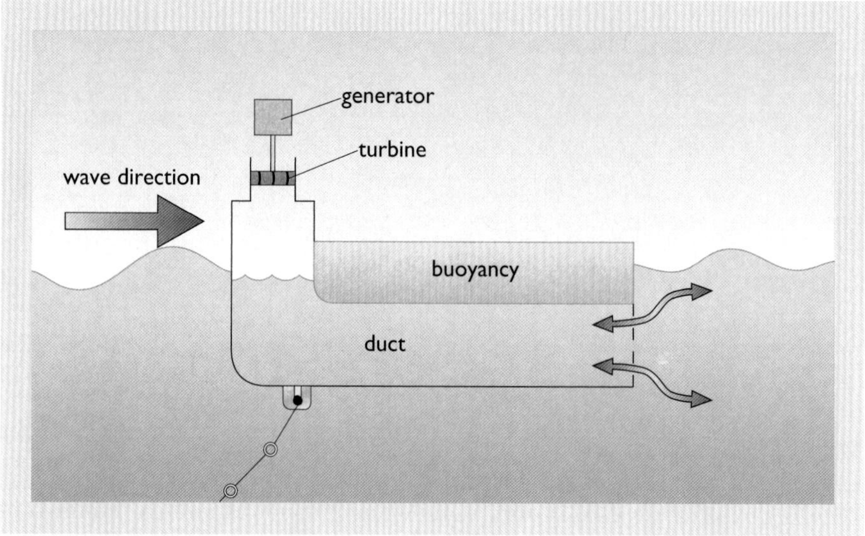

Figure 8.20 The Backward Bent Duct Buoy (BBDB)

The Whale has been tested both at model scale and as a full-scale prototype. A rather massive structure (50 m in length, 1000 tonnes in weight, rated at 110 kW) is required to provide a reasonably stable frame of reference for the Whale, but since the concept incorporates uses such as a breakwater and leisure provision in addition to the generation of electricity, the research team believes that the Whale will prove cost-effective (CADDET, 1999). Tests commenced in September 1998, in Gokasho Bay, Japan, where the typical wave conditions were found to be $H_s = 1$ m; $T_e = 5–8$ s; mean power density = 4 kW m^{-1}. The mean power output was 6–7 kW, in energy periods of 6–7 s, giving an overall efficiency of some 15% (Washio, 2001).

The BBDB concept has been tested at model scale in China, leading to a projected electricity cost of 6p per kWh (Masuda *et al.*, 2000). Recently a comparison of physical and numerical models for different types of BBDB device was carried out in Ireland – details of the results can be found in Lewis *et al.* (2003).

Floating Wave Power Vessel (FWPV)

A FWPV, based on ship construction, was anchored in 50–80 depth of water 500 m from the Shetland coast (Scotland) in 2002. This device has been developed by the Swedish company Sea Power International as part of the Scottish Renewables Order scheme. It is designed to have a maximum power output of 1.5 MW, producing about 5.2 million kWh per year at a cost of 7p per kWh (Lagström, 2000). This device (Figure 8.21) functions by capturing the water from waves that run up its sloping front face. The captured water is returned to the sea via a standard Kaplan hydroelectric turbine (see Chapter 5). In many respects this device may be compared to a floating version of the TAPCHAN. See http://www.seapower.se/ [accessed 3 December 2003] for more details of this project.

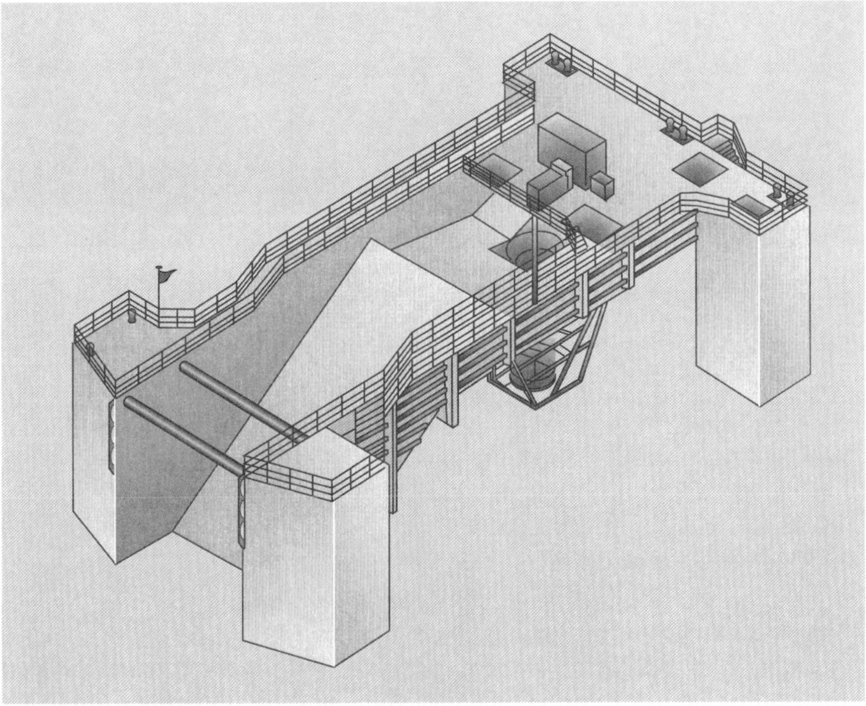

Figure 8.21 The Floating Wave Power Vessel (FWPV)

Clam

The circular Clam developed at Coventry University in the UK in the 1980s consists of twelve interconnected air chambers, or cells, arranged around the circumference of a toroid (see Figure 8.22a), with Wells turbines in each cell. At full scale this would be 60 metres or so in diameter, and would be deployed in deep water (40–100 m). Each cell is sealed against the sea by a flexible reinforced rubber membrane. Waves cause the movement of air between cells. Air, pushed from one cell by the incident wave, passes through at least one of the twelve Wells turbines on its way to fill other cells. As the air system is sealed, this flow of air will be reversed as the positions of wave crest and trough on the circle change. Figure 8.22(b) overleaf shows a cross section through the Clam.

Figure 8.22 (a) A prototype of the circular Clam under test

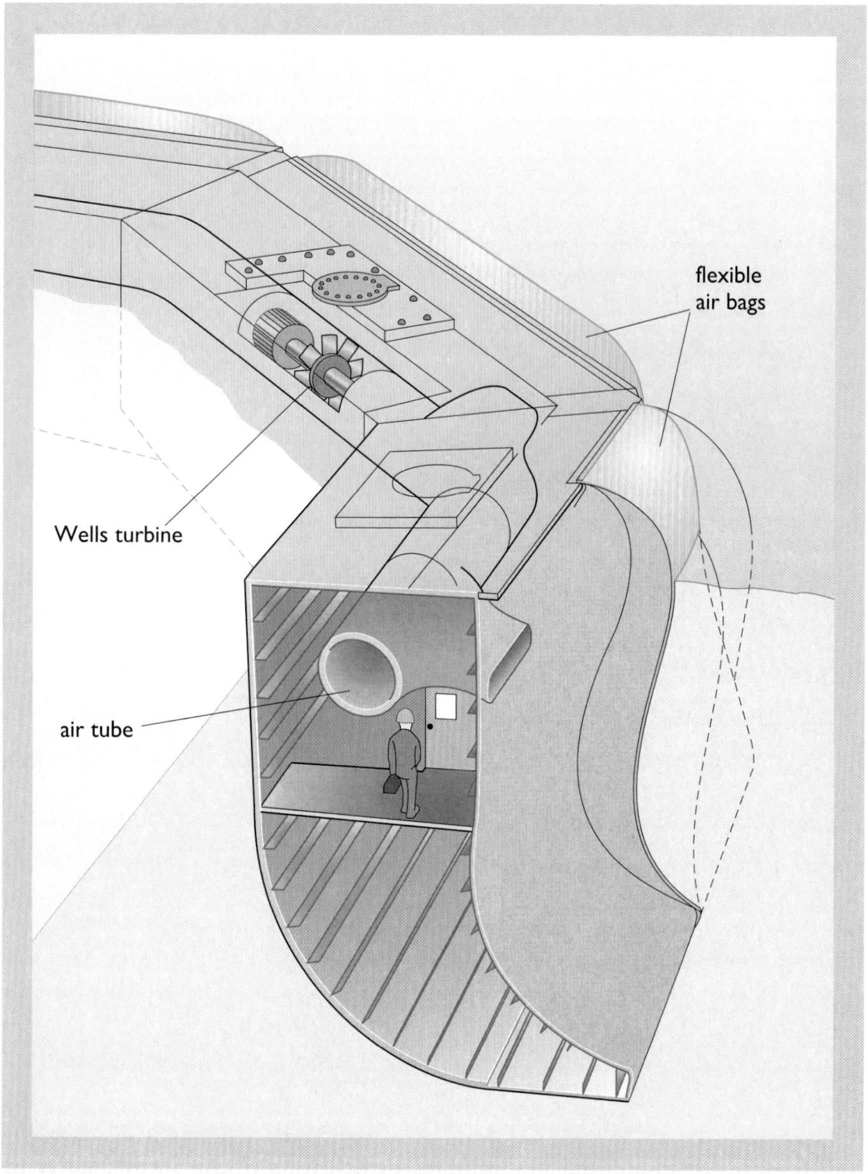

(b) Cross section of the proposed circular Clam

Duck

The Edinburgh Duck concept (Figure 8.23), conceived by Professor Stephen Salter at Edinburgh University in the UK in the 1970s, was originally envisaged as many cam-shaped bodies linked together on a long flexible floating spine which was to span several kilometres of the sea. The spine would be oriented close to the principal wave front making the Duck largely a terminator. The Duck was designed to match the orbital motion of the water particles, as discussed earlier. This matching can be nearly 'perfect' at one wave frequency and the efficiency in long waves can be improved by control of the flexure of the spine through its joints. The concept is theoretically one of the most efficient of all wave energy schemes, but it will take some years to develop fully the engineering necessary to utilize

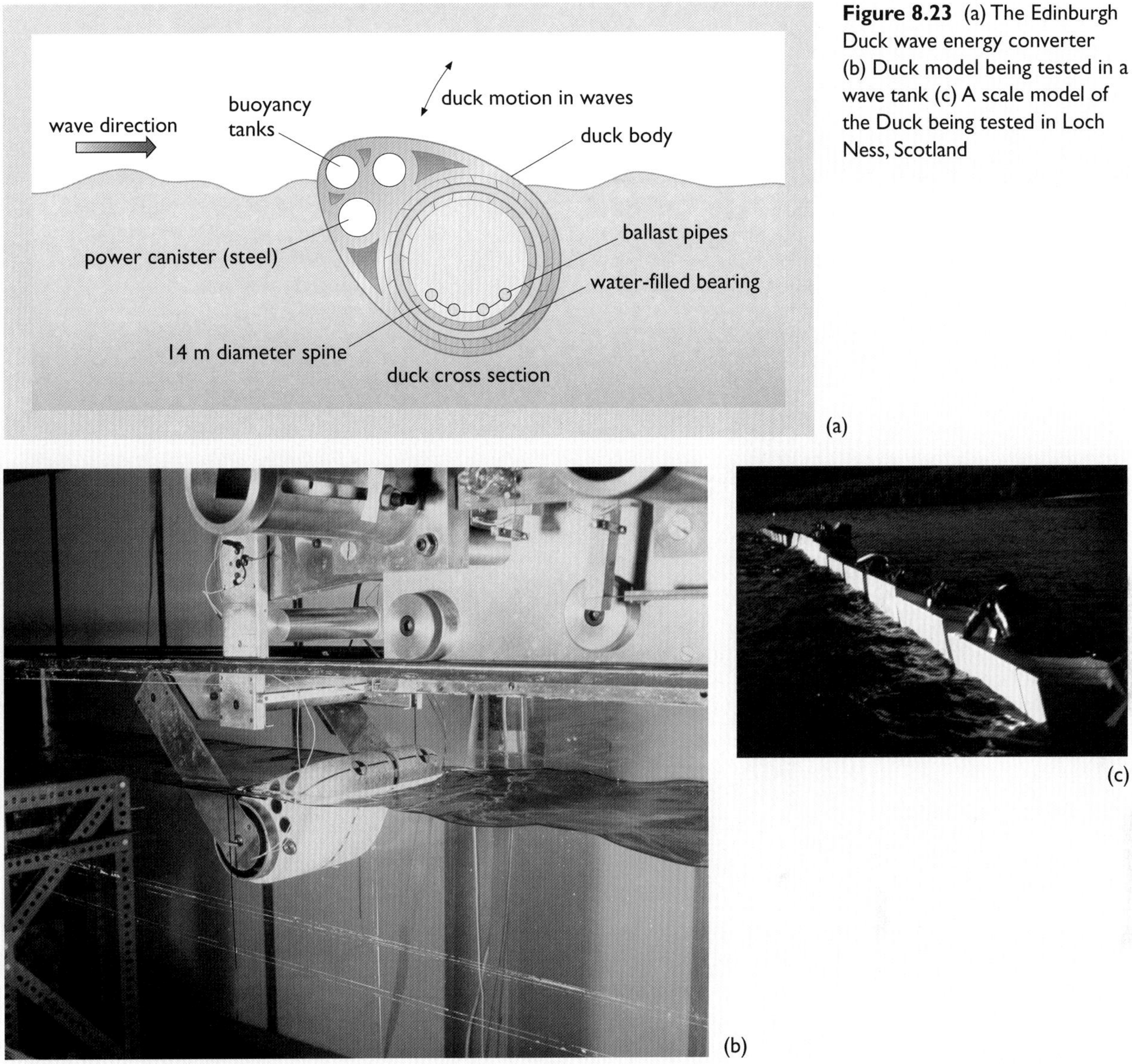

Figure 8.23 (a) The Edinburgh Duck wave energy converter (b) Duck model being tested in a wave tank (c) A scale model of the Duck being tested in Loch Ness, Scotland

this concept at full scale. One of the difficulties is that of extracting energy from a 'randomly' rocking body.

Pelamis

The Pelamis, or 'sea snake', is a device being developed for deployment offshore by the UK firm Ocean Power Delivery Ltd (OPD).

Pelamis has its ancestry in the Edinburgh Duck, consisting of a number of cylindrical sections hinged together, though in this case these are arranged as an attenuator device and are therefore the active components. The wave-induced motion of the cylinders is resisted at the joints by hydraulic rams that pump high-pressure oil through hydraulic motors via smoothing accumulators, and the hydraulic motors drive electrical generators to produce electricity.

Survival has been a key feature of the development programme; the Pelamis is capable of inherent load shedding, which means that the spine is not subjected to the full structural loadings that would otherwise be imposed on it during a storm. As an attenuator it sits down the waves rather than across them – see Figure 8.24 – and so becomes detuned in long storm waves, where the waves are much longer than the device.

Figure 8.24 Prototype of the Pelamis under test

A 750 kW device will be 150 meters long, 3.5 metres in diameter, and composed of five modular sections. OPD makes use only of existing technology. Power will be linked to the grid via sub-sea power cables. OPD has tested models on several scales and has signed a 15-year Power Purchase Agreement under the Scottish Renewables Order (1999) to deliver electricity at less than 7p per kWh. The device was due to be launched in 2003. OPD anticipates costs reducing to under 3p per kWh in the next few years as it refines the design, and claims that with further control of the power take-off a price of 2p per kWh is achievable (Yemm, 2000). For further information see http://www.oceanpd.com/ [accessed 3 December 2003].

Tethered devices

Float systems, with the main body of the structure floating on the surface, but moored to the seabed via a pump, are now attracting some attention. These can act as point absorbers which draw in energy from a greater width of water than their own physical diameter. In theory a perfect wave energy point absorber could capture the energy from a wave front with a length equal to $\lambda / 2\pi$ metres. For example, a wave with period of 6 seconds would have a wavelength of between 56 m and 72 m, depending on whether the water is deep or shallow, and so a perfect point absorber would absorb the energy from a width of about 10 m (between about 9 m and 12 m). In reality, however, the capture width is much less due to the limitations of the vertical amplitude of the motion of the absorber. The mathematical explanation of this is outlined by Nielsen and Plum (2000) who go on to report experimental results. These devices are said to have a **capture width ratio**, i.e. apparent diameter/physical diameter, greater than 1. Figures 8.29 and 8.30 later in the chapter show examples of float systems. The concept of **latching**, or holding the float under water for a second or so before allowing it to follow the wave, has been developed to maximize the energy capture by permitting a large amplitude of motion of the float, which is needed for optimal performance (Falnes and Lillebekken, 2003).

Hose Pump Wave Energy Converter

The **Hose Pump Wave Energy Converter** has been developed over almost 20 years by Technocean in Sweden and is intended to pump sea water from an array of hose pumps fixed to the seabed (Figure 8.30). The hose pump consists of a reinforced vertical rubber cylinder, anchored to the seabed and attached to a float on the surface. Careful choice of the geometry of the reinforcing cords, by winding them as spirals up the tube, means that the sea water inside the tube will be pressurized when the tube is stretched by the float as it moves upwards in response to a wave crest. The sea water is then pumped out of the tube and up to a storage reservoir on land. A Pelton wheel (see Chapter 5) extracts energy from the water as it is released from the reservoir back to sea. The tube is refilled with water via a non-return valve.

Interproject Service Convertor

The **Interproject Service (IPS) Converter** was developed by Interproject Service AB, Sweden, in the early 1980s and tested at a 1:10 scale in a lake and as a full-scale prototype at sea. It consists of a long buoy with a tube open at both ends attached underneath. A piston inside the tube is linked to the buoy and power is extracted by the interaction of the buoy and the water in the tube. A new configuration, the sloped IPS, embodies a number of attractive features, particularly by operating at an angle to the vertical giving it additional energy capture. By employing a hydraulic accumulator the device can effectively provide reasonably smooth output and so offers the prospect of firm power for slender electricity networks or small islands (Salter and Lin, 1995). For more information see IPS's website at http://www.ips-ab.com/ [accessed 3 December 2003].

The AquaBuoy system shown in Figure 8.25 is a development based on both the IPS and the hose pump wave energy converter. An experimental programme aimed at installing a 1 MW scheme consisting of units rated at about 100 kW off the coast of Washington in the US is being conducted by AquaEnergy Group.

Figure 8.25 The AquaBuoy system, based on the IPS and hose pump concepts

Wave energy research and development activity around the world

Japan

Japan has had a substantial wave energy research programme, with many teams working on a variety of projects. Some of these, such as the Whale, BBDB and Pendulor, have already been discussed here. Further details can be found in Miyazaki (1991); Miyazaki and Hotta (1991); Kondo (1993); Washio *et al.* (2001); and ISOPE (2002).

Norway

TAPCHAN and the multi-resonant OWC have already been discussed (see Box 8.1 and the subsection on fixed devices earlier in this section). Important work is continuing on various ways of improving wave energy capture, including the use of latching to maximize energy output (mentioned above).

Twin, or duplex, linked OWCs are also being studied in Norway as they may have a wider frequency response than single OWCs. In 1998, another Norwegian concept, the **ConWEC** system was announced, in which a float oscillates up and down inside a cylindrical structure. The energy in the waves is converted to mechanical energy by a pump with its piston rigidly connected to the float. Sea water can thus be pumped to a reservoir and returned to the sea via a turbine. This scheme resembles a fixed OWC but has the advantages of including latching to maximize performance, and storage for output smoothing (CADDET, 1999).

Traditionally, all of Norway's electricity has come from renewable hydropower, so until quite recently Norway had little interest in deploying wave energy technology for its own use. However, as mentioned in Section 8.1, it was keen to develop an export market for wave energy technology, and thus it continued to invest in research and development. In recent years electrical supply has been gradually falling below demand (the level was 99.4% in 1997) because of public opposition to the construction of new hydro plants, and by 1998 Norway was a net importer of 3.7 TWh from Denmark and Sweden. Concern about CO_2 emissions from bridging this gap with fossil-fuel burning power stations could therefore create the basis for a Norwegian wave energy market.

United Kingdom

In addition to the UK wave energy projects that have already been mentioned – such as the 'designer gully' LIMPET OWC, the Edinburgh Duck and Pelamis – other UK wave energy research projects which have been costed include the **Bristol Cylinder** and the **Pitching and Surging FROG** (PS FROG).

The Bristol Cylinder concept consisted of a large cylinder held underwater but allowed to orbit in a wave-like manner, and hence absorb energy from the waves. The engineering difficulties associated with resisting such large forces but permitting large movements make this an unlikely contender for future full-scale deployment.

The Pitching and Surging FROG, developed by Professor French and his colleagues at Lancaster University, is a reaction wave energy converter which achieves energy-absorbing behaviour by the movement of internal inertial mass (McCabe *et al.*, 2003). Hence, it is a compact structure which does not require a large spine to provide a stable frame of reference (Figure 8.26).

A bottom-mounted OWC, termed **OSPREY 1** (Ocean Swell Powered Renewable Energy) was launched by the Scottish company Wavegen in 1995 (Figure 8.27), but Hurricane Felix destroyed the steel shell before it could be properly ballasted into location. At the time of writing the company is planning another version of the device, OSPREY 2000 for deployment off the Irish coast (see http://www.wavegen.co.uk [accessed 3 December 2003]. Wavegen is also planning a combined wave–wind scheme called **WOSP**, which would have a total installed capacity of 3500 kW.

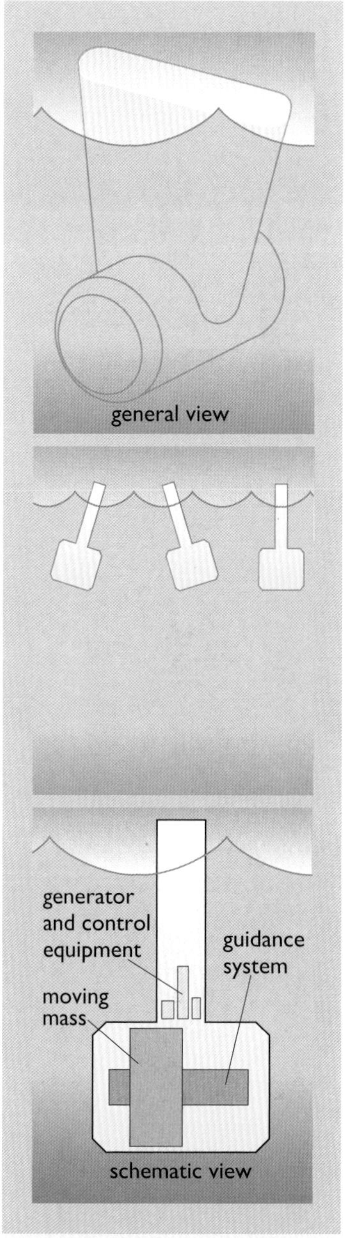

Figure 8.26 The Pitching and Surging FROG wave energy converter

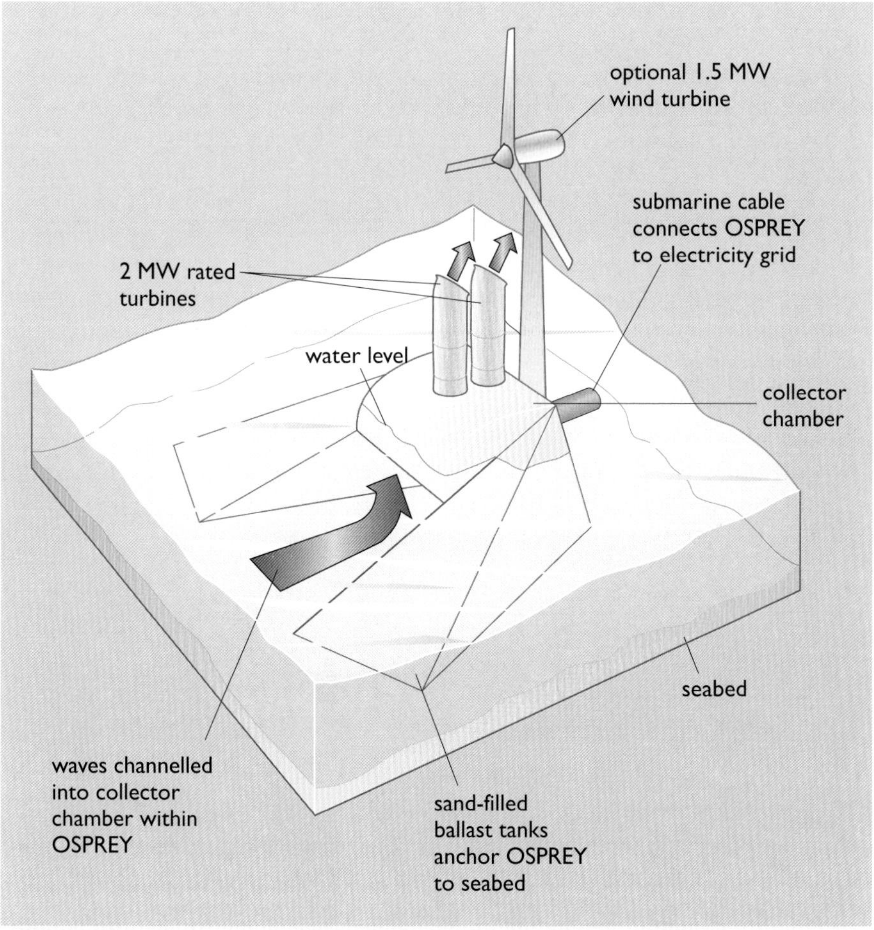

Figure 8.27 The Wavegen OSPREY 1

China

Wave energy activity is developing in China. Much of the Chinese work is linked to Japan, either in concept or by the exchange of ideas and staff. Some of the work concentrates on navigation buoys, some on theoretical modelling. China is known to have built Backward Bent Duct devices and to have tested a small OWC, in addition to the numerous wave-powered navigation buoys that it has deployed. The shoreline 3 kW OWC, in a natural gully on Dawanshan Island in the Pearl River, was so successful that it is currently being upgraded with a 20 kW turbine. Experiments on the BBDB concept have been conducted at the Guangzhou Institute of Energy Conversion (Masuda *et al.*, 2000; Xianguang *et al.*, 2000). For more up-to-date information on Chinese developments see ISOPE (2002).

India

Trials of a multi-resonant OWC device, installed in a breakwater and employing a Wells turbine of 2 metres in diameter, driving a 150 kW generator (Figure 8.28), commenced off the Trivandrum coast in 1991 (Ravindran *et al.*, 1995). The double benefits of the breakwater – the provision of a harbour and the generation of electricity – mean that the function of electrical production does not have to justify the total capital cost, and so the generated cost of electricity may be more economically attractive. The device is estimated to be capable of delivering an average of 75 kW during the monsoon period from April to November, and 25 kW from December to March. A power conversion system incorporating two units has been installed so that in the higher power period both units can be operational.

Since the annual average wave power density along the Indian coast is only between 5 and 10 kW per metre, it is perhaps surprising to see such research and development activity. However, many more harbours are planned on the Indian coastline and this has led to consideration of the potential to incorporate OWC wave energy converters.

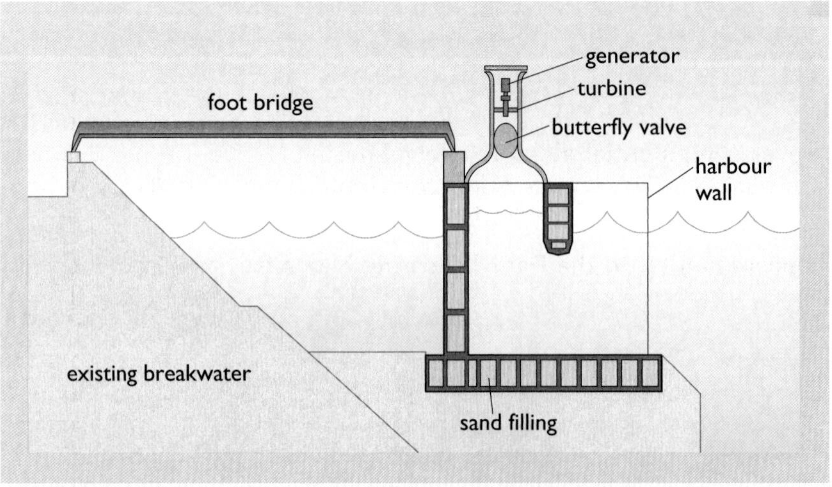

Figure 8.28 Cross-section through the Indian breakwater OWC

Denmark

Denmark has had an ambitious wave energy programme, with substantial support from successive coalition governments. However, the change of government in 2000 led to the curtailment of much of the Danish wave programme. A summary of Danish wave power research can be found in Meyer and Nielsen (2000).

Research effort in Denmark includes work on a tethered buoy system, as illustrated in Figure 8.29. The floating buoy responds to wave activity by pulling a piston in a seabed-mounted unit. This piston pumps water through a submerged turbine. An array of these buoys could be deployed and arranged to have an integrated output. There have been some difficulties with leaking seals on the prototype, but these should be overcome with further development.

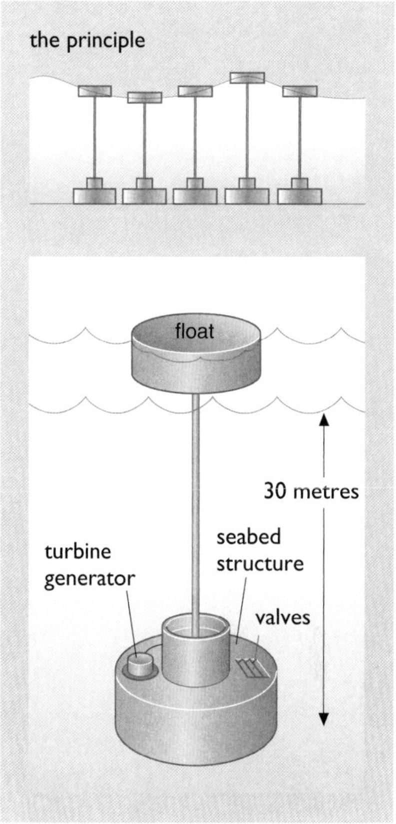

Figure 8.29 Tethered buoy wave energy converter

Several other projects have been assessed as model or prototype developments, including the **Waveplane**, the **Wave Dragon** and the **Swan DK3**.

The Waveplane is a wedge-shaped device which directs incoming waves of varying frequency into a trough in a spiral configuration, creating a vortex which is used to drive a turbine. Tests using a 1:5 scale model have been under way in Mariager Fjord in Jutland since May 1999.

The Wave Dragon is effectively a floating TAPCHAN-type device – waves run up a tapered channel and a head of water collects in a reservoir. This water returns to the sea via a set of simplified Kaplan hydroelectric turbines. A full-scale version was launched near to the University of Aalborg in March 2003, and this delivered its first power to the grid in June 2003. More turbines were installed at sea in September 2003, demonstrating that maintenance at sea is possible (Sorensen *et al.*, 2003). Finally, the Swan DK3 is based on the L-shaped Backward Bent Duct Buoy concept initially developed in Japan and then China. For more details see Meyer and Nielsen (2000).

Sweden

The potential along the Swedish coast is about 5–10 TWh per year (an average power of 0.57–1.1 GW, representing 3–7% of demand). In addition, however, the potential along the neighbouring Norwegian coast is put at around 3.0–3.5 GW, which could contribute some 12–15% of Sweden's electricity demand via the Nordic grid.

A concept similar to the Danish floating buoys has been investigated in Sweden, but using a reinforced rubber hosepipe as both the tether and pumping mechanism (see Figure 8.30). The reinforcing cords in the hose are arranged at a carefully chosen angle to the main axis of the hose. As the buoy rises with a wave, the hose is stretched and the cord angle changes in such a way that it causes the internal volume of the hose to be reduced. This raises the pressure of the working fluid (sea water) contained within the hose. The working fluid is thus pumped into a high-pressure reservoir where it is subsequently used to generate electricity.

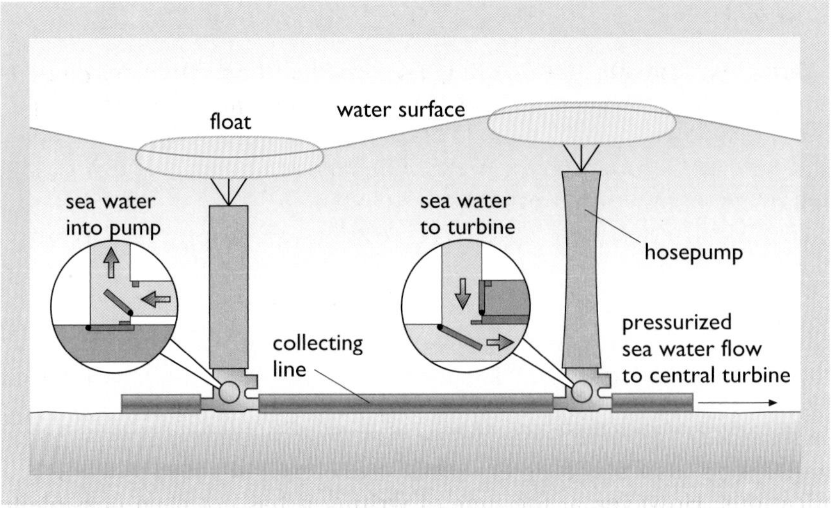

Figure 8.30 Swedish hose pump wave energy converter

In 1992 Sea Power International installed a floating wave power vessel (FWPV) for testing off the west coast of Sweden. This steel vessel resembles a floating TAPCHAN in that waves run up a sloping ramp and are collected in a raised internal basin. The water flows from the basin back into the sea via low-head turbines. This device is not sensitive to tidal range, and by varying its ballast the device can be tuned to different wave heights. As mentioned earlier, in 2002 Sea Power International also installed a 1500 kW pilot version 500 metres off the west coast of Shetland (Lagström, 2000).

Sea Power International also plans to introduce floating wave power vessels adapted to produce hydrogen gas (FWPVHY). Using this system, the wave energy would be stored as hydrogen for use on demand in fuel cells (see Chapter 10, Box 10.3). Sea Power International estimates that in five years it should be possible to have five FWPVHY plants commissioned. These could each deliver 1 340 000 m^3 of hydrogen. For more information see http://www.seapower.se/pres_eng.pdf [accessed 22 December 2003].

Portugal

An OWC with a capacity of 500 kW has been installed on the island of Pico, part of the Azores, in the north Atlantic. This is located on the sea bottom, close to the rocky shoreline. It is in a slightly sheltered bay which helps to protect it from the excesses of the largest storms. The scheme consists of a 12-metre-by-12-metre concrete structure built *in situ* on the rocky sea floor spanning a natural harbour, and equipped with a horizontal-axis Wells turbine-generator capable of producing an average annual output of 124 kW, but rated for a maximum instantaneous value of 525 kW. The Wells air turbine is designed to operate between 750 and 1500 rpm, has eight blades of constant chord and a diameter of 2.3 m (Falcao, 2000). A second turbine with pitch-controlled blades is to be operated as a contrast to the fixed-pitch one (Taylor, 2002).

Netherlands

The Dutch **Archimedes Wave Swing** developed by Team Techniek consists of a number of interconnected mushroom-shaped chambers located just below the sea surface. These chambers are topped with moveable floats or hoods. The system consists of a cylindrical, air-filled chamber (the 'floater'), which can move vertically with respect to the cylindrical 'basement', which is fixed to the seabed (CADDET, 1996). This movement is generated by the changes in buoyancy of the air within the floater as waves pass over the top. The resulting changes apply forces to a tether which actuates a hydraulic system and generator.

In the late 1990s Team Techniek was constructing a 2 MW pilot scheme in Romania for deployment in Portugal. This was supported financially and technically by a large Dutch utility and several industrial companies. Team Techniek was also planning an 8-MW three-chamber design for future applications. However, at the time of writing it has not been possible to find any further information or establish the fate of this project.

Korea

Baek Jae Engineering of Korea has designed a prototype wind–wave energy scheme. This design has many novel features, in particular a floating lattice structure fabricated from plastics and composites. The new aspects of the design are intended to reduce the overall capital cost of the scheme by minimizing the non-productive wave loading on the device and thus allowing a less robust and cheaper construction material to be used. The design is at an early stage of development, i.e. at pre-prototype phase. An independent assessment has been carried out which indicated that, if research in these areas is successful and the device lives up to its early promise, it is likely to be economically competitive with a range of electricity generation technologies (both conventional and from renewable energy sources) if deployed in energetic wave climates such as those of Western Europe. The most recent reports available indicate that Baek Jae Engineering is continuing to develop the design and is intending to progress to testing a prototype in the near future (Thorpe, 1999).

Indonesia

In the late 1990s the Norwegian company Indonor AS started building a 1.1 MW TAPCHAN on the Indonesian island of Java. The contract was sealed by the former Norwegian Prime Minister Gro Harlem Brundtland on her visit to Indonesia in 1996 (Gemini, 1996). The structure consists of 7-metre deep channel which narrows from 7 meters wide down to 25 cm over its 60-metre length. The bay where it is being built has its own natural basin with an area of 7500 m^2 capable of holding water 4 metres above sea level. This, along with a low tidal range, provides perfect conditions for the TAPCHAN. The total cost was expected to be about £6 million and the predicted cost of electricity exported from the plant was 5.5p per kWh.

At the time of writing it has not been possible to determine the outcome of this project.

Republic of Ireland

The west coast of the Republic of Ireland is particularly suitable for the deployment of both shore-mounted and offshore wave energy converters. The overall wave resource around the west and south coasts of Ireland is estimated to be 25 GW, amounting to a gross energy output of 219 TWh per year or nearly 10 times the current electrical demand for Ireland (23.5 TWh in 2001) (Ó'Gallachóir, 2001).

Research in Ireland has concentrated on OWCs, and on other forms of self-rectifying air turbines as alternatives to the Wells turbine.

A 40-metre long prototype of a device called the **McCabe Wave Pump** was deployed off the coast of Kilbaha in Ireland a few years ago (see Figure 8.31). This device consists of three narrow rectangular steel pontoons, which are hinged together across their beam, which points into the incoming waves. The pontoons move relative to each other and energy is extracted from this motion by linear hydraulic rams mounted between the pontoons near the hinges. The output is intended for use in applications such as desalination, as well as electrical generation. Kraemer *et al.* (2000) report on experimental and theoretical results of the motion of a McCabe wave pump. Modelling of the power take-off system for a hinged barge wave energy converter is reported by Nolan *et al.* (2003).

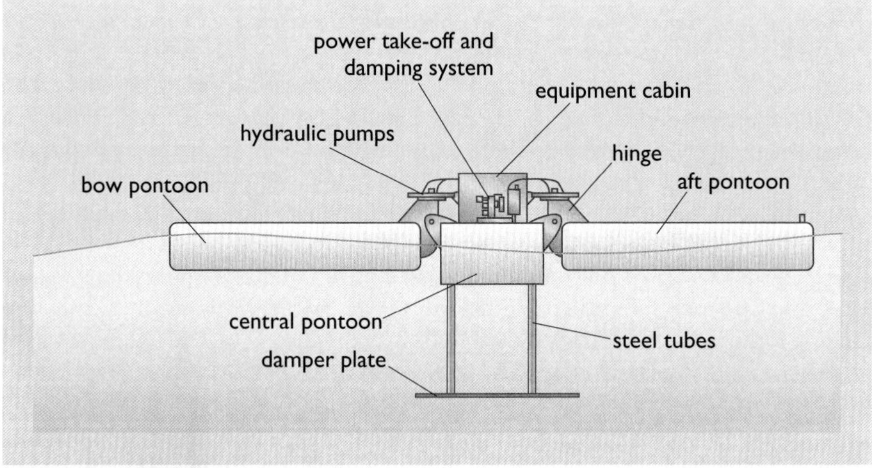

Figure 8.31 The McCabe wave pump

United States

A small amount of work on wave energy has been carried out in the United States. Government support has been modest, but commercial organizations have promoted several concepts to the stage of preliminary design and model testing, particularly from the late 1990s to the present. These have included a scheme based on the OWC, as well as the **Seamill** concept, which resembles an OWC but has a float on top of the internal water surface. The motion of this float is used to provide hydraulic power, which drives a turbine and hence generates electricity.

Another concept, the **Wave Energy Converter** developed by Ocean Power Technology (OPT) consists of a modular buoy-based system which drives generators using mechanical force developed by the vertical movement of the device. Each module is relatively small, permitting low-cost regular maintenance, leading to expected lifetimes of at least 30 years and hence delivered energy at 2 to 3p per kWh for 100 MW schemes, and 5 to 7p per kWh for 1 MW plants (Taylor, 2000).

A unit of the OPT system has been extensively tested on a large scale off the coast of New Jersey, USA. The first commercial schemes are being built for an electrical utility in Australia, for the US Navy and for the State of New Jersey, with a number of other schemes in the pipeline.

AquaBuoy, a derivative of the IPS concept, is being deployed off the Washington coast – see the subsection on tethered devices above.

Australia

Energetech Australia Pty is developing a new wave energy concept, an advanced shoreline OWC. This uses a novel, variable-pitch turbine to improve efficiency and a parabolic wall to focus the wave energy on the OWC collector. It is designed for use in harbours or rocky outcrops where the water at the coastline is deep. As wave fronts approach they are amplified up to three times at their focal point by the parabola shaped collector, before entering the 10-metre-by-8-metre OWC structure.

Figure 8.32 The Energetech Australia Pty wave energy concept

A 500 kW scheme is being constructed at Port Kembla, New South Wales, and was due to be completed by the end of 2003. This is a joint venture between the developer and a local power retailer, Integral Energy, which expects to purchase over 1 GWh per year from the scheme. The prototype is expected to produce electricity at about AU$0.05 per kWh (around 2p per kWh). A contract is being drawn up with another utility for a second device in a European country (CADDET, 2000).

8.6 Economics

Reducing operation and maintenance costs is the key to successful economic implementation of wave energy stations. The *capital cost* per kW of establishing a wave-energy run power station is likely to be at least twice that of a conventional station running on fossil fuels; and the *load factor* is likely to be much lower than a conventional station due to the variability of the wave climate. Therefore, wave energy costs can only be competitive if the *running costs* are significantly below those for a conventional station. Naturally the 'fuel' or wave energy costs are zero, leaving the operation and maintenance costs as the determining factor. Schemes will therefore have to be reliable in their energy conversion and robust enough to survive the wave climate for many years. This means schemes designed for long lifetimes and with small numbers of moving parts (to minimize failures). The oscillating water columns and TAPCHAN schemes are good examples of what is required.

The total capital investment required for wave energy schemes is dependent on overall average efficiencies and on location. Many of the devices detailed here have average efficiencies of around 30%. Frequency response characteristics and limitations of swept volume and survival when operating in very energetic seas are responsible for the generally low overall efficiency. The capital cost is typically around £1000 per installed kW, although the cost of particular schemes may vary markedly from this.

The European 5th Framework Programme fixed a production cost target of 10p per kWh for energy delivered into the grid, and many wave energy concepts now have predicted costs that would satisfy this criterion. Figure 8.33 provides a rough historical guide to the costs of electricity produced by, or predicted for, a variety of wave energy designs. In the 1970s projected costs were 20p per kWh or more. Understandably, the UK Government of the day was reluctant to continue to fund further research and development; but it was the government's own remit to evaluate only the potential of large 2 GW schemes that produced these high costs. Large schemes are technically demanding because of the high structural loads imposed by the north Atlantic wave climate. As time has passed there have been

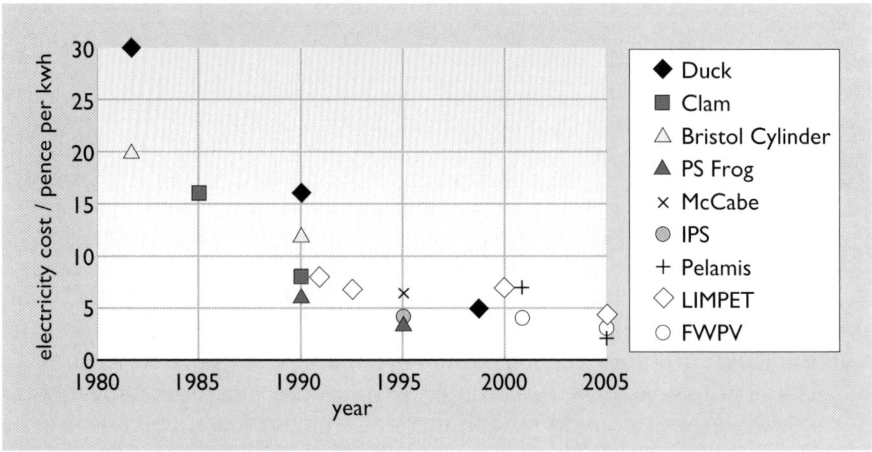

Figure 8.33 Evolution of the cost of electricity from UK wave power devices

improvements in design, performance and construction techniques, together with rationalization of some of the problems and a move to smaller schemes. Smaller schemes are technically simpler and less financially risky, and hence the capital costs and insurance costs are reduced, with commensurate reduction in produced energy costs. Looking back to the 1970s and 1980s, with hindsight we can appreciate that the UK Government's brief to wave energy teams to design 2 GW schemes was extremely over-ambitious – analogous to expecting someone to design a Boeing 747 in the early days of aviation without going through the evolution of the biplane, single seater monoplane, jet engine, etc.!

It must be emphasized here that the electricity costs shown in Figure 8.33 have been gleaned from a wide variety of sources – official, unofficial, developers estimates and fully assessed costings – and hence the individual values cannot be taken as precise or accurate. The figure does, however, show an obvious downward trend in wave energy costs and an extrapolation to well below 5p per kWh seems reasonable.

All three of the 1999 Scottish Renewables Order (SRO) schemes (Pelamis, LIMPET and the Swedish FWPV) have been contracted to supply electricity for 15 years at less than 7p per kWh. The inventor of the Pelamis 'sea snake', Richard Yemm, says that he expects the unit cost of electricity from future versions of Pelamis to be under 3p per kWh and possibly below 2p per kWh.

Development teams, naturally, expect to learn from their early prototypes in order to reduce constructional and operational costs. Some commentators have made the point that no other energy technology (coal, oil, nuclear, wind) has ever started commercial production from such a low unit cost. If wave energy costs follow the example of wind power then a rapid reduction can be expected. As wave energy is considered to be environmentally benign, it is likely to become an attractive commercial and political proposition and this should result in an extensive installation programme with wave farms deployed in many locations.

Wave energy technology has moved into the commercial world and several developers are anxious to demonstrate prototypes before executing ambitious plans to deploy multiple devices generating electricity (or other outputs) at favourable prices. Coupled with incentives for avoided carbon dioxide emissions, the economic prospects for commercial wave energy exploitation appear to be good.

BOX 8.5 A European test centre

With funding of £5.65 million from the UK Government, the European Marine Energy Centre (EMEC) off Billia Croo, near Stromness in the Orkney Islands, was due to open in late 2003. This centre will solve many of the problems that have dogged the construction of prototype wave-power generators in the past. Storm-proof moorings and armoured cables will facilitate the simple and cheap installation and testing of new devices. EMEC can accommodate up to four machines at a time, so designers can directly compare devices under identical conditions, giving them a chance to spot small design 'tweaks' that will improve the efficiency of energy generation. See http://www.scotland.gov.uk/ [accessed 6 January 2004] for more information.

8.7 Environmental impact

Wave energy converters may be among the most environmentally benign of energy technologies because:

- They have little potential for chemical pollution. At most, they may contain some lubricating or hydraulic oil, which will be carefully sealed from the environment.
- They will have little visual impact except where shore-mounted.
- Noise generation is likely to be low – generally lower than the noise of crashing waves. (There might be low-frequency noise effects on cetaceans, but this has to be confirmed.)
- They should present a small (though not insignificant) hazard to shipping.
- They should present no difficulties to migrating fish.
- Floating schemes, since they are incapable of extracting more than a small fraction of the energy of storms, will not significantly influence the coastal environment. Of course, a scheme such as a new breakwater incorporating a wave energy device will provide coastal protection, and may result in changes to the coastline. Concrete structures will need to be removed at the end of their operating life.
- Near-shore wave energy schemes will release an estimated 11 g of CO_2, 0.03 g of SO_2 and 0.05 g of NO_X for each kWh of electricity generated (Thorpe, 1999), making them very attractive in comparison to the conventional UK electrical generating mix of coal, gas and nuclear plants (see the companion volume, Boyle *et al.*, 2003, Chapters 13 and 14). Thus wave energy can make a significant contribution in meeting climate change and acid rain targets.

8.8 Integration

The electrical output from a wave energy scheme can be used directly but it is much more likely that the electricity will be fed into a grid. A discussion of the electrical issues surrounding integration can be found in Beattie (1997).

Wave energy for isolated communities

If the grid is small and serves a small, remote community, great care must be taken in integrating the electrical output from any wave energy scheme into the grid. This is because the output from the scheme will fluctuate wildly (except in the case of schemes such as the TAPCHAN) and may cause swings in voltage or frequency on the grid unless precautions are taken (Figure 8.34).

Many small communities currently depend on diesel generators for their electricity. The diesel generator is best run at a constant output close to its design capacity – say 50 kW. Therefore, if this diesel unit is the sole source of electricity, the load from the grid should always be matched to 50 kW. Clearly, the consumers will cause the load to vary as they switch appliances on and off. To cope with this, a ‘dump’ load can be incorporated into the

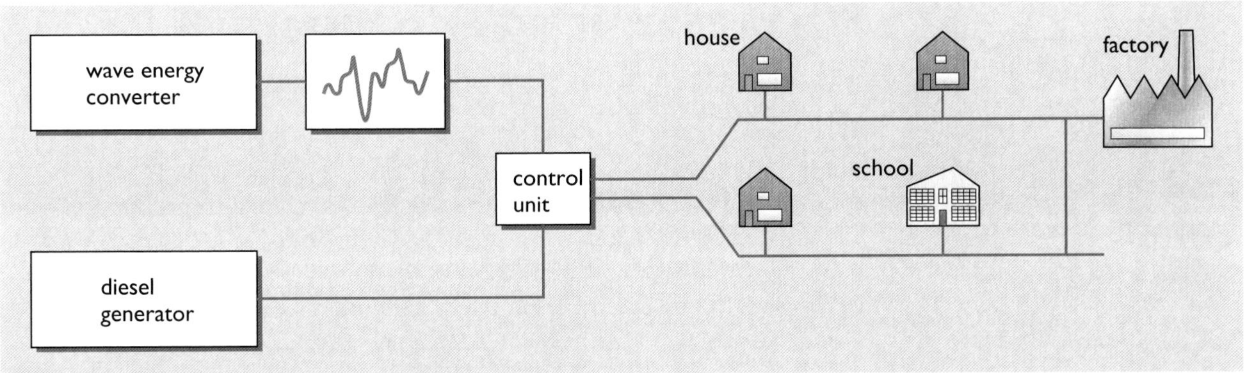

Figure 8.34 Integration of a wave energy/diesel system

system. The diesel output voltage of 240 V a.c. rises if the load falls much below 50 kW – for instance at night when the demand is low. The grid line voltage also rises, and in order to protect other appliances and make use of the diesel output, a voltage sensor set at 250 V allows electricity to be sent to a 'dump' load consisting of electrical space and water heaters. When other appliances are switched on again, the line voltage falls and the dump load is disconnected.

Similarly, the incorporation of a varying electrical output from a wave energy scheme into the grid can partially be accommodated by the use of such dump loads to stabilize the grid voltage (Figure 8.35).

By careful overall design of an integrated scheme, a remote community could enjoy significant gains in electricity supply from a wave energy scheme. The reduction in diesel oil consumption would be substantial, and since it is costly to transport diesel oil to remote locations, the cost savings could be large.

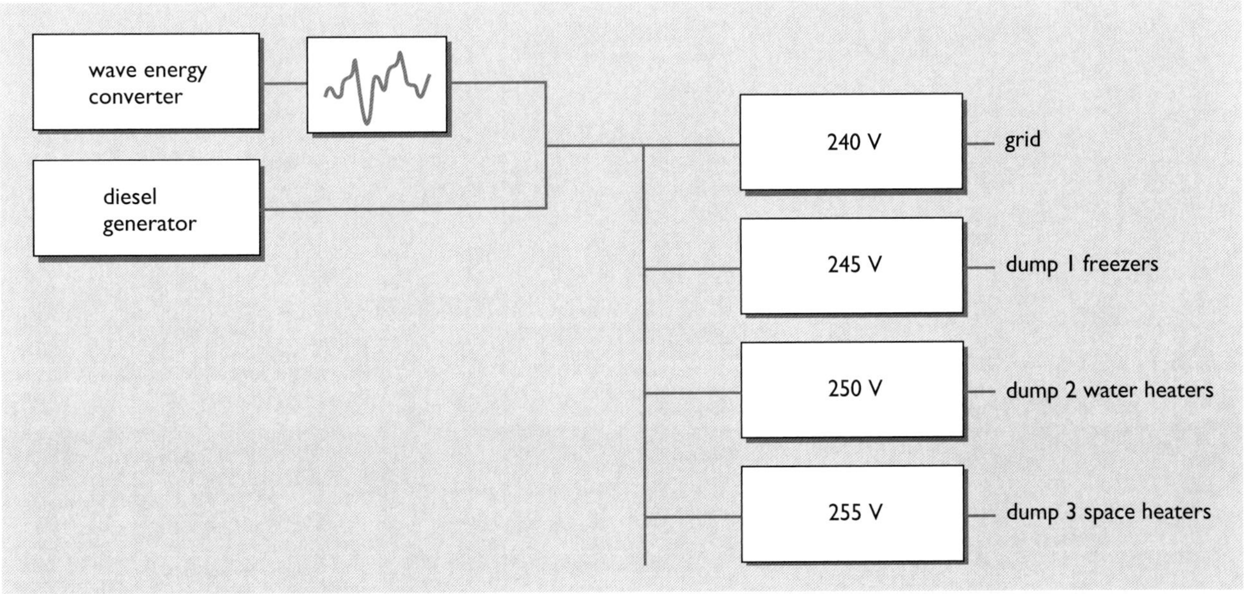

Figure 8.35 Use of voltage sensors and dump loads

Wave energy for large electricity grids

When the electrical outputs of several wave energy units are added together, the total output will be generally smoother than for a single unit. If we extend this to an array of several hundred floating devices, then the summed output will be smoother still. In addition, any fluctuations in output will be less important if the electricity is to be delivered to large national systems like those of the UK, where in most locations the grid is 'strong' enough to absorb contributions from a fluctuating source. Figure 8.36 illustrates a typical scheme.

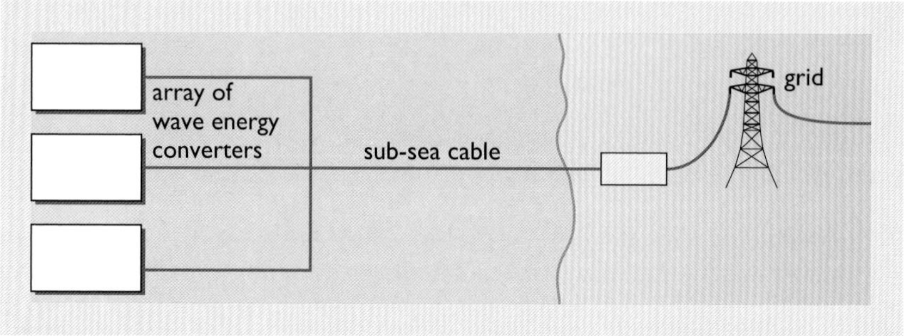

Figure 8.36 Electrical connections for an array of wave energy devices

Finally, although we have dwelt upon short-term fluctuations of seconds or minutes, the wave resource also varies on a day-to-day and season-by-season basis. For those countries in the north-east Atlantic such as the UK, Denmark and Norway, the seasonal variation is favourable, as the more energetic waves appear in the bad weather of winter when the electrical demand is greatest (Figure 8.37) (see for example, Rugbjerg *et al.*, 2000 and Petroncini and Yemm, 2000).

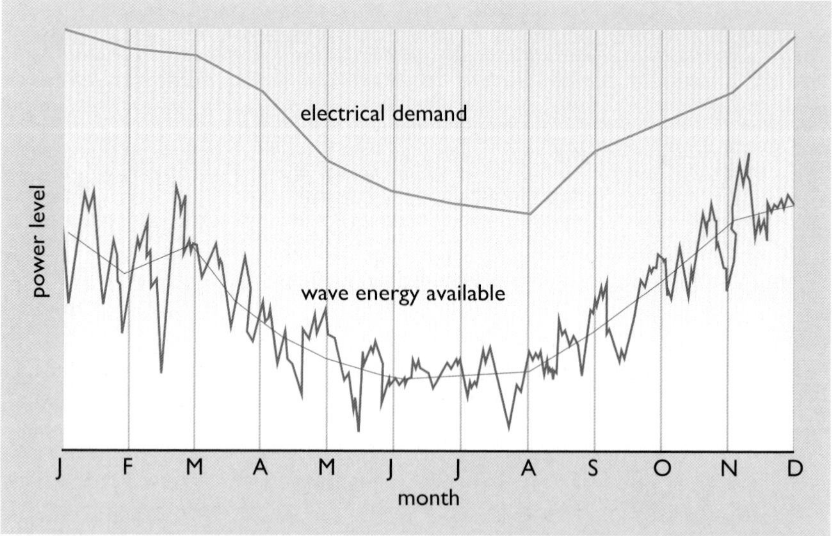

Figure 8.37 Seasonal availability of wave energy and electrical demand for the UK

8.9 Future prospects

Whilst the UK is fortunate in having a good wave climate, the political climate has not always favoured the technology. Attitudes are changing very quickly, however, prompted by the need to address global climate change, the issue of long-term resource security of fossil fuels and by the increasingly competitive economics of wave energy. UK teams conducted much of the early work, but as this chapter records many other countries are now very active in the development of wave energy systems. Commercial involvement has had an important impact and we can now see commercial schemes or concepts emerging from countries such as The Netherlands, Norway, Australia, Sweden, Denmark and the US, as well as the UK.

In addition, the generation of electricity is not the only option for the delivered energy. Desalination, coastal protection, water pumping, mariculture, mineral recovery from sea water and hydrogen generation are among the benefits being developed.

The development of wave energy technologies has been a long process, but the economics of current designs are potentially quite attractive. Some of the developers predict much lower unit costs in the future when their particular technology has matured, either by refinement of the design to improve performance, or by the benefit of scale of production. Proving the long-term survival and cost effectiveness of designs should make the prospect of wave energy stations on a large scale a real possibility. We may well see converters deployed in large numbers to harvest the considerable wave energy that is present at some locations. These devices will probably have evolved from one or more of these described in this chapter.

References

Beattie, W. C. (1997) 'From wave to wire – control techniques for acceptable power', *Wave Power: An Engineering and Commercial Perspective*, IEE Colloquium.

Boake, C. B., Whittaker, T. J. T. and Folley, M. (2002) 'Overview and initial operational experience of the LIMPET wave energy plant', *The Proceedings of the Twelfth (2002) International Offshore and Polar Engineering Conference, Volume I*, Kitakyushu, Japan, ISOPE.

Boyle, G. A., Everett, B. and Ramage, J. (2003) *Energy Systems and Sustainability*, Oxford, Oxford University Press in association with The Open University.

CADDET (1996, 1999, and 2000) Newsletter of the Centre for the Analysis and Dissemination of Demonstrated Energy Technologies, CADDET website, http://www.caddet-re.org.

Duckers, L. J. (1998) 'Wave power developments', *Renewable Energy World*, James and James, vol. 1, no. 3, pp. 52–57.

ETSU (1985) *Wave Energy: The Department of Energy's R&D Programme 1974–1983*, ETSU Report R26.

Falcao, A. F. de O. (2000) 'The shoreline OWC wave power plant at the Azores', *Fourth European Wave Energy Conference*, Aalborg, Denmark, pp. 42–48.

Falnes, J. and Lillebekken, P. M. (2003) 'Budal's latching controlled buoy type wave power plant', *Fifth European Wave Energy Conference*, Cork, Ireland.

Falnes, J. and Løvseth, J. (1991) 'Ocean wave energy', *Energy Policy*, vol. 19, pp. 768–775.

Garratti, G., Lewis, A. and Howett, D. (eds) (1993) 'Wave energy R&D', *Workshop Proceedings*, Cork, Ireland, CEC, EUR15079EN.

Gemini (1996) 'Norwegian wave-power to Java', NTNU (Norwegian University of Science and Technology) website, http://www.ntnu.no/gemini/1996-04/7.html [accessed 11 January 2004].

Heath, T., Whittaker, T. J. T. and Boake C. B. (2000) 'The design, construction and operation of the LIMPET wave energy converter (Islay, Scotland)', *Fourth European Wave Energy Conference*, Aalborg, Denmark, pp. 49–55.

House of Commons Select Committee on Science and Technology (2001) *Wave and Tidal Energy*, Seventh Report Session 2000–2001, HC291.

ISOPE (2002) *The Proceedings of the Twelfth (2002) International Offshore and Polar Engineering Conference, Volume I: Ocean Resources and Engineering*, Kitakyushu, Japan, ISOPE.

Kondo, M. (1993) *Proceedings of International Symposium on Ocean Energy Development*, Muroran Institute of Technology, Japan.

Kraemer, D. R. B., Ohl, C. O. G. and McCormick, M. E. (2000) 'Comparisons of experimental and theoretical results of the motions of a McCabe wave pump', *Fourth European Wave Energy Conference*, Aalborg, Denmark, pp. 211–218.

Lagström, G. (2000) 'Sea Power International – Floating Wave Power Vessel (FWPV)', *Fourth European Wave Energy Conference*, Aalborg, Denmark, pp. 141–146.

Lewis, A. W., Gilbaud, T. and Holmes, B. (2003) 'Modelling the Backward Bent Duct Device – B2D2 – a comparison of physical and numerical methods', *Fifth European Wave Energy Conference*, Cork, Ireland

Masuda, Y., Kuboki, T., Xianguang, L. and Peiya, S. (2000) 'Development of terminator type BBDB', *Fourth European Wave Energy Conference*, Aalborg, Denmark, pp. 147–152.

McCabe, A. P., Bradshaw, A., Widden, M. B. *et al.* (2003) 'PS FROG MK 5: an offshore point absorber wave energy converter', *Fifth European Wave Energy Conference*, Cork, Ireland.

Meyer, N. I. and Nielsen, K. (2000) 'The Danish wave energy programme second year status', *Fourth European Wave Energy Conference*, Aalborg, Denmark, pp. 10–18.

Miyazaki, T. (1991) 'Wave energy research and development in Japan' *Oceans 91 Symposium*, Honolulu, Hawaii.

Miyazaki, T. and Hotta, H. (eds) (1991) *Third Symposium on Ocean Wave Energy Utilization*, Tokyo, Japan (in Japanese).

Nielsen, K. and Plum, C. (2000) 'Point absorber – numerical and experimental results', *Fourth European Wave Energy Conference*, Aalborg, Denmark, pp. 219–226.

Nolan, G., O'Cathain, M., Ringwood, J. V. and Murtagh, J. (2003) 'Modelling, simulation and validation of the power take-off system for a hinge barge wave energy converter', *Fifth European Wave Energy Conference*, Cork, Ireland.

Ó'Gallachóir, B. P. (2001) 'Technological advances in power generation from renewable sources of energy', *Proceedings of Energy Ireland*, Dublin (reviewed in Martin, Pierce 'Stimulating the energy market', *The Engineers Journal*, vol. 55, no. 8, p. 51).

Petroncini, S. and Yemm, R. W. (2000) 'Introducing wave energy into the renewable energy marketplace', *Fourth European Wave Energy Conference*, Aalborg, Denmark, pp. 33–41.

PIU (2001) 'PM announces step change in support for renewable energy', Press Release, 2001, 10 Downing Street website, http://www.number10.gov.uk/ [accessed 11 January 2004].

Ravindran, M. *et al.* (1995) 'Indian wave energy programme: progress and future plans', *Proceedings of the Second European Wave Power Conference*, Lisbon, November 1995.

Ross, D. (1995) *Power from the Waves*, Oxford, Oxford University Press.

Rugbjerg, M., Nielsen, K., Christensen, J. H. and Jacobsen, V. (2000) 'Wave energy in the Danish part of the North Sea', *Fourth European Wave Energy Conference*, Aalborg, Denmark, pp. 64–71.

Salter, S. and Lin, C. P. (1995) 'The sloped IPS wave energy converter', *Second European Wave Power Conference*, Lisbon.

Sorensen, H. C., Christensen, L., Hansen, L. K. *et al.* (2003) 'Development of Wave Dragon from scale 1:50 to prototype', *Fifth European Wave Energy Conference*, Cork, Ireland.

Taylor, G. W. (2000) 'The history, current status, and future prospects for the modular OPT wave power system', *Wave Power: Moving Towards Commercial Viability*, IMechE Seminar Publication, pp. 69–76.

Taylor, J. (2002) Personal communication.

Thorpe, T. W. (2001) 'The UK market for marine renewables', *All-Energy Futures Conference*, Aberdeen.

Thorpe, T. W. (1992) *A Review of Wave Energy*, ETSU Report R72.

Thorpe, T. W. (1998) *Overview of Wave Energy Technologies*, AEAT-3615 for The Marine Foresight Panel.

Thorpe, T. W. (1999) *A Brief Review of Wave Energy*, ETSU Report R122.

Washio, Y., Osawa, H. and Ogata, T. (2001) 'The open sea tests of the offshore floating type wave power device "Mighty Whale" – characteristics of wave energy absorption and power generation', *Ocean 2001 Symposium*.

Watabe, T., Yokouchi, H., Kondo, H. *et al.* (1999) 'Installation of the new Pendulor for the second stage sea test', *Proceedings of the 9th International Offshore and Polar Engineering Conference*, Brest, France, pp. 133–138.

Whittaker, T. J. T. *et al.* (1997) 'The Islay wave power project – an engineering perspective', *ICE Proceedings, Water Maritime and Energy*, no. 124, pp. 189–201.

Xianguang, L., Peiya, S., Wei, W. and Niandong, J. (2000) 'The experimental study of BBDB model with multipoint mooring', *Fourth European Wave Energy Conference*, Aalborg, Denmark, pp. 173–185.

Yemm, R. W. (2000) 'The history and status of the Pelamis wave energy converter', *Wave Power: Moving Towards Commercial Viability*, IMechE Seminar Publication 2000-8, pp. 51–59.

Chapter 9

Geothermal Energy

by Geoff Brown and John Garnish

9.1 Geothermal energy – an overview

The mining of geothermal heat

In the continuing search to find cost-effective forms of energy that neither contribute to global warming nor threaten national security, geothermal energy has become a significant player. Geothermal is the only form of 'renewable' energy that is independent of the sun, having its ultimate source within the earth. It is a comparatively diffuse resource; the amount of heat flowing through the earth's surface, 10^{21} joules per annum (J a^{-1}), is tiny in comparison with the massive 5.4×10^{24} J a^{-1} solar heating of the earth which also drives the atmospheric and hydrological cycles (see Figure 1.26). Fortunately, there are many places where the earth's heat flow is sufficiently concentrated to have generated natural resources in the form of steam and hot water (180–250 °C), available in shallow rocks and suitable for electricity generation. These are the so-called 'high-enthalpy' resources (see Box 9.1).

The techniques for exploiting the resources are very simple in principle, and are analogous to the well-established techniques for extracting oil and gas. One or more boreholes are drilled into the reservoir, the hot fluid flows or is pumped to surface and is then used in conventional steam turbines or heating equipment.

Obviously, electricity is a more valuable end-product than hot water, so most attention tends to be focused on those resources capable of supporting power generation, i.e. hot enough to make electricity generation economic. By 2000 world electrical power generating capacity from geothermal resources had reached almost 8 gigawatts electrical (8 GW_e), a small but significant contribution to energy needs in some areas (Table 9.1). About a further 16 gigawatts thermal (16 GW_t) is also being harnessed in non-electrical 'direct use' applications, principally space heating, agriculture, aquaculture and a variety of industrial processes. Many of these applications occur outside high-enthalpy regions where geological conditions are nevertheless suitable to allow warm water (less than 100 °C) to be pumped to the surface; these are 'low-enthalpy' resources. In 2000, these installations supplied a total of 45 000 GWh.

BOX 9.1 Enthalpy

Enthalpy is defined as the heat content of a substance per unit mass, and is a function of pressure and volume as well as temperature.

Geothermal resources are usually classified as 'high enthalpy' (water and steam at temperatures above about 180–200 °C), 'medium enthalpy' (about 100–180 °C) and 'low enthalpy' (<100 °C). The term 'enthalpy' is used because temperature alone is not sufficient to define the useful energy content of a steam/water mixture. A mass of steam at a given temperature and pressure can provide much more energy than the same mass of water under the same conditions. The distinction is important to geothermal practitioners so the term has entered into general use.

For the purposes of this chapter, however, it is usually sufficient to think of temperature and enthalpy as going hand in hand.

Table 9.1 Geothermal power generation around the end of the twentieth century

Nation	1995 MW_e	2000 MW_e	2005 (est. MW_e)
Argentina	0.67	0	n/a
Australia	0.17	0.17	n/a
China	28.78	29.17	n/a
Costa Rica	55.0	142.5	161.5
El Salvador	105.0	161	200
Ethipoia	0	8.52	8.52
France	4.2	4.2	20
Guatemala	0	33.4	33.4
Iceland	50.0	170	186
Indonesia	309.75	589.5	1987.5
Italy	631.7	785	946
Japan	413.7	546.9	566.9
Kenya	45.0	45	173
Mexico	753.0	755	1080
New Zealand	286.0	437	437
Nicaragua	70.0	70	145
Philippines	1227.0	1909	2673
Portugal	5.0	16	45
Russia	11.0	23	125
Thailand	0.3	0.3	0.3
Turkey	20.4	20.4	250
USA	2816.7	2228	2376
Totals	**6833**	**7974**	**11 414**

(from Huttrer, 2000)

In almost all these situations heat is being removed faster than it is replaced, so the concept of 'heat mining' is appropriate. Although geothermal resources are non-renewable on the scale of human lifetimes and, strictly, fall outside the remit of this book, they are included because they share many features with the true renewable resources. For example, geothermal is a natural energy flow rather than a store of energy like fossil or nuclear fuels. At one time, it was thought that many high-enthalpy resources were indeed renewable, in the sense that they could be exploited indefinitely, but experience of declining temperatures in some producing steam fields and simple calculations of heat supply and demand show that – locally – heat is being mined on a non-sustainable basis (see Box 9.2). Nevertheless, and especially as techniques become available for extracting energy from rocks that do not support natural water flows (the so-called Hot Dry Rock concept: see section 9.3), the volumes of rock that are potentially exploitable are so large in comparison to an individual reservoir that – to use an agricultural analogy – the resource could be 'cropped' on a rotational basis. Once a particular zone has been depleted, further boreholes could be drilled to a deeper layer, or a few kilometres further away, and production continued. After three or four such operations, each with a lifetime of 20–30 years, the original zone would have regenerated to economically exploitable levels.

The source of heat

Heat flows out of the earth because of the massive temperature difference between the surface and the interior: the temperature at the centre is around 7000 °C. So why is the earth hot? There are two reasons: first, when the earth formed around 4600 million years ago the interior was heated rapidly as the kinetic and gravitational energy of accreting material was converted into heat. If this were all, however, the earth would have cooled within 100 million years; of much greater importance today is the second mechanism: the earth contains tiny quantities of long-lived radioactive isotopes, principally thorium 232, uranium 238 and potassium 40, all of which liberate heat as they decay (see Book 1, Chapter 11). These radiogenic elements are concentrated in the upper crustal rocks. Cumulative heat production from these radioactive isotopes (approximately 5×10^{20} J a^{-1} today) accounts for about half the surface heat flow, though the exponential decay laws for radioactivity imply that heat production was about five times greater soon after the earth formed. Heat is transferred through the main body of the earth principally by convection, involving motion of material mainly by creep processes in hot deformable solids. This is a very efficient heat transport process resulting in rather small variations of temperature across the depth of the convecting layer. Closer to the surface, across the outer 100 km or so of the earth, the material is too rigid to convect because it is colder, so heat is transported by conduction and there are much larger increases of temperature with depth (i.e. larger 'thermal gradients': see Section 9.4). This rigid outer boundary layer, or shell, is broken into a number of fragments, the **lithospheric plates**, which move around the surface at speeds of a few centimetres a year in concert with the convective motions beneath (Figure 9.1). Only this last point is of direct relevance to geothermal exploitation; our ability to drill into the earth is restricted to the upper few kilometres, so we have to look for mechanisms and locations where the earth's interior heat is brought within our reach.

From our point of view, it is at the boundaries between plates, mainly where they are in relative extension or compression, that heat flow reaches a maximum. Here, the heat energy flowing through the surface averages around 300 milliwatts per square metre (300 mW m^{-2}) as compared with a global mean of 60 mW m^{-2}.

However, along plate margins heat flow is even more concentrated locally because rock material reaches the surface in a molten form, resulting in volcanic activity that is often spectacular. Storage of molten, or partially molten, rock at about 1000 °C just a few kilometres beneath the surface strongly augments the heat flow around even dormant volcanoes. These heat flows result in high thermal gradients, really the hallmark of high-enthalpy areas, which are further enhanced in the upper regions by induced convection of hot water. Over geological periods of time, this high heat flow has resulted in large quantities of heat being stored in the rocks at shallow depth, and it is these resources that are mined by geothermal exploitation and commonly used for electricity generation. In areas of lower heat flow, where convection of molten rock or water is reduced or absent, temperatures in the shallow rocks remain much lower, so any resources will be suitable only for direct use applications. So we see that high enthalpy resources, including all those currently exploited for geothermal electric power (see Figure 9.1), are confined to volcanically active plate margins or localized hot spots like the Hawaiian islands. Boiling mud pools, geysers and volcanic vents with hot steam are characteristic features of such geothermal areas.

.... incipient plate boundaries ===== divergent boundaries ● major high-enthalpy geothermal energy producing areas

- - - - convergent boundaries ——— strike-slip boundaries (sideways motion)

Figure 9.1 Map of the earth's lithospheric plates indicating the relative speeds of motions by the lengths of the arrows (generally 1–10 cm per annum). Large dots indicate major high-enthalpy geothermal energy-producing areas

BOX 9.2 Extraction and recharge rates

The following simple calculations show the order of magnitude discrepancy between commercial geothermal extraction rates and thermal recharge by the earth's natural heat flow. They demonstrate that it is *not the heat flow but the heat store that is being exploited,* even in high-enthalpy areas.

(a) The currently exploited area of Tuscany, northern Italy, totals about 2500 km². This is a generous estimate which ignores the fact that the active fields are only a small sub-set of this. The average heat flow is about 200 mW m^{-2} so the total heat flow through this surface is

$$\{2500 \times 10^6\ (m^2) \times 200 \times 10^{-3}\ (W\ m^{-2})\} = 500\ MW_t$$

The region currently supports generating capacity of >400 MW_e; at a mean generating efficiency of 13%, this requires 3000 MW_t. Moreover, steam supplies have been proved to support an electricity generating capacity of 1500 MW_e, which would require an input of approximately 12 000 MW_t. So the ratio of forecast commercial production rate to thermal recharge is at least

$$\frac{12000\ MW_t}{500\ MW_t} \text{ or } 24:1.$$

(b) In the Imperial Valley of California, USA, a commercial lease of 4 km² is expected to support generating capacity of some 40 MW_e, (which will require at least 250 MW_t). Assuming an average heat flow of 200 mW m^{-2}, the thermal recharge is less than 1 MW_t, giving a ratio of

$$\frac{250\ MW_t}{1\ MW_t} \text{ or } 250:1.$$

(c) The well field in Krafla, Iceland, covers an area of some 50 km². It currently generates some 60 MW_e, implying heat extraction of some 400 MW_t. To recharge the field on a sustainable basis, the heat flow would have to be 8 W m^{-2}. Again, this is much higher than could be expected even in this very active area. Of course, there is recharge by the earth's natural heat flow but a thermally depleted reservoir may take many tens or even hundreds of years to regenerate. However, in regions of large geothermal resources this is generally a minor problem. NOTE: MW_t denotes megawatts of thermal energy; MW_e denotes megawatts of electrical energy

Historical perspective

The historical exploitation of geothermal resources dates back to Greek and Roman times, with early efforts made to harness hot water for medicinal, domestic and leisure applications. Roman spa towns in Britain generally sought to exploit natural warm water springs with crude but reliable plumbing technology. The early Polynesian settlers in New Zealand, who lived for 1000 years undisturbed by European influence until the eighteenth century, depended on geothermal steam for cooking and warmth, and hot water for bathing, washing and healing. Indeed, the healing properties of geothermal waters are renowned throughout the modern world and have important medical benefits.

By the nineteenth century, progress in engineering techniques made it possible to observe the thermal properties of underground rocks and fluids, and to exploit these with rudimentary drills. In Tuscany, where the indigenous geothermal fluids were exploited as a source of boron from the eighteenth century onwards, natural thermal energy was used in place of wood for concentrating and processing the solutions. An ingenious steam collection device, the *lagone coperto* ('covered pool'), sparked a rapid growth in the Italian chemical industry, resulting in flourishing international trade. The generation of electrical power started in 1904, fostered by Prince Piero Ginori Conti, and 1913 saw the arrival of the first 250 kW_e power plant at Larderello, marking the start of new industrial activity (Figure 9.2). Today the Larderello power station complex (Figure 9.3) has a capacity exceeding 700 MW_e and a rebuilding programme is in progress that will take the capacity to some 1200 MW_e.

Figure 9.2 The Larderello field prior to development for geothermal power production

Figure 9.3 The Larderello 3 station, which produces 120 MW from six turbine units. The geothermal pipeline network consists of an inner steel pipe lagged with asbestos fibre and covered with aluminium plate

The Wairakei field in New Zealand was the second to be developed for commercial power generation, but not until the early 1950s, and it was followed closely by the Geysers field in northern California where electricity was first generated in 1960. With an installed capacity that peaked at 2800 MW_e in the early 1990s, the Geysers field is still the most extensively developed in the world, though it will be overtaken soon by the Philippines. The Geysers field, in fact, illustrated the dangers of uncoordinated exploitation; the field was exploited by several different companies, without co-operation or re-injection of extracted fluids. The result was a decline in steam pressure and a reduction in output capacity by several hundred megawatts. It was this decline that accounted for the very low net growth in world capacity during the second half of the 1990s. Fortunately, the problem was recognized in time, and reinjection is now practised widely; steam pressures and volumes are now recovering.

With the notable exceptions of Italy, the most volcanically-active European country, and Iceland, which lies on the volcanic ridge of the central Atlantic, the chief geothermal nations are clustered around the Pacific rim. Japan, the Philippines and Mexico have shared in recent technological developments; the installations in El Salvador and Nicaragua are strategically vital to the economies of those nations, and several other countries, notably Costa Rica, Ecuador and Chile, have recently joined the list of geothermal electricity producers.

Meanwhile, schemes making direct use of geothermal heat for district heating and agricultural purposes have advanced, with Japan, China, the former Soviet republics (mainly Kamchatka in the Far East, and the Georgia and Dagestan regions between the Caspian and Black Seas), Hungary and Iceland being the major producers. France developed substantial heating systems in the 1970s and 1980s, but the most significant event within continental Europe has been the opening up of the formerly centrally-planned economies of eastern Europe. Many of these countries – notably the former East Germany, Poland, Romania, Hungary and Slovenia – possess good low-enthalpy aquifers (aquifers are defined in Section 9.2 below) With the increasing drive to minimize the environmental effects of their outdated and dirty solid fuel fired heating systems, they have begun to develop district heating schemes along the lines of those in Paris. (To be fair, it should be said that a few schemes had already been developed independently in these countries before the opening up of eastern Europe, but progress was hampered by the lack of high grade materials for pipelines, pumps and heat exchangers that could withstand the corrosive effects of brines in poorly maintained systems).

Finally, there has been a quiet revolution in direct use in the past few years, especially in Europe and the USA, with the widespread adoption of ground source heat pumps (see Figure 9.11 in Section 9.3). GSHPs differ in a number of ways from conventional installations, most notably in that the scale is that of the single dwelling rather than the multi-user district heating scheme. They extract heat at only 12–15 °C from depths of 100–150 m and do not depend on the presence of water-bearing rocks at that depth. Consequently, they can be exploited almost anywhere. There are now tens of thousands of units in operation in Europe (notably in Switzerland, Germany and Sweden) and over 400 000 units in the US, a figure that is increasing by over 40 000 annually. They are particularly valuable when cooling is required as well as heating.

BOX 9.3 Case study: The Southampton Geothermal District Heating Scheme

The city of Southampton is located in one of several areas in the United Kingdom known to be underlain by rock strata containing hot water (Figure 9.6(a)). The Southampton scheme demonstrates how it is possible to make direct use of the hot water beneath the city. This scheme was not designed to produce electricity for the grid – it simply serves the city by providing a supply of hot water. Though very small in comparison with the developments around Paris, it provides a useful case study within the UK of a small-scale geothermal scheme that actually works.

A single geothermal well, a kind of borehole, was drilled in 1981 to a depth of just over 1800 m beneath a city centre site in Southampton (Figure 9.4). It was an exploration borehole drilled as part of the former Department of Energy's research programme, with an agreement that the City Council would develop a district heating scheme if the well were successful. Near the bottom of the hole, a 200 million year old geological formation known as the Sherwood Sandstone, containing water at 70 °C, was encountered. This particular formation is both porous and permeable, meaning that it holds water and can transmit considerable volumes. It therefore behaves like a giant rigid sponge, and is a good geothermal **aquifer** (see Section 9.2). The fluid itself contains dissolved salts and, as in most geothermal areas, is more accurately described as **brine**. Within the aquifer the brine is pressurized and so rises unaided to within 100 m of the surface. However, a turbine pump, located at 650 m depth in the well, brings the hot brine to the surface where its heat energy is exploited.

The brine passes through coils in a heat exchanger where much of its heat energy is transferred to clean water in a separate district heating circuit. The heated 'clean' water is then pumped around a network of underground pipes to provide central heating to radiators, together with hot water services, in the Civic Centre, Central Baths and several other buildings within a 2 km radius of the well. The cooled geothermal working fluid (brine) is discharged via drains into the Southampton marine estuary. (Nowadays, it is almost universal practice to reinject the spent fluid into the reservoir.) Nevertheless, in this particular case the scheme is seen as environmentally acceptable, and it is saving the equivalent of over a million cubic metres of gas (or 1000 tonnes of oil) a year.

More recently, the geothermal heat supply – originally 1 megawatt thermal (1 MW_t) – has been increased to 2 MW_t using heat pumps, and this is capable of satisfying the base load demand. However, during periods of higher demand, fossil fuel boilers augment the plant's heat output to a maximum of 12 MW_t. A further technical aspect is the inclusion of a diesel generator which supplies electrical power to the various circulation pumps and the monitoring equipment. Heat from the generator is fed directly into the district heating scheme and any surplus power is sold to the local electricity company.

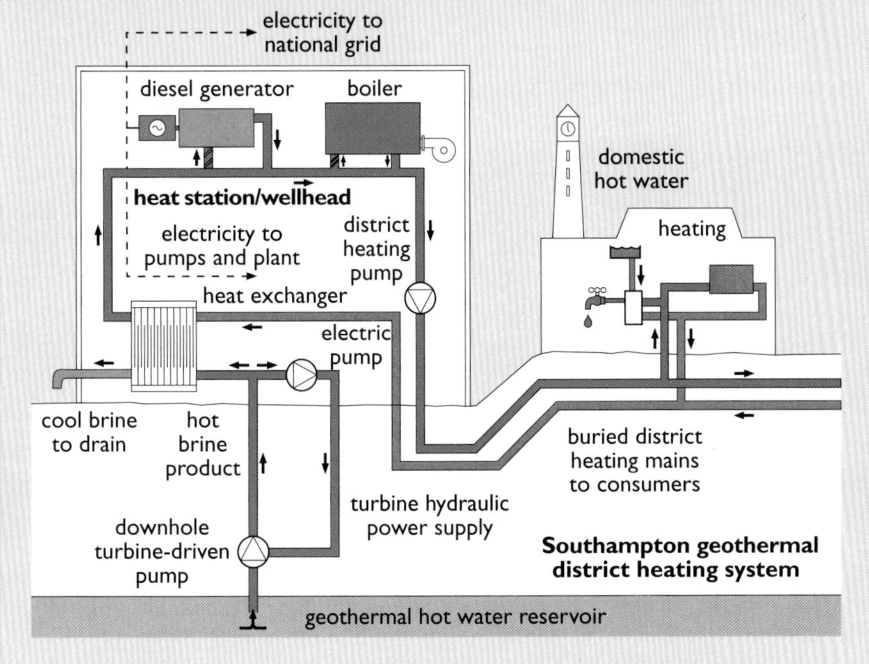

Figure 9.4 The Southampton geothermal district heating system: technology schematic

9.2 The physics of geothermal resources

Primary ingredients

Geothermal resources of most types must have three important characteristics, as shown in Figure 9.5: an **aquifer** containing water that can be accessed by drilling; a **cap rock** to retain the geothermal fluid; and a **heat source.**

First, what is an aquifer? Natural aquifers are porous rocks that can store water and through which water will flow. **Porosity** refers to the cavities present in the rock, whereas the ability to transmit water is known as **permeability.** A geothermal aquifer must be able to sustain a flow of geothermal fluid, so even highly porous rocks will only be suitable as geothermal aquifers if the pores are interconnected. In Figure 9.6, rocks (a) and (c) are porous and likely to be highly permeable, whilst (b) and (d) have low porosity and permeability. Example (e), however, has low

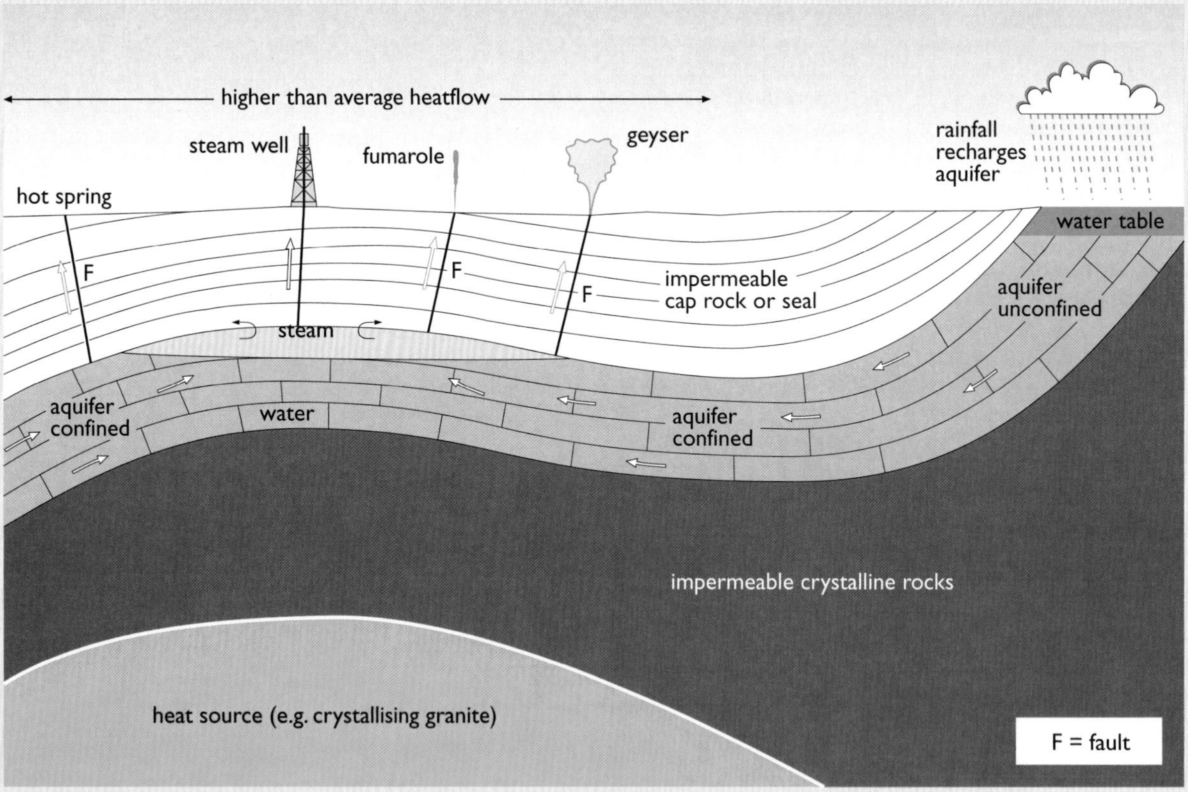

Figure 9.5 Simplified schematic cross-section to show the three essential characteristics of a geothermal site: an aquifer (e.g. fractured limestone with solution cavities); an impermeable cap rock to seal the aquifer (e.g. clays or shales); and a heat source (e.g. crystallising granite). Steam and hot water escape naturally through faults (F) in the cap rock, forming fumaroles (steam only), geysers (hot water and steam), or hot springs (hot water only). The aquifer is unconfined where it is open to the surface in the recharge area, where rainfall infiltrates to keep the aquifer full, as indicated by the water table just below the surface. The aquifer is confined where it is beneath the cap rock. Impermeable crystalline rocks prevent downward loss of water from the aquifer

permeability despite its high porosity whereas cavities developed in (f) by dissolution of more soluble components give high porosity and permeability. Permeability due to fracturing ('fracture permeability'), as in (g), is particularly important in many geothermal fields.

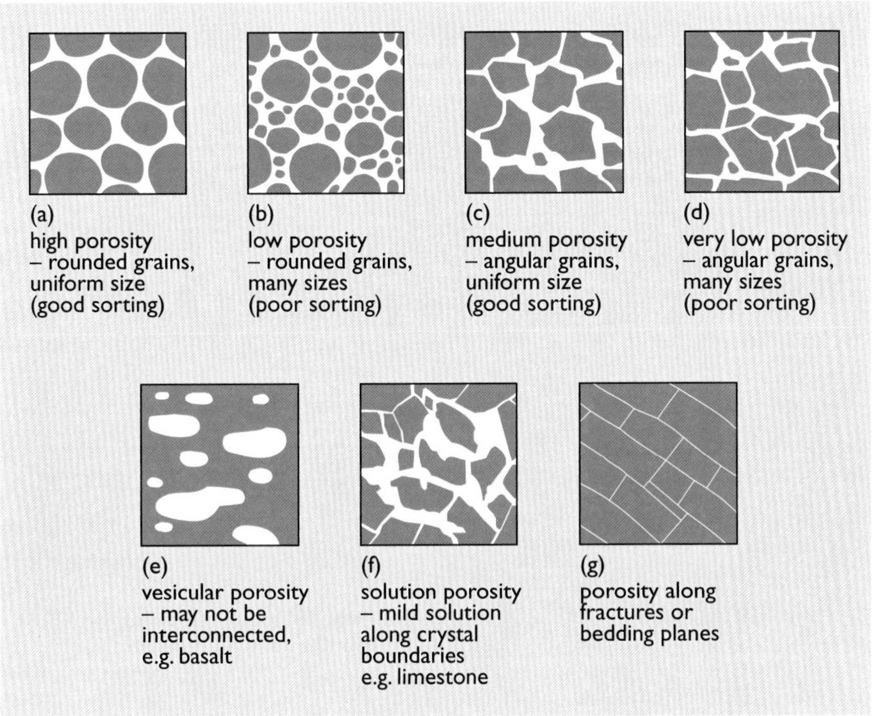

Figure 9.6 The relationship between grain size, shape and porosity in sedimentary rocks, especially sandstones (a–d); vesicular porosity in crystallised lava flows due to gas bubbles (e); and solution porosity resulting from rock dissolution, especially where acid groundwaters attack limestone (f). Porosity also develops in rocks along original planes of weakness, especially bedding planes and fractures (joints and faults) (g)

A good measure of the permeability of a rock is its hydraulic conductivity (K_w). Darcy's Law states that the speed (v) of a fluid moving through a porous medium is proportional to the hydraulic pressure gradient causing the flow:

$$v = K_w \frac{H}{L} \qquad (1)$$

Here, H is the effective head of water driving the flow, and is measured in metres of water. The pressure gradient, or hydraulic gradient (H/L) is the change in this head per metre of distance L along the flow direction.

The volume of water (Q) flowing in unit time through a cross-sectional area A m^2 is v times A. So Darcy's Law may also be written:

$$Q = AK_w \frac{H}{L} \qquad (2)$$

and K_w (the hydraulic conductivity) may be interpreted as the volume flowing through one square metre in unit time under unit hydraulic gradient. Some values of hydraulic conductivity for different rocks are given in Table 9.2.

Table 9.2 Typical porosities and hydraulic conductivities

Material (m day^{-1})	Porosity (%)	Hydraulic conductivity
Unconsolidated sediments		
Clay	45–60	$<10^{-2}$
Silt	40–50	10^{-2}–1
Sand, volcanic ash	30–40	1–500
Gravel	25–35	500–10 000
Consolidated sedimentary rocks		
Mudrock	5–15	10^{-8}–10^{-6}
Sandstone[(a)]	5–30	10^{-4}–10
Limestone[(a)]	0.1–30	10^{-5}–10
Crystalline rocks		
Solidified lava[(a)]	0.001–1	0.0003–3
Granite[(b)]	0.0001–1	0.003–0.03
Slate	0.001–1	10^{-8}–10^{-5}

(a) The larger values of porosity and hydraulic conductivity apply to heavily fractured rocks and, for limestones, may also reflect the presence of solution cavities (see Figure 9.6(f))

(b) Granite is a coarsely crystalline rock that has cooled down slowly from a melt at depth in the earth. Such rocks are generally non-porous and impermeable, but contain many natural fractures and acquire limited permeability

Notice that the highest values of K_w occur in coarse-grained unconsolidated rocks, such as the ash layers which are particularly common in volcanic areas, but that values are also quite high in some limestones and sandstones. These are aquifer rocks, with high permeability. It should be remembered also that fracture permeability is often important in geothermal aquifers (see Figure 9.6(g)), and is central to the Hot Dry Rock (HDR) concept (see below).

In a confined aquifer such as in Figure 9.5 the fluid pressure beneath the extraction point is high because there is a relatively impermeable rock, or seal, to prevent fluid escaping upwards. Such a '**cap rock**' is essential if a steam field is to develop. Mudrocks, clays and unfractured lavas are ideal. The importance of cap rocks was demonstrated in the early 1980s during exploration for geothermal resources in a very obvious place, the flanks of Vesuvius volcano. Only small amounts of low pressure fluid were discovered because the volcanic ashes that form its flanks are apparently quite permeable throughout. Given time, alteration of the uppermost deposits or overlying sediments by hot water and steam can create clays or deposit salts in pore spaces, so producing a seal over the aquifer, and in this way many geothermal fields eventually develop their own cap rocks. For this reason, however, the youngest volcanic areas, like Vesuvius, are not necessarily the most productive from a geothermal viewpoint.

The third prerequisite for exploitable geothermal resources is the presence of a heat source. In high-enthalpy regions, abundant volcanic heat is available, but in low-enthalpy areas the heat source is less obvious. In such regions there are two main types of resource: (a) those located in deep **sedimentary basins** where aquifers carry water to depths where it becomes warm enough to exploit, and (b) those located in **hot dry rocks** where natural heat production is high but an artificial aquifer must be created by fracturing the rocks in order that the geothermal resource may be exploited. Let us now look at each type of resource in more detail.

Volcano-related heat sources and fluids

The heat supply for a high-enthalpy field is usually derived from a cooling and solidifying body of magma (partially molten rock), which need not necessarily be centred directly beneath the geothermal field (Figure 9.7). It may seem surprising that much of the magma rising beneath a volcano is not erupted but instead reaches only a level of neutral buoyancy at which its density is the same as that of the surrounding rocks. Two factors conspire to halt the rising magma: first the pressure of overlying rocks reduces as the magma ascends; this promotes the separation of liquid magma from its dissolved gases which are lost, increasing the density of the remaining magma; second, shallower rocks are inherently less dense than rocks at

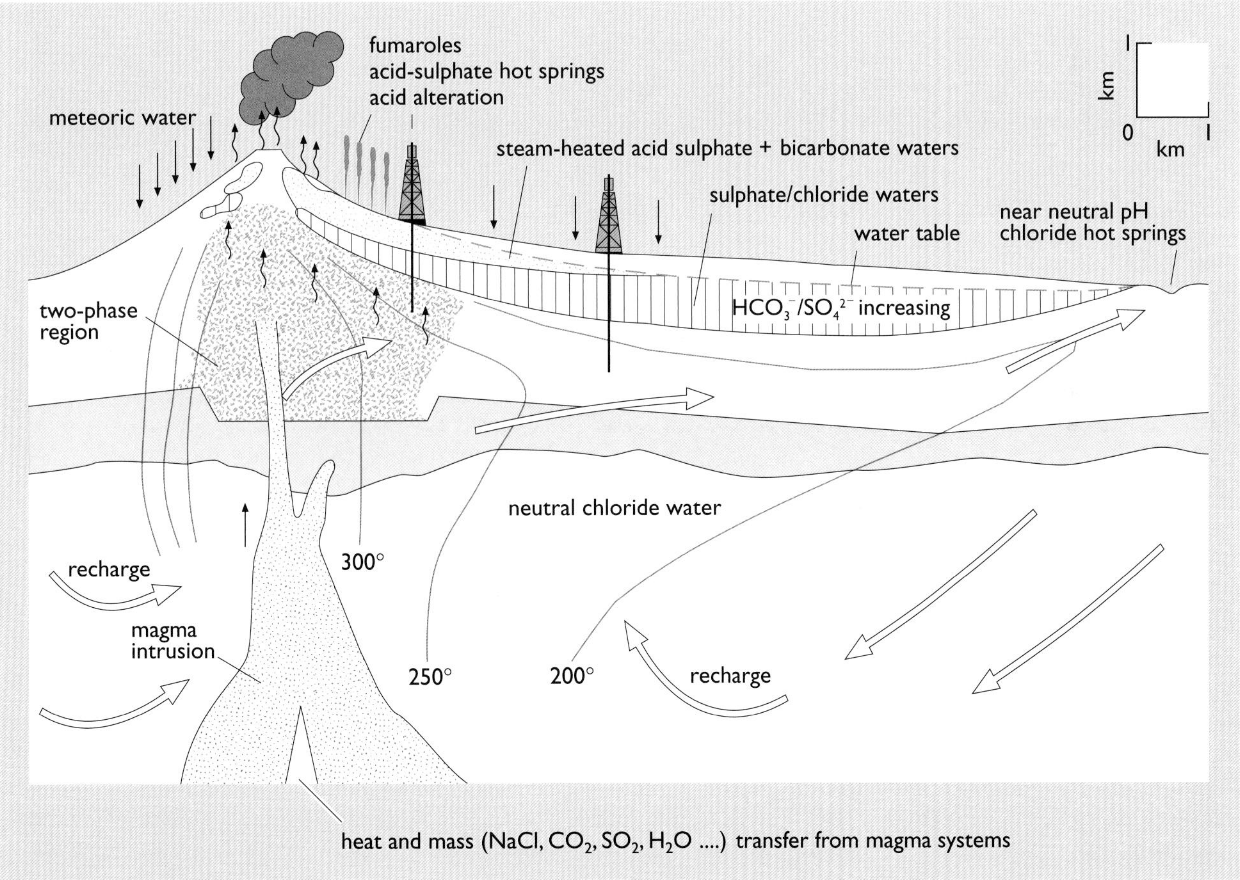

Figure 9.7 Conceptual model of a typical volcanic geothermal system in which meteoric (rain) waters percolate deep into the volcanic superstructure where they are heated by a body of magma which loses dissolved gases as it rises, and forms an intrusion (see text). The hot aqueous fluid rises and may reach the point at which water boils to form steam, producing a two-phase (steam + water) zone. The hydraulic gradient causes the geothermal fluid to migrate through any permeable rocks in the volcano flank. Here the fluids may be accessed by drilling (see drill rig symbols); the chemistry of the fluid changes during migration due to mixing with CO_2-saturated rain water (see Box 9.2)

greater depth, usually because they are less compressed. So, whereas volcanic eruptions are driven by exceptionally high gas pressures, many magmas form 'intrusions', coming to rest and crystallizing beneath the surface at 1–5 km depth.

In the 1980s, experiments were undertaken in the USA with the ultimate aim of drilling directly into or very close to magma bodies where temperatures may be up to 1800 °C, and to harness geothermal power by cycling water through their outer margins. In preparation, the US Magma Energy Program succeeded in drilling into the molten Kilauea lava lake (Hawaii) and ran successful energy extraction experiments. In any drilling operation, the drill bit is cooled and lubricated by circulating fluid (usually mainly water, but referred to as '**mud**'), which also lifts cuttings to the surface. In this case, as was expected, the circulating water solidified a thin shell of lava around the drill bit. The resulting tube of solid but thermally fractured rock acted as a heat exchanger, with heat being transferred to the drill hole by convecting magma. While a useful test, this is still a long step from drilling into a live magma chamber with high-pressure dissolved gases, and no further work has been undertaken in recent years.

A close encounter with magma occurred during the development of the Krafla field in northern Iceland when, in 1977, rising magma reached the depth of a borehole at 1138 m and three tonnes of magma was erupted through the hole in 20 minutes! Quite by chance, the development history of this field was dogged by a series of eruptions in 1975 and 1984, the first at Krafla for over 250 years, but progress improved once the eruptions ceased and the field now supports 60 MW_e of power generation.

Several of the world's most advanced geothermal sites (for example in northern Italy and the western USA) are located in extinct volcanic areas. Fortunately for geothermal exploitation, because rocks are such good insulators, magmatic intrusions may take millions of years to cool to ambient conditions. Such intrusions, therefore, continue to act as a focus for 'hot fluid', or hydrothermal convective cycles in permeable strata as in Figure 9.7. The nature of the resource then depends on the local conditions of pressure and temperature in the aquifer, and this determines the extraction technology and the profitability of the site.

The range of pressures and temperatures in which we are interested geothermally lies typically between 100 and 300 °C, below the critical point

BOX 9.4 Pressure and depth

As noted in earlier chapters, hydrostatic pressure increases by about 1 atmosphere for an increase of 10 metres in depth. It follows that a geothermal aquifer 1 km thick will produce a pressure increase of 100 atmospheres (100 bar). (A pressure of 1 bar is approximately equal to 1 atmosphere.)

The SI unit for pressure is the pascal or, more appropriately for the high pressures in geothermal systems, the megapascal (MPa). One megapascal is approximately 10 atmospheres, and the 20 MPa mentioned in the text is thus 200 atmospheres.

Many geothermal aquifers also contain a steam zone and, since steam has a much lower density than water, the **vapourstatic** increase in pressure with depth (i.e. the pressure due to the weight of the column of steam) is much smaller than the hydrostatic increase. Note that it is not strictly correct to refer to water vapour as steam. Steam is condensed droplets of water (as in clouds), which is why it can be seen. Water vapour is an invisible gas. However, steam is often used as a synonym for water vapour, though the meaning should generally be clear from the context.

at which liquid water and water vapour become indistinguishable (though conditions in some fields exceed 400 °C), and in the pressure range up to about 20 megapascals (20 MPa).

In simple terms, high-enthalpy systems are subdivided into vapour-dominated and liquid-dominated, depending on the main pressure-controlling phase (i.e. steam or liquid water) in the reservoir. Vapour-dominated systems are the best and most productive geothermal resources, largely because the steam is dry (free of liquid water) and is of very high enthalpy. Where reservoir rocks are at pressures below hydrostatic, which promotes steam formation (perhaps 3–3.5 MPa at depths down to 2 km), there must be some barrier to direct vertical groundwater infiltration. The Larderello field in Figures 9.2 and 9.3 is of this type.

In contrast, liquid-dominated systems are at higher than hydrostatic pressures, exceeding 10 MPa at depths below 1 km (because at 1 km depth the hydrostatic pressure is about 100 bars, i.e. 10 MPa, see Box 9.4). Production of electricity from liquid-dominated systems benefits from the higher fluid pressures at depth, and water may 'flash' into steam as it crosses the boiling point curve (towards lower pressures) *en route* to the surface. However, the steam is often wet and of lower enthalpy, which adds to the technical problems for electricity production. The famous Wairakei field in New Zealand is liquid-dominated but, typically for such systems, has developed a two-phase zone as pressures have fallen during exploitation. Fortunately, the groundwater zone has a relatively low permeability which suppresses the tendency for natural venting of steam over most of the Wairakei area.

The heat source in sedimentary basins

An important key to understanding many geothermal resources is the heat conduction equation:

$$q = K_T \frac{\Delta T}{z} \tag{3}$$

This is analogous to Darcy's Law, but here q is the one-dimensional vertical **heat flow** in watts per square metre ($W\ m^{-2}$), ΔT is the temperature difference across a vertical height z, and $\Delta T/z$ is thus the **thermal gradient.** The constant K_T relating these quantities is the **thermal conductivity** of the rock (in $W\ m^{-1}\ °C^{-1}$) and is equal to the heat flow per second through an area of 1 square metre when the thermal gradient is 1 °C per metre along the flow direction.

Values of K_T for most rock types are quite similar, in the range 2.5–3.5 $W\ m^{-1}\ °C^{-1}$ for sandstones, limestones and most crystalline rocks. However, mudrocks (clays and shales) are the exceptions, with lower values of 1–2 $W\ m^{-1}\ °C^{-1}$. These are also among the most impermeable rocks (Table 9.2), so mudrocks contribute two of the essential characteristics for geothermal resources: they act as impermeable cap rocks and as an insulating blanket, enhancing the geothermal gradient above aquifers in regions of otherwise normal heat flow.

So, even under conditions of average heat flow (60 $mW\ m^{-2}$), it is possible to obtain temperatures of 60 °C within the top 2 km of the earth's crust.

Box 9.5 demonstrates how the differing insulating properties of rocks influence the way the temperature varies with depth. To maintain the same vertical heat flow, low-conductivity rocks require a steeper temperature gradient than a relatively good conductor, and are accordingly important in augmenting temperatures at depth.

This has led to exploration programmes aimed at locating natural waters in areas of thick sedimentary rock sequences containing mudrocks and

BOX 9.5 Thermal gradient and heat flow

Consider the situation where there is a steady upward flow of heat through the top few kilometres of the earth's crust. We can use Equation 3 to relate this flow to the temperature at any depth if we know the thermal conductivity of the rock.

If, for instance, the temperature is found to be 58 °C at a depth of 2 km (2000 metres) and the surface temperature is 10 °C, the temperature gradient is

$$(58 - 10) / 2000 = 0.024\ ^\circ\text{C m}^{-1}$$

and if the thermal conductivity of the rock is 2.5 W m^{-1} $°C^{-1}$, the heat flow rate is

$$2.5 \times 0.024 = 0.060\ \text{W m}^{-2}$$

or 60 milliwatts per square metre.

Suppose, however, that this same 60 mW flows up through several layers with different thermal conductivities. Equation 3 tells us that the thermal gradient must be different in each layer, with the temperature changing most rapidly through the layer with the lowest conductivity, as in Figure 9.8 below.

We can check that the diagram shows the correct temperatures by using Equation 3 to calculate the temperature gradient for each layer and comparing this with the gradient read from the graph:

Layer 1

The calculated gradient is 0.060 / 2.5 = 0.024 °C m^{-1}

The measured gradient is (34.5 – 10.0) / 1000 = 0.0245 °C m^{-1}

Layer 2

The calculated gradient is 0.060 / 1.5 = 0.040 °C m^{-1}

The measured gradient is (54.5 – 34.5) / 500 = 0.040 °C m^{-1}

Layer 3

The calculated gradient is 0.060 / 3.0 = 0.020 °C m^{-1}

The measured gradient is (64.5 – 54.5) / 500 = 0.020 °C m^{-1}

Within the precision of the data therefore, the temperatures shown are consistent with a heat flow rate of 60 mW per square metre through each layer. Comparing this case (Figure 9.10) with the uniform rock considered above, it is obvious that the presence of the thin layer with low thermal conductivity has appreciably enhanced the temperature at a depth of 2 km.

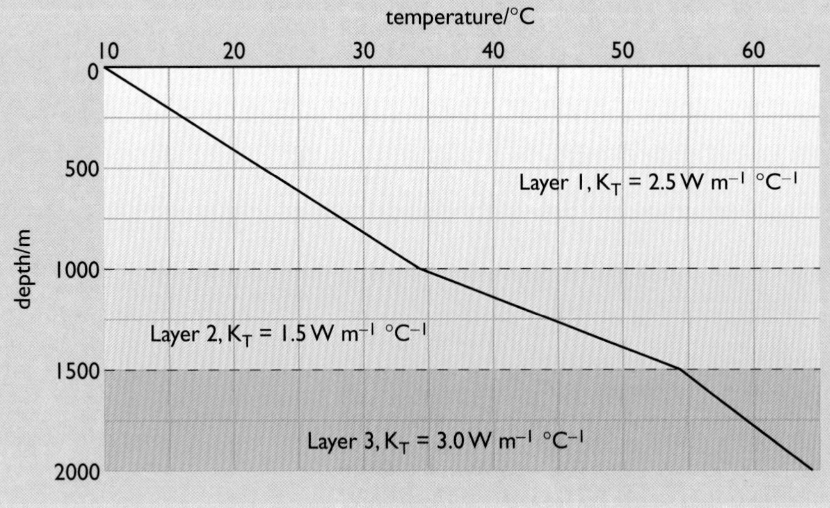

Figure 9.8 Variation of temperature with depth across three zones of differing thermal conductivity, K_T

permeable limestones or sandstones. For example, the Paris area is at the centre of a 200 km wide depression in the crystalline basement rocks. Exploration for hydrocarbon resources in the 1960s and 70s was extremely successful in locating hot water, between 55 °C and 70 °C at depths of 1–2 km, but found very little oil or gas! While low enthalpy water resources are unsuitable for power generation (no high pressure steam can be produced, and temperatures are too low to permit an acceptable generation efficiency), they can be of considerable benefit in meeting demands for low grade heat (space heating, etc.). However, to be economic they must be located close to a heat load. The Paris area is ideal in this respect. Similar resources occur in some of the sedimentary basin areas of the UK, such as beneath the Yorkshire-Lincolnshire coast and in Hampshire, where the Southampton geothermal scheme operates (see Box 9.3). However, most are remote from suitable heat loads.

Finally, there are two extensions of the criteria discussed above which make some sedimentary basin resources more attractive:

1 Large-scale non-electrical applications of geothermal energy worldwide arise in basins where the background heat flow is above average. The geological reasons for the association of high heat flow with sedimentary basins are not altogether surprising: stretching processes within the earth's outer plate layer induce thinning which can radically raise the heat flow as well as creating a surface depression on which sedimentation occurs. Beneath the south Hungarian Plain, for example, geothermal gradients as high as 0.15 °C m^{-1} have been recorded and 120 °C water occurs at 1 km depth.

2 In other areas, larger sedimentary thicknesses may occur. For example, high-pressure fluids at temperatures of 160–200 °C occur at 3–5 km depth in the Gulf of Mexico, southern Texas and Louisiana. Because of chemical processes occurring as a result of the depth and temperature of burial and the efficient sealing of the aquifers by impermeable rocks, pressures greatly exceed hydrostatic and 100 MPa has been recorded in local pockets of fluid. The fluids are highly saline brines with trapped gas, especially methane. These so-called 'geopressured brines' are a potentially important geothermal resource for power generation, a resource that has remained untapped to date but on which there is intermittent government funding for research in the USA. The great advantage of geopressured resources is that they offer three kinds of energy: geothermal heat, hydraulic energy (because of the high pressure), and the large quantities of methane that are found dissolved in the fluid.

Geothermal waters

Most of the foregoing has been concerned with the source of geothermal heat. Exploitation of the heat, however, requires that the geothermal water be brought to surface, and that brings with it a different set of problems. Water that has been in contact with rock for long periods (and geothermal waters can be thousands or even millions of years old) contains dissolved minerals. Hot water tends to be more reactive than cold water, so geothermal waters can often contain around 1% of dissolved solids. Typically, these will be carbonates, sulphates or chlorides, and dissolved silica becomes significant where waters have been in contact with rocks above 200 °C. For this reason, geothermal fluids are often called 'brines'. Dissolved gases are also common, especially at higher temperatures. Techniques are available to deal with all of these, but it is essential that they be taken into consideration at the design stage of the plant. With correct design, these contaminants can all be handled and disposed of without either operational or environmental difficulty. If they are ignored, however, or the plant designed before the water has been properly characterized, the entire system can fail within a matter of months. The various techniques that can be used are beyond the scope of this chapter, but some examples are quoted in Section 9.3.

Why are there hot dry rocks?

Our attention now turns from sedimentary strata to the underlying crystalline 'basement'. The term 'hot dry rock (HDR) resources' refers to the heat stored in impermeable (or poorly permeable) rock strata and to the process of trying to extract that heat. What is required is the creation of an artificial heat exchanger zone within suitably hot rocks. Because rocks are (by normal standards) poor conductors of heat, very large heat transfer surfaces (of the order of square kilometres) are required if heat is to be extracted at useful rates. This can be achieved by enhancing the natural fracture system that occurs in all such crystalline rocks. Water is circulated through the enhanced zone so that heat may be extracted, ideally to generate steam and, hence, electrical power. Although the technology to create suitable arrangements for reproducible heat recovery has not yet been perfected, in theory at least HDR technology could be applied over a significant proportion of the earth's surface.

Because drilling is expensive, with costs rising exponentially with depth, only the top 6 km of the earth's crust is generally used in calculating geothermal energy potential (though some hydrocarbon and research drilling has gone as deep as 15 km). Given current technical and economic constraints on drilling depths, a minimum geothermal gradient of around 0.025 $°C\ m^{-1}$ is required if development is to be economical. With a typical thermal conductivity of 3 $W\ m^{-1}\ °C^{-1}$, this requires (from Equation 3) a heat flow of 75 $mW\ m^{-2}$, only a little above the earth's average. In practice, however, it is customary to look for rocks with much higher heat flows (as at experimental sites in the USA, Japan and France). Granite bodies are ideal targets, because such rocks occupy large volumes of the upper crust and they crystallized from magmas that had naturally high concentrations of the chemical elements with long-lived radioactive isotopes – uranium, thorium and potassium. Here we reach a situation in which heat flow through the earth's surface itself is augmented because of the heat production within certain shallow crystalline rocks. If, in addition, a layer of poorly conducting sedimentary rocks overlies the granite, its 'blanketing' effect will increase the temperature gradient and make higher temperatures available at shallower depths.

9.3 Technologies for geothermal resource exploitation

Resources in high-enthalpy steam fields

The first stage in prospecting for geothermal resources in volcanic areas involves a range of geological studies aimed at locating rocks that have been chemically altered by hot geothermal brines, and finding surface thermal manifestations such as hot springs or mud pools. Investigations of fluid chemistry and, increasingly, the release of gases through fractured rocks allow assessment of the composition and resource potential of trapped fluids. These studies provide the first clues to the likely presence and location of exploitable resources. However, geophysical prospecting techniques, particularly resistivity surveying and other electrical methods designed to detect zones with electrically conducting fluids (i.e. brines), are probably the most effective for precise location of buried geothermal resources. Once a suitable geothermal aquifer has been located, exploration and production wells are drilled using special techniques to cope with the much higher temperatures and, in some cases, harder rock conditions than in oil and water wells. Since fluid pressures in the aquifer range up to about 10 MPa, the driller must ensure that the 'mud' used is dense enough to counteract these pressures and avoid 'blow out', where an uncontrollable column of gas is discharged. The well is lined with steel tubing that is cemented in place, leaving an open section or a perforated steel casing at production depths. As each set of casing has to be inserted through its predecessors as the well depth increases, the well diameter decreases with depth from perhaps 50 cm near the surface to 15 cm at production depths. A well-head with valve gear is welded to the steel casing at ground level. This allows the well to be connected to a power plant via the network of insulated pipes that are a familiar sight in geothermal areas (as in Figure 9.3).

Technologies for electrical power generation depend critically on the nature of the resource – not just the fluid temperature and pressure but also its salinity and content of other gases, all of which affect plant efficiency and design. The size of any power station is determined by the economics of scale; conventional coal- or oil-fired stations are typically a few hundred megawatts per unit. A typical geothermal unit, by contrast, is usually 30–50 MW_e. This is because the amount of steam delivered by one well is usually sufficient to generate only a few MW_e, and wells are linked across the field and back to the station by pipeline. Above a certain capacity, the cost of the pipelines is such that it is cheaper to develop a separate station in another part of the field.

Given the fact that most of the costs of the electricity derived from geothermal resources are accounted for by the need to pay back the capital investment, with day-to-day operating costs being relatively minor (and insensitive to output variation), there is a great incentive to maximize the efficiency with which the relatively low grade heat (by power generation standards) is converted into useful energy. Today there are several hundred installations operating worldwide and these include four main types, described below.

Dry steam power plant

As the name implies, this type of system (Figure 9.9 (a)) is ideal for vapour-dominated resources where steam production is not contaminated with liquid. The reservoir produces superheated steam, typically at 180–225 °C and 4–8 MPa, reaching the surface at several hundred km h^{-1} and, if vented to the atmosphere, sounding like a jet engine at close proximity. Passing through the turbine, the steam expands, causing the blades and shaft to rotate and hence generating power. Temperatures up to 300–350 °C and correspondingly greater pressures are increasingly being exploited, leading to greater efficiency in electricity production.

In the simplest form of power plant, a 'back-pressure' unit, the low-pressure exhaust steam is vented directly to the atmosphere. Although such units are simple, they are also very inefficient; their main use is as temporary

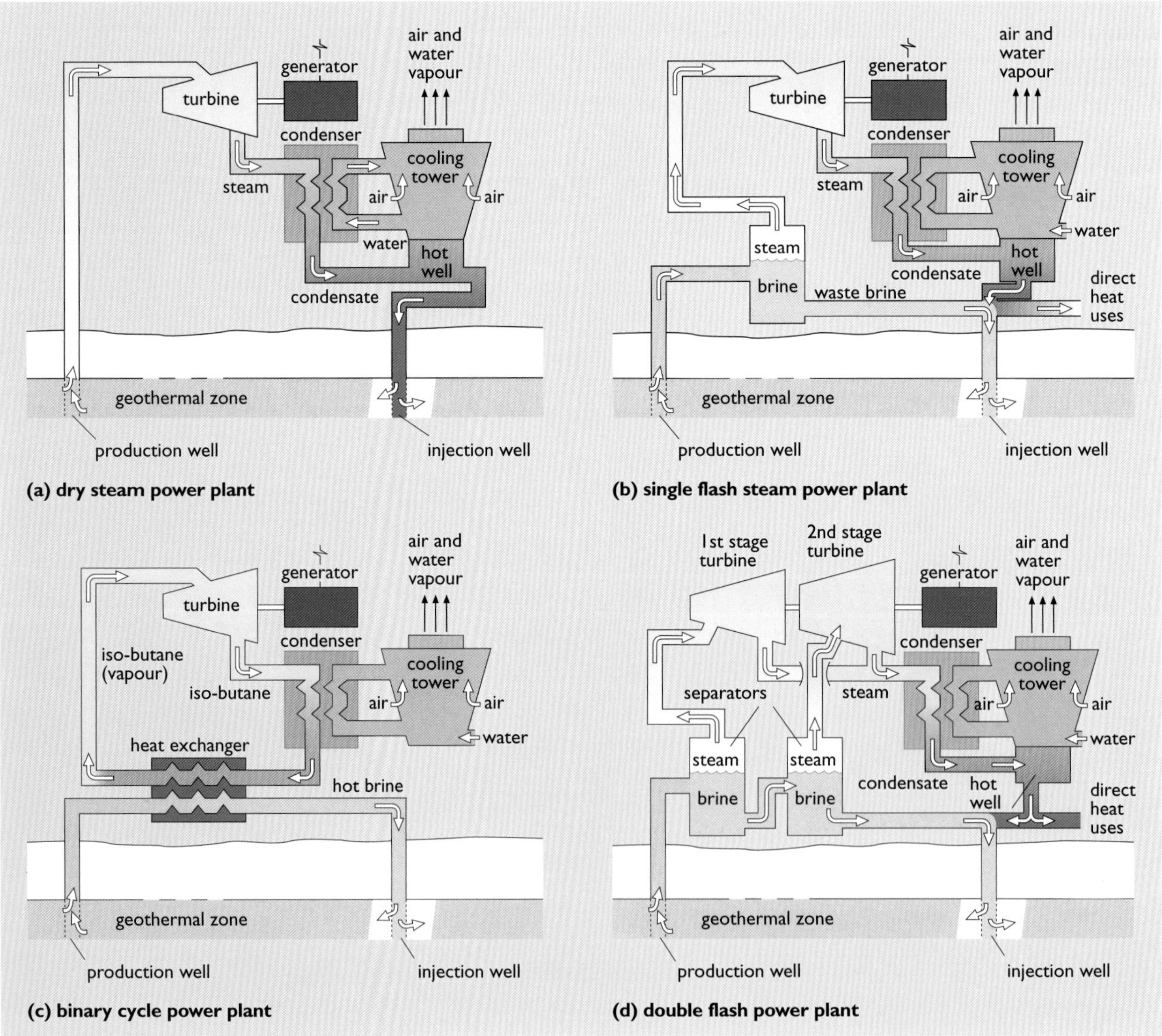

Figure 9.9 Simplified flow diagrams (a–d) showing the four main types of geothermal electrical energy production

transportable units during the development of a new field. Once the steam supply is ensured, normal practice is then to install 'condensing' plant as shown in Figure 9.9(a). These achieve greater efficiency by condensing the exhaust steam to liquid, thus dramatically increasing the pressure drop across the turbine because liquid water occupies a volume roughly 1000 times less than the same mass of steam. Of course, the cooling towers generate 'waste heat' in just the same way as conventional coal- and oil-power stations (or, indeed, any heat engine). At temperatures typical of geothermal fluids, efficiencies are low and, despite the use of high temperature superheated steam, rarely exceed 20%. Nevertheless, whereas a 1960s plant required almost 15 kg steam per saleable kWh in optimum conditions, modern dry steam plant with higher temperature steam and better turbine designs can achieve 6.5 kg per kWh, so a 55 MW_e plant requires 100 kg s^{-1} of steam.

Plant efficiency, and therefore profitability, is strongly affected by the presence of so-called 'non-condensable' gases such as carbon dioxide and hydrogen sulphide in the geothermal fluid. When the turbine exhaust gases are cooled, achieving a suction effect on the turbine as the water condenses into liquid at around 100 °C, gases that do not similarly condense cause higher residual pressures at the back end of the turbine. A small percentage of such gases reduces suction efficiency and so impacts on the economics of the system; for this reason, many geothermal plant are fitted with gas ejectors. However, the ejectors themselves require either a steam supply or electric power from the turbine-generator and, consequently, reduce output. Non-condensable gases have an additional economic impact: it is no longer acceptable in most places to vent them to the atmosphere, so they must either be trapped chemically or reinjected with the waste water to avoid pollution, and both options entail additional costs.

In general, dry steam plant is the simplest and most commercially attractive. For that reason, dry steam fields were exploited early and have become disproportionately well known. In fact, only the USA and Italy have extensive dry steam resources, though Indonesia, Japan and Mexico also have a few such fields. (Elsewhere, and even in these countries, liquid-dominated fields are far more common.). While in some areas it is common practice to reinject the spent fluid, this was not undertaken in the largest field, the Geysers in the USA, until falling fluid pressures led to a recognition among the various private operators that the field was being over-exploited. There is now an agreed reinjection policy to make the resource more sustainable; some 70% of the mass of produced steam is now reinjected, and field production is rising again. (Not all of this fluid comes from the condensed steam from the turbines, because much is evaporated from the cooling towers. Instead, in a new and imaginative environmental development, treated sewage water is piped 48 km from a local community – thereby solving an unrelated local disposal problem as well as helping to maintain the geothermal reservoir.)

Single flash steam power plant

Here (Figure 9.9 (b)) the geothermal fluid reaching the surface may be steam (the water having 'flashed', i.e. vaporized, within the well as pressure dropped during ascent) or hot water at high (close to reservoir) pressure.

In the first case, a separator is installed simply to protect the turbine from a massive influx of water should conditions change. However, it is often better to avoid flashing in the well because this can lead to a rapid build-up of scale deposits as minerals dissolved in the fluid come out of solution, leading to the plugging of the well. For this reason, the well is often kept under pressure to maintain fluid as liquid water. To deal with hot, high-pressure water requires complex equipment designed to reduce the pressure in a controllable manner and induce flashing so that steam may be separated. Again, a conventional condensing steam turbine is at the heart of the plant, but lower steam pressures and temperatures (0.5–0.6 MPa, 155–165 °C) are common, so the plant typically requires more steam per kWh than would be required in a dry steam plant, say around 8 kg per kWh. Moreover, the bulk of the fluid produced, often up to 80%, may remain as unflashed brine which is then reinjected unless there are local direct use heating applications available. In general, therefore, reinjection wells must be available for fluid disposal both at single flash plants and at plants incorporating the newer types of technology described below.

Binary cycle power plant

This type of power plant (Figure 9.9(c)) uses a secondary working fluid with a lower boiling point than water, such as pentane or butane, which is vaporized and used to drive the turbine. It is more commonly known as an Organic Rankine Cycle (ORC) plant. Its main advantage is that lower-temperature resources can be developed where single flash systems have proved unsatisfactory. Moreover, chemically impure geothermal fluids can be exploited, especially if they are kept under pressure so that no flashing ever takes place. The geothermal brine is pumped at reservoir pressure through a heat exchange unit and is then reinjected; the surface loop is closed and no emissions to the environment need occur. Ideally, the thermal energy supplied is adequate to superheat the secondary fluid. (Note: a superheated fluid is a liquid above the normal boiling point, usually prevented from boiling by increasing the pressure.) For temperatures below about 170 °C, higher generating efficiencies are possible than in low-temperature steam flash plants. About 100 units of this type are in operation today. A disadvantage is that keeping the geothermal fluid under pressure and repressurizing the secondary fluid can consume some 30% of the overall power output of the system because large pumps are required. Large volumes of geothermal fluid are also involved; for example, the Mammoth geothermal plant in California (Box 9.8) uses around 700 kg s^{-1} to produce 30 MW_e.

Double flash power plant

Recently, several attempts have been made to develop improved flashing techniques, particularly to avoid the high capital costs and parasitic power losses (e.g. circulating pumps for the secondary fluid) of binary plant. Double flash (Figure 9.9(d)) is ideal where geothermal fluids contain low levels of impurities and so the scaling and non-condensable gas problems that affect profitability are at a minimum. Quite simply, unflashed liquid remaining after the initial high-pressure flashing flows to a low-pressure tank where another pressure drop provides additional steam. This steam is mixed with the exhaust from the high-pressure turbine to drive a second turbine (or a second stage of the same turbine), ideally raising power output by 20–25% for only a 5% increase in plant cost. Even so, extremely large fluid volumes are required. The East Mesa plant in southern California, for example, commissioned in 1988, uses brine at 1000 kg s^{-1} from 16 wells to generate 37 MW_e; i.e. around 5 times as much fluid as for similar dry steam plant (though temperatures would be much higher in the latter case).

Future developments

As the geothermal industry continues to expand, there will be a need to develop technologies that can produce geothermal power from a variety of resources that are less ideal than dry steam. Increasing use is being made of geothermal fluids which are at either lower temperature than (but similar pressure to) those in dry steam fields, or at the same or higher temperature and much higher pressure. These are essentially liquid-dominated resources, albeit of high enthalpy, and they exist in large volumes. Inevitably, variants on the binary and double flash systems will continue to be developed; they are at the leading edge of current research. More recently, greater use has been made of the produced fluids by operating combined cycles, using an Organic Rankine Cycle to extract further work from the main turbine exhaust or the separated water.

A number of other approaches are being developed to increase the efficiency of generation from lower temperature fluids, using either different working fluids or new power cycles such as the Kalina cycle. The latter uses as working fluid an ammonia-water mixture, the composition of which varies throughout the cycle. Capital costs are expected to be significantly lower than for ORC and the net generating efficiency up to 40% higher; at an inlet temperature of 130 °C the net efficiency is estimated at over 13% (58% of the theoretical Carnot efficiency). The early demonstration plant are still showing teething problems, but the operators remain optimistic.

Resources for direct use geothermal energy

Some of the countries that are exploiting geothermal resources for non-electrical purposes have chosen to develop these direct use applications in areas flanking the main steam fields. Japan, New Zealand, Iceland and Italy are obvious examples, where wet steam or warm water at a range of temperatures is readily available for industrial, domestic and leisure applications. In this section, however, we are concerned principally with the low-temperature resources found in sedimentary basin areas, several of which have been developed across central Europe. Drilling techniques resemble those discussed earlier, but the process is generally less hazardous since the geothermal fluid is found under much lower pressure and temperature conditions than in hot steam fields, and pumps are usually required to bring the fluid to the surface at adequate flow rates. However, the hot water is usually too saline and corrosive to be allowed directly into heating systems, so once again corrosion-resistant heat exchangers are widely used. The secondary circuit might be a vast greenhouse complex using both aerial and underground pipes, or it might be a domestic heat load with a combination of underfloor and radiator pipes. The dense multi-storey apartment blocks of the Paris suburbs are ideal heat loads for these local resources.

The French led the development of these 'low-enthalpy' resources in Europe. Over the last 30 years no less than 55 group heating schemes were installed in the Paris Basin, with several more in south-western France. At the design stage, a twin production and reinjection borehole system would be planned on the basis of supplying between 3 and 5 MW_t of heat energy (25–50 l s^{-1} of water at 60–70 °C) over a lifetime of 30–50 years. The spacing of the wells must be designed to maintain high fluid pressures by reinjection while avoiding the advance of a 'cold front' (i.e. fluid at reinjection temperatures) towards the production well until capital costs are paid back, and this means that flow conditions in the aquifer need detailed study. A typical layout for a twin production well scheme, with a schematic of the heat transfer technology, is shown in Figures 9.10(a) and (b). In Figure 9.10(b) note the interesting application of heat pumps to enhance the system efficiency by reducing the reinjection temperature. Heat pumps work on the same principle as refrigerators, but here produce a concentrated high temperature output. Of course, they consume electrical energy but, in the example shown in Figure 9.10(b), they enable the number of heated apartments to be doubled.

Although the French group heating schemes have been a great practical success, their economic benefits were only marginal at times of low oil prices, increasing availability of natural gas and high interest rates. For this reason, development work stopped during the period 1989–92. A few schemes were abandoned, for technical or economic reasons, but some 45 remain in operation and new and extended schemes are being planned. Currently, they produce a saving of over 200 000 tonnes of oil (or equivalent in other fossil fuels) per year in an area which, 30 years ago, had no obvious geothermal potential. The same concept applies to the analogous UK scheme in Southampton.

In Germany, reunification has had a significant effect on the way in which geothermal developments have occurred; the better (though still low temperature) resources tend to be concentrated in the eastern part of the country. Although a few schemes existed before reunification, with the freeing of capital following unification these are really beginning to take off. Several large scale district heating schemes are already in operation and even more under active development. By the end of 1999 direct thermal use of geothermal energy in Germany amounted to an installed thermal power of roughly 397 MW_t. Of this total, 27 major centralized installations accounted for approximately 55 MW_t. Ground source heat pumps (see below) are estimated to contribute an additional 342 MW_t and a substantial number of new schemes, both large and small, are under development.

Ground source heat pumps

Between 1992 and 2000 the capacity of direct heat installations worldwide increased from 4000 MW_t to 16 000 MW_t. In fact, the increase in large installations like those described above has been quite modest (for the same reasons of high interest rates and low fossil fuel prices) and the bulk of the increase has been accounted for by a new type of geothermal installation, the ground source heat pump (GSHP).

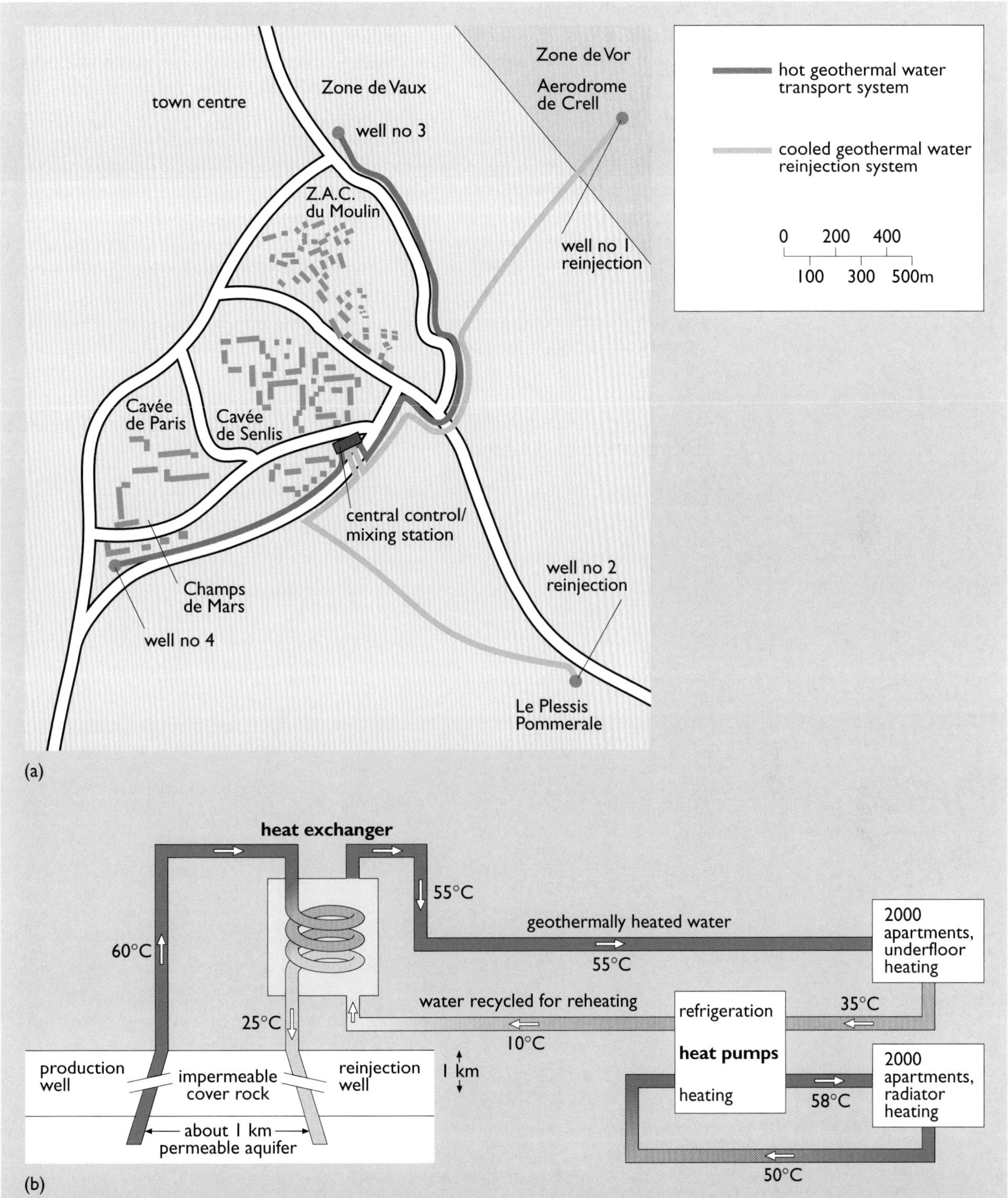

Figure 9.10 The Creil district heating scheme installed in 1976 north of Paris: (a) Map showing location of two production and two reinjection wells, and the various apartment blocks served by the system. (b) This shows how the geothermal heat is exchanged to a secondary freshwater circuit. The circuit is used first to heat 2000 apartments by underfloor heating; the residual energy is then boosted using heat pumps to provide radiator heating in a further 2000 apartments. Note that the main function of the heat pumps is to lower the reinjection temperature and so extract more heat from the geothermal fluid rather than to raise the production temperature

The general arrangement is illustrated in Figure 9.11(a). Unlike other geothermal techniques, this one relies on heat transfer by conduction from the walls of the borehole, not on the extraction of groundwater. The heat available from a well 100–150 m deep is only a few kW_t, but that is sufficient for a single domestic installation, and boreholes of this depth are cheap enough to make the installation competitive with conventional heating systems. A simple loop of pipe is inserted in the well and grouted in place. A heat transfer fluid (usually water) circulates in the loop and transfers heat from the surrounding sub-soil to a heat pump. More than 15 years' experience has shown that a few kilowatts can be extracted throughout the heating season; the sub-soil temperature drops by a few degrees but regenerates over the remainder of the year (Figure 9.11(b)). Table 9.3 shows the rate at which energy can be extracted – on a quasi-continuous basis – from typical 100–150 m deep holes.

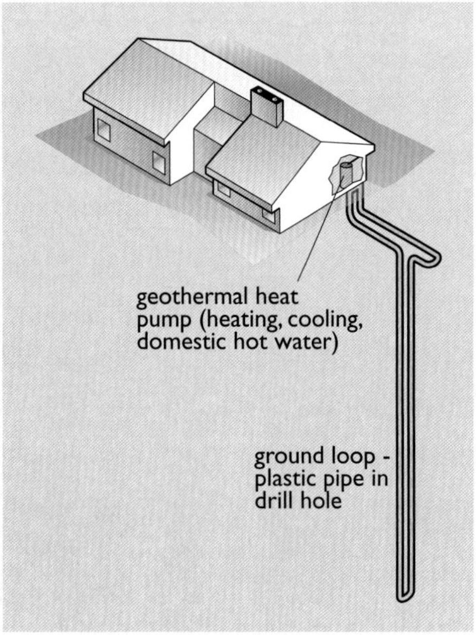

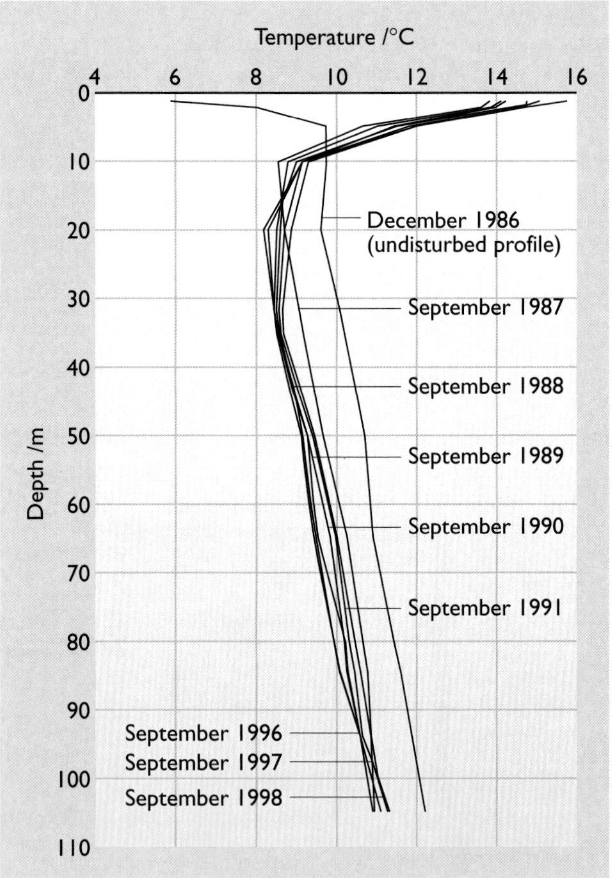

Figure 9.11 (a) The geothermal heat pump (GHP) concept used to extract heat from warm shallow groundwater to supply a single domestic dwelling. In the winter heat is removed from the earth and delivered in a concentrated form via the heat pump. Because electricity is used, in effect, to increase the temperature of the heat, not to produce it, the GHP can deliver three to four times more energy as heat than the energy content of the electricity it consumes; (b) Measured data from a Swiss GSHP installation. After an initial drop, the ground temperature below the solar-dominated surface zone (see Box 9.6) recovers to the same value year after year. This appears to be a truly renewable operation

Table 9.3 Borehole heat exchanger performance in different rock types

Rock type	**Thermal conductivity ($W\ m^{-1}\ K^{-1}$)**	**Specific extraction rate ($W\ m^{-1}$)**	**Energy yield per metre of borehole ($kWh\ m^{-1}\ a^{-1}$)**
Hard rock	3.0	max. 80	135
Unconsolidated rock, saturated	2.0	45–50	100
Unconsolidated rock, dry	1.5	max. 30	65

Source: Rybach and Eugster, 1998

BOX 9.6 Solar or geothermal?

The question is often asked whether GSHPs are truly geothermal devices, or whether they are really solar units. The answer depends on the configuration of the heat exchanger loop. All the GSHPs discussed in this chapter are 'borehole' units, with the heat exchanger located some 100 m or so below ground, and these are true geothermal systems. There is another type of GSHP, however, that uses 'ground coils', horizontal loops of piping buried just beneath the surface. These latter *do* derive their energy from the daily solar input; they are cheap to install but their performance varies on a seasonal basis according to the solar input.

This difference in performance again has its origins in the poor thermal conductivity of soils. The thermal pulse from the daily and seasonal solar input penetrates very slowly, and for practical purposes virtually disappears at depths greater than 15–20 m (Figure 9.12). At greater depths, the temperature is controlled almost entirely by the geothermal heat flow (except where it may be influenced by local groundwater flow).

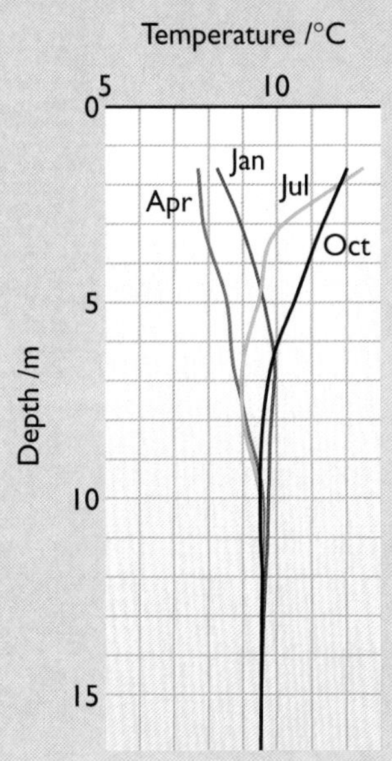

Figure 9.12 Ground temperatures typical of northern Europe, as a function of depth for different periods of the year

If reversible heat pumps are used, the same systems can provide cooling in summer with the added advantage of helping to recharge the underground and so increasing the sustainable heat extraction from the system. It should be emphasized that, unlike other geothermal systems, GSHPs are ideally suited to the domestic scale with a single module providing just a few kW_t.

Because the presence of groundwater is not a prerequisite, this technology can be applied almost anywhere. The types of buildings that are using ground source heating and cooling in this manner range from small, utility or public housing, through to very large (MW sized) institutional or commercial buildings. This technology can offer up to 40% reductions in CO_2 emissions against competing technologies. Better still, if the electricity to drive the heat pump is supplied from non-fossil sources, then there should be no CO_2 emissions associated with heating and cooling a building.

The concept was developed independently in the US and Europe and, although Sweden and Switzerland have installed many thousands of units to provide winter heating in houses (the growth rate in Switzerland is 10% per year), the activity in USA and Canada in the last fifteen years has overtaken the European installation rate. The number of installations in North America is now approaching half a million.

Recently, several large scale arrays have been installed in the US to feed larger complexes where suitable supplies of deep geothermal water are not available. In the largest development to date, 4000 units – each with its own borehole – have been established on a US Army base in Louisiana to provide heating and cooling. Peak electrical demand has dropped by 6.7 MW_e compared with the previous installation, gas savings amount to 2.6 TJ a^{-1} and – perhaps most telling of all – service calls have dropped from 90 per day in summer to almost zero.

Installation of such units has begun only recently in UK; there are probably around 70 installations. A geotechnical consulting group has been heating its offices in this way since 1996 (practising what it preaches!). In 1998 a 4-borehole system was fitted in the new Health Centre at St. Mary's, Isles of Scilly, using a 25 kW reversible heat pump to supply hot water, heating and cooling to the building. Since then, there have been a number of larger, non-domestic installations in the UK (for details, see Curtis, 2001).

Hot dry rock technology

All conventional geothermal systems (except, perhaps, GSHP) rely on the presence of water circulating through the rock to extract heat and bring it to the surface. However, even in a good aquifer more than 90% of the heat is contained in the rock rather than in the water. Moreover, the vast majority of rocks are poorly permeable at best and the occurrence of an exploitable geothermal reservoir is a rarity. On the other hand, heat exists everywhere, and the amount of energy stored within accessible drilling depths (say, down to 7000 m) is colossal. One cubic kilometre of rock (which is about the scale of a geothermal reservoir) will provide the energy equivalent of 70 000 tonnes of coal if cooled by 1 °C. A recent report expressed the potential in another way:

> The U.S. Geological Survey has calculated the heat energy in the upper 10 kilometres of the earth's crust in the U.S. is equal to over 600 000 times the country's annual non-transportation energy consumption. Probably no more than a tiny fraction of this energy could ever be extracted economically. However, just one hundredth of 1% of the total is equal to half the country's current non-transportation energy needs for more than a century, with only a fraction of the pollution from fossil-fuelled energy sources.
>
> (McLarty *et al.*, 2000)

All over the world, temperatures around 200 °C are accessible under a high percentage of the landmass; if this store of heat could be exploited, it would give almost every country the opportunity to generate electricity from an indigenous and (for all practical purposes) renewable resource. It is this prospect that has motivated a number of countries to spend over $300 million over the past 30 years to find a way to exploit the resource.

The fundamental problem is that, as mentioned in Section 9.2, rocks are very poor conductors of heat and, to extract energy at a rate sufficient to pay back the high cost of the boreholes needed to reach these depths, very large heat transfer surfaces are needed – of the order of several square kilometres! There is now a consensus that the only practical way of achieving this figure is to work with nature, exploiting the fact that most deep rocks contain extensive networks of natural fractures. The focus has been on learning how to stimulate and manage these fracture networks to support a useful and controllable flow of water between boreholes. The project at Rosemanowes, Cornwall, played a key role in this (Box 9.7).

The principal costs in the construction of an HDR system lie in drilling wells deep into hard crystalline rocks and creating the artificial heat exchanger. We are dealing with rocks at higher temperatures and under higher stresses, at 3–6 km depth, than in most conventional geothermal areas. The familiar concept of twin production and injection boreholes provides the basis of HDR system designs, but here drilled into relatively hard granite and terminating several hundred metres apart. A suitable heat exchange surface is then created by opening pre-existing joints (fractures). This is done by using a variant of an oil industry technique known as hydro-fracturing, which consists of pumping water down the borehole at increasing pressure until fractures in the rock are opened. The progressive development of the opening fractures is followed by listening to and locating the sound of rock surfaces moving over one another. This is known as microseismic monitoring.

If the second borehole has already been drilled, it may be necessary to repeat the stimulation in that hole in order to link up with the first zone. Alternatively, the second hole may be drilled after the first stimulation to intersect the stimulated zone. A closed circuit water circulation through the fracture system is thereby generated. The trick has been in learning precisely how to control the injection conditions in each hole to ensure that water can flow through the system with a minimum of resistance. At the same time, the stimulated zones around each hole must link up in such a way that water losses are minimized, because water losses mean wasted pumping costs (and in many regions water itself is a valuable commodity). Water is circulated down the injection well, through the HDR reservoir and up the production well to a heat exchanger and turbo-generator where the thermal energy is converted to electricity. (Lower-temperature district heating schemes are also under consideration, but the high capital costs of such an operation require a large local market for the heat produced).

Significant progress has been made in the field of 'Hot Dry Rock' (HDR) during the past few years. It has also been recognized that 'Dry' is a misnomer; very few basement rocks have proved to be completely dry. Water has been found in fractured basement even at the deepest levels in the exploration boreholes in the Kola Peninsula, Russia (15 km) and in the Black Forest, Germany (>8 km). There is a gradual acceptance of the term 'enhanced geothermal systems' (EGS), but the phrase 'hot dry rock' is too catchy to disappear easily. The overall technology has been redefined as '*any system in which reinjection is necessary to maintain production at commercially useful levels*'. This redefinition also emphasizes the continuity which exists in the spectrum of reservoir permeabilities.

The pioneering work took place at Fenton Hill in New Mexico in the 1970s and 80s where the Los Alamos National Laboratory (LANL) developed two systems at temperatures of 200 °C and >300 °C. The Fenton Hill project proved the principle in 1979, when a 60 kW_e ORC generator operated for a month on the produced water. The operating parameters were very far from those that would be needed in a commercial system, however, and a number of teams have worked in the intervening years to understand and refine the techniques required. Notable among these activities was the full-scale (but deliberately low temperature) rock mechanics experiment at Rosemanowes Quarry in Cornwall, which laid many of the foundations for our current understanding of the behaviour of natural fracture systems and how they might be managed (see Box 9.7).

BOX 9.7 **Rosemanowes**

During the 1980s much HDR research focused on developing rock fracture technologies, principally at Fenton Hill, New Mexico, USA, and at the Camborne School of Mines (CSM) site, a disused quarry at Rosemanowes, Cornwall (see Figure 9.13) in the UK. Based on lessons learnt from Fenton Hill, the Rosemanowes project was begun in 1978 with the aim of developing a better understanding of the rock mechanics issues while avoiding the problems of dealing simultaneously with high temperatures. It was intended to reproduce all the necessary characteristics of a commercial-scale system with the sole exception of temperature (and hence depth). Consequently, the wells at Rosemanowes were drilled to only 2 km, less than half the depth required for any commercial electricity production, simply as a rock mechanics experiment.

Two wells were drilled initially, deviated at depth by up to 35° so that they intersected the natural joints and fissures (which are more or less vertical at this depth). Under the conditions prevailing in the Cornish granite, the stimulated zone was expected to develop vertically, so the deviated sections of the two wells were arranged one above the other with vertical spacing of several hundred metres. The key stage, stimulation, involved pumping large volumes of water or low viscosity gel, via the lower section of the injection well, at a pressure high enough to induce the required permeability between the two wells.

Using this approach, hydraulic connections were established at about 2 km depth. Because the main stimulated zone did not develop in the expected direction, however, a third well was drilled subsequently to intersect the zone. The resulting heat exchange system was subjected to long-term circulation test, during which time detailed seismic monitoring was used to pinpoint the position and development of fracturing in three dimensions. The tests revealed that the natural fractures *could* be stimulated by hydraulic pressure, although artificial fractures are difficult to induce except in close proximity to the wells. While good circulation results were achieved, together with significant advances in understanding of the detailed rock mechanical processes, several problems remained.

First, despite a number of different stimulation operations, the system maintained a higher resistance to flow than was desirable. This is expressed in terms of **impedance,** defined here as the pressure difference across the two boreholes divided by the production flow rate. It has been found that there is a critical pressure (about 10 MPa for most systems so far examined) above which runaway shear failure occurs in the fracture network and water losses become uncontrollable. So, for a flow rate of 100 l s^{-1} this would limit the acceptable impedance to <0.1 MPa l^{-1} s^{-1}. This is just one of the target parameters that have been set for a successful HDR system (see Table 9.6). The best results from Cornwall were a maximum flow rate of 24 l s^{-1} and a minimum impedance of 0.5 MPa l^{-1} s^{-1}, and under these conditions the production temperature fell from 70 °C to 55 °C during a three year test from 1985 to 1988. This indicated that 'short-circuits', where cooling advanced much more rapidly than is desirable, had been driven between the wells as a result of the repeated stimulation experiments. So, a second problem is that effective heat transfer surfaces must be increased to give commercial lifetimes of 20–30 years. A third, related, problem is that of water loss which, for economic purposes, should be kept below 10%. Initial losses were very high, especially during high pressure operation, although it proved possible eventually to reduce these to around 10%.

Despite these technical difficulties, the Rosemanowes experiment increased our fundamental understanding of these processes, and provided the confidence and incentive for the European teams to move on to the Soultz project (see main text).

Figure 9.13 (Left) Deep drilling, to about 2 km depth for hot dry rock geothermal research at the Rosemanowes granite quarry, near Penryn (outside Truro), Cornwall, in 1981; (Right) Natural permeable fissure in otherwise solid impermeable granite from the quarry face at Rosemanowes (Length of scale at centre of photograph = 1 m)

The countries principally involved in HDR research (US, Japan, UK, France, Germany and, most recently, Switzerland) worked closely together throughout this period; this type of research makes great demands on both money and expertise and is aimed at what should prove to be a generally applicable technology, so it is an ideal subject for international collaboration. Following this logic, during the 1980s the various teams in UK, France and Germany agreed to pool their resources and, with the support of the European Commission, to develop a single experimental site at Soultz-sous-Forêts in the Upper Rhine Valley. The aim was to build on the results derived from the work in Cornwall, but at a site where higher temperatures could be achieved at shallower depths.

The Soultz site, like Fenton Hill, benefits from the blanketing effect of 1000 m of sedimentary rock above the crystalline basement. The geothermal gradients through the sediments average 0.08–0.1 °C m^{-1}, falling to 0.028–0.05 °C m^{-1} in the crystalline basement beneath. Temperatures at a given depth in the basement are therefore higher than in the granites of south-west England. Teams from France, Germany, UK, Italy and Switzerland work together on the site, which now has two deep boreholes (>3800 m, 170 °C) sited 450 m apart, one borehole to 5000 m (200 °C), and several shallower boreholes which are used for the microseismic monitoring system. Geologically, the site is located in the Upper Rhine Graben, where E-W tensional forces have stretched the crust and caused the granitic basement to subside. The basement is heavily fractured and even supports a small amount of fluid flow.

After 15 years of operation, the project at Soultz, co-ordinated and part-funded by the European Commission, is now recognized as the world leader in developing HDR technology. It has been shown that by careful control of the pressure and density of the stimulation fluid the stimulated zone can be persuaded to develop in a pseudo-horizontal direction so that it can be accessed by a conventional arrangement of two or more boreholes deviated in opposite directions. In late 1997, after several years of testing and hydraulic stimulation of the fracture system at 3.5 km depth, a 4 month circulation test was carried out. 25 kg s^{-1} of water was circulated on a continuous basis between the two deep wells. The system operated in a closed loop, with the heat produced (ca. 10 MW_t) being dumped via a heat exchanger. The overall loss rate was zero and no make-up water was required. In a two-well system like this, such a result was possible only because a down-hole pump was used in the production well, altering the sub-surface pressure field to ensure that losses from the injection well could be balanced by input from the natural in-situ fluids. Tracers added to the injected fluid proved that circulation was occurring; production also declined very rapidly when reinjection was stopped, both observations demonstrating that this is a true HDR system under the above definition. Of equal importance was the finding that the overall system impedance was less than 0.2 MPa l^{-1} s^{-1}, closer to the targets than any previous project (see Table 9.6). This was achieved by careful stimulation of the natural system, based on understanding developed over the past 20 years of collaborative research.

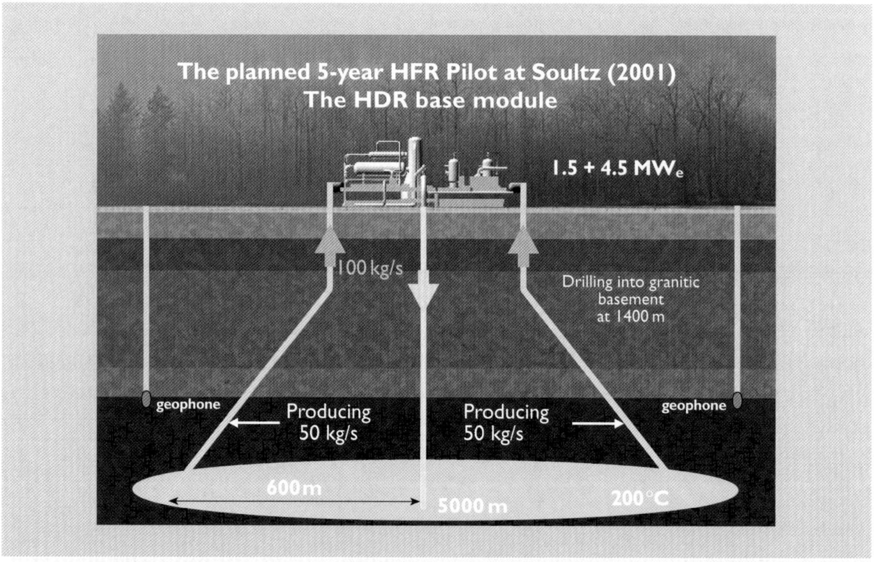

Figure 9.14 The planned 5-year HFR Pilot at Soultz. Note: the term HFR refers to Hot Fractured Rock – a specific type of hot dry rock

In the light of these results, a pilot plant is now being constructed at Soultz. To give better control of pressure distribution in the reservoir and thereby minimize the potential for fluid loss, however, it will be based on a minimum of 3 wells rather than 2. The aim is for the first electricity to be generated around the end of 2004. If that pilot plant leads to a successful commercial demonstration, the work of all these teams over the past 25 years will have given the world a potentially huge new energy resource, available in virtually all countries and capable of contributing significantly to power production needs without the emission of greenhouse gases or other pollutants.

Even before that, many of the lessons will have been applied to enhancing the output of natural high-enthalpy systems (hence the alternative – and probably more accurate – name of 'enhanced geothermal systems'). An idea of the potential was given in the paper cited earlier:

> Experts estimate that up to 6 GW in the U.S. and 72 GW worldwide could be produced with current *[i.e. conventional geothermal]* technology at known hydrothermal sites. With enhanced technology, these estimates increase to 19 GW and 138 GW.
>
> (McLarty *et al.*, 2000)

9.4 Environmental implications

Significant environmental concerns associated with geothermal energy include those to do with site preparation, such as noise pollution during the drilling of wells, and the disposal of drilling fluids, which requires large sediment-settling lagoons. Noise is also an important factor in high enthalpy geothermal areas during well-testing operations when steam is allowed to escape, but once a field comes into production noise levels rarely exceed those of other forms of power plant. Accidents during site development are rare, but a notable exception in 1991 was the failure of a well originally drilled in 1981 at the Zumil geothermal station on the flanks of Santiaguito volcano in Guatemala. Hundreds of tonnes of rock, mud and steam were blown into the atmosphere when the well 'blew its top', apparently because of gravitational slippage of the ground beneath the site.

Longer-term effects of geothermal production include ground subsidence, induced seismicity and, most important, gaseous pollution. In dry steam fields, where the reservoir pressures are relatively low and the rocks are self-supporting, subsidence is rare (as at the Geysers and Larderello). But significant reduction of the higher pressures in liquid-dominated systems, for example due to inadequate fluid reinjection, can induce subsidence, usually on the millimetre to centimetre scale (although maximum subsidence of 3 metres has occurred at Wairakei as a result of early exploitation without reinjection). Reductions in reservoir pressure can also have an adverse effect on natural manifestations (geysers, hot springs) which are a common accompaniment of high-enthalpy fields and often important to the local tourist industry. Such concerns have severely restricted the development of power generation in Japan.

The question of whether there is induced seismicity around geothermal sites has been much debated, and it must be recognized that most steam fields are located in regions already prone to natural earthquakes. There is evidence that fluid injection lubricates fractures and increases pressures, creating small earthquakes (microseismicity), especially when reinjection is not at the same depth as the producing aquifer (mainly for reasons of fluid disposal). However, in cases where reinjection is designed to maintain reservoir pressures, seismicity is not greatly increased by geothermal production.

Geothermal 'pollutants' are chiefly confined to the non-condensable gases: carbon dioxide (CO_2), with lesser amounts of hydrogen sulphide (H_2S) or sulphur dioxide (SO_2), hydrogen (H_2), methane (CH_4) and nitrogen (N_2). In the produced water there is also dissolved silica, heavy metals, sodium and potassium chlorides and sometimes carbonates, depending on the nature of the water-rock interaction at reservoir depths. Today these are almost always reinjected and this also removes the problem of dealing with waste water. Traditionally, geothermal fields have received a bad press on account of their association with the 'rotten eggs' smell of H_2S. However, this and other gaseous products of old leaking plant have now been reduced so that the environmental impact of thermal production is at a minimum. Modern plants are fitted with elaborate chemical systems to trap and destroy H_2S. Interestingly, the level of atmospheric H_2S over the Geysers field is now lower than that produced from hot springs and geysers before geothermal developments began. Nevertheless, the image of polluting geothermal systems has slowed developments at several new sites. For example, environmental legislation covering the Miravalles plant, located on the periphery of a rainforest conservation area in northern Costa Rica, delayed completion of the plant for four years. A project on Mount Apo on Mindanao Island in the Philippines was turned down by the World Bank and the Asian Development Bank on social and environmental grounds. Objectors claimed that 111 hectares of forest would be threatened, 28 rivers polluted and a national park destroyed.

BOX 9.8 Environmental concerns

The Mammoth plant in California, (Figure 9.15) brought on line in the late 1980s, illustrates contemporary reactions to environmental concerns at the planning stage. The plant lies just outside a popular ski resort and exploits a shallow reservoir with abundant but relatively impure geothermal fluid. A fluid volume of about 900 kg s^{-1} is used to produce 40 MW_e in a binary cycle operation. The geothermal fluid is kept under pressure, requiring expensive pumping so that it does not flash, and the pollutants stay in solution. Because the producing reservoir has high sustainable flow rates, reinjection is directed considerably beneath the reservoir and is done solely to avoid pollution. The Mammoth plant also incorporates an air cooling system for the post-turbine isobutane gas, so avoiding all forms of visible gas emission from the plant. This is a costly form of condensation system and environmental restrictions on the height of the cooling towers mean that on warm days condensation is inefficient, thus reducing the electrical power output. The plant is coloured green to blend with the landscape and a screening embankment with trees reduces the visible impact from the Mammoth Lakes ski resort.

Despite all these environmental restrictions, this low-level, totally self-contained and emission-free geothermal plant makes a profit through fixed price contracts with the Southern California electric utility company.

Figure 9.15 The Mammoth ORC Geothermal plant

The position over emissions of carbon dioxide, an important greenhouse gas, is rather more complicated. Geothermal reservoirs often contain significant quantities of CO_2, so emissions from the power plant will also be higher in CO_2 than might have been expected. On the other hand, exploitation of the field often reduces natural emissions. Leaving aside that possible benefit, however, a recent survey carried out by the International Geothermal Association (IGA, 2002) shows a wide variation in CO_2 emissions from existing plant, ranging from 4 g per kWh to 740 g per kWh. The weighted average is 122 g per kWh. Typical CO_2 emission rates from fossil-fired power stations range from 460 g per kWh for the most up-to-date natural gas fired combined cycle plant to 960 g per kWh for the best coal-fired stations (see the companion text, Chapter 13, Table 13.10).

Overall, the facts are now quite plain – geothermal developments have a net positive impact on the environment compared with most conventional energy systems because of their much smaller pollution effects. In producing the same amount of electrical power, modern geothermal plants emit significantly less CO_2 than do the cleanest fossil fuel units. Figures for the acid rain gas SO_2 and for particulates are less than 1% and 0.1% respectively. In terms of social developments, geothermal plant requires very little land, taking up just a hectare or so for plant sizes of 100 MW_e or more. Geothermal drilling, with no risk of fire, is safer than oil or gas drilling and, although there have been a few steam 'blow out' events, there is far less potential for environmental damage from drilling accidents. In direct use applications, geothermal units are operated in a closed cycle, mainly to minimize corrosion and scaling problems, and there are no emissions. So while the acidic briny fluids are corrosive to machinery such as pumps and turbines, these represent technological challenges rather than environmental hazards.

There is an interesting social footnote, which emphasizes that geothermal energy developments are not always entirely benign when viewed at short range. Not surprisingly, drilling of direct use geothermal wells in the heart of the Paris suburbs led to conflicts over noise pollution. During the preparation of a new site in 1980, for example, an enormous poster was hoisted on an apartment block which nicely summarized the Parisian attitude: '*Oui à la géothermie, non aux nuisances*'. It is worth remembering, however, that once the drilling and installation is complete, there is often little or nothing to see, no noise and no emissions. All the equipment is contained within the boiler room of the apartment blocks or in a small building little larger than a garage.

9.5 Economics and world potential

On an international scale, geothermal is one of the most significant 'renewable' energy resources. Its strength in this respect is that it can provide firm, predictable power on a 24 hour per day basis.

This high availability distinguishes geothermal from many other renewables, and results in significantly greater amounts of energy supplied for a given installed capacity.

Table 9.4 Performance of typical geothermal power plant

	Italian 60 MW	Italian 20 MW	Japanese 50 MW
Year	**1999**	**1999**	**1/4/97–31/3/98**
Installed capacity (MW_e)	60	20	50
Maximum load (MW_e)	55	17	48.3
Annual produced electricity (MWh)	462 845	142 248	361 651
Hours of operation of plant	8748	8483	8112
Load factor (%)	96.1	95.5	85.5
Availability factor (%)	99.9	96.8	92.6

(from IGA, 2001)

There was quite spectacular growth in geothermal installed capacity of approximately 14% p.a. following the oil embargoes of the early 1970s, at a time when conventional generating capacity grew at between 0 and 3% p.a. Stabilization of oil prices brought the growth rate down to about 8% p.a. by the early 1990s and cheap natural gas together with liberalization of electricity markets reduced rates to 3% in more recent years. This trend seems now to have reversed, driven in part by environmental concerns over fossil fuel usage. The underlying trend is again close to 10% p.a., though it has been masked in recent years by the downturn in production of the Geysers.

The Geysers field (Table 9.5) provides a useful illustration of historical economic improvements engendered by falling plant costs as planned lifetimes and operational returns have increased. These *costs* may be compared with wholesale electricity *prices* which, for most of the developed world, are generally in the region of US¢5–9 per kWh (*IEA, 1998*).

Table 9.5 The Geysers field: improvements in cost-effectiveness

	Capital cost per kW of capacity	Operational / maintenance costs (US¢/kWh)	Overall energy production cost (US¢/kWh)
1981	$3000	4.0	8.5
1991	$2600	2.2	5.7

The underlying trend in these figures has continued as advanced drilling, exploration and conversion technologies have been introduced, so that production costs (calculated on the same basis) would now be about US¢4 per kWh. The data for the Geysers field have been complicated, however, by the production problems already mentioned and the need to introduce reinjection, which has added to capital costs. Nevertheless, the climate for geothermal development continues to improve. In New Zealand, for instance, where the electricity industry was privatized during the 1990s and different fuels now have to compete on equal terms with hydroelectricity, a new privately-owned 55 MW_e plant was commissioned in 1996 and a second in 1999. Sharing of experience and R&D costs among the different operators will be a vital factor in achieving targets. The obvious economic advantages of high-enthalpy resources in providing a good return on capital have stimulated loan investments by international agencies such as the World Bank in geothermal developments, especially in Central and South American countries. But perhaps the greatest economic gain to society in general lies in the 100 million barrels of oil a year which is already being saved. Moreover, in some areas, geothermally generated steam is cheap (e.g. $2.50 per tonne in Iceland compared with $15.00 per tonne from oil-fired boilers).

Although geothermal resources make a significant contribution in some high-enthalpy areas (e.g. 8% of energy usage in Costa Rica, 1.5% in Italy, 10% in New Zealand), the total amount of geothermal electricity produced in 1998 (some 46 TWh) accounted for only about 0.35% of global electricity consumption. Yet the long-term potential is much higher, especially in volcanically active countries, and may be realized as the technology improves.

The economics of lower-grade geothermal resources are much more marginal and depend on local political and economic conditions, such as the availability and price of fossil fuels, the willingness of governments to invest in new energy concepts, the degree of environmental awareness and the related tax incentives to promote 'clean' energy commercially.

The economics of future HDR developments are speculative at present, and will remain so until the technology is fully demonstrated. The best estimates, derived from recent progress in reservoir development and reductions in drilling cost, and for sites with a mean temperature gradient of 35–40 °C km^{-1}, are that electricity might be produced for about ¤0.10–0.15 per kWh in the early pilot plant, reducing to half these figures for a multi-module commercial system. There is an important caveat, however: these estimates come from **cost models** not **financial analysis**, and the distinction is important. **Financial analysis** (see Appendix 2, and the companion text, Chapter 12) can be applied to an existing operation or to a technology that is proven; all the steps in the process are known, and the costs of each step can be calculated. This information can be used to derive the break-even cost of the product in an unambiguous way. **Cost modelling**, on the other hand, is usually applied to an unproven technology – often one that is still being developed It examines the possible costs of each step in terms of assumptions about the performance of the step itself and each preceding step. It says only that '*...if the technology performs in this way, then the cost will be...*'. The result is only as good as the initial assumptions. Cost modelling is a useful tool for setting the targets that various elements of the technology must achieve, or for establishing which aspects of the research offer the best opportunities for improvement, *but it does not predict prices*. This is a common error, and one that HDR research has suffered from, particularly in the UK.

Used correctly, however, such analyses can be very useful. Cost analyses of this type have been carried out for two-well systems by all the research teams involved in HDR, and resulted in general agreement on the target parameters to be achieved for a two-well HDR reservoir, aiming to produce 200 °C water for power generation over a 20 year life:

Table 9.6 Target parameters for a 2-well HDR system

Flow rate	75–100 kg s^{-1}
Effective heat exchange area	> 2×10^6 m^2
Accessible rock volume	> 2×10^8 m^3
Impedance *	0.1 MPa l^{-1} s^{-1}
Water losses	< 10%

* Note: strictly, impedance is not a constant, but varies with flow rate (pressure). The specified figure is the resistance to flow at the operating rate

Until recently, none of the projects had come close to achieving simultaneously the required flow, impedance and loss rate. In 1997/8, however, the Soultz project demonstrated a closed loop circulation for 4 months with zero losses and impedance close to the necessary target. It is on the basis of extrapolating the findings from this experiment that the previously mentioned cost estimates are derived. It is still worth repeating, however, that these figures come from cost modelling, not financial analysis. Although we can make reasonable estimates of the capital costs of HDR schemes, we still do not know for certain how to construct a good reservoir, so we can only base our estimates on what its performance and properties ought to be. Predictions of HDR costs are arguably premature until further technological developments have provided better reservoir performance data.

9.6 Geothermal potential in the United Kingdom

Sedimentary basin aquifers

As in many other countries, it was the oil crisis of the mid-1970s that spurred geothermal resource evaluation in the UK. By 1984, new maps of heat flow (Figure 9.16(a)) and of promising geothermal *resource* sites (Figure 9.16(b)) had been produced. Three radiothermal granite zones stand out with the highest heat flow values, but significant heat flow anomalies also occur over the five sedimentary basins identified, partly because these are regions of natural hot water upflow. Many shallow heat flow boreholes were drilled during this period, together with the four deep exploration well sites of Figure 9.16(b) and Table 9.7.

In each case the main aquifer is the permeable Lower Triassic Sherwood Sandstone (named after its most notable outcrops in the East Midlands).

Table 9.7 Characteristics of deep exploration well sites

Location	Completion	Well depth (m)	Bottom temp.	Main aquifer depth (m)	Temperature
Marchwood	Feb 1980	2609	88 °C	1672–1686	74 °C
Larne	July 1981	2873	91 °C	960–1247	40 °C
Southampton	Nov 1981	1823	77 °C	1725–1749	76 °C
Cleethorpes	June 1984	2092	69 °C	1093–1490	44–55 °C

The shallower intersections with this aquifer at Larne and Cleethorpes are at a rather low temperature for geothermal exploitation but have reasonably high flow rates because of the large aquifer thickness. Unfortunately, the other two wells, which intersect the aquifer at a better temperature, produce rather low flow rates because of the restricted vertical height of good aquifer rock. The yield is reduced not just because the sedimentary sequence is thinner than in the Southampton area, but also because much of the Sherwood Sandstone proved to be more highly cemented and therefore less permeable. The resources are nevertheless substantial (Table 9.8).

Table 9.8 Potential UK geothermal energy resources at different temperatures

Basin	Potential resource at 40–60 °C (10^{18} J)	Potential resource at >60 °C (10^{18} J)
East Yorks and Lincs	26.2	0.2
Wessex	2.8	1.8
Worcester	3.0	–
Cheshire	8.9	1.5
Northern Ireland	6.7	1.3
TOTALS	**47.6**	**4.8**

Assumptions behind these estimates are that the 40–60 °C resource would be exploited with the use of heat pumps, producing a reject temperature of 10 °C, whereas the >60 °C resource would not, giving a 30 °C reject temperature. The latter resources could be doubled with heat pumps. For comparison, current UK electrical energy demand is around 350 TWh per year (1.3×10^{18} J per year, or about 30 GW_e as equivalent continuous power), so we are considering here quite large resources of renewable energy, but as heat rather than electricity.

So why are geothermal aquifers not being exploited much more widely?

The problem is not just one of marginal economics and geological uncertainty, but rather the mismatch between resource availability and heat load, itself a function of population density. More than half the UK resources are located in east Yorkshire and Lincolnshire, essentially rural areas lacking the concentrated populations of the Paris basin. While electricity can readily be transported over long distances from source to market, hot water is more of a problem. Transmission distances for the latter under UK conditions are likely to be restricted to just a few kilometres so the resources are likely to remain undeveloped. It is interesting to note, however, that the distance limitation arises from cost rather than heat loss; some Icelandic pipelines exceed 50 km in length with a temperature drop of only about 1 °C.

Hot dry rocks

Of the three principal granite zones in the eastern Highlands, northern England and south-west England, the latter is characterized by the highest heat flow, as shown in Figure 9.16(a). However, large areas of the more northerly granite masses are covered by low thermal conductivity sedimentary rocks and so temperatures will be higher at depth than if the granite bodies came to the surface (Figure 9.8). Substantial areas of Cornwall

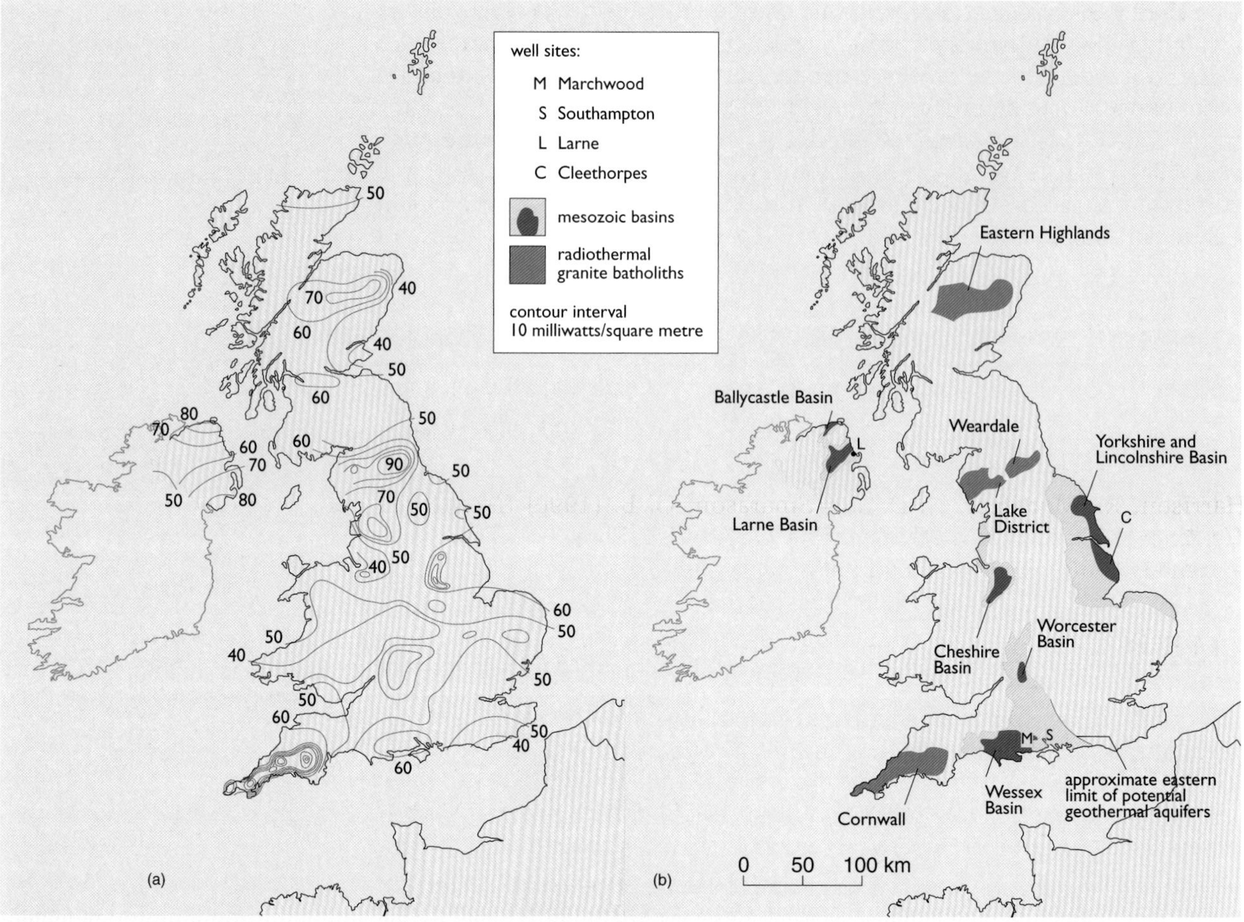

Figure 9.16 (a) Heat flow map of the UK based on all available measurements to 1984 compiled and published by the British Geological Survey; (b) Distributions of radiothermal crystalline rocks (granites) and major sedimentary basins likely to contain significant geothermal aquifers in the UK. (Sites of four major wells already drilled are indicated in the key

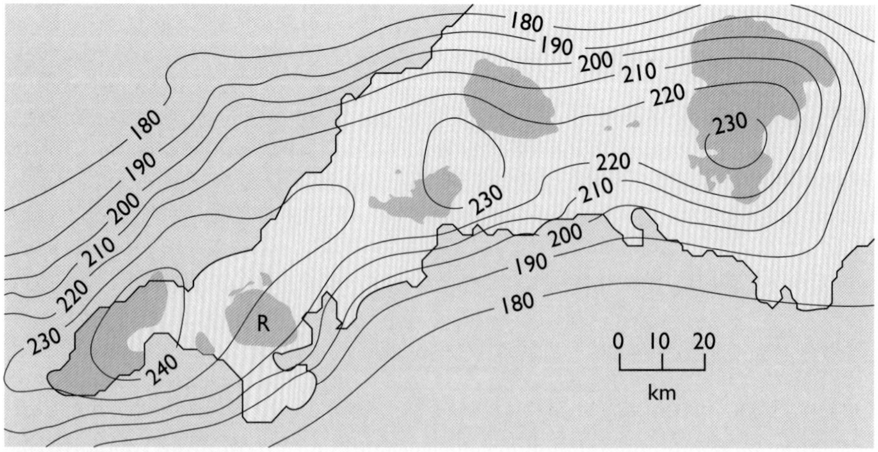

Figure 9.17 Projected temperatures (°C) at 6 km depth beneath south-west England. Granite bodies that crop out at the surface are shaded
(R = Rosemanowes)

and Devon are projected in Figure 9.17 as having temperatures above 200 °C at 6 km depth and it has been estimated that the HDR resource base in south-west England alone might match the energy content of current UK coal reserves. One estimate suggests that 300–500 MW_e (about 10^{16} J a^{-1}) could be developed in Cornwall over the next 20–30 years with much more to follow later. If and when the technology has been successfully demonstrated at Soultz, it seems probable that developments in Cornwall will follow soon after.

References

Curtis, R. (2001) 'Earth energy in the UK', *Proc. International Geothermal Days 'Germany 2001' conference,* Bad Urach. Available in PDF format on www.uni-giessen.de [accessed 23 September 2003].

Harrison, R., Mortimer, N. D. and Smarason, O. B. (1990) *Geothermal Heating: A Handbook of Engineering Economics,* Pergamon Press. [An extremely comprehensive account with particular focus on technology and economics; good general introduction chapter.]

Huttrer, G. W. (2000) 'The status of world geothermal power generation 1995–2000', *WGC 2000*, pp.1–22.

IEA (1998) *Key World Energy Statistics,* International Energy Agency annual publication.

IGA (2002) 'Geothermal power generating plant: CO_2 emission survey', *IGA News*, no. 49, July–September 2002.

IGA (2001) 'Performance indicators for geothermal power plant', *IGA News*, no. 45, July–September 2001.

Lund, J. W. and Freeston, D. H. (2000) 'World-wide direct uses of geothermal energy 2000', *WGC 2000*, pp. 23–38.

McLarty, L., Grabowski, P., Entingh, D. and Robertson-Tait A. (2000) 'Enhanced geothermal systems R&D in the United States', *WGC 2000*, pp. 3793–96.

Rybach, L. and Eugster, W. (1998) 'Reliable long-term performance of BHE systems and market penetration – the Swiss success story', *Proc. 2nd Stockton International Geothermal Conference*, Richard Stockton College, New Jersey, March 1998. [See also Rybach, L. and Eugster, W. (2000) 'Sustainable Production from Borehole Heat Exchanger Systems', *WGC 2000*, p. 825–30, for an extended and updated analysis.]

Rybach, L., Mégel, T. and Eugster, W. J. (2000) 'At what time scale are geothermal resources renewable?', *WGC 2000*, pp. 867–72.

Further reading

The world wide web

A great deal of information on geothermal resources and their exploitation is now available on the World Wide Web. The following sites are particularly useful, and offer further links:

International Geothermal Association (IGA) – http://iga.igg.cnr.it [accessed 23 September 2003]. This site offers an e-mail enquiry service, allowing questions about geothermal energy to be directed to a large cross-section of the IGA membership. A quarterly newsletter, *IGA News*, is also available on this site.

IGA European Branch – www.geothermie.de/iga european branch forum.htm [accessed 23 September 2003].

Geothermal Resources Council (GRC) – www.geothermal.org/ [accessed 23 September 2003]. A US-based association for geothermal researchers and practitioners, closely linked to the IGA.

GeoHeat Centre – geoheat.oit.edu/ [accessed 23 September 2003]. Based at the Oregon Institute of Technology, GHC focuses on direct uses of geothermal energy. It publishes a useful quarterly newsletter, many articles from which can be downloaded in PDF format from the GHC web site. Of particular interest to European readers was the content of vol. 22, no. 2 (June 2001): 'Geothermal Use in Europe'.

Ground Source Heat Pumps: www.earthenergy.co.uk/casestudies.html [accessed 23 September 2003].

European Hot Dry Rock Project (Soultz-sous-Forêts): www.soultz.net/ [accessed 23 September 2003].

Conference proceedings

IGA and GRC organize jointly a World Geothermal Congress every 5 years. WGC 2000 was held in Japan in May–June 2000. The Proceedings (*Proceedings World Geothermal Congress 2000, Kyushu – Tohoku, Japan, May 28 – June 10, 2000)* have been published on CD-ROM which presents some 670 papers in PDF format (available from the IGA Secretariat, c/o ENEL/ERGA, Via Andrea Pisano 120, I-56100 Pisa, ITALY; e-mail: igasec@.enel.it). In the references given above, this conference is designated *WGC 2000*.

Background material

Armstead, H. C. H. and Tester, J. W. (1987) *Heat Mining*, Chapman and Hall. An authoritative text addressing HDR, based mainly on experiences in the USA.

Downing, R. A. and Gray, D. A. (eds) (1986) 'Geothermal Energy: The Potential in the United Kingdom', London, HMSO. A comprehensive account of a full decade of UK research into aquifers and HDR including much more technical data than it is possible to provide in this section.

Garnish, J. D., Vaux, R. and Fuller, R. W. E. (1986) 'Geothermal Aquifers – summary of research on UK aquifers – Department of Energy R & D Programme, 1976–1986', *ETSU Report R-39*, London, HMSO.

Popovski, K. and Sanner, B. (eds) (2001) *International Geothermal Days 'Germany 2001'*, comprising the Proceedings of an International Workshop on Balneology and 'Water' Tourist Centres, an International Course on Geothermal Heat Pumps and an International Seminar on Hot Dry Rock Technology. Available on CD-ROM (ISBN 0-567-98766-4) and in PDF format on the University of Giessen web site at: www.uni-giessen.de [accessed 23 September 2003].

Chapter 10

Integration

by Bob Everett and Godfrey Boyle

10.1 Introduction

In the year 2000, renewable energy contributed about 18% of the world's primary energy needs (see Chapter 1). This proportion is likely to increase substantially throughout this century as concerns about climate change and declining fossil fuel reserves are addressed. The current energy situation in the UK and several other countries, each with different circumstances, has been described briefly in Chapter 1, and in much more detail in the companion text *Energy Systems and Sustainability*.

In Chapters 2–9, we have reviewed each major renewable-energy source in turn, with estimates of the contributions that each could make to future needs. Modern industrial society demands energy in extremely large quantities and in many different forms. In order to meet these requirements, a vast, world-wide network of energy supply and distribution systems has been built up. To what extent will these existing energy networks need to be modified and supplemented if intermittent renewables are to make an increasing contribution? Can renewables deliver our energy, not only in significant *amounts* and at an acceptable *price*, but also in the right *form*, at the right *time*, and in the right *place*? What are the factors that currently make it difficult for renewable-energy sources to compete with conventional sources? What can governments do to promote their use?

These are some of the questions we shall be addressing in this final chapter. We shall return to many of the topics discussed in earlier chapters, and try to draw some of the threads together. In general, we concentrate on the situation in the UK, before looking further afield.

Firstly, we look briefly at the existing energy systems and how they currently supply our various demands (Section 10.2). Next, in Section 10.3, we discuss the notion of a 'resource' or 'potential' for each renewable-energy technology and how practical and economic constraints impose limits on the contributions that each might make. Sections 10.4 and 10.5 look at these technologies in more detail, investigating the extent to which they can deliver energy in suitable forms, in the right place and at the right time. This includes the difficult problem of integrating electricity from renewable-energy supplies into the national grid systems. In Section 10.6, we consider some possible system solutions, including the adoption of hydrogen as an energy carrier and a transport fuel.

Figure 10.1 The existing energy infrastructure has replaced traditional renewable energy technologies, such as this windmill in the Netherlands - how can new ones be integrated?

Section 10.7 is concerned with the problem of balancing the costs of renewable energy (and energy conservation) against estimates of the 'external' environmental and social costs of conventional energy forms. Then, Section 10.8 looks at the methods that the UK and other governments have used to promote renewable energy over the last 30 years. Finally, Sections 10.9–10.10 describe some 'energy scenarios' that have been constructed by governmental and non-governmental organizations in order to explore the various possible patterns of energy supply and demand in the future – and in particular the roles that renewables might play in them.

10.2 The existing UK energy system

Energy flows

Renewable energy has to fit into the existing patterns of energy use. In the current UK system, delivered energy forms are of five main kinds:

- *liquid fuels*: almost entirely oil and its derivatives – petroleum (gasoline), diesel, kerosene, etc.,
- *gaseous fuels*: mostly methane ('natural gas'), plus some 'bottled' gas (propane and butane),
- *solid fuels*: almost entirely coal and its derivatives, but with small contributions from fuel wood, refuse incineration and straw burning,
- *electricity*: almost all from fossil-fuelled or nuclear power stations, with a small contribution from hydroelectricity and other renewable sources,
- *heat*: although this is only done to a limited extent in the UK, some energy is distributed to buildings directly in the form of hot water or steam. This is provided by centralized boilers in district heating (DH) (also known as community heating) schemes, or waste heat from power stations in combined heat and power (CHP) schemes. It can also include heat from geothermal sources.

These forms of energy flow into the various distribution networks and are delivered to final consumers in the main energy-using sectors of the economy, normally categorized as *domestic*, *services*, *industry* and *transport*.

Within each of these sectors, this delivered energy is used to provide the *energy services* that we actually need. These can be categorized under the broad headings of heat, motive power and electricity-based services.

Heat is required in many forms – from tepid water to superheated steam – for washing, cooking, space heating and industrial processing. It can be provided either by burning fuels close to where the heat is needed, by piping heat in from a more distant CHP or DH plant, or by using electricity, either in resistance heaters or in more sophisticated devices like microwave ovens.

Motive power is needed, obviously, for transport (cars, trucks, buses, trains, ships and aircraft, etc.) and to drive machinery. In most cases, the form of delivered energy currently used to supply it is oil. However, a significant proportion of energy for motive power is in the form of electricity – some of it for electric railways, but a great deal in the almost unnoticed but ubiquitous form of power for electric motors. These are now used in enormous numbers for everything from watches and video recorders to lifts and heavy industrial machinery.

Electricity-Based Services include heat, lighting and motive power, each of which can also be provided by other forms of delivered energy.

Nevertheless, electricity is *essential* for those systems that simply cannot function with any other energy form – not just computers and communications systems but also more specialized electro-chemical processes such as the manufacture of aluminium or chlorine.

Distribution

It is worth dwelling a little on the changing nature of our energy distribution systems. Most of our energy demand actually occurs within the relatively small areas of our major cities. Many of these are built in sheltered inland valleys, deliberately to *avoid* the worst excesses of wind and wave energy. Many large industrial towns grew up around their fuel supplies, coal or wood. Some faded to obscurity when their fuel ran out. Others survived by importing fossil fuels.

At the end of the nineteenth century, the UK was a major producer and exporter of coal. Most European cities ran mainly on coal, brought by rail or by sea. It was burned immediately where it was needed, or was converted into electricity or 'town gas' in local plants. The precise forms in which we make use of electricity (as a 230 volt AC supply) and gas (as a low pressure piped supply) had largely been fixed by the beginning of the twentieth century. Yet, at that time, oil as we know it today scarcely existed as a commercial commodity.

Today coal, oil, natural gas and electricity are distributed not just country-wide but internationally. Coal, once the major fuel in the UK, is now mainly just a fuel for power stations. Some is still locally mined, but much is imported by ship from countries such as Australia and South Africa.

The UK was a major importer of oil until the 1970s when discoveries in the North Sea made it not merely self-sufficient but a net exporter. There is an extensive internal network of pipelines conveying oil around the country. North Sea production is now declining and it is likely that the country will again become a net oil importer after 2010, requiring oil to be brought in by sea and pipeline.

Until the 1960s, the UK gas supplies were 'town gas' made from coal. Then the large discoveries of natural gas in the North Sea revolutionized the industry. Hundreds of small production plants were closed down and a complete new infrastructure of long-distance distribution pipes was built to bring in the North Sea gas. Although its higher energy density meant that all gas boilers and cookers had to be fitted with new burners, it also meant that the energy carrying capacity of the existing pipework was doubled. This was useful since there was an immediate rise in sales of gas-fired central heating systems. Natural gas is *the* fuel of the moment and probably will remain so for the next few decades. Its low carbon content compared to coal has led to the construction of a whole new generation of gas-fired power stations, which has increased demand even further.

Looking further afield, there is an impressive Europe-wide system of pipelines for gas distribution (Figure 10.2), stretching from gas fields in Algeria in the south to Norway in the north, from Siberia and the Caspian Sea in the east to the Republic of Ireland in the west. Natural gas can also be liquefied and transported by ship.

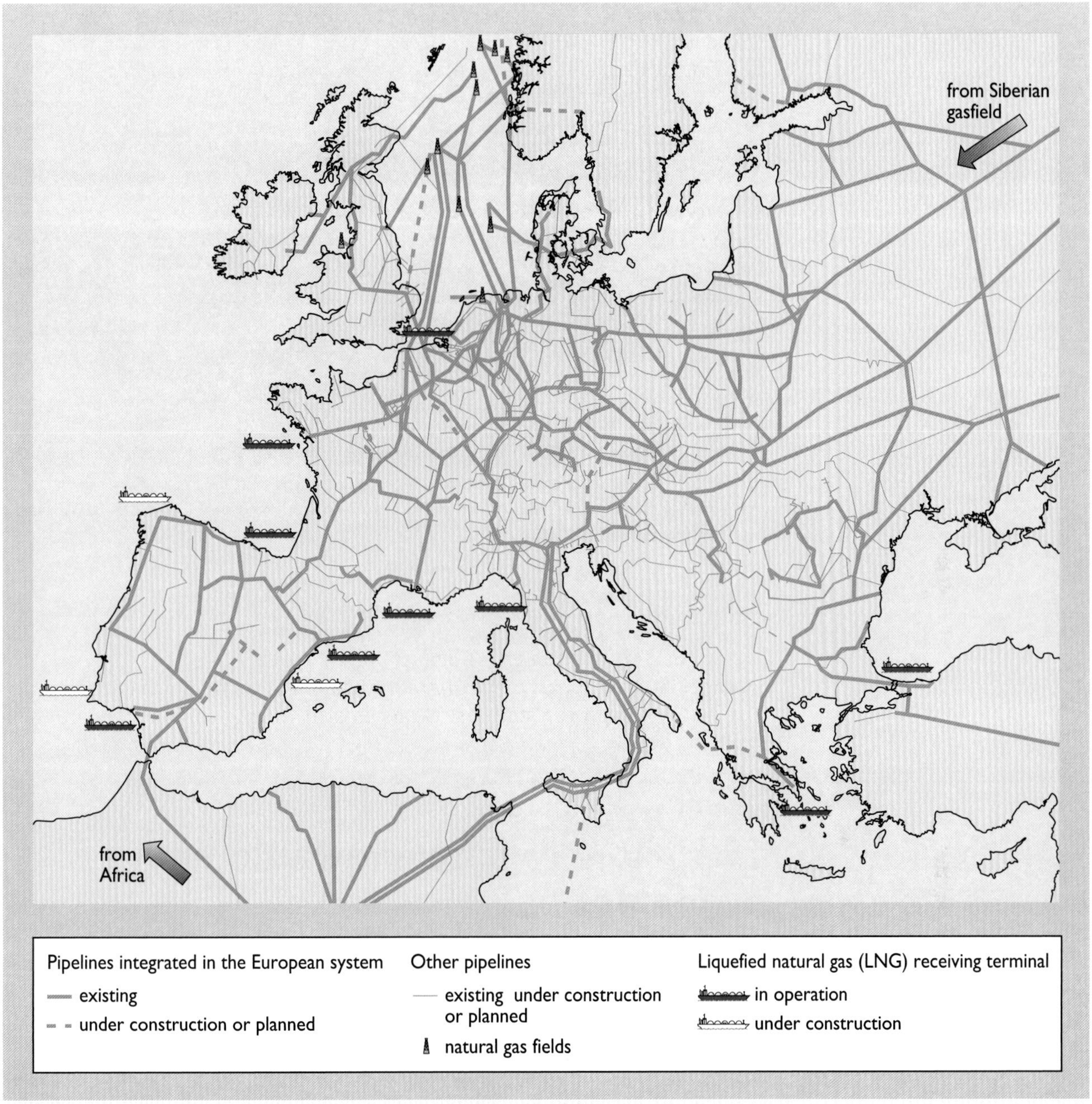

Figure 10.2 The Europe-wide network of natural gas pipelines distributes supplies from many sources including the North Sea, North Africa and Siberia. In addition, liquefied natural gas (LNG) and liquefied petroleum gas (LPG) are transported by ship (source: Eurogas 2001)

This gas infrastructure is continuing to grow rapidly. The European Commission has estimated (CEC, 2001) that European gas consumption will grow by 60% by 2020. It is likely that this will coincide with a period of decline of North Sea gas production and that from 2005 the UK will be start becoming a net importer of gas (PIU, 2002). The Commission estimates that in excess of 200 billion euros will have to be spent in new infrastructure to supply natural gas from other sources, mostly to the east. A major pipeline such as that shown in Figure 10.3 can cost £1 million per kilometre to lay.

Figure 10.3 A high-pressure gas pipeline being laid. Such pipes can carry several gigawatts of power, i.e. a similar amount to that carried by a 400 kV electricity power line (photo courtesy of Transco)

Turning to electricity, today it is mostly generated in very large plants, usually of 500 MW or more output, and in the UK it is rare to find one really close to a major city. The output is distributed to consumers via the National Grid, a network of high-voltage cables that covers the UK, with links to neighbouring countries. Since they are likely to be major carriers of renewably-generated electricity in the future, we will look at the National Grids of the UK and the Republic of Ireland in more detail later.

10.3 How much renewable energy is available?

Let us now consider what contribution could be made by renewable energy, particularly in the UK. It is easy to make sweeping statements about the very large natural energy flows available. These form the total 'available resource'. Defining how much energy a particular source or a specific technology might actually supply in practice requires looking at the various constraints, technical, social and economic, on its use. These whittle down the 'resource' or 'potential' towards something more realistic. A possible ranking of such constraints is given in Box 10.1. The 'economic potential' at the end is likely to be far smaller than any notion of a total available resource (see Figure 10.4).

In recent years, the UK Department of Trade and Industry (DTI) has carried out many detailed assessments of the future potential for different renewable-energy technologies. Future electricity supplies are of particular interest as the UK government has set a target of a 10% contribution by 2010. UK electricity demand in 2001 was 340 TWh, of which renewables only supplied about 10 TWh (under 3%). Large-scale hydro produced 3.8 TWh and wind power only about 1 TWh. The remainder came from biofuels (DTI, 2002b).

BOX 10.1 Resource size terminology

Firstly, we must make a distinction between finite *fossil fuel reserves* (as discussed in the companion text) and *renewable-energy resources*. World oil reserves, for instance, are the total quantity of oil known to be in place and accessible under specified conditions. Such a 'total' would obviously be meaningless for wind power or solar energy, and for these and the other renewables, it is customary to specify their potential ***annual*** *contributions*. It is important to bear in mind this distinction when comparing the potential contributions from the fossil and renewable sources.

There are many definitions of the annual **resource** or **potential** of a specific form of renewable energy, differing mainly in the extent to which they take into account the practical and economic limitations to its use. The following terms have all featured in recent UK energy literature.

Available resource (or **total resource**) – the total annual energy delivered by the source; for example the total energy carried by ocean waves or the wind, or the total incident solar energy.

Technical potential (also referred to as **accessible resource**) – the maximum annual energy that could be extracted from the accessible part of the Available Resource using current mature technology. Although this could change with technological advances, it may ultimately be limited by basic laws of physics (determining for instance the properties of wind turbines or the efficiency of heat engines.) It is also limited by basic accessibility constraints due to:

- practical difficulties such as the presence of roads, buildings and lakes,
- institutional restrictions and the need to avoid such areas as National Parks, Sites of Special Scientific Interest (SSSIs), Areas of Outstanding Natural Beauty (AONBs), etc.

Practicable potential (also referred to as the **practicable resource**) – the technical potential reduced by taking into account:

- constraints on using or distributing the energy – such as transportation problems, access to the electricity grid or problems of intermittent supply,
- further limitations on land or technology use due to public acceptability. It may be difficult to quantify these, since they may only become apparent when planning permission is sought and environmental objections are expressed.

Economic potential – the amount of the technical potential that is economically viable. Any judgement about this requires the specification of an acceptable energy price and a discount rate that sets the cost of borrowing money for investment.

These kinds of estimates, together with assumptions about economic conditions, underlie the various energy scenarios described later in this chapter.

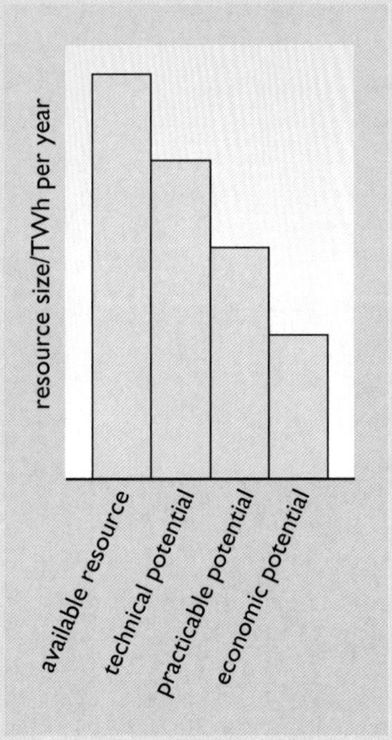

Figure 10.4 A notional ranking of renewable-energy resource sizes – the 'practicable' resource or 'economic' resource may be considerably less than the 'available' resource.

Table 10.1 below shows a simplification of a 1998 view of what might be prospects for renewable electricity generation looking forward to the year 2025. The table excludes large scale hydro because it is considered to be fully developed. Many of the figures are large – the estimated total technical potential for offshore wind power alone is ten times the total UK electricity consumption in 2001. Even taking into account many practical constraints, the total estimated 'practicable potential' from all the sources listed is over 250 TWh yr^{-1}. We will look at the nature of these constraints later.

Table 10.1 Estimates of renewable-energy resource and cost in 2025 for the UK

Technology	Technical Potential TWh yr⁻¹	Practicable Potential TWh yr⁻¹	Cost pence per kWh	Economic potential at this cost TWh yr⁻¹
Building-Integrated PV	266	37	7.0	0.5
Offshore Wind	~3500	100	2.5–3.0	100
Onshore Wind	317	8*–58	<3.5	58
Biomass (energy crops)	'large'	'large'	4.0	33
Wave	600+	50	4.0	33
Tidal Stream	36	1.8	7.0	1.8
Small Hydro	40	3	7.0	1.8
Municipal Solid Waste	13.5	6.5	7.0	6.5
Landfill Gas	7	7	2.5	7

Source: Chapman and Gross, 2001

* the source authors, writing in 2002 (PIU, 2002), felt that the DTI's 1998 figure of 58 TWh yr^{-1}, did not take enough constraints into account and suggested a more modest figure of 8 TWh yr^{-1}

The table also considers economic constraints. Some renewable-energy technologies are undoubtedly cheaper than others; their output has to be compared with the costs of electricity from conventional generation. In 2001 electricity from combined cycle gas turbine (CCGT) plant cost 2–2.3p/kWh and that from nuclear plant 3–4p/kWh (PIU, 2002). Wind turbines sited in the windiest places will produce electricity more cheaply

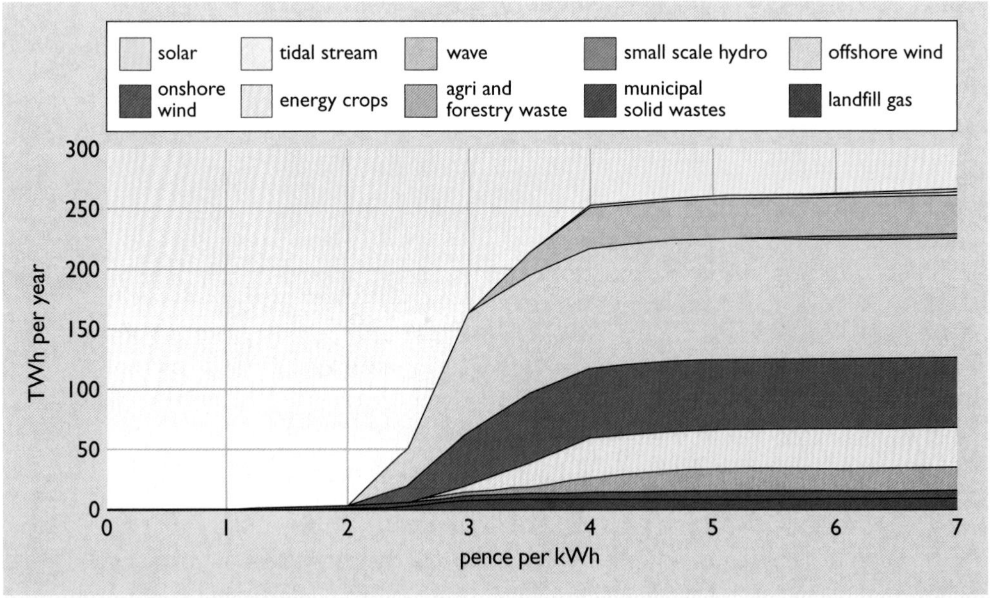

Figure 10.5 Renewable electricity generation: resource-cost curve for the UK in 2025, based on an 8% discount rate (source: ETSU, 1999)

than those in more sheltered locations. Wave power is likely to be more expensive than wind power. Photovoltaic power could be even more expensive still. We do not just have a 'potential for renewables' in isolation: rather, it is a 'potential available at less than a particular price'. The more we are prepared to pay, the greater the potential resource and the wider the choice of options.

Figure 10.5 shows a resource-cost curve for the technologies listed in Table 10.1. The resource is minimal below an electricity price of 2.0p/kWh, but increases rapidly between 2.0p/kWh and 4.0p/kWh and then flattens out. Although it extends to 7.0p/kWh a willingness to pay more than 4.5p/kWh does not significantly increase the resource size – it is limited by *practical* considerations. The bulk of the PV resource does not appear because it costs more than 7.0p/kWh. Note that other technologies, such as tidal barrages have not been included in this particular assessment.

Drawing up such a curve requires making assumptions about the nature of the costing, in particular the discount rate used for borrowing money. The 8% figure used here is more in line with current Treasury thinking than figures of 15% often used in the 1990s. A low discount rate acts in favour of many capital-intensive renewable-energy schemes. This topic is discussed further in the companion text *Energy for a Sustainable Future*.

10.4 Are renewable-energy supplies available *where* we want them?

So the potential contribution of renewables to UK energy needs could be very large. Let us take our first practical constraint – can renewables supply our energy *where* we want it?

It would be helpful if such supplies were located at the points of maximum energy demand, the major cities. Some can be. Solar thermal or photovoltaic panels can be fitted to the roofs of buildings. Passive solar measures are actually a part of the buildings themselves.

Turning to energy crops, these are likely to be available in rural areas, but perhaps not those where food production is seen as more important. Fuels such as wood and forestry wastes need to be gathered and then transported from where they are grown, possibly in remoter country areas, to where they will be used, probably in towns. As we have seen in the Bioenergy chapter, wood has only half the energy density of coal, so transporting it can be a problem – although modern Scandinavian techniques can reduce wood to anything from small chips or pellets to a fine powder. Reducing wood to a dense 'bio-oil' can create a fuel that is more convenient to transport and store, though at the expense of about 15% of its initial energy content.

Municipal solid wastes (MSW), forestry and agricultural wastes are widely used as fuels. In combined heat and power (CHP) plants they are used to make both electricity and heat. In district or community heating plants, they are just used for heat production.

In all these cases, the heat can be distributed through city-wide heating networks as is done in Denmark (see Figure 10.6). Such networks can also be used to distribute geothermal heat where it is accessible, such as in Southampton in the UK.

Figure 10.6 District heating pipes being laid in Denmark. These pipes, produced by Logstor Rør, have a single casing which contains both the flow and return water pipes and some insulation (photo: Logstor Rør)

Biofuels with the potential to be used as petroleum substitutes, such as vegetable oils and sugars, need to be harvested and transported to processing plants, as does the fuel for their processing. Once in the right form they can then be inserted into the existing distribution chain for petrol and diesel fuels. Since transport fuels tend to have a higher energy density and also a higher value to society than fuels for heat, the problems of their distribution are perhaps not so significant. Integrating large-scale renewable electricity supplies into the existing distribution systems is less straightforward and will probably need careful planning.

The present electricity grid

New electricity generating renewables will have to feed into the existing National Grid structures that have been built up over the past 70 years (see Figure 10.7). In England and Wales, the National Grid has developed to distribute electricity to the major cities from the gigawatt-scale coal-fired stations close to the pits in Yorkshire and Wales. Extra links have been added from nuclear stations, most of which are located close to the sea for

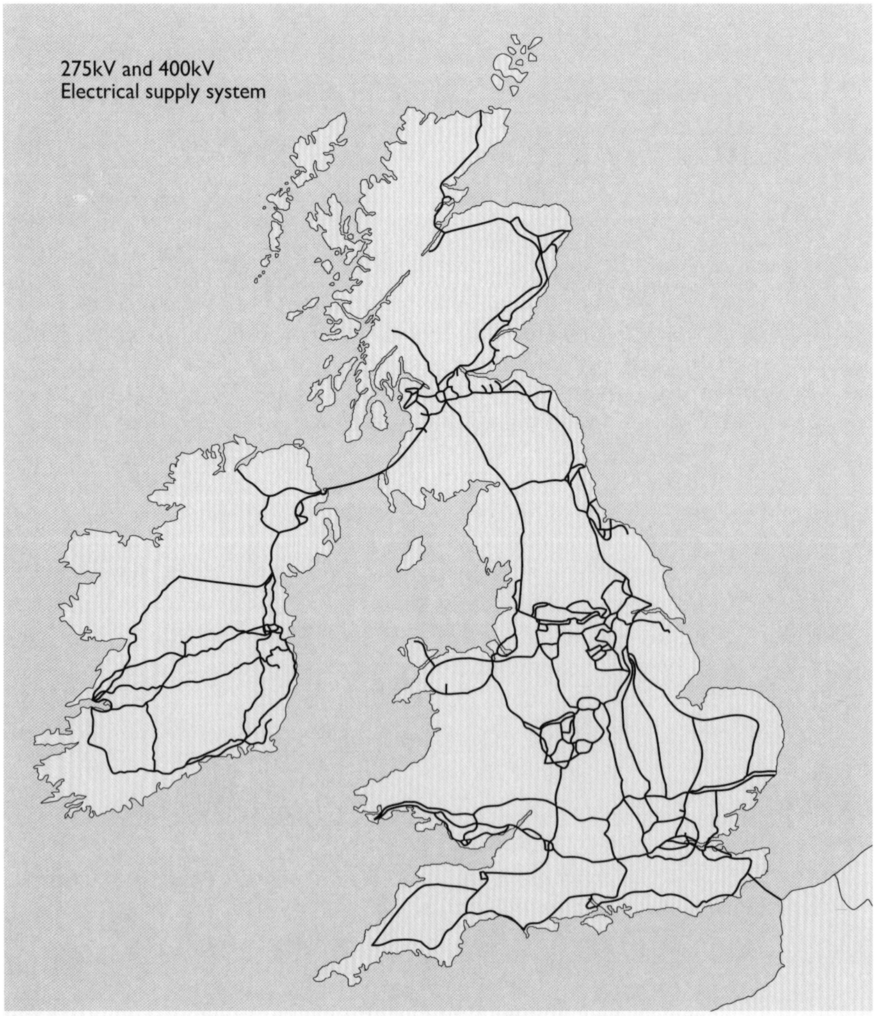

Figure 10.7 The existing UK and Irish electricity grids

access to cooling water, and there is a cross-Channel link to France. The load is becoming increasingly concentrated in the south of England. The electricity system of Scotland is relatively self-contained, with the demand being met by a mix of coal, nuclear and hydroelectric plants, yet it is linked to both that of England and Wales and that of Northern Ireland. This, in turn, is linked to that of the Republic of Ireland. There Dublin, on the *east* coast, forms a major load, in part supplied by power stations running on imported coal and oil sited on the *west* coast.

Wave, wind and tidal power

Figure 10.8 summarizes some of the information from earlier chapters on the locations of the best wind, wave and tidal energy sources in the British

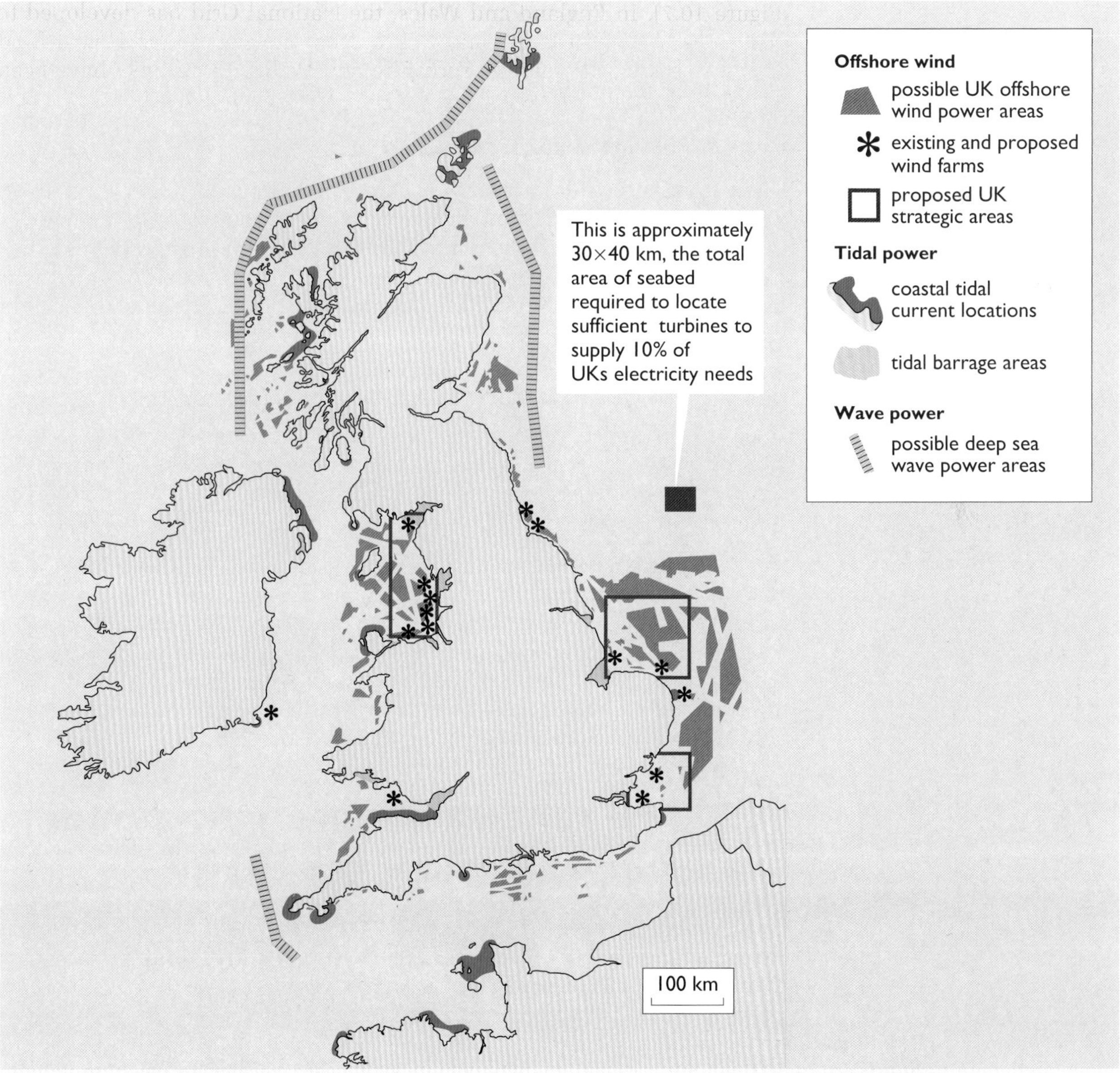

Figure 10.8 Geographical distributions of offshore wind, wave and tidal energy resources in the British Isles

Isles. As can be seen, these electricity-generating sources are not uniformly distributed. Both the UK and the Irish Republic have enormous potential for renewable-energy supplies. Comparison of this map with Figure 10.7 raises the question of how the existing National Grids will need to evolve in order to match the new energy sources to the loads.

The best **wave** power resources are in Scotland, the Atlantic coast of Ireland and the south-west of England. The best **onshore wind** resources are in Scotland, Wales, Cornwall, the north and west of England and the west of Ireland. All of these are areas currently under rapid development.

The prime areas for **offshore wind** development are the shallow waters of the North and Irish Seas. As shown in Figure 10.8, possible areas are necessarily restricted by shipping lanes. Yet it would only require an area of sea 30 km by 40 km to supply 10% of the UK's electricity needs. The DTI have identified three 'strategic areas' for development: in Morecambe Bay on the west coast, and around the Wash and the Thames Estuary on the east coast. These are all conveniently close to existing major grid links. In Ireland, a major offshore wind farm has been approved for construction at Arklow Bank, south of Dublin. This project alone could provide more than 10% of the Republic's electricity.

The potential for **tidal barrages** is concentrated on a few large estuaries, particularly the Severn, though 'lagoon' type structures could be built in more open sea areas. The potential for **tidal current** devices is in similar estuary locations, but there are many possibilities around prominent headlands, notably the 'Race of Alderney' between Alderney and Cap la Hague on the French coast.

As Figure 10.7 shows, the UK and Irish grids mostly do not run in a concentrated 'point to point' manner. They are built as a 'mesh' and some lines are apparently duplicated. This is deliberate, to allow the system to continue operating despite the failure of any one given line. This gives flexibility of operation and potentially allows the insertion of extra sources into the existing system.

There are other pressures on the development of National Grids. The European Union would like to develop the European electricity market. From its perspective the 2 GW link from France to England and the 500 MW link from Scotland to Northern Ireland have been identified as 'bottlenecks', restricting free trade and pushing up prices. National Grids are also a key to dealing with the time variability of many renewable-energy supplies, as will be discussed in Section 10.5 below.

10.5 Are renewable-energy supplies available *when* we want them?

Our demand for energy is not constant. It varies widely over the day, the week and the year. We need more energy for heating buildings in winter than in summer. As a result, the UK consumes three times as much natural gas in a typical December as it does in a summer month.

Generally, in the 'developed' countries, there are few constraints on our demand for energy. Our electricity supply systems are organized in such a way that power is virtually certain to be available whenever we turn on a switch. Gas is always there waiting for us to turn on the cooker. Every major highway has petrol stations at regular intervals ready to serve us when we drive in.

Complex infrastructures have been put into place to enable supply to meet demand. A greater need for, say, gas will set off a whole series of pumps distributing supplies from gas-holder to gas-holder through a nationwide set of mains. Any failure of the infrastructure to deliver energy on demand – be it gas, electricity, heating oil or petrol – causes a consumer outcry.

Biofuels have many of the advantages of fossil fuels. Most of them can be easily stored and used on demand. However, most other renewable sources are intermittent. Wind, wave and solar energy are dependent on weather conditions. They are not completely 'firm' supplies – that is, we cannot absolutely guarantee that they will be available when we need them. Hydro power is also dependent on weather conditions but its availability is improved by the built-in storage provided by the reservoirs that form part of most hydro installations. Tidal power is intermittent, but entirely predictable.

We'll look at the characteristics of these difficulties first with renewable sources of heat and then with the more difficult topic of electricity.

Renewables as heat suppliers

Heat is a relatively easy topic to consider. Individual solar water heaters can be mounted on the roofs of houses or other buildings. They effectively reduce the demand for other forms of heating and do so mainly in the summer.

As pointed out above, a wide range of biofuels and household and agricultural wastes, together with fossil fuel sources, can be used to produce heat for district or community heating networks. Such systems also allow the input of heat from solar sources, heat pumps, geothermal aquifers, or even surplus wind power. Although there have been demonstration schemes in which heat is stored 'interseasonally', from summer to winter, they are expensive. Most heat stores associated with community heating schemes, such as that shown in Figure 10.9, hold only sufficient supplies for a day or so. Using them frees the system designers from concerns of how the heat demand of the local buildings varies over the day, but they do need to think about how it varies over the year.

Figure 10.9 The Pimlico Accumulator in central London holds enough hot water to supply 4000 homes for a day

Municipal solid waste is not a fuel that it is desirable to store for long periods. Whatever the means used to turn it into heat and electricity – mass burning, gasification or pyrolysis – it is best to do so steadily year-round. Other biofuels can be stored and used to meet winter space heating demands.

Experience in many countries has shown that it is economically viable to insulate existing buildings to levels where their space heating demands are dramatically reduced. They can then be heated easily with relatively low-temperature district heating systems and the problems of winter demands are not so serious.

Integrating renewable electricity

At present (2003), this is a matter of considerable concern. In order to meet the UK government's target for 10% of electricity supplies to come from renewable sources by 2010, it is estimated that between 6–8 GW of renewable-energy supply will need to be connected. Meeting further targets of 20% or more could require several tens of gigawatts of capacity.

To understand the formidable problems involved in coping with inputs of such sizes it is worth describing the existing system in some detail. As we will see, to a large extent the practical resource limitations of *where* and *when* overlap.

Electricity: how the current UK system works

Electricity is a much more flexible and valuable commodity than heat, but it suffers from one problem – it is difficult to store. Basically, it has to be generated immediately to suit the demand. In the UK, as in many other countries, electricity demand varies enormously over the year, and hour by hour throughout the day. Figure 10.10 shows sample national daily demand patterns for summer and winter.

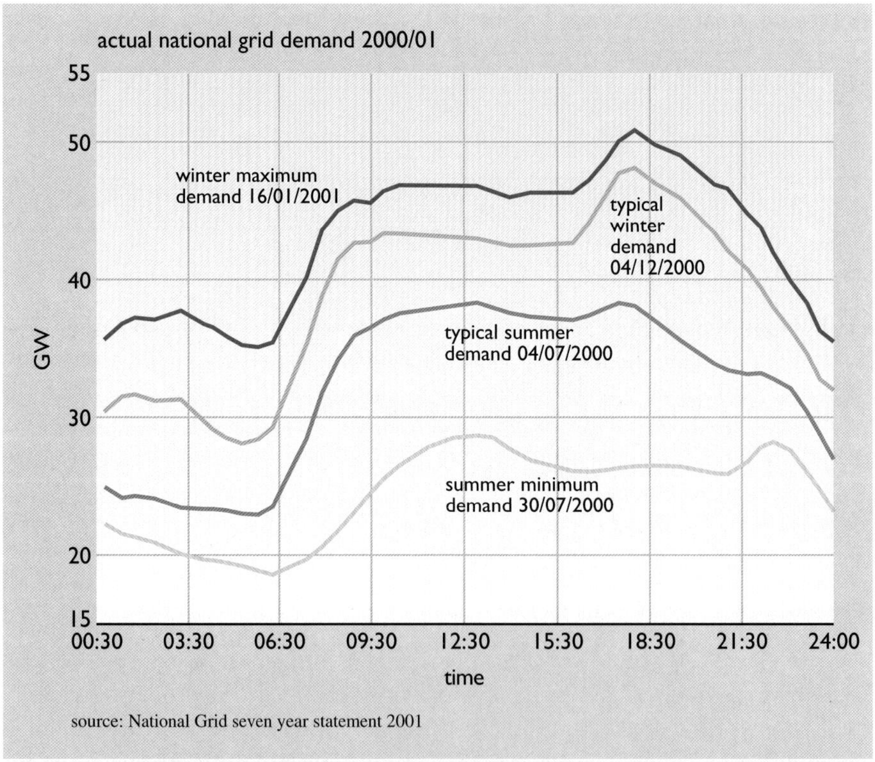

Figure 10.10 Variation of electricity demand on the UK National Grid for some sample days during 2000 and 2001 (source: National Grid Co.)

At night, demand is relatively low (in summer below 20 GW), but it picks up very rapidly early in the morning and flattens out during the day. In winter months there is a pronounced peak (sometimes reaching over 50 GW) in the early evening.

We can look at electricity demand on such a large aggregated scale because of the existence of the National Grid. Although we use electricity in an intermittent manner, turning on a light here and a heater there, on average it usually adds up to a smoothly varying demand that the grid system can cope with relatively easily. This is known as **diversity of demand**.

The history of the National Grid's development has been described in detail in the companion text *Energy Systems and Sustainability*. When its construction was initiated in the 1930s, electricity was generated locally in hundreds of small separate power stations with low average generation efficiencies. The first phase of the grid was completed in 1934, initially to link together the most efficient power stations within *regions* of the UK, allowing mutual backup and coverage for peak demand. These all had to operate using alternating current (AC) at a common frequency, chosen to be 50 cycles per second (50 Hz). Hundreds of small inefficient stations were closed down.

After experiments in 1937, it was found that it was safe to connect together all the regional systems of England, Scotland and Wales into a single *national* network. The output of individual stations was controlled centrally in order to optimize the overall system performance. It has largely been run as such since 1938.

At that time, many of the power stations and local distribution networks were privately owned. Although the UK industry was nationalized in 1947 and then privatized in 1989, this has not affected the *national* nature of the distribution system.

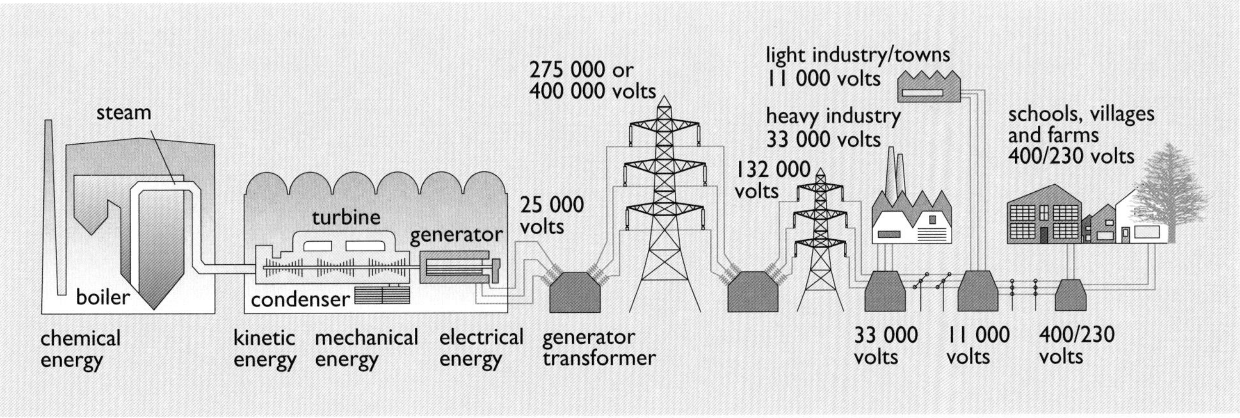

Figure 10.11 Diagrammatic representation of the current UK electricity system. Electricity is generated mostly in large power plants, most of them like the one shown on the left, transmitted over long distances at 275 V or 400 kV by the National Grid Company and distributed at 132 kV or lower by the local network operating companies

The grid has been repeatedly strengthened and reinforced since 1934. The main links are now at 275 kV and 400 kV with lines capable of carrying 2 GW or more. The overall philosophy has been one of large generating plant, justified on the basis of economies of scale, centralized control, and distribution outwards and downwards in voltage to the consumer (i.e. from left to right in Figure 10.11). At the lowest voltage, the system consists of buried cables in the streets capable of supplying both 400 V, 3-phase AC and 230 V, single phase AC (the normal for domestic consumers).

In practice, the grid provides enormous flexibility of operation. The demands of Oxford can be met from a power station in Yorkshire or a storage plant in Wales. Across Europe, the electricity systems operate on the common frequency of 50 Hz. Because they are linked across national borders, power plants are able to share loads or export surpluses of electricity.

Matching supply and demand

At present in the UK, the total demand pattern is met by a mixture of generating plant, owned by different companies, which competes in a centrally controlled **power pool**, a competitive market in electricity. Electricity is bought in from all the competing power stations and then distributed via the National Grid to consumers. The precise terms of trading are governed by a set of **trading arrangements.** The current set is known as **NETA** (New Electricity Trading Arrangements), and covers just England and Wales, but this is scheduled to be shortly replaced by a revised version, **BETTA** (British Electricity Trading and Transmission Arrangements), which will include Scotland.

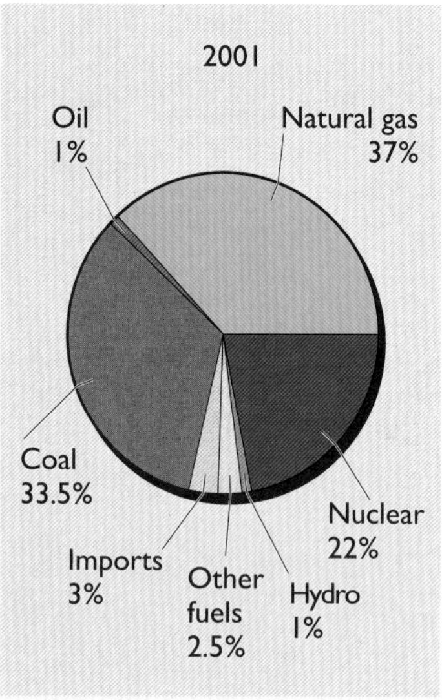

Figure 10.12 The UK generation fuel mix in 2001

The current NETA arrangements encourage fierce competition between generating companies. Which of these generates at any particular time depends on the bulk prices of gas and coal and the availability of various nuclear stations. Figure 10.12 shows the generation mix for 2001. This can perhaps be simplistically described as a three-way split between coal, gas and nuclear power, with small inputs from other sources.

The ability of different plants to provide for a changing load depends on their scale. A large coal-fired or nuclear power station may take 24 or even 36 hours to reach full output from cold. A smaller combined cycle gas turbine (CCGT) station may be able to produce some power within an hour but take 8 hours to reach full output. Following a changing demand can be done by running stations at part load, but this reduces their overall efficiency (and increases their CO_2 emissions per kilowatt of electricity generated). It is best if they are run continuously at full output.

However, demand is not conveniently constant. There can be rapid changes in demand for several reasons.

- The daily increase in human activity in offices and factories can increase demand in the morning by 12 GW in the space of two hours.
- Simultaneously showing the same TV programmes over the whole country can produce a synchronization of behaviour of consumers. It can send a large proportion of the population rushing to use their electric kettles at the same time, leading to a **demand pickup** of over 1 GW in a matter of minutes.
- The sudden breakdown of a large power station or failure of a major transmission line may mean that generating capacity of 600 MW or more could suddenly disappear, again requiring a response in a matter of minutes.

The flexibility of hydroelectricity can go a long way to meeting these rapid changes in demand. The traditional way of dealing with them in countries with limited hydro resources was to keep 'spinning reserve' – having large fossil-fuelled or nuclear power stations generating electricity at part load, but with sufficient steam up to cope with any sudden increase in demand.

However, running stations in this manner reduces their efficiency and wastes fuel. So, over the years a range of other technologies has been brought into use to cope with this problem (see Box 10.2).

BOX 10.2 Matching electricity supply to short-term demand fluctuations

Electricity has to be generated on demand and the voltage and frequency of the AC supply have to be held within relatively tight limits. A range of technologies is used to meet rapid and possibly unexpected increases in demand.

Pumped storage plants

These have been described in the Hydroelectricity chapter. At times of low demand, surplus electricity is used to pump water into high level reservoirs. At times of sudden peak demand, they can use the stored potential energy of the water to generate electricity with as little as 10 seconds' notice. The UK currently has three such plants, two in Wales and one in Scotland. Their combined peak output power is over 2 GW, about 5% of the UK's typical winter electricity demand. Other countries have similar schemes. Typically, such plants have an overall storage efficiency of 70–80%.

Gas turbine and diesel 'peaking' plants

These can be run up to full power in half an hour or less. Small 'open cycle' gas turbines, lacking the steam stage of the CCGT, run on natural gas or light heating oil and typically supply between 10 MW and 100 MW of electricity. Diesel generators of around 1 MW output can be brought on-line in minutes. The disadvantage of both types is that they consume fossil fuel and are less thermally efficient than larger 'base-load' power stations.

Compressed air energy storage (CAES)

In practice, a large part of the power of a gas turbine is used to compress the air before it reaches the combustion chamber (see Chapter 8 of the companion text). A CAES plant uses off-peak electricity to compress air and store it in an underground cavern. At times of peak demand, this is fed to a peaking gas turbine reducing its gas consumption by more than 60%. The first commercial plant was a 290 MW unit built in Germany in 1978 using storage in a salt mine. A 2.7 GW plant is planned for construction in Ohio. This will compress air to over 100 atmospheres pressure, storing it in an existing limestone mine nearly 700 metres under ground.

Rechargeable batteries

Lead acid batteries These use solid electrodes made of lead and have been used by electricity utilities for peaking power and emergency backup since the late nineteenth century. The largest is currently a 40 MWh system in California. Lead acid technology is limited by the number of cycles that a battery can be put through before it needs replacing

Sodium sulphur batteries These use a positive electrode of molten sulphur and a negative electrode of molten sodium. The chemicals combine to produce sodium polysulphides and electricity. The battery has to be kept at 300 °C for the reaction to take place. When the battery is recharged, the elemental sulphur and sodium are regenerated. A number of MW scale plants have been built for electricity utilities in Japan.

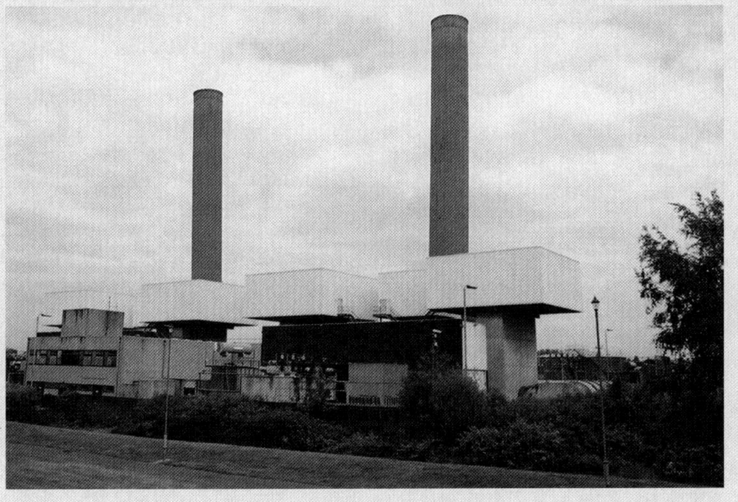

Figure 10.13 A 132 MW Open Cycle Gas Turbine (OCGT) plant in North London. It has two turbines that can run on natural gas or light heating oil

Flow Batteries Unlike most conventional batteries where the key active chemicals are solid, in a flow battery they are liquids and can be stored in tanks separately from the battery itself. This is similar to the hydrogen fuel cell (see Box 10.3) which is a 'gas battery'. A system known as 'Regenesys', using liquid sodium bromide and sodium polysulphide is being demonstrated in a 15 MW, 120 MWh plant at Little Barford Power Station in the UK. Other flow batteries in the MW class, based on Vanadium or Zinc Bromide, are being developed for use in Japan.

Which is best?

Balancing the electricity grid of an entire country requires large amounts of power and appreciable amounts of stored energy. At present only pumped storage systems have power ratings of over 1 GW and the capacity to supply this for more than an hour or so, but such systems require suitable sites in mountainous regions. Peaking gas turbines can be installed almost anywhere, but consume fossil fuel in a relatively inefficient manner. CAES systems can supply 100 MW or more, but require the special geology suitable for underground high pressure air storage.

Rechargeable batteries can potentially be installed anywhere, but at present are only available in ratings of less than 50 MW and are typically only used to supply that power for periods of an hour or less. They have the advantage of very rapid response – fractions of seconds rather than minutes. They are more likely to be used to absorb short surges and to correct control instabilities in local distribution systems. They may also act as 'starting batteries' to allow a power station to recover from a grid failure and perform what is known as a *black start*.

Figure 10.14 The 12 MW Regenesys flow battery under construction at Little Barford in Bedfordshire in 2001. The two tanks contain the electrolytes (photo courtesy Regenesys Technologies)

Connecting the renewables

Where do the renewables fit into all this? The answer depends on the particular source and the extent to which the timing and quantity of its output matches demand.

Hydroelectricity

Large scale hydro power is perhaps the most desirable of all renewable electricity sources from the point of view of flexibility of supply. Water can be stored in reservoirs for months or even years, yet the generators can be wound up to full power and turned off again in minutes. In the UK, most plants are in the range 100 kW to 100 MW and are connected at voltages of 11 kV or above. Elsewhere in Europe, such as in Norway and Sweden, hydro power plays a major role, with the output, and in some cases the pumped storage capacity, sold across international borders.

Although hydropower can be used as a solution to short term variations in demand, there may be long term year-to-year variations in its potential depending on rainfall.

Biofuel plant

Generation plants using MSW, waste wood or landfill gas are relatively small, typically in the output range 100 kW to 50 MW. As such, they are likely to be connected to the system at 11 kV or 33 kV and run fairly continuously. Many smaller plants are unmanned and run under automatic control. Apart from breakdowns, their output is highly predictable and, as such, their electricity is every bit as valuable as that from larger power stations.

Solar power

In the UK, photovoltaic systems are most likely to be at the kilowatt scale and connected locally at the 230 V or 400 V level. Naturally, they only produce electricity during the day and their output will be higher in summer than in winter. Given their relative expense, they are only likely to make up a small proportion of any renewable electricity mix for the UK in the near future.

This is not the case in sunnier places such as California or Greece where the electricity demand peaks not in winter, but with increasing air conditioning loads in summer. This makes solar electricity more desirable. The large-scale SEGS plants, described in the Solar Thermal chapter, generate at the multi-megawatt scale and are connected to the electricity grid at high voltage. Although no new SEGS plants have been built for many years, it is likely that any future schemes would take the form of a solar-assisted gas fired power station, and fit quite normally into the existing infrastructure.

Multi-megawatt PV schemes have been proposed for Greek islands, where electricity is normally supplied by diesel generators. Although the latter are relatively inefficient when run at part load (leading to high electricity prices) they are very flexible in operation. When the sun shines, the PV cells provide the power. In the evening, the diesel generators take over.

Wind and wave power

With the current large-scale development of onshore and offshore wind farms, there is much interest in exactly how much wind-generated electricity can be absorbed by the existing infrastructure. Although wave power is much less developed, it shares the same basic problems.

Modern individual wind and wave generators are likely to have power ratings of between 50 kW and 5 MW. Current wind farms and future large wave power devices could have total outputs in excess of 100 MW. Thus they are likely to be connected to the grid within the local distribution network at voltages of 11 kV, 33 kV or even higher.

The output of wind and wave power generators is not perfectly predictable (although detailed weather forecasting can help) but it pays to use the output when it is there, since there are no fuel costs.

Typically, a 1 MW wind turbine will produce 300–400 kW on *average*. It will produce full output on a windy day but nothing on a calm one. At modest wind speeds, its output may vary considerably from minute to minute. Modern large turbine designs incorporate sophisticated power electronics that can reduce this variability.

Similarly, the output of a wave power device is dependent on the variable intensity of the waves. On a stormy day it will run at full power, on a flat calm one it will produce nothing, and on an intermediate day, its output will be variable.

As these sources do not usually incorporate significant amounts of energy storage, using them means reducing the output of some other plant (usually fossil-fuelled) at fairly short notice when they become available and increasing it again when their output drops.

If such sources are widely spaced, then just as *diversity of demand* adds up to a smoothly varying total demand on the National Grid, so **diversity of supply** can also smooth out the local variations of output of various renewable-energy sources. When the wind stops blowing in Scotland, it may still be blowing in Wales.

However, there is still a fundamental question of what should be done on 'the day with no wind power'. It has been argued (for example, Laughton, 2002) that when a high pressure weather system covered the whole of the British Isles, there would be no output from wind turbines, and consequently there would need to be 100% backup from other sources. This might require extra expenditure on Open Cycle Gas Turbine plant (such as shown in Figure 10.13), or other power plant might have to be kept in reserve. All this would create extra 'system costs' over and above those of the normal generation of electricity.

The counter-argument to this is that the electricity supply system needs to have wide-ranging backup anyway to cope with diversity of demand, demand pickup and large-scale faults. The *Energy Review,* produced by the UK Cabinet Office, concluded that small amounts of generation from intermittent sources such as wind power create insignificant extra system costs, but that these increase with the proportion of electricity from intermittent sources. They are less than 0.1p/kWh for 10% electricity from intermittents, under 0.2p/kWh for 20% and under 0.3p/kWh for 45% (PIU, 2002).

A further more detailed study (ILEX, 2002) investigated possible future mixes of wind power and other generation. It concluded that large amounts of intermittent wind power would require increased expenditure on backup. In addition, not surprisingly, accessing large amounts of wind power in the areas with the best resource (such as Scotland) would require expenditure in extra grid links. A mix of wind power located in England and Wales (i.e. close to the main loads) and biomass-fuelled plant distributed throughout Britain gave the lowest extra system costs.

A more surprising integration problem is what to do on 'the *night* with *too much* wind power'. The voltage and frequency stability of the National Grid has to be kept with in tightly defined limits. Traditionally this has been ensured by the flywheel inertia and stored steam pressure of enormous 600 MW coal and nuclear generating sets. Yet on a summer night with low demand and plenty of wind, it is possible that all of these would have been turned off and the system would then be *wholly powered by renewables*. In this situation, it has been suggested that the task of stabilization should be given to the large pumped storage plants on the system, which would be busily occupied absorbing yet more *surplus* wind power.

This problem has already been encountered in Denmark, where wind power now (2003) provides almost 20% of the country's electricity. New offshore wind farms are being designed to with equipment to maintain the frequency stability of the network, and the possibility of dumping surplus wind power as heat into the country's district heating systems is being investigated.

In the Irish Republic, the proposed 500 MW offshore wind farm at Arklow Bank may not pose serious intermittency problems since it is conveniently located only 40 km from the country's pumped storage plant in the Wicklow mountains.

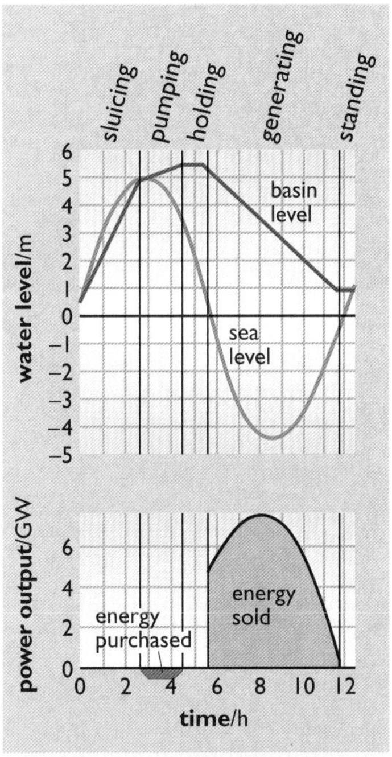

Figure 10.15 A tidal power station such as the Severn Barrage would produce bursts of power of up to 8 GW in magnitude and 6 hours long. These could be difficult to integrate into the UK system

Tidal power

As Chapter 6 has described, tidal power is intermittent, but highly predictable. A scheme such as the proposed Severn Barrage, generating only on the ebb tide, could produce a pulse of power of up to 8 GW about six hours long every 12.4 hours (see Figure 10.15).

It would, in theory, be necessary to schedule a number of conventional power stations to shut down when the tidal one is about to start generating; but given their thermal inertia, this would not be very fuel efficient. It might therefore be better to have a large double (or even triple)-basin scheme in this location with one basin kept 'high' and the other 'low' and with flexible generation and pumping between them. Although expensive, this could produce something closer to 'firm' renewable electricity and could supply pumped storage, short-term backup and grid stability control where it is needed, in the south of England.

The position with tidal current turbines seems more promising but, at present, it is only possible to estimate their performance. Theoretically, such a turbine is likely to produce power in proportion to the cube of the speed of the water flowing through it (this is analogous to the performance of wind turbines). The output will peak every 6.2 hours, on the incoming tide and again on the ebb tide.

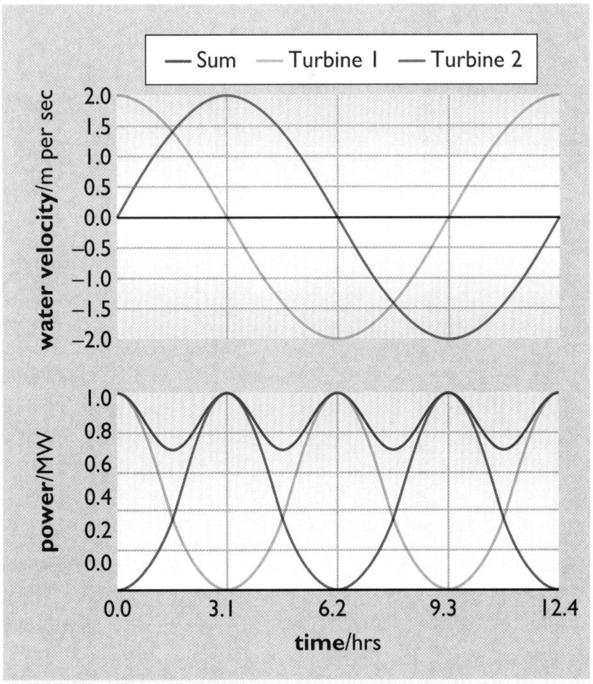

Figure 10.16 Theoretically, two tidal stream turbines at widely spaced locations could in combination produce an almost continuous supply of electricity

The output of a single turbine might consist of a rather unpromising sequence of pulses of power three to four hours long. However, the output of *two* identical devices in different locations where the times of high tides differed by about 3 hours (such as Portland Bill in Dorset and Dover in Kent) could add up to an almost constant supply (see Figure 10.16). In practice, it is likely that such turbines would be deployed at a large number of different locations. In total, they might produce a supply that varied little over the day, but would have long monthly cyclic variations with the spring-tide to neap-tide changes in tide amplitude. The price that might have to be paid for any large-scale deployment would be a strengthening of the National Grid to make use of the differing times of the tides at different locations.

10.6 Some system solutions

To a certain extent, there are some common solutions to the *where* and *when* of deploying renewable energy: stronger electricity grids, better demand management, the use of embedded generation, and in the longer term, the development of the hydrogen economy.

Grid strengthening

The existing grid has grown up around the power stations of the past. Connecting large amounts of power from new renewable sources will undoubtedly require strengthening the National Grid. This will not come cheaply. Typically, an overhead 400 kV line costs about £150 per MW per kilometre (ETSU/PB Power, 2002). On this basis one capable of carrying 2 GW would cost about £300 000 per kilometre. Moreover, a new line from, for instance, Scotland to England would almost certainly face environmental objections. Placing sections of the cable underground would be extremely expensive, with prices of up to £5 million per km.

An alternative scheme using high voltage DC underwater cables through the Irish Sea might turn out to be cheaper. In a similar manner to the existing 2 GW link under the English Channel, this would require relatively expensive AC/DC converters at each end connected by twin cables costing only £1 million per kilometre. A 'point to point' link 200 km long is estimated to cost about £790 million. This could be extended (at a price) as part of a wider 'grid mesh' designed to harvest and distribute renewable energy from a wide area.

The UK is not alone in considering such grid strengthening. The electricity systems of Denmark, Norway, Sweden and Finland are now operated as a single market, the Nordpool. Denmark is a small country and has strong grid links to Sweden and Norway. This has allowed it to develop its wind power in the knowledge that it can rely on the flexible backup of over 40 GW of Swedish and Norwegian hydropower. Most of Danish electricity comes from its coal-fired CHP plants, which can cover for shortages in hydropower during dry years. The Nordic Grid Master Plan (Nordel, 2002) suggests further north-south grid expansion and even a possible link from Norway to England.

Demand management

It would be convenient for fossil-fuelled, nuclear and most renewable electricity supplies if electricity demand varied as little as possible over the day and night. This can be encouraged through the use of 'off-peak' electricity tariffs. Electricity for heating (or other) purposes is supplied at cheaper rates at night and the heat can then be stored through to the next day. This allows existing power stations to continue running into the night, rather than having to build extra ones to cope with higher peak demands. In the UK, only a relatively small amount of electricity is used for heating purposes. In France, with its high reliance on nuclear power, the proportion is much higher. There, the effect of the promotion of off-peak electricity in flattening out the winter demand profilehas been quite considerable (see Figure 10.17).

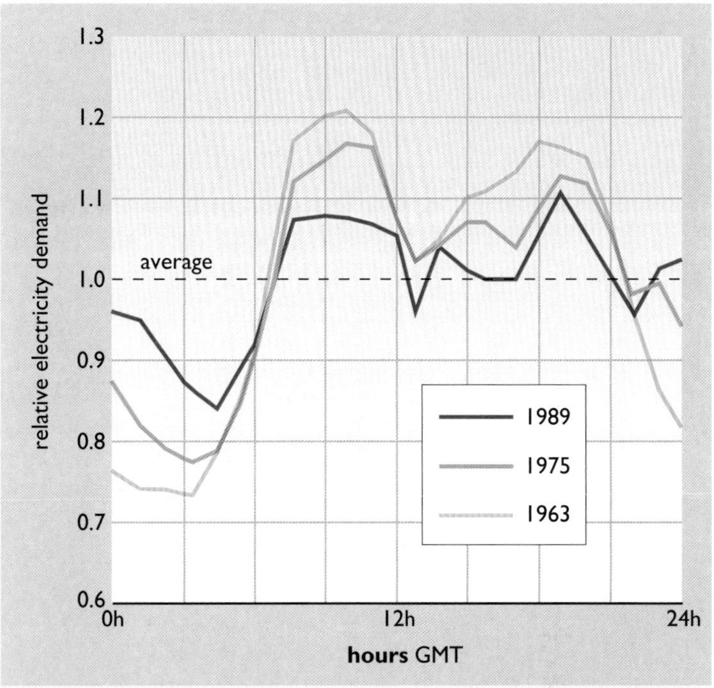

Figure 10.17 Typical daily electricity demand profiles for France for the years 1965, 1975 and 1989. Over the years, the practice of offering cheap electricity at night has flattened the daily demand profile (source: P. Careme, Electricité de France)

Taking this further, there are many electrical loads that are not immediately needed. One example is large-scale water pumping which could be placed under remote control. At times of any impending shortage of renewable electricity supply, pumps could be turned off. At present the mechanisms for such 'load-shedding' are not well developed, perhaps because any inability to meet all requests for supply could be interpreted as some kind of 'failure' of a market-based system.

PV, micro-CHP and emergency generators

It is often argued that large numbers (possibly millions) of small **embedded** generators (i.e. at the low voltage end of the grid) could cut the need for large scale electricity grids and provide backup for wind generators.

Small-scale generators are designed to operate completely automatically. They will connect and disconnect from the grid in response to their *local* energy needs or circumstances. The information technology to allow remote 'scheduling' of such devices (i.e. placing them under the same kind of control as large power stations) undoubtedly exists, but at present the operating procedures (Engineering Recommendations) that cover this only apply to larger generators. A recent report acknowledged that these need to be amended to cover smaller generators. This report (EA Technology, 2000) also estimates that there is already 20 GW of emergency backup from diesel generators in UK hospitals and other large buildings. At present, they only come into operation when the grid has collapsed, and give a restricted supply of power within the buildings they serve. In theory, much of this plant could be modified to operate in parallel with the grid providing backup capacity at a lower cost than purchasing new Open Cycle Gas Turbine (OCGT) peaking plant. This is an area of continuing research of considerable importance.

Hydrogen – the fuel of the future?

Hydrogen has been widely advocated as an 'energy carrier' for the future. Its use as a fuel has many advantages:

- it can act as a store of renewable energy from season to season,
- it can provide a transport fuel not dependent on the world's declining reserves of oil,
- the only by-products of its combustion are water and a very small amount of nitrogen oxides, and even the emissions of these can be reduced to zero if fuel cells (see Box 10.3) are used.

Hydrogen is already used in large quantities as a feedstock for the chemical industry, mainly in the manufacture of fertilizers. Currently, it is mainly produced by steam 're-forming' of natural gas (methane) which necessarily also produces carbon dioxide:

$$\begin{array}{ccccccccc} 2H_2O & + & CH_4 & \rightarrow & CO_2 & + & 4H_2 \\ \text{steam} & + & \text{methane} & \rightarrow & \text{carbon dioxide} & + & \text{hydrogen} \end{array}$$

One possibility is a hydrogen economy using such fossil-fuel sources together with **carbon capture and sequestration**. (See the companion text for a detailed account.) However, renewable or 'solar' hydrogen can be produced without CO_2 by-products, in a number of ways:

- by the **electrolysis** of water using electricity from non-fossil sources. If direct current electricity is passed between two electrodes immersed in water, hydrogen and oxygen can be collected at the electrodes. This process could be used to produce hydrogen from renewable electricity virtually anywhere: solar plants in the deserts of Africa, wind power in the north of Scotland, or geothermal energy or hydropower in Iceland.
- by the **gasification** of biomass. This is described in the Bioenergy chapter. Large amounts of hydrogen can be produced leaving a residue of high-grade carbon for chemical purposes. This carbon, is of course likely to end up as CO_2, but will be re-absorbed as long as the biomass is sustainably grown.
- by the **thermal dissociation** of water into hydrogen and oxygen using concentrating solar collectors (probably in desert areas). To do this directly would require very high temperatures, over 2000 °C, but with more complex processes using extra chemical compounds the same result may be achievable at temperatures of under 700 °C. These processes have not yet been developed on a commercial scale.

Other techniques are under investigation, including the use of photoelectrochemical cells that produce hydrogen directly from water via artificial chemical photosynthesis.

Using hydrogen as a fuel is well understood. 'Town gas' produced from coal before the arrival of natural gas consisted mainly of a mixture of hydrogen and carbon monoxide. Space rocket motors run on a mixture of liquid hydrogen and liquid oxygen.

When burned, 1 kilogram of hydrogen will produce 120 MJ of heat, assuming that the resulting water is released as vapour. Although this is nearly three times the energy per unit *mass* of petrol or diesel fuel, hydrogen has the disadvantage of being a gas, with a low energy per unit *volume* at atmospheric pressure. It can be stored in a number of forms:

- as a gas in pressurized containers, typically at around 300 atmospheres. These containers obviously have a weight penalty,
- by absorbing it into various metals, where it reacts to form a metal 'hydride': the hydrogen can be released by heating,
- as a liquid, although this requires reducing its temperature to –253 °C and the use of highly insulated storage. Natural gas (methane) is already widely shipped in liquid form, but this only requires temperatures of –162 °C.

Hydrogen can also be pumped through pipelines. Here it has a disadvantage: at atmospheric pressure, its energy density is only 10 MJ/m^3, about a quarter of that of natural gas. Although this would limit its use as a direct substitute in the existing heating network, a simple answer to its initial deployment might be to add a modest proportion to the existing natural gas flows, effectively reducing their overall carbon content. This would have to be limited to 15–20% hydrogen by volume before the modification of existing burners or other end-use technologies would be required.

It has been suggested that countries with plentiful supplies of renewable energy, such as Iceland, which has large untapped reserves of hydroelectricity and geothermal energy, could give up the use of fossil fuels entirely. It might be possible to convert all its road vehicles to run on hydrogen. As a step in this direction, three hydrogen buses using fuel cells are being deployed on the Reykjavik bus system and a high pressure hydrogen filling station was opened in April 2003. A variety of hydrogen-powered vehicles are now under test in a number of other countries world-wide.

Figure 10.18 A hydrogen-powered bus. The Proton Exchange Membrane (PEM) fuel cell and high-pressure hydrogen tanks are mounted on the roof. It is propelled by a electric motor mounted at the rear (photo courtesy Ballard Power Systems)

BOX 10.3 **Fuel cells**

As discussed in Chapters 1 and 2, all heat engines are inherently limited in the efficiency with which they can convert heat into motive power, and hence into electricity if the engine is driving a generator. Normally, most of the energy in the input fuel emerges as 'waste' heat (although of course this can often be harnessed and put to good use).

The fuel cell (Figure 10.19) enables hydrogen and oxygen fuel to be converted to electricity at potentially a higher efficiency than could be achieved by burning them in a heat engine. A fuel cell is in principle a battery where the active elements are not solids (such as the lead and lead dioxide in a car battery) or liquids (as in the Regenesys plant mentioned above) but gases. Indeed, when the fuel cell was first invented by Sir Charles Grove in 1839 he called it a 'gas battery'.

The principle of operation of the fuel cell is similar to electrolysis but in reverse: gases such as hydrogen and oxygen (or air) are pumped in and DC electricity is the output.

The only by-products are water and there are virtually no pollutants. There is some waste heat, but much less than in most combustion-based generation systems, and there are no moving parts. As with other types of battery, the voltage from an individual cell is low, typically about 0.7–0.8 volts, and multiple cells are connected in series to get a useful working voltage.

Despite its early invention, development work did not really start until the late 1950s. Now there is a whole range of different types:

The Alkaline Fuel Cell (AFC) was the first to be developed, for the US Gemini and Apollo space programmes in the 1960s. Although simple with low manufacturing costs, they are limited by the need to remove any CO_2 from the air supply to prevent contamination of the potassium hydroxide electrolyte.

The Solid Polymer Fuel Cell (SPFC) which is being developed as two forms:

- the Proton Exchange Membrane Fuel Cell (PEMFC) – a strong candidate for transport and portable power applications. It has been demonstrated in cars and buses and is available in sizes of up 250 kW.
- the Direct Methanol Fuel Cell – another candidate for transport applications but still at the development stage. Although methanol is poisonous, it is easier to handle as a fuel than hydrogen.

The Phosphoric Acid Fuel Cell (PAFC) is the most developed of the fuel cell types and is available commercially, usually as 200 kW unit packaged with a steam reformer to allow it to run on natural gas. There has been significant PAFC development in Japan, where more than 100 plants ranging in size from 50 kW to 11 MW output are operating.

Several other types are under development.

Currently, many fuel cell designs require the use of small quantities of 'noble metals', such as platinum, as catalysts. Although most of the metal in spent fuel cells can be recycled, if their use became very widespread, there might be catalyst supply problems.

Typical fuel cell electricity generation efficiencies are currently in the range 40–60%. While this is better than the 25–30% that might be expected from small reciprocating engines running on natural gas or hydrogen, it is still only just competitive with the 45–55% efficiencies of large CCGT power stations. Fuel cell developers are aiming to produce devices with costs competitive with more conventional plant and with higher efficiencies. For the moment, the strength of the fuel cell concept lies in its lack of noise, low pollution at the point of use and flexibility as a 'gas battery' since, like other batteries, its power output can be changed very rapidly, often within fractions of a second.

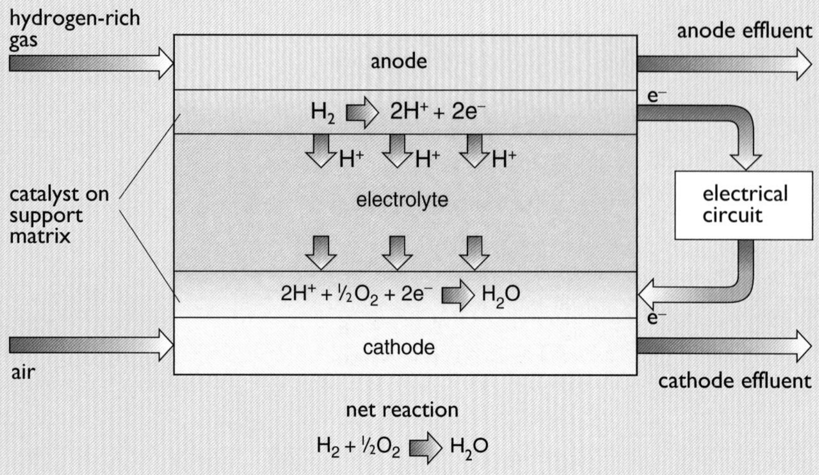

Figure 10.19 Operation of the phosphoric acid fuel cell (PAFC)

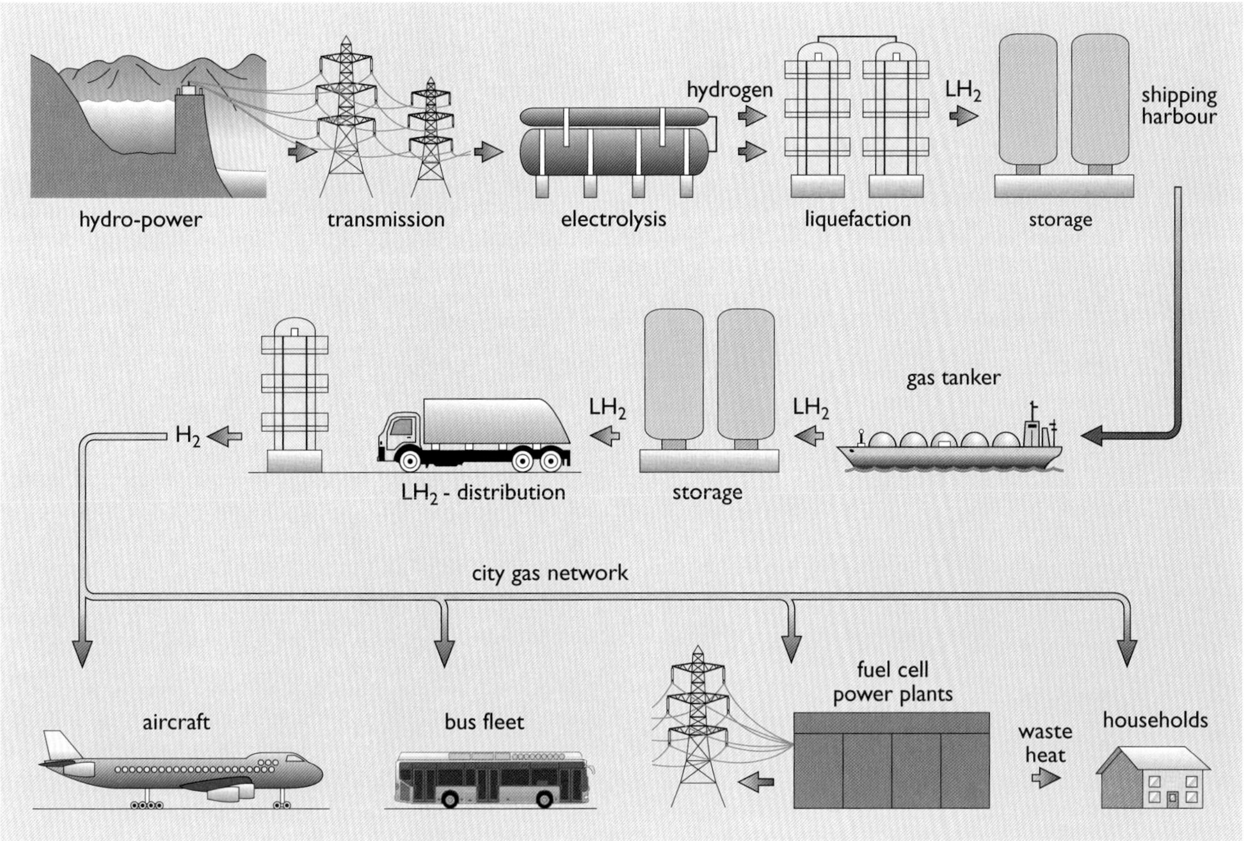

Figure 10.20 A possible future hydrogen economy. Many possible sources of renewable energy could be tapped to manufacture hydrogen. This could be shipped to the consumers and used in a variety of ways

Any practical scheme for the large-scale use of hydrogen would require many steps, such as those illustrated in Figure 10.20, each of which would require large capital investments.

For example, surplus hydropower could be electrolysed to produce hydrogen, which would then be stored, either as high pressure gas or as a low temperature liquid, before being shipped to its destination in special tankers. At the receiving end there would need to be further storage facilities and provision to distribute the hydrogen by road tanker or pipeline. Finally, there are the individual end uses:

- in homes for cooking and heating – possibly as a mixture with natural gas,
- in fuel cell power stations (of all sizes) to generate electricity and useful heat,
- in road vehicles using fuel cells or hydrogen-fuelled internal combustion engines,
- and even ultimately in jet aeroplanes.

Bringing such a vision to the current energy marketplace will require a high premium for hydrogen's carbon-free and pollution-free qualities, since it will initially be in competition with natural gas and its well-developed technologies. Even taking low prices for surplus renewable energy and long-term prospective prices for processing and distributing the hydrogen, its cost to the end user is likely to be several times higher than current gas

Table 10.2 Comparisons of costs of supply of natural gas and renewably-generated hydrogen

Component of cost	Current (2000) natural gas supply cost in p kWh^{-1}	Prospective hydrogen alternative cost in p kWh^{-1}
Intermittent renewable-energy source		2.0
Electrolysis		0.5
Compression and Storage		0.5
Distribution to industrial consumers		0.3
Total cost to industrial consumers	**0.7**	**3.3**
Extra distribution costs to residential consumers	1.0	1.5
Total cost to residential consumers	**1.7**	**4.8**

Source: ICCEPT, 2002

prices. Table 10.2 gives an illustrative breakdown of the costs of supplying natural gas under the current conventional supply system (2000 figures) and a possible future renewable hydrogen alternative.

Taking this further, the ICCEPT report from Imperial college, London, cited in Table 10.2 suggests that with prospective future fuel cell prices of £500/kW, hydrogen-produced *electricity* might cost about 5.8p/kWh to industrial consumers. This would represent an increase of 60% on conventionally industrial generated electricity prices in 2000. Yet this potential supply would have to co-exist in competition with other sources of electricity, renewable and nuclear. It may well prove cheaper to invest in stronger power grids to transport renewable electricity directly, and in other forms of storage, than to invest in hydrogen technology. Only time will tell.

In the short-to-medium term, it seems likely that transport will be the major developing user of hydrogen, because of (a) the difficulty of finding other viable alternatives to petrol and diesel fuel and (b) the fact that these fuels are highly taxed already. For example, in 2000, petrol in the UK cost about 8.5p/kWh of which about 75% was tax (DTI, 2002c). Even given the production costs listed above, a hydrogen alternative could be sold at a similar price and still have room for taxation.

10.7 Balancing economic options

Renewables and conservation

In practice, there are even wider choices to make, not just between different renewable supplies, or between renewables and conventional supplies, but between these and extra investment in improved energy efficiency. This is most important because conservation and efficiency measures:

- are usually more cost-effective than many renewable supply options,
- narrow the gap between demand and the supply of renewable energy.

We can extend the idea of a resource–cost curve to mixtures of conservation/efficiency and renewable supply, as in the Swedish example shown in Figure 10.21. Note that this is plotted slightly differently to Figure 10.5 earlier – the resource on the *x*-axis and the price on the *y*-axis.

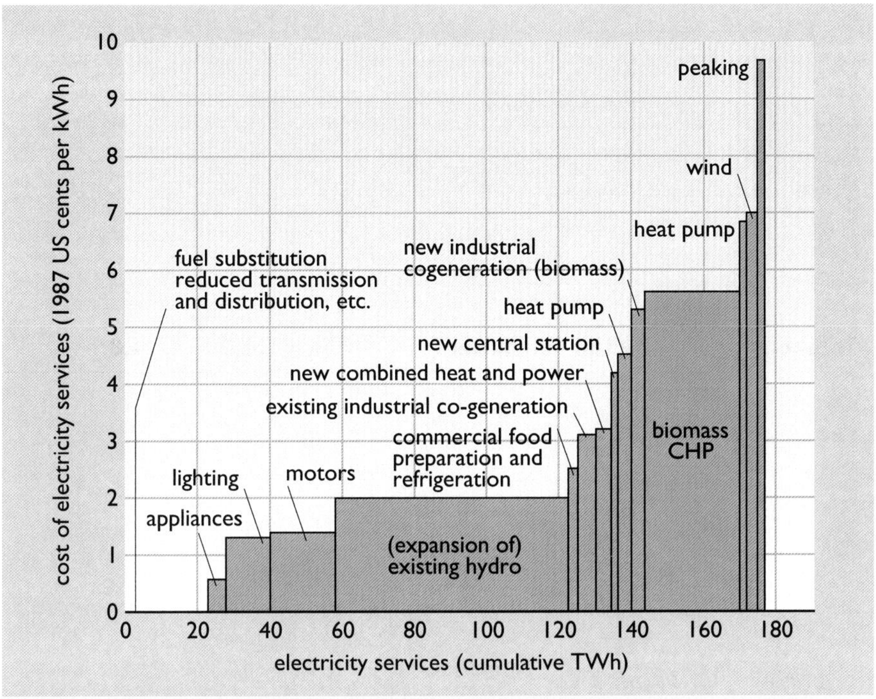

Figure 10.21 Resource-cost chart for investments in electricity conservation measures and/or new electricity supply technologies in Sweden (Note: In 1987, 1 US cent = 0.6 pence) (source: Bodlund et al., 1989)

What is striking about such a chart is the diversity of options available. Many conservation options cost little or nothing. At a given energy price there may be choices between investments in biomass, onshore wind, more efficient refrigeration or lighting. This diversity is important to consider when we add in the difficulties of deciding between fossil fuel, nuclear and renewable-energy supplies on environmental grounds. In practice, it is rare to see such a chart drawn up, since 'providing supply' and 'cutting demand' are usually the realms of completely separate government departments.

Balancing economic and environmental considerations

This is perhaps the most difficult area. There are several benefits of using renewable-energy supplies in preference to conventional sources:

- they cut carbon dioxide emissions,
- they decrease a country's reliance on imported fuel and add to diversity of energy supply,
- they cut emissions of acid rain pollutants, sulphur dioxide and nitrogen oxides.

However, all renewable-energy sources have some environmental consequences; their benefits have to be seen in relation to the alternatives. The problems of each individual technology have been discussed in the preceding chapters and a detailed comparison made in the companion text *Energy Systems and Sustainability.*

It is unlikely, for example, that there will be much further development of large-scale hydroelectricity within the UK or the EU, aside from re-powering existing schemes. The problems of flooding large areas of land and the possibility of methane emissions from rotting vegetation are sufficient to rule out further expansion. Although this technology is counted as part of total renewable-energy targets, it is omitted from most promotion schemes.

Similarly, the mass burn of municipal solid waste (MSW) is widely used and contributes to national renewable-energy totals, yet it faces opposition because of fears of dioxin emissions. This, too, is omitted from many promotion schemes, although the gasification and pyrolysis of waste, which might have lower emissions, are included.

Other forms of biomass also need to be burned cleanly to avoid air pollution. In the extreme, the Indian government has a policy of promoting LPG and kerosene to *discourage* the use of firewood as a cooking fuel in cities.

The intensive cultivation of energy crops is being encouraged although it may require the use of fossil fuel in the production of fertilizers, and the harvesting and transport of the produce.

Putting actual costs to the relative benefits and disbenefits of different technologies is difficult. Energy normally has a market-place price in terms of £ per GJ or pence per kWh. What is missing are the 'external costs', those borne by society at large. The EU, under its ExternE study, has attempted to evaluate these for various electricity generation technologies (see Table 10.3). It has taken into account a wide range of factors, including health risks and environmental damage, and expressed them in euro cents per kWh. These are, of course, the extra costs over and above the normal market price of electricity, which as an EU average, is about 4 euro cents per kWh.

Table 10.3 Estimated external costs of electricity generation from various primary energy sources: European Union average

Primary Energy Source	External Costs Euro cents per kWh
Coal	5.7
Gas	1.6
Biomass	1.6
PV Solar	0.6
Hydro	0.4
Nuclear	0.4
Wind	0.1

1 euro cent (2001) = 0.6 p

Source: European Commission (2001)

However, this study specifically excluded global warming costs resulting from the emission of greenhouse gases. For nuclear power and the renewable energy technologies listed above, these are small. However, for coal and gas-fired generation we can provide figures to be added to those in the table above by looking at the likely carbon dioxide abatement costs under the proposed EU carbon trading arrangements (described in Section 10.8).

When coal or gas are used to produce electricity with current technologies, there are marked differences in the total emissions of the greenhouse gases, CO_2, methane and nitrous oxide. The effects of these can be summarized in terms of the equivalent amount of CO_2 produced per kWh, as in Table 10.4.

We can go further by assigning a possible abatement cost to these amounts under future EU carbon trading arrangements. It is likely that when trading begins in 2005 the price for permission to emit 1 tonne of CO_2 will be about 5 euros, or 0.5 euro cents per kg (Christiansen, 2003).

The external costs produced by the ExternE study are obviously subject to considerable uncertainties, and the 'market price' of CO_2 will no doubt

Table 10.4 Likely CO_2 abatement costs for electricity generation

	kg CO_2 equivalent produced per kWh generated	**CO_2 abatement cost at 5 euros per tonne/Euro cents per kWh**
Modern coal plant (including FGD and low NO_x burners)	1.1	0.55
Combined cycle gas turbine	0.5	0.25

Note: FGD = flue gas desulphurization

Emission data source: Eyre, 1990

change. However, there are large cost differences between the total external and CO_2 abatement costs for coal-fired electricity generation (over 6 euro cents per kWh), those for gas-fired generation (1.85 cents per kWh), and those for renewables and nuclear power (0.1–1.6 cents per kWh). This sets the context for the various promotion mechanisms for renewable energy, which will be described below.

Renewable energy and planning permission

Obtaining planning permission for renewable-energy projects can often be a time consuming and expensive task. In the UK, large projects, above 50 MW, have to obtain consent from central government, usually the Department of Trade and Industry. For most other smaller projects, below 50 MW, planning consent has to be obtained from local authorities, which operate within established town and country planning procedures.

In addition, projects 'which are likely to have a significant effect on the environment by virtue ... of their nature, size or location' require the submission of an **environmental impact statement** setting out the developer's assessment of a project's likely environmental effects and the measures for modifying or mitigating them.

To help local authorities adjudicate on specific issues, in 1993, the Department of Environment (now DEFRA) published a 'Planning Policy Guidance Note' on renewable energy (PPG 22). It strongly supports the concept of renewable-energy development and urges authorities to look sympathetically at proposals for the generation of power by alternative means to fossil fuels. It requires them to balance these with local environmental and planning concerns in the context of the government's overall policy on renewables. Given that planning issues have become a significant problem for some projects, particular wind farms, a revised version of PPG 22 is planned.

In addition, the government has asked the Regional Assemblies, the next layer up in the local government system, to produce planning guidance notes at their level – in effect, regional plans, including targets, set in the wider context of the UK's overall renewable-energy targets. The idea is that the regional plans and targets may, in turn, be used to shape local plans and policies.

The UK also has to put relevant EU legislation into effect, although in most cases it has been possible to do this through adjustments to existing UK legislation.

In the case of very large projects, such as nuclear power stations, major Public Inquiries, chaired by a government-appointed Inspector, have sometimes been set up. It is possible that in the case of a very large renewable-energy project, such as the Severn Barrage, the government would choose to set up a Public Inquiry. It would then consider the verdict of the Inquiry Inspector before reaching a decision on whether or not to proceed with the project. However, in 2001, radical changes were proposed in a Green Paper on Planning, which, if accepted, could lead to the introduction of new 'fast track' procedures for major projects of national significance, with planning consent, in effect, being decided by Parliament.

10.8 Promoting renewables

Renewable energy can be promoted in many ways. The simplest is through publicity, ensuring that potential users are fully informed of the latest technologies. Once under way, some technologies, such as wind power, have become self-publicising. Photos of wind turbines now adorn the annual reports of most energy-related companies who may be anxious to be seen to be 'green'. However, the potential of other technologies, such as some biofuels, may not be widely known.

Then there are educational initiatives. For example, the original text of the first edition of this book was based on material from an Open University resource pack on Renewable Energy for Tertiary Education, sponsored by the UK Department of Trade and Industry (DTI). The availability of the Internet, and its search engines, has helped enormously. The DTI at www.dti.gov.uk [accessed 25 June 2003], the Danish Wind Industry Association at www.windpower.org [accessed 25 June 2003] and the Energy Efficiency and Renewable Energy section of the US Department of Energy at www.eere.energy.gov [accessed 25 June 2003] have extensive information on renewable energy together with links to other more specific sites.

Supporting research and development

Since 1990, UK expenditure on research and development (R & D) for renewable energy has run at between £10–20 million per year. Many of the experimental projects funded have been described in the previous chapters. It is important to appreciate that support for R & D is very different from subsidising the large scale commercial deployment of new systems which may still be somewhat experimental. In the latter case, the sums of money involved can run into hundreds of millions of pounds per year (see Financial Incentives, below).

A modest amount of R & D investment can have significant results. In the 1980s, when a large number of small Danish companies were developing wind turbines to sell to California, the Danish Risø laboratory provided test facilities and certification procedures. These set the basis for reliable products and for the rapid expansion of the Danish turbine manufacturing industry which is now the largest in the world.

Targets

Governments can also promote renewable energy by setting targets. The European Commission in its 1997 White Paper (European Commission, 1997) asked for the contribution from renewables to EU primary energy consumption to be increased from 6% to 12% of EU by 2010. This proposal has been reinforced by a more recent Green Paper on Security of Supply (EC, 2000). The increase is to be achieved by the deployment of by 2010 40 GW of wind power, 3 GW of photovoltaics and an expansion in the use of small hydro, biomass and solar thermal collectors. As part of this, the UK has adopted a target of obtaining 10% of its electricity from renewable sources by 2010. A further EU directive proposes a target of 5.75% for the share of biofuels in the transport sector by this time.

Legislation and building regulations

Another important role task is the specification of standards and codes for the use of new forms of renewable energy – for example the Engineering Recommendations (Engineering Recommendation G77, 2000) that cover the connection of photovoltaic panels to the grid. Planning procedures can also be simplified. As mentioned above, documents such as the Planning Policy Guidance Note 22, can give advice and encouragement on planning for renewable energy. The difficult process of setting up an offshore wind farm can be simplified by the establishment of a 'one stop shop', where permission can be obtained from the large number of different authorities involved.

A more forceful approach is to include renewable energy in building codes. At present, the UK building regulations concentrate on energy conservation. In other countries, such as Greece or Israel, there are requirements that new housing should have solar collectors for water heating.

Financial incentives

In practice, most countries rely on various financial incentives. These can be summarized as:

- Exemption from energy taxes
- Capital Grants for renewable-energy schemes
- Auctions of supply contracts for renewable energy as used in the UK Non-Fossil Fuel Obligation (NFFO) scheme
- Renewables Obligations – obligations on electricity suppliers to purchase a specific proportion of renewable energy
- Renewable Energy Feed-In Tariffs (REFIT) – fixed premium prices for electricity from renewable sources.

Energy taxes

Energy is currently very cheap. In the UK, it is almost at an all-time low in relation to earnings, a topic which is discussed in the companion text *Energy for a Sustainable Future*. Other countries, notably Denmark and Japan, have higher taxes and energy prices. The imposition of energy taxes is fraught with difficulties relating to the effects on industrial production, commercial transport and public opinion. Despite this, high levels of taxation on petrol and diesel fuel are accepted as normal in most European countries.

In 2001, the UK government introduced its Climate Change Levy (CCL) for energy supplies to most companies (but not to domestic consumers). This is a tax of 0.43p per kWh on electricity, 0.15p per kW on gas and coal and 0.07p per kWh on LPG. Energy supplies from *new* renewable sources or 'high quality' CHP are exempt. This acts as a stimulus for the take-up of renewables and energy efficiency improvements and should raise approximately £1 billion per year. Despite its name, the CCL does not exempt nuclear power, despite the fact that it does not produce a significant amount of greenhouse gases. The exclusion of domestic consumers avoids problems of increasing the number of households in 'fuel poverty'. In addition, companies that use large amounts of electricity, such as the aluminium industry, are exempt, but they do have to make a commitment to making emission savings.

Capital grants

These have also been widely used to promote renewable energy. In the UK, they are available on a competition basis for offshore wind developments, planting energy crops, and a new scheme for solar thermal and PV systems has been introduced.

Auctions of renewable-energy supply contracts – NFFO

The main mechanism for supporting renewable-energy projects in the UK between 1990 and 2002 was funding from the **Non-Fossil Fuel Obligation (NFFO)** and its associated fossil fuel levy. This levy on electricity consumers was set up in 1989 at the time of the privatization of the electricity industry primarily to provide a subsidy for *nuclear* power. In 1990/91, it raised a total of £1.175 billion, though the amounts declined in subsequent years. Only a small proportion of this money was used to subsidise renewable energy.

From time to time, the government issued successive rounds of Non-Fossil Fuel Orders. These were auctions of limited numbers of contracts for the supply of renewable energy within a series of technology bands (wind, small-scale hydro, landfill gas, etc). The contracts gave each project a guaranteed market over a period long enough to recover its investment costs. They were awarded to those bids with the lowest electricity prices submitted within each band.

Initially, given the inclusion of nuclear power, it was not clear whether the European Commission would allow the UK to continue the fossil fuel levy beyond 1998. As a result, NFFO contracts approved before 1993/94 were also set to expire in 1998. Since the capital costs for these projects had to be recouped within a very short period, they had to be given a high price for their electricity, anything from 6p to 11p per kWh. This has given the unfortunate impression that renewable energy is expensive.

Subsequently the Commission raised no objection to support for renewables over a longer period and from 1994 onwards, contracts were awarded for up to 15 years. This resulted in substantially lower prices, ranging from around 3.5p/kWh for electricity from landfill gas, through to 3–6p/kWh for power from wind farms, to around 8.5p/kWh for electricity produced through the gasification of forestry wastes. The original NFFO arrangements applied only to England and Wales, but comparable schemes were set up in 1993 for Scotland and Northern Ireland. The Republic of Ireland and some other countries have also adopted a similar approach.

By the end of 2002 some 3.6 GWdnc (declared net capacity) of projects had been contracted of which just over 1 GWdnc were operational. The use of the 'dnc' terminology (see Box 10.4) allows rough comparisons with the annual amounts of electricity produced by conventional generation capacity. The mix of operational technologies includes 106 MWdnc of biomass, 436 MWdnc of landfill gas, 223 MWdnc of municipal and industrial waste and 214 MWdnc of wind power. The latter is equivalent to about 500 MW when expressed in terms of the full rated power of the turbines.

BOX 10.4 Declared net capacity

Under the UK NFFO scheme, the power ratings of renewable-energy equipment are specified in 'MW dnc' or **Declared Net Capacity**. Elsewhere they are simply described in terms of their 'rated capacity' in 'MW', or 'MW_{pk}' for PV systems. This can cause confusion.

The 'rated capacity' of a generating plant (sometimes called the 'nameplate' rating, i.e. the figure written on the generator together with the manufacturer's name) is the maximum output that can be sustained continuously. Declared net capacity normally describes the *net average power output*. For large fossil-fuelled plants this takes into account the electricity used to run their own pumps and internal services, so their average *net* power output or *capacity*, is slightly less than 100% of their *rated capacity*.

For renewable-energy technologies, this terminology was extended to take into account the intermittency of the source, allowing some comparison of the likely annual electricity outputs of different types of conventional and renewable-energy plant. For example:

Hydro and biofuel plant	dnc = Rated capacity
Wind turbines	dnc = 0.43 × Rated capacity, i.e. 1 MW rated = 430 kW dnc
Wave and tidal plant	dnc = 0.33 × Rated capacity, i.e. 1 MW rated = 330 kW dnc

Renewables obligation

In 2002, the NFFO system was replaced by a **Renewables Obligation**, a statutory duty placed on electricity supply companies. Each year this requires that they obtain a certain proportion of their electricity from renewable sources. How exactly they do this is left to the market. Each year this proportion will be increased towards the 10% target for 2010. The supply companies can pass any extra costs on to the consumers, but a ceiling of 3p/kWh has been set on these. Given that the extra costs apply only to a maximum of 10% of electricity supplied, this implies that, at worst, the average consumer might end up paying about 4% extra on their bill 2010.

At present (2003) renewable energy in the UK is promoted by a range of methods. Table 10.5 summarizes those that currently apply in England, Wales and Scotland.

Table 10.5 Summary of incentives for renewables

Source	10% Target	Renewables obligation	CCL exemption	Capital grants
Landfill and sewage gas	✓	✓	✓	
Energy from waste – mass burning	✓		✓	
Large hydro >10 MW	✓			
Small hydro <10 MW	✓	✓	✓	
Onshore wind	✓	✓	✓	
Offshore wind	✓	✓	✓	✓
Agricultural and forestry residues	✓	✓	✓	
Energy crops	✓	✓	✓	✓
Wave power	✓	✓	✓	
Photovoltaics	✓	✓	✓	✓

Source: adapted from DTI (2000)

Guaranteed prices – renewable energy feed-in tariff (REFIT)

Another approach, which has been used to considerable effect in Denmark, Germany and Spain, has been to offer guaranteed high purchase prices for electricity generated from renewable sources – a renewables feed-in tariff or REFIT. For example, in Denmark in 2002, an offshore wind turbine could expect to receive the normal market price for electricity plus an extra 0.453 DKK per kWh (about 4p per kWh).

Which system is best?

From a government perspective, it is best if renewable-energy targets can be met with the minimum of bureaucracy. While capital grants seem simple, it is desirable that payments are related to performance to the annual output rather than the rated capacity. The subsidy schemes in California in the 1980s, where payments were based simply on capital costs, sometimes led to wind farms of dubious mechanical reliability being set up.

Governments would also like to minimize the use of taxpayers' money. This favours 'market solutions' such as the Renewables Obligation. On the other hand, from a renewable-energy investor's point of view, firm prices with the minimum of paperwork are desirable. Both NFFO and REFIT offer this, though the NFFO process required the preparation of a competitive bid, with the possibility of it being rejected.

Payments under both NFFO and REFIT are based on the amount of electricity actually produced, giving a strong incentive to renewable-energy generators to maintain output. However, the competitive nature of the NFFO bidding has been criticized for leading to wind farms in the UK being built in the areas with the highest wind speeds, which are also the most environmentally sensitive. This has created much opposition and many schemes have been refused planning permission.

Supporters of the competitive NFFO system have stressed that there has been a 'convergence' between the premium prices paid and the normal market prices, so that eventually renewables will become commercially viable without further support.

A report by the European Environment Agency (EEA, 2001), noted that 80% of new wind energy output in the EU between 1993 and 1999 had been created in the three countries that operated REFIT funding systems. By 2002, Germany had installed some 10 000 MW of wind capacity while the UK, with a much better wind resource had only installed around 500 MW. In Denmark, the simplicity of the REFIT system has encouraged the smaller developer. More than 80% of the country's wind turbines are owned by wind energy co-operatives or individual farmers (Krohn, 2002).

Future EU policies

A persistent problem for renewable-energy support mechanisms has been the conflict between two major goals of the European Union: the protection of the environment and promotion of a pan-European free market in energy. Under this, ideally, any consumer should be able to purchase energy from any supplier within the EU. This would make it difficult to levy individual *national* energy taxes. There have been court challenges in both Denmark and Germany as to whether or not measures designed to promote energy conservation and renewable energy are legal under the free energy market regulations. At present (2003) the situation can only be described as confused.

The European Commission is attempting to move towards policies based on **carbon trading**, i.e. where some, at least, of the 'environmental benefit' of renewable energy is also a tradable commodity. The UN Framework convention on Climate Change negotiated at Kyoto in 1997 set targets for emission reductions for greenhouse gases for each participating country or region. In practice, different countries have different opportunities for reducing their emissions. Countries that can achieve savings in emissions above their individual quotas can sell carbon credits in the form of emission permits to those who cannot. The Commission has now approved EU-wide trading to start in 2005 in order to meet CO_2 emission targets for individual member countries.

The Commission has also set up a **Renewable Energy Certificate System** (RECS) which it intends to develop. The general idea is that a renewable electricity plant, for example, has two products. The electricity is sold at the market price, by the MWh, on the conventional electricity market; and its 'environmental benefit' is sold in the form of certificates in a separate market. Although such ideas are in their infancy now, they could become an incentive for the take-up of renewable-energy technologies in the future.

A further problem is the proper integration into the EU Common Agricultural Policy (CAP) of policy on the growing of biofuels. Since the end of the Second World War, most western European countries have given their farmers subsidies to increase agricultural production. The food shortages of the 1950s have given way to embarrassing production surpluses. Recent attempts to reduce these subsidies have included a policy under which farmers are paid to 'set aside' part of their land on which food will not be grown. This has encouraged some farmers to look at alternative sources of income, such as wind farming or growing biofuels. The reform of CAP in 2003 now allows UK farmers to grow biofuels on 'set-aside' land.

10.9 Energy scenarios: Danish examples

Future energy scenarios for the UK have been mentioned briefly in Chapter 1 and are discussed in more detail in the companion text. Scenarios can be influential in shaping national policy as shown by the example of Denmark. (Some of the finer details of Danish energy use have been described in the companion text.)

Briefly, during the 1960s, Denmark's energy use expanded rapidly in an era of cheap oil (see Figure 10.22). By 1972, it had become almost totally reliant on imported oil and was badly hit by the OPEC oil price rises in 1973 and 1979, which created considerable political embarrassment.

The situation encouraged energy researchers from Danish universities to produce, in 1983, an 'alternative energy scenario' (AE83). It outlined a programme aimed at cutting oil and coal imports to zero by 2030, to be achieved by an increase in the proportion of renewables (wind, solar and biofuels) and a sharp reduction in total energy demand. It also did not include nuclear power, which was very unpopular in Denmark at the time. The projection, also shown in Figure 10.22, contrasted with the official 1981 Danish government projection (EP81), which suggested a continued growth in energy demand (see Norgaard and Meyer, 1989).

In practice, the government solved its immediate problems in several ways.

(a) it pursued a policy of energy conservation, imposing high energy taxes and new regulations to encourage the insulation of buildings and promote the use of CHP. The resulting cuts in energy use for space heating have been quite spectacular. Between 1972 and 1985, the total area of heated building floor area increased by 30%, but the total amount of energy used to heat it *decreased* by 30%.

(b) it switched its power stations from oil-firing to coal-firing. Consequently, national reliance on oil dropped from 93% in 1972 to only 43% twenty years later in 1992.

(c) it developed its own oil and gas resources in the North Sea. The results started to appear in the early 1980s and by 1997 Denmark had become a net energy exporter.

(d) it made considerable efforts to start building up its renewable resources – particularly biomass and wind power.

The fact that actual Danish energy consumption up to 2001 has not followed either the AE83 projection or the government EP81 projection confirms the point that scenarios are only pictures of what *could happen* rather than what *will happen*.

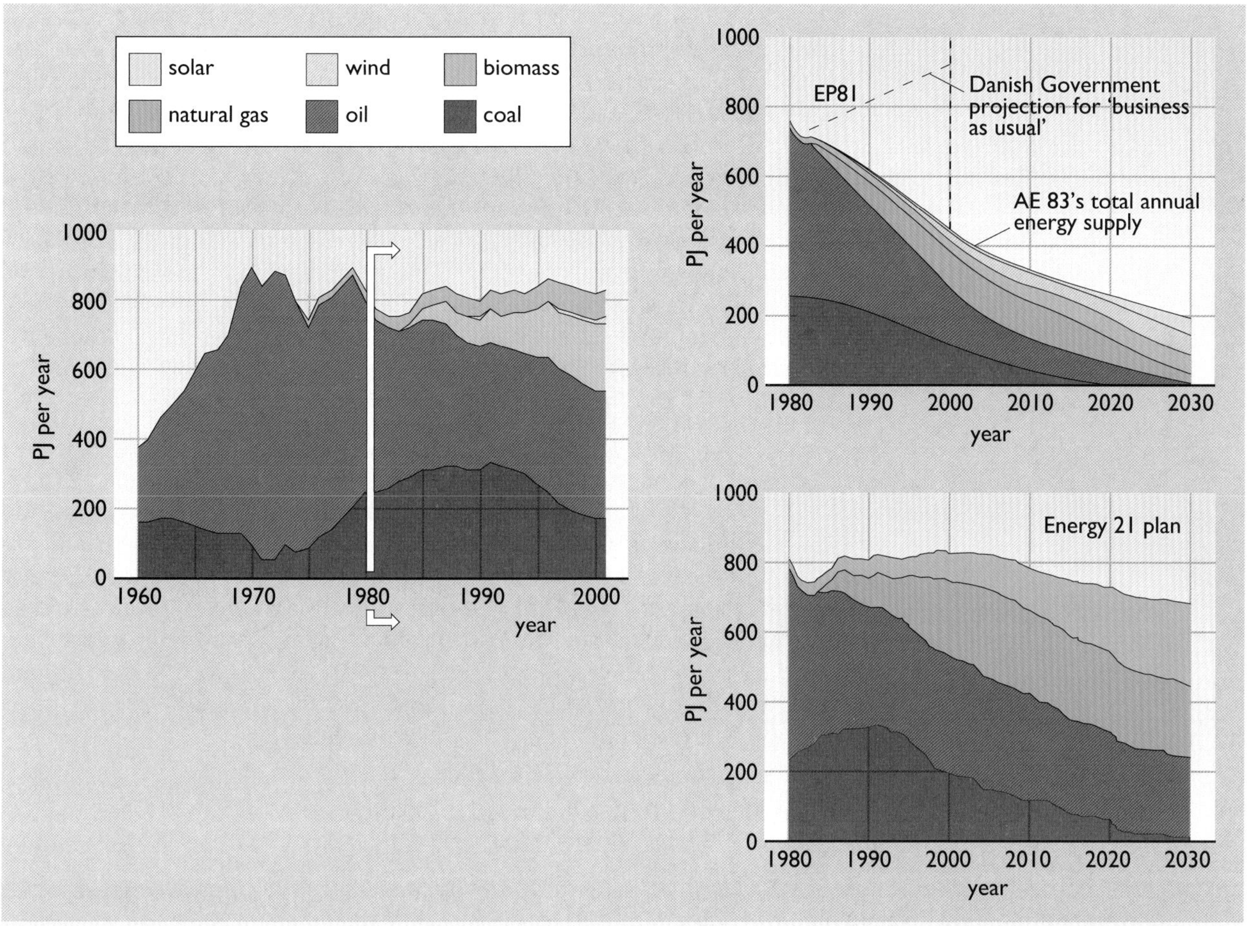

Figure 10.22 (Left) Primary energy consumption in Denmark, 1960–2001, compared with (Right Top) the 1983 Alternative Energy scenario (AE83), and the Danish Government's 1981 'business as usual' projection (EP81) and (Right Bottom) their Energy 21 scenario (sources: Eurostat, 1993; Ministry of Environment and Energy, 1996; Danish Energy Agency, 2003; Norgaard and Meyer, 1989)

There are several lessons to be drawn from Danish energy policy during the 1980s:

(a) Primary energy use and GDP do not necessarily go hand in hand – between 1977 and 1999, Danish GDP increased by nearly 50%, yet primary energy consumption hardly increased at all (Dal and Jensen, 2000)

(b) Actually reducing *total* energy demand may be difficult – despite high taxes and progressive policies, motor fuel consumption increased by almost 60% and electricity demand by nearly 70% over the same period.

(c) Renewable energy use can be promoted effectively. By 2001, its contribution was 90 PJ per year, a *larger* amount than that suggested in the AE83 scenario.

By the mid–1990s, the focus of energy policy had shifted from self-sufficiency to climate change. Denmark is a low-lying country and any future sea level rises would have serious consequences. In 1996, the centre-left government adopted a policy that national CO_2 emissions should be cut by 20% from their 1988 levels by 2005 and by 50% by 2030. This was set out in their *Energy 21* plan (Ministry of Environment and Energy, 1996). Under this scenario, total primary energy use would fall only slightly and the use of coal would be replaced by increased use of natural gas and renewable energy. It was suggested that by 2030 a half of Denmark's electricity could come from renewable energy. In order to implement this policy, there were continued high energy taxes and support for renewable energy under a REFIT scheme.

But time moves on. In 2001, a new centre-right government was elected committed to tax reform. Since by 2002 Denmark already had 2600 MW of wind power, against an official target of only 1500 MW by 2005, (Krohn, 2002) the new government has taken the attitude that the country's environmental initiative is 'ahead of schedule'. The *Energy 21* plan has been dropped, orders for three offshore wind farms postponed and many tax incentives for energy efficiency scrapped. It is basing its policy to deal with future CO_2 emission quotas on carbon trading. Although the subsidy cuts have caused much dismay amongst the Danish renewable-energy community, the government's latest energy forecast (DEA, 2003) still expects the amount of renewable energy to increase slowly.

10.10 Global scenarios

If renewables were to be given high levels of government support and a substantial share of global energy investment in the future, could they make a major contribution to world energy needs in the long term? A wide variety of studies of long-term energy futures at global level have been prepared by various governments, corporations and non-governmental organizations in recent years. As we shall see, most of them envisage an increasingly important role for renewable energy.

International Energy Agency projections

Looking into the medium-term future, to 2020, one of the most widely-quoted set of energy projections is published at regular intervals by the International Energy Agency (IEA). The Agency's *World Energy Outlook* (IEA, 2000 and subsequent years) suggests that by 2020 world energy demand is likely to have risen to around 550 EJ (c.13 000 Mtoe) (see Figure 10.23). Of this, approximately 23% (c. 3000 Mtoe) is likely to be supplied by coal, 27% (c.3500 Mtoe) by gas, 42% (c. 5500 Mtoe) by oil, 4% (c. 500 Mtoe) by nuclear energy, 2% (c. 300 Mtoe) by hydropower and another 2% (c. 300 Mtoe) by other renewables, including modern biofuels (but excluding non-commercially traded energy principally traditional biofuels). Clearly, the IEA envisages renewables still playing only a minor role in the energy system of 2020 – though it does envisage a much bigger role for them in the following decades.

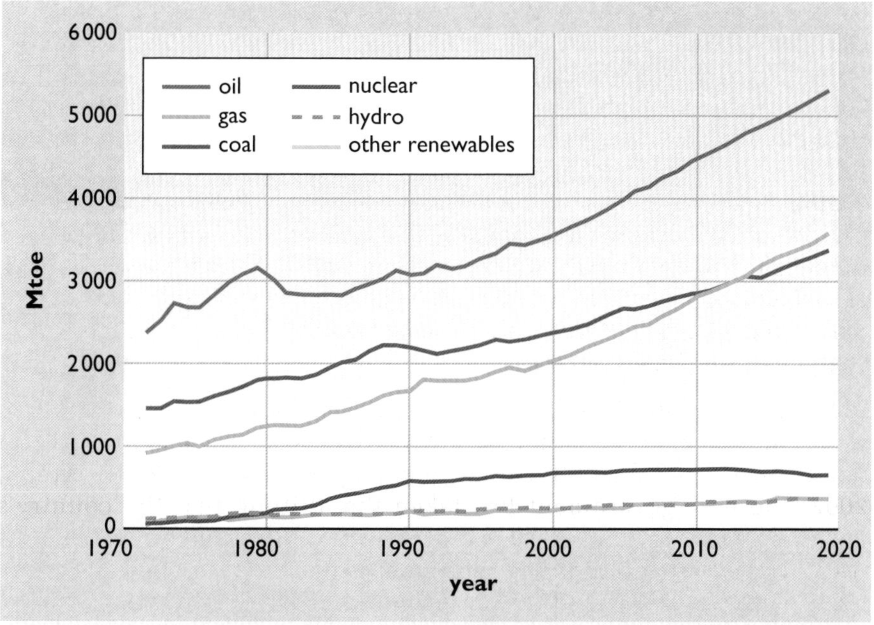

Figure 10.23 World Primary Energy Supply by Fuel, 1971–2020. Note: 1000 Mtoe is equivalent to 42 Exajoules (EJ). Note also that the two lines representing hydro and other renewables, respectively, are almost coincident (source: International Energy Agency, 2001)

The World Energy Council Scenarios

A comprehensive set of long-term world energy scenarios for the twenty-first century has been produced in collaboration between the World Energy Council, an association of the world's major energy companies, based in London, and the International Institute for Applied Systems Analysis (IIASA), a 'think tank' based in Austria. For simplicity, the scenarios will be referred to here as the World Energy Council scenarios. The following brief summary of them is based on a report prepared as part of the United Nations 'World Energy Assessment' exercise (Goldemberg *et al*, 2000).

The WEC/IIASA team produced a set of three basic 'families' of scenarios, which they call 'cases'. The first, Case A, is called 'high growth' and includes three variants: A1, 'ample oil and gas'; A2, 'return to coal'; and A3, 'non-fossil future'. The second, Case B, which has no variants, is called 'middle course' and is designed to represent a path to the future involving only gradual changes. The third, Case C, is called 'ecologically driven' and includes two variants: C1, 'new renewables'; and C2, 'renewables and new nuclear'.

The three cases are illustrated in Figure 10.24 and the main scenario characteristics for 2050 are summarized in Table 10.6.

The WEC's scenarios envisage the share of renewable energy rising rapidly, to around 22% by 2050 in the Middle Course scenario, to between 22–30% in the High Growth scenarios, and to 37–39% in the Ecologically Driven scenarios.

Table 10.6 The World Energy Council Scenarios: Characteristics of the three cases for the world in 2050 compared with 1990

	Base year: 1990	**Case A**			**Case B**	**Case C**	
		(A1)	(A2)	(A3)		(C1)	(C2)
Primary energy, Gtoe	9	25	25	25	20	14	14
Primary energy mix, percent							
Coal	24	15	32	9	21	11	10
Oil	34	32	19	18	20	19	18
Gas	19	19	22	32	23	27	23
Nuclear	5	12	4	11	14	4	12
Renewables	18	22	23	30	22	39	37
Resource use 1990 to 2050, Gtoe							
Coal		206	273	158	194	125	123
Oil		297	261	245	220	180	180
Gas		211	211	253	196	181	171
Energy sector investment, trillion US$	0.2	0.8	1.2	0.9	0.8	0.5	0.5
US$/toe supplied	27	33	47	36	40	36	37
As a percentage of GWP	1.2	0.8	1.1	0.9	1.1	0.7	0.7
Final Energy, Gtoe							
Final energy mix, percent							
Solids	30	16	19	19	23	20	20
Liquids	39	42	36	33	33	34	34
Electricity	13	17	18	18	17	18	17
Other[a]	18	25	27	31	28	29	29
Emissions							
Sulphur, MtS	59	54	64	45	55	22	22
Net carbon, GtC[b]	6	12	15	9	10	5	5

Note: Subtotals may not add due to independent rounding

[a] District heat, gas and hydrogen

[b] Net carbon emissions do not include feedstocks and other non-energy emissions or CO_2 used for enhanced oil recovery.

Note: GWP is annual Gross World Product

The Shell Scenarios

In 1995, planners at Shell International Petroleum published the results of a detailed study of the long-term future prospects for the world's energy system (see Shell International, 1995 and Herkstroter, 1997).

In it, they created two scenarios sketching out two possible evolutionary paths that the world's energy system might follow during the twenty-first century.

The first Shell scenario, 'Sustained Growth', envisages world energy demand increasing from the current level of around 400 EJ to approximately

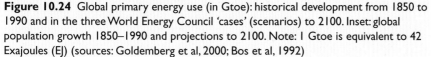

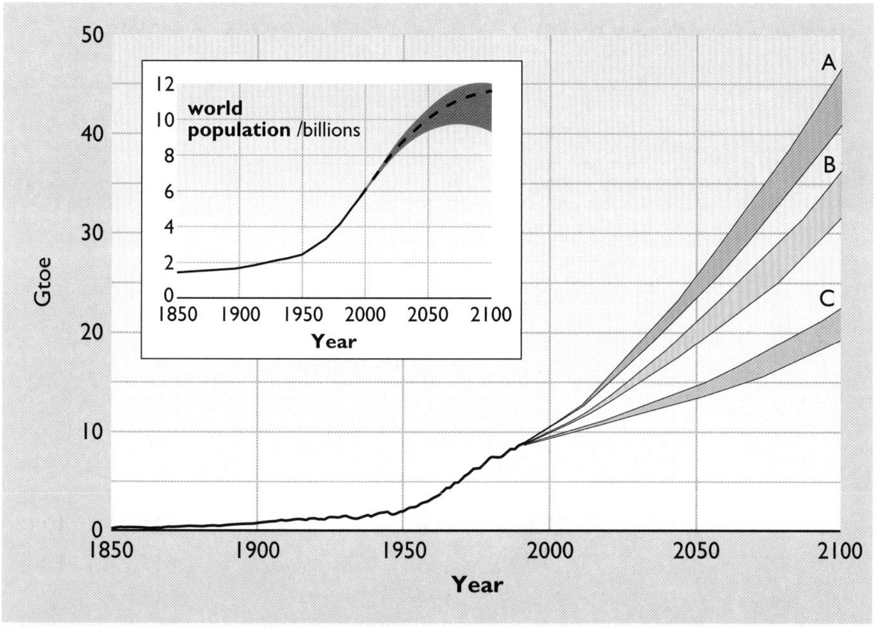

Figure 10.24 Global primary energy use (in Gtoe): historical development from 1850 to 1990 and in the three World Energy Council 'cases' (scenarios) to 2100. Inset: global population growth 1850–1990 and projections to 2100. Note: 1 Gtoe is equivalent to 42 Exajoules (EJ) (sources: Goldemberg et al, 2000; Bos et al, 1992)

1500 EJ by 2060, in the absence of any special efforts being made to introduce energy efficiency improvements. World GDP growth averages 3% per annum and world energy demand grows on average at 2% per annum. By the middle of the next century, the world's sources of energy are considered likely to be more diversified than at present, with new renewables supplying around 50% of global commercial energy demand and fossil fuels beginning to decline – not so much because of absolute limitations on supply but because renewables have become more competitive.

In the second Shell scenario, called 'Dematerialization', the world economy becomes more frugal in its use of materials and energy. Technological developments in currently-unrelated areas converge to create breakthroughs in the efficiency of products and processes. Nevertheless, energy demand rises at around 1% per annum to around 1000 EJ by 2060, with the renewables' contribution slightly lower than the 50% envisaged in 'Sustained Growth' because of the lower overall demand.

In 2001, Shell International went on to publish two new long-term energy scenarios, called Dynamics as Usual, and Spirit of the Coming Age (Shell International, 2001). 'Dynamics as Usual' envisages 'an evolutionary progression from coal to oil to gas to renewables (and possibly nuclear)...', whereas 'Spirit of the Coming Age' is more revolutionary, envisaging 'the potential for a hydrogen economy – supported by developments in fuel cells, advanced hydrocarbon technologies and carbon dioxide sequestration' (Shell, 2001).

Figure 10.25 and Table 10.7 show Shell's analysis of the market shares of various fuels in the world fuel mix for the past (1850 to 2000), and its projections of the likely future shares from 2000 to 2050 in its Dynamics as Usual scenario. This envisages overall world demand rising to some 852 EJ by 2050, with the market share of coal, oil and traditional biofuels continuing to decline as the twenty-first century progresses. The share of nuclear power stays relatively constant, the share of gas increases until around 2030 and then declines slightly, the share of hydro increases and then levels-out, and the share of biofuels and 'new' renewables steadily grows. By 2050, the combined market share of all renewables, including 'new' renewables, hydropower, new and 'traditional' biofuels, has risen to around 33%.

Shell's projections for its Spirit of the Coming Age scenario are shown in Figure 10.26 and Table 10.7. Here, world demand by 2050 rises to 1121 EJ, substantially higher than in Dynamics as Usual. The share of solid and liquid fuels steadily declines, but the share of gaseous fuels, in the form of either natural gas or hydrogen, steadily grows, as does the share of electricity from hydro, nuclear and 'new' renewables. By 2050, the share of all renewables reaches around 30% of total demand.

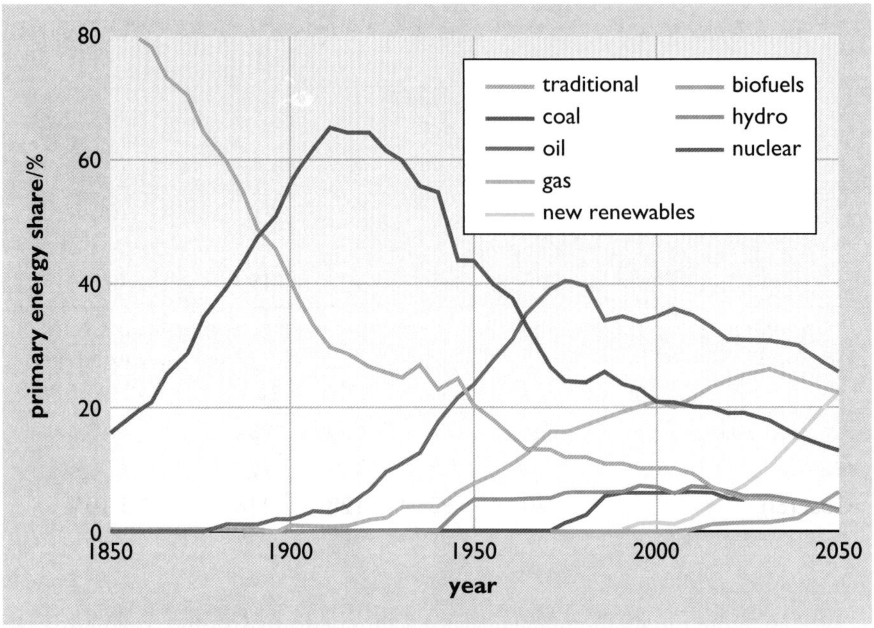

Figure 10.25 In Shell's 'Dynamics as Usual' Scenario, energy supplies continue to evolve from high- to low-carbon fuels and towards electricity (from increasingly distributed sources) as the dominant energy carrier, driven by demands for security, cleanliness and sustainability (source: Shell, 2001)

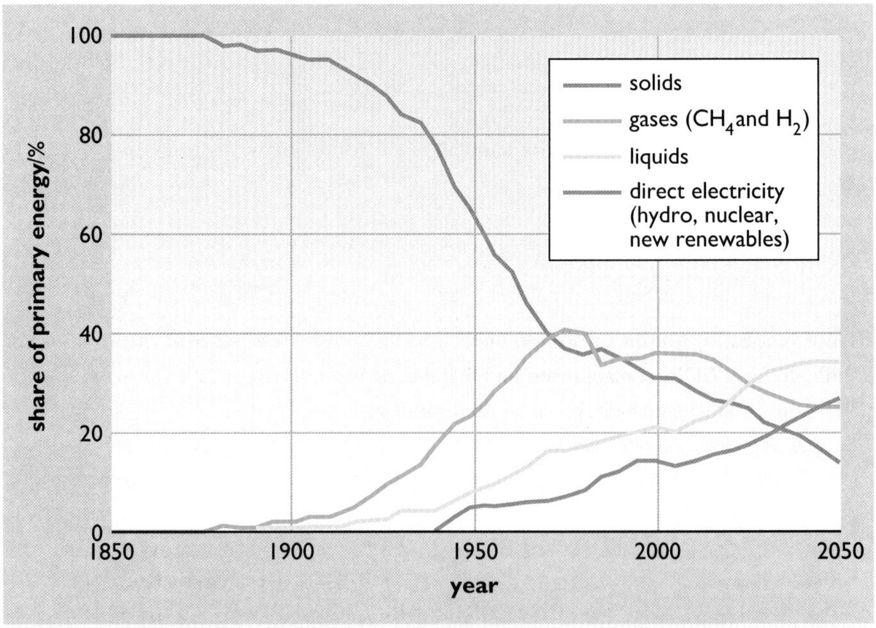

Figure 10.26 In Shell's 'Spirit of the Coming Age' scenario, energy supplies continue to evolve from solids through liquids to gas (first methane then hydrogen), supplemented by direct electricity from renewables and nuclear, and by long-term sources of hydrogen gas (source: Shell, 2001)

Table 10.7 Summary of the main quantitative characteristics of Shell's 'Dynamics as Usual' and 'Spirit of the Coming Age' scenarios

	1975	2000	2025	2050	Annual growth rate 1975–2000	Annual growth rate 2000–25	Annual growth rate 2025–50
World Population (billion)	4	6	8	9	1.50%	1.00%	0.60%
World GDP (trillion 2000 $ PPP*)	23	49	108	196	3.10%	3.20%	2.40%
Dynamics as Usual	**1975**	**2000**	**2025**	**2050**	**Annual growth rate 1975–2000**	**Annual growth rate 2000–25**	**Annual growth rate 2025–50**
Primary Energy (EJ)	256	407	640	852	1.90%	1.80%	1.20%
Oil (EJ)	117	159	210	229	1.20%	1.10%	0.30%
Coal (EJ)	70	93	128	118	1.10%	1.30%	–0.30%
Coal CH_4/H_2† (EJ)	0	0	4	16	0.00%	0.00%	5.90%
Natural Gas (EJ)	47	93	167	177	2.70%	2.40%	0.20%
Nuclear (EJ)	4	29	35	32	8.10%	0.80%	–0.40%
Hydro (EJ)	17	30	41	39	2.40%	1.30%	–0.30%
Biofuels (EJ)	0	0	5	52	0.00%	10.20%	10.10%
Other Renewables (EJ)	0	4	50	191	8.70%	11.20%	5.50%
Spirit of the coming age	**1975**	**2000**	**2025**	**2050**	**Annual growth rate 1975–2000**	**Annual growth rate 2000–25**	**Annual growth rate 2025–50**
Primary Energy (EJ)	256	407	750	1121	1.90%	2.50%	1.60%
Oil (EJ)	117	159	233	185	1.20%	1.60%	–0.90%
Coal (EJ)	70	93	150	119	1.10%	1.90%	–0.90%
Coal CH_4/H_2† (EJ)	0	0	6	97	0.00%	0.00%	11.60%
Natural Gas (EJ)	47	93	220	300	2.70%	3.50%	1.30%
Nuclear (EJ)	4	29	46	84	8.10%	1.90%	2.40%
Hydro (EJ)	17	30	49	64	2.40%	2.00%	1.10%
Biofuels (EJ)	0	0	7	108	0.00%	11.80%	11.80%
Other Renewables (EJ)	0	4	38	164	8.70%	9.90%	6.00%

Note: Nuclear, hydro, wind solar and wave contributions are expressed as thermal equivalents.

* PPP denotes GDP estimate made on a Purchasing Power Parity basis

† Denotes methane or hydrogen manufactured from coal

Source: Shell International, 2001

In both of Shell's 2001 scenarios, the overall shares of renewable energy projected for 2050 are around one third, significantly lower that the share of roughly 50% projected in their 1995 predecessors. This change illustrates how different value judgements made in the preparation of scenarios, whether by corporations, governments or non-governmental organizations, can make a major difference to the outcomes. In Shell's case, its commitment to expansion of natural gas in the medium term seems to have played a key role in this re-evaluation of the relative roles of renewables and fossil fuels.

As Philip Watts, then chairman of Shell group's committee of managing directors, stressed in his introduction to the 2001 report, 'Expanding the use of gas is perhaps the most important immediate way of responding to the climate threat, as well as of improving air quality.'

The Greenpeace fossil-free energy scenario

Perhaps the most optimistic scenario describing the prospects for renewables by the end of the twenty-first century is the 'fossil-free energy scenario' developed for Greenpeace International by the Stockholm Environment Institute in the early 1990s (Lazarus *et al.*, 1993). Like the WEC and UNCED scenarios, the Greenpeace scenario is largely based on conventional assumptions about population increases and economic growth. The authors state that they have reservations about these assumptions on environmental and social grounds, but have retained them in order to allow their scenario to be compared with others.

In the fossil-free energy scenario (FFES), world primary demand increases by a factor of approximately 2.5, to around 1000 EJ (approximately 24 000 Mtoe) by the end of the twenty-first century (Figure 10.26). All fossil and nuclear fuels are assumed to be entirely phased out by the end of the period, and replaced by a mixture of solar, wind, biomass, hydro and geothermal energy. Improvements in energy efficiency, to be achieved using 'market or near-market' technologies, would result in overall primary energy demand levelling out and then falling slightly around 2030, before rising again towards the end of the century.

The main renewable-energy technologies contributing to the supply mix by 2100 in the Greenpeace scenario include co-generation of electricity and heat (CHP) from biomass wastes, the use of fuel cells for electricity (and heat) production, and increasing use of wind turbines, photovoltaics and solar thermal electric power generation. Hydrogen, produced by

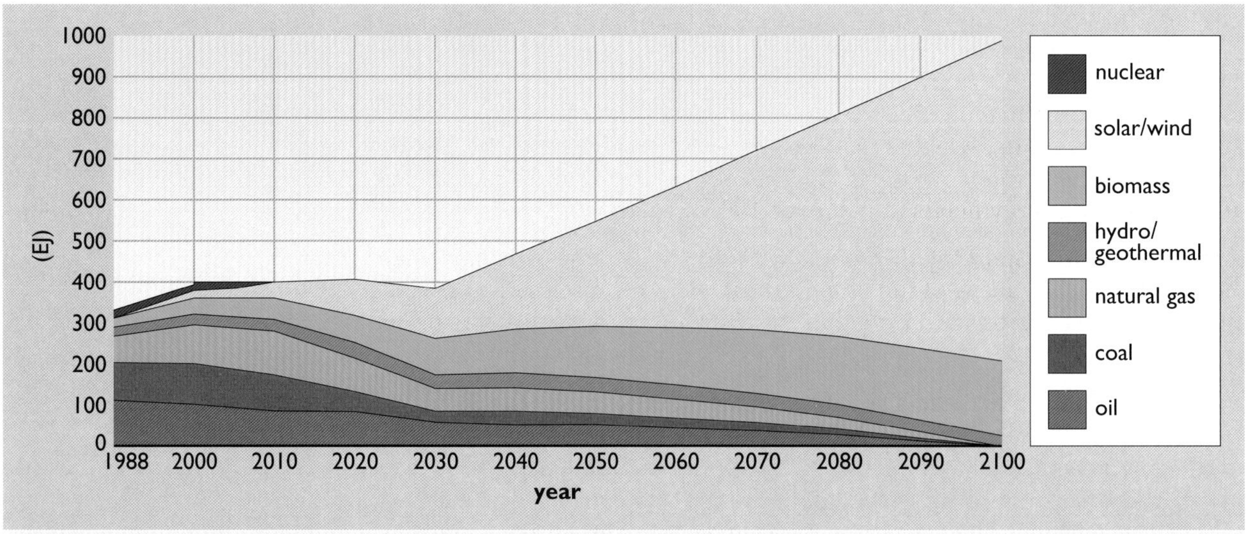

Figure 10.27 The Greenpeace Fossil-Free Energy Scenario. World energy demand increases by a factor of approximately 2.5. All fossil and nuclear fuels are phased out and replaced by a mixture of solar, wind, biomass, hydro and geothermal sources by 2100 (source: Lazarus et al, 1993)

electrolysis from solar/wind sources or from biomass, provides an increasing proportion of transport fuel and a means of storing power from intermittent sources.

10.11 Conclusions

Greenpeace's scenario of a totally fossil-free energy future may seem Utopian to many. Clearly, the technical, social and political challenges involved in phasing-out fossil and nuclear fuels and replacing them entirely with renewables are immense.

On the other hand, there seems little doubt that world population, and the world economy, will continue to grow very substantially during the twenty-first century. An accompanying rise in global primary energy use seems extremely difficult (even if not technically impossible) to avoid. Moreover, if a substantial share of this additional energy is not to be supplied by renewables, it will have to come from fossil or nuclear fuels, with all the familiar environmental and/or resource depletion concerns that the use of these sources entails.

The extent to which renewables prove successful in increasing their share of world-wide energy supplies will depend on many factors. These include: the extent to which investment in research and development and large-scale production can bring about efficiency improvements and cost reductions; the outcome of debates about the environmental and social costs of conventional sources and the extent to which these costs are reflected in energy prices; the future patterns of world economic and population growth, and their effect on the level of demand for various forms of energy; the impact of these considerations on the priorities of governments; and the environmental and social acceptability of renewables to the public.

However, our overall conclusion is that renewables are likely to play a greatly-increased role in future energy supplies. Even in the World Energy Council's 'High Growth' scenarios, the renewables' contribution to world energy needs by 2050 is projected to rise to between 22–30%. The WEC's 'Ecologically-driven' scenarios envisage a renewables contribution by 2050 of nearly 40%. And if history should prove the even more optimistic Greenpeace projections to be more accurate, renewables could be supplying *all* the world's energy needs in just over 100 years' time.

Meanwhile, we hope that this book will not only help to promote an improved understanding of the potential of the renewable energy sources, but also play a part in facilitating their deployment on a world-wide basis, as countries progress towards a sustainable world economy during the twenty-first century.

References

Bodlund, B. *et al.* (1989) 'The challenge of choices: technology options for the Swedish electricity sector' in Johansson *et al* (eds) *Electricity – Efficient End Use*, Lund University Press.

Christiansen, A. C. (2003) *Viewpoint: Lessons learned in 2002 and future challenges,* Pointcarbon Europe Weekly, Jan 10, 2003, downloadable from Pointcarbon.com [accessed 26 June 2003].

European Commission (2001) *ExternE: Externalities of Energy,* DG 12, Luxembourg, Science Research and Development.

Eyre, N. J. (1990) *Gaseous Emissions due to Electricity Fuel Cycles in the UK*, Harwell, Oxon, ETSU.

Blomen, L. M. J. (1989) 'Fuel cells: a review of fuel cell technology and its applications' in Johansson *et al* (eds) *Electricity – Efficient End Use*, Lund University Press.

Chapman, J., and Gross, R. (2001) *Technical and economic potential of renewable energy generating technologies: Potentials and cost reductions to 2020,* Working Paper for UK Cabinet Office Performance and Innovation Unit (now Strategy Unit) Energy Review, downloadable from www.strategy.gov.uk/2002/energy/workingpapers.shtml [accessed 26 June 2003].

Commission of the European Communities (2001) *European Energy Infrastructure*, COM (2001) 775 final, downloadable from europa.eu.int/comm/ [accessed 26 June 2003].

Dal, P. and Jensen, H. S. (2000) *Energy Efficiency in Denmark,* Danish Energy Ministry.

Danish Energy Authority (2003) *Energy consumption forecast 2003,* downloadable from www.ens.dk [accessed 26 June 2003].

Department of Environment (1993) Planning Policy Guidance Note No.22 – Renewable Energy, HMSO.

Department of Trade and Industry (1995) *Energy Projections for the UK*, Energy Paper 65, HMSO.

Department of Trade and Industry (2000) *New and Renewable Energy: Prospects for the 21st Century: The Renewables Obligation Preliminary Consultation*', downloadable from www.dti.gov.uk [accessed 26 June 2003].

Department of Trade and Industry (2002a) *Energy Flow Chart 2001*, HMSO downloadable from www.dti.gov.uk [accessed 26 June 2003].

Department of Trade and Industry (2002b) *NFFO Fact Sheet 11*, downloadable from www.dti.gov.uk [accessed 26 June 2003].

Department of Trade and Industry (2002c) *Energy in Brief, December 2002*, HMSO downloadable from www.dti.gov.uk [accessed 26 June 2003].

EEA (2001) 'Renewable energies: success stories', *Environmental issue report No 27,* Copenhagen, EEA. Available as pdf from: reports.eea.eu.int/environmental_issue_report_2001_27/en/Issues_No_27_content.pdf [accessed 10 December 2003].

EA technology (2000) *Overcoming Barriers to Scheduling Embedded Generation to Support Distribution Networks*, ETSU report K/EL/00217/REP, downloadable from www.dti.gov.uk [accessed 26 June 2003].

Engineering Recommendation G77 (2000) *UK technical guidelines for inverter connected single phase photovoltaic (pv) generators up to 5 kVA*, Electricity Association, online at: www.pv-uk.org.uk/reference/grid_con/g77oview.htm [accessed 27 June 2003].

ETSU (1999) *New and Renewable Energy: Prospects in the UK for the 21st Century – Supporting Analysis*, HMSO.

ETSU / PB Power (2002) *Concept Study – Western Offshore Transmission Grid*, downloadable from www.dti.gov.uk [accessed 26 June 2003].

Eurogas (2002) *Annual Report 2001*, downloadable from www.eurogas.org [accessed 26 June 2003].

European Commission (1997): *Energy for the Future – Renewable Sources of Energy – White Paper for a Community Strategy and Action Plan*, COM (97)599 final (26/11/1997) downloadable from www.europa.eu.int/comm/energy/en/com599.htm [accessed 26 June 2003].

European Commission (2000) *Green Paper – Towards a European strategy for the security of energy supply*, (COM (2000) 769 final) downloadable at www.europa.eu.int/comm/energy_transport/en/lpi_lv_en1.html [accessed 26 June 2003].

Eurostat (1993) *Energy Statistical Yearbook 1992*, The Statistical Office of the European Communities.

Goldemberg, J. (ed.) (2000) *World Energy Assessment: Energy and the Challenge of Sustainability*, New York, United Nations Development Programme, United Nations Department of Economic & Social Affairs and World Energy Council.

Herkstroter, C. (1997) *Contributing to a Sustainable Future – the Royal Dutch Shell Group in the Global Economy*, paper presented at Erasmus University, Rotterdam, March 17th 1997, London, Shell International, 9 pp.

ICCEPT (2002) *Assessment of Technological Options to Address Climate Change*, Imperial College, London, downloadable from www.iccept.ic.ac.uk [accessed 26 June 2003].

ILEX (2002) *Quantifying the system costs of additional renewables in 2020*, downloadable from www.dti.gov.uk/energy/developep_080scar report_v2_0.pdf [accessed 26 June 2003].

International Energy Agency (2001) *World Energy Outlook*, IEA, Paris, p. 421.

Johansson, T. B., Bodlund, B. and Williams, R. H. (eds) (1989) *Electricity – Efficient End Use*, Lund University Press.

Johansson, T. B., Kelly, H., Reddy, A. and Williams, R. (eds) (1992) *Renewable Energy: Sources for Fuels and Electricity*, Washington DC, Island Press, 1160 pp.

Krohn, S. (2002) *Wind Energy Policy in Denmark Status 2002*, Danish Wind Industry Association, www.windpower.dk (see under publications), [accessed 26 June 2003].

Lazarus, M. *et al.* (1993) *Towards a Fossil Free Energy Future: The Next Energy Transition*, Boston, Stockholm Environment Institute.

Millborrow, D., (2001) *Penalties for Intermittent Sources of Energy*, Working Paper for UK Cabinet Office Performance and Innovation Unit (now Strategy Unit) Energy Review, downloadable from www.strategy.gov.uk/2002/energy/workingpapers.shtml [accessed 26 June 2003].

Ministry of Environment and Energy (1996) *Energy 21*, accessible at Danish Energy Authority at: www.ens.dk/graphics/publikationer/energipolitik_uk/e21uk/contents.htm [accessed 26 June 2003].

Nordel (2002) *Nordic Grid Master Plan*, downloadable from Eng.Elkraft-System.dk (look under Publications) [accessed 26 June 2003].

Norgaard, J. S. and Meyer, N. I. (1989) 'Planning implications of electricity conservation: the case of Denmark' in Johansson *et al* (eds) *Electricity – Efficient End Use*, Lund University Press.

PIU (2002) *The Energy Review*, UK Cabinet Office Performance and Innovation Unit (now Strategy Unit), downloadable from www.strategy.gov.uk [accessed 26 June 2003].

Shell International (1995) *Evolution of the World's Energy System, 1850–2060*, London, Shell International.

Shell International (2001) *Energy Needs, Choices and Possibilities: Scenarios to 2050*, London Shell International, 60 pp.

Further information

Electricity Storage Association at: www.electricitystorage.org [accessed 26 June 2003].

US Department of Energy at: www.eere.energy.gov/hydrogenandfuelcells/ [accessed 26 June 2003].

Icelandic New Energy at: www.newenergy.is [accessed 26 June 2003].

International Energy Agency (IEA) at: www.iea.org/ [accessed 26 June 2003].

World Energy Council (WEC) at: www.worldenergy.org/ [accessed 27 June 2003].

International Institute for Applied Systems Analysis (IIASA). Summary of WEC-IIASA scenarios at: www.iiasa.ac.at/cgi-bin/ecs/book_dyn/bookcnt.py [accessed 27 June 2003].

World Energy Assessment details at: www.undp.org.seed.eap.activities.wea [accessed 27 June 2003].

Shell 2001 Scenarios downloadable from: www2.shell.com/home/media-en/downloads/51852.pdf [accessed 27 June 2003].

The Intergovernmental Panel on Climate Change (IPCC) home page at: www.ipcc.ch [accessed 27 June 2003]. *Special Report on Emissions Scenarios* at: www.grida.no/climate/ipcc/emission/index.htm [accessed 27 June 2003].

Appendix

This Appendix is designed as a quick reference source for 'energy economics' calculations and for some of the basic physical concepts and units used in energy calculations.

A1 Investing in renewable energy

Introduction

The sun, wind and tides may all be freely available, but the equipment to harness their energy has to be paid for. There are several ways of looking at the economics of renewable energy technologies. We can only give a brief summary of the main techniques here. A more detailed description will be found in the companion text *Energy for a Sustainable Future*.

The cost of energy from any system, whether it is fossil-fuelled, nuclear or renewably-powered has to take into account four elements: capital costs – and the cost of borrowing the money; fuel costs; operation and maintenance (O & M) costs and decommissioning costs.

For most renewable energy systems, the capital costs are significant and the most important component in any calculation. For example, for hydro plants they can be around £1500/kW of capacity in contrast to under £300/kW for a competing Combined Cycle Gas Turbine power station.

However, for most renewable energy systems the fuel costs are zero. Some bioenergy projects do have positive fuel costs resulting from the need to grow, harvest and transport it. For others, such as Municipal Solid Wastes schemes, the fuel costs may even be negative, since any other method of waste disposal costs money. Similarly, O & M and final decommissioning of most renewable energy schemes are usually low.

Payback time

For the simplest of calculations we can take the capital cost, the energy output and a competing energy price to calculate a 'payback time', the number of years taken to recover the capital outlay (see Item 1 in the wind turbine example in Box A1.1).

Simple Annual Method

Alternatively, we can calculate the cost per unit energy on an annual basis. In the simplest method, the capital is first **annuitized**, i.e. considered to be repaid in equal annual amounts over the project lifetime. The average O & M costs and fuel costs (if any) are added to these annual payments, and the average cost of energy is then:

$$\text{cost per unit of energy} = \frac{\text{annual capital repayment} + \text{average running costs}}{\text{average annual energy output}}$$

Item 2 in Box A1.1 gives an example.

BOX A1.1 Economics of a small wind turbine project ignoring discounting

The capital cost of a small wind turbine generator is £20 000, its estimated operating and maintenance costs are £200 per year* and it is expected to produce an output of 50 000 kWh per year for 25 years.

1 Ignoring the O & M costs calculate the payback time, given a competing electricity price of 5p kWh^{-1} (i.e. 5 p per kWh).

Value of annual electricity generation = 50 000 × 5p/100 = £2500

Payback time = capital cost/value of annual output = 20 000/2500 = **8 years**

2 Calculate the cost of the electricity generated in pence per kilowatt-hour assuming that this is simply the total cost per year (including O & M) divided by the output per year.

Cost of capital spread over 25 years = £20 000/25 = £800 yr^{-1}

Operating and maintenance costs = £200 yr^{-1}

Total costs = £1000 yr^{-1}

Overall cost per kWh = 100 000p/50 000 kWh = **2.0 p kWh^{-1}**

*Often referred to as 'per annum', yr^{-1} or a^{-1}.

Using discounted cash flow

In practice, these calculations are too simplistic because a pound earned or spent tomorrow is not worth the same as a pound today, for a number of reasons.

Firstly, there is our **time preference for money**. Given a choice, most people would rather have a pound today than a pound in the future. Put another way, we would need to be offered a pound plus some additional sum, say x%, next year to forgo the use of one pound today.

Then we have the ability to lend out money and charge interest. We can forgo the use of a pound today in order to have a pound plus an additional sum in the future. If the interest rate is high enough and the time long enough, this sum can be appreciable. For example, if we invested £100 at a 10% rate of interest per annum, we would expect to be able to withdraw £260 in 10 years time. We can say that £100 has a **future value** of £260. Put another way, the **present value** of £260 in ten years' time at an interest rate of 10% is only £100.

We are 'discounting' future payments, saying that sums of money in the future can be expressed in terms of smaller sums today. This concept leads to a technique of economic appraisal known as **discounted cash flow** (DCF) analysis. Much of this now takes the form of standard computer spreadsheet functions. Generally, the terms 'discount rate' and 'real interest rate' tend to be used interchangeably.

Inflation is another factor that needs to be included. This can be thought of a 'disease of money' progressively eroding its value. In order to adjust for its effects, we should use a 'real' interest rate, rather than the purely monetary one. This is simple enough:

Real interest rate = monetary interest rate – rate of inflation

Before the era of computers and spreadsheets, banks used simple pre-calculated tables to work out repayments on borrowed money. This approach is still useful today. Table A1.1 shows the annual repayments on £1000 of borrowed capital – these can also be calculated using the standard spreadsheet PMT (Payment, Month, Term) function.

Generally, a loan is agreed to be paid back over a certain number of years. Financial institutions are often unwilling to consider loans spread over more than 25 years because of the essential uncertainty about the future. Yet the working life of a hydroelectric station plant may be in excess of a hundred. It is worth remembering that the capital repayment time (to pay off a loan) may be shorter than the working lifetime of a scheme.

Table A1.1 Annuitized value of capital costs (annual repayment in £ per £1000 of capital) for various discount rates and capital repayment periods

Capital repayment period/years[1]	Real discount rate /%						
	0	2	5	8	10	12	15
5	200	212	231	250	264	277	298
10	100	111	130	149	163	177	199
15	67	78	96	117	131	147	171
20	50	61	80	102	117	134	160
25	40	51	71	94	110	127	155
30	33	45	65	89	106	124	152
40	25	37	58	84	102	121	151
50	20	32	55	82	101	120	150
60	17	29	53	81	100	120	150

1 This is not necessarily equal to the total physical lifetime of the project.

Using this table, we can revisit the previous example. Using a discount rate of 10% and a project lifetime of 25 years, we find that the original estimate of 2.0 p kWh^{-1} is an underestimate and arrive at a much higher figure of 4.8 p kWh^{-1} (see Box A1.2).

BOX A1.2 Cost of electricity from a wind turbine – Including discounting.

Calculate the cost of the electricity from the wind turbine described in Box A1.1, assuming a discount rate of 10% and a project lifetime of 25 years.

First, we find the annuitized value of the capital cost. From Table A1.1 the annual repayments on £1000 over 25 years at 10% per year = £110 per year.

For a borrowed sum of £20 000 they will be 20 × £110 = £2200.

Note that this is much higher than the repayment figure of £800 used in Box A1.1. That figure can be derived using the annuitized value of £1000 over 25 years at 0% per year = £40 per year. 20 × £40 = £800.

Our total annual costs now read:

Cost of capital spread over 25 years = £2200 yr^{-1}

Operating and maintenance costs = £200 yr^{-1}

Total costs = £2400 yr^{-1}

Overall cost per kWh = 240 000p/50 000 kWh = **4.8 p kWh^{-1}**

This approach is fine if the construction time is short, and the O & M payments and output are uniform over the life of the project. For projects such as large-scale hydro and tidal schemes, which may take many years to build, and be subject to periodic refurbishment, a full 'Net Present Value' calculation is required. This process involves six steps:

1 Itemise the capital and running costs for each year of the project life
2 Calculate the separate Present Values of all these annual costs using an appropriate discount rate, and sum them to give a Net Present Value (Net Present Value, NPV, is a standard spreadsheet function)
3 Itemise the output for each year over the project lifetime.
4 Calculate the Net Present Value of all these annual outputs, expressed usually in kWh.
5 Calculate the unit cost in pence per kWh as:

$$\frac{\text{Net Present Value of costs (pence)}}{\text{Net Present Value of output (kWh)}}$$

It may seem a little odd to discount the value of the output in kWh, but it has a value, just like money. It is just that we cannot assign its magnitude until the calculation is done. A sample calculation for a proposed nuclear power station will be found in the companion text.

Choice of Discount Rate

Many renewable energy technologies involve high capital costs, a characteristic also shared by nuclear power. As such, their economics are critically dependent on the cost of capital and choice of discount rate.

Past UK evaluations of the economics of renewable energy projects have used discount rates of 8% and even 15%, these being taken as the expectations of earnings in the private sector. However, there are many who suggest that these figures are too high (see the companion text for some discussion of this). In the 'Green Book' produced by the UK Treasury (HM Treasury, 2003) it is suggested that public sector projects should use a figure of 6% and even this figure is broken down into separate 'allowances':

1 A 'social time preference rate' (STPR) of 3.5%, a figure for the time preference of money for society as a whole. It suggests that even this figure should be reduced for long term projects, for example to 2.5% for projects with a life expectancy of around 100 years.
2 An allowance for the risk in a project – the probability that it may fail either financially, or through accident.
3 An allowance for 'optimism bias' in project costings – the fact that they are usually made with a rosy view of the future and usually do not make allowance for construction delays and cost over-runs.

Finally, the calculations above only cover the most basic financial analysis. There are other issues of subsidies to promote renewable energy and of the 'external costs' of pollution, etc. These are described in Chapter 10 and in the companion text.

Reference

H. M. Treasury (2003) *Appraisal and Evaluation in Central Government (The Green Book)*, HMSO, downloadable at http://greenbook.treasury.gov.uk/ [accessed 11 November 2003].

A2 Units

Orders of magnitude

Discussions of energy quantities may involve numbers whose magnitudes range from the very *large* to the very *small*. These can be expressed by using powers of ten and exponents, or by using prefixes.

Powers of Ten

The number two million can be written as:

$$2\ 000\ 000 = 2 \times 10 \times 10 \times 10 \times 10 \times 10 \times 10 = 2 \times 10^6$$

The quantity 10^6 is called *ten to the power six*, and the 6 is known as the **exponent**.

The method can also be used for very small numbers, with the convention that one tenth (0.1) becomes 10^{-1}; one hundredth (0.01) becomes 10^{-2}, etc. The separation of two atoms in a typical metal, for instance, might be about 0.13 of a billionth of a metre, which is 0.000 000 000 13 m. In more compact form, it becomes 0.13×10^{-9} m.

Prefixes

The powers of ten provide the basis for the prefixes used to indicate multiples (including sub-multiples) of units. Table A2.1 shows the most common prefixes in decreasing order.

Table A2.1 Prefixes

Symbol	Prefix	Multiply by	... which is
E	exa-	10^{18}	one quintillion
P	peta-	10^{15}	one quadrillion
T	tera-	10^{12}	one trillion
G	giga-	10^{9}	one billion
M	mega-	10^{6}	one million
k	kilo-	10^{3}	one thousand
h	hecto-	10^{2}	one hundred
da	deca-	10	ten
d	deci-	10^{-1}	one tenth
c	centi-	10^{-2}	one hundredth
m	milli-	10^{-3}	one thousandth
μ	micro-	10^{-6}	one millionth
n	nano-	10^{-9}	one billionth

Units and Conversions

The most common units of energy are the joule, the kilowatt-hour, the tonne of oil equivalent (toe) and tonne of coal equivalent (tce), i.e. the energy content of a tonne of an 'average' sample of these fuels.

The basic unit of power is the watt, which is a rate of energy conversion of 1 joule per second. Thus 1 watt-hour = 3.6 kJ. The kilowatt is commonly used for generators and motors but the horsepower is still often used for car engines. There are also many traditional units of mass, length, speed, area and volume still in use.

The tables below give conversion factors between the units most commonly encountered in this book and other energy literature.

Table A2.2 Energy

	MJ	GJ	kWh	toe	tce
1 MJ =	1	0.001	0.2778	2.4×10^{-5}	3.6×10^{-5}
1 GJ =	1000	1	277.8	0.024	0.036
1 kWh =	3.60	0.0036	1	8.6×10^{-5}	1.3×10^{-4}
1 toe =	42 000	42	12 000	1	1.5
1 tce =	28 000	28	7800	0.67	1

Table A2.3 Power

Rate	Joules... per hour	Joules... per year	Kilowatt-hours per year	Oil equivalent per year	Coal equivalent per year
1 kW	3.6 MJ	31.54 GJ	8760	0.75 toe	1.1 tce
1GW	3.6 TJ	31.54 PJ	8.76×10^9	0.75 Mtoe	1.1 Mtce

Table A2.4 Other Quantities

Quantity	Unit	SI equivalent	Inverse
mass	1 lb (pound)	= 0.4536 kg	1 kg = 2.205 lb
	1 t (*tonne*)	= 1000 kg	1 kg = 10^{-3} t
length	1 ft (foot)	= 0.3048 m	1 m = 3.281 ft
	1 yd (yard)	= 0.9144 m	1 m = 1.094 yd
	1 mi (mile)	= 1609 m	1 m = 6.214×10^{-4} mi
speed	1 km hr^{-1} (kph)	= 0.2778 m s^{-1}	1 m s^{-1} = 3.600 kph
	1 mi hr^{-1} (mph)	= 0.4470 m s^{-1}	1 m s^{-1} = 2.237 mph
	1 knot	= 0.5144 m s^{-1}	1 m s^{-1} = 1.944 knots
area	1 acre	= 4047 m^2	1 m^2 = 2.471 x 10^{-4} acre
	1 ha (hectare)	= 10^4 m^2	1 m^2 = 10^{-4} ha
volume	1 litre	= 10^{-3} m^3	1 m^3 = 1000 litre
	1 gal (UK)	= 4.546×10^{-3} m^3	1 m^3 = 220.0 gal
	1 gal (US)	= 3.785×10^{-3} m^3	1 m^3 = 264.2 gal (US)
energy	1 eV (*electron-volt*)	= 1.602×10^{-19} J	1 J = 6.242×10^{-18} eV
power	1 HP (*horse power*)	= 745.7 W	1.341×10^{-3} HP

Reference

The Royal Society (1975) *Quantities, Units and Symbols*, London, The Royal Society.

Acknowledgements

Grateful acknowledgement is made to the following sources for permission to reproduce material within this product.

Chapter 1

Figures

Figure 1.3: Laherrère, Jean. *Forecasting Future Production for Past Discovery,* OPEC Seminar, 28 September 2001. Courtesy of the author; Figure 1.5: *UK Energy in Brief,* December 2002, DTI. Crown copyright material is reproduced under class licence number C01W0000065 with the permission of the Controller of Her Majesty's Stationery Office and the Queen's Printer for Scotland; Figure 1.6b: International Panel on Climate Change (2001) *Climate Change 2001: The Scientific Basis,* Cambridge University Press. Summary for Policymakers and Technical Summary at www.ipcc.ch/pub/reports.htm.

Chapter 2

Figures

Figures 2.4, 2.6, 2.22, 2.36 and 2.40: Courtesy of Bob Everett; Figure 2.23: Courtesy of ARCON Solvarme A/S, Denmark; Figure 2.24: Courtesy of M. G. Davies; Figure 2.26: © Derek Taylor; Figures 2.31 and 2.32: Courtesy of Commission of New Towns and J. Doggart; Figure 2.35: © Hockerton Housing Project; Figure 2.38: Guildhall Library; Figure 2.39: © David Parker/Science Photo Library; Figure 2.42: Courtesy of Biblioteca Cuidades para un futuro más sostanible, Madrid; Figure 2.43: Courtesy of Sandia National Laboratories; Figures 2.44 and 2.45: © Hank Morgan/Science Photo Library; Figure 2.46: Courtesy of DOE/NREL; Figure 2.49: Schlaich Bergermann und Partner; Figure 2.50: Courtesy of Heeley City Farm. Trainee featured, Adam Ashman.

Chapter 3

Figures

Figure 3.1: © Bibliotheque Nationale de France; Figure 3.3: © Popperfoto; Figure 3.4: © NASA/Science Photo Library; Figures 3.5 and 3.6: *Trends on Photovoltaic Applications*, IEA Photovoltaic Power Systems Programme, August 2003; Figure 3.10: Green, M. (1982) *Solar Cells*, Prentice-Hall; Figure 3.16: © AP Photos; Figure 3.23 left: © Martin Bond/Science Photo Library; Figure 3.23 middle: © Kaj R. Svensson/Science Photo Library; Figure 3.23 right: © James King-Holmes/Science Photo Library; Figure 3.24: Courtesy of Susan Roaf; Figure 3.25: Courtesy of Solar Century; Figure 3.28: © Martin Bond/Science Photo Library; Figure 3.29: © Colin Cuthbert/Science Photo Library; Figure 3.30: © Tony Craddock/Science Photo Library; Figure 3.31: © Martin Bond/Science Photo Library; Figure 3.33: © Martin Bond/Science Photo Library; Figure 3.34: © Martin Bond/Science Photo Library.

Tables

Table 3.1: Jardine, C. N. and Lane, K. 'PV-compare: energy yields of photovoltaic technologies in northern and southern Europe', *Photovoltaic Science, Applications and Technology,* Proceedings of Joint Meeting of UK Solar Energy Society and PVNET, Loughborough University, 2003.

Chapter 4

Figures

Figures 4.4 and 4.20: Courtesy of Coppice Resources Ltd.; Figure 4.5: © Peter Nixon; Figure 4.6 and page 120 top: Courtesy of EPR Ely, Ltd.; Figure 4.11: Courtesy of SELCHP; Figure 4.12: Courtesy of Reprotech (Pebsham) Ltd.; Figure 4.14: Courtesy of Bob Everett; Figure 4.15: Courtesy of Farmatic Biotech Energy UK; Figure 4.18: Courtesy of Jonathan Scurlock; Figure 4.16: *ETSU Project Summary Unit 192: Biofuels/MSW Digestion* (1990) Crown copyright, reproduced by permission of the controller, HMSO.

Chapter 5

Figures

Figure 5.1: Hill, G. (1984) *Tunnel and Dam: the Story of the Galloway Hydros*, South of Scotland Electricity Board (now Scottish Power); Figure 5.2: © 2000 Scottish Power UK plc; Figure 5.3: Courtesy of www.dee-ken-fishing.org; Figure 5.11: © Reuters; Figure 5.12: © Calvin Larsen/Science Photo Library; Figure 5.16: Courtesy of Gilkes, www.gilkes.com; Figures 5.17 and 20: Courtesy of Bob Everett Figure 5.26: © Sue Cunningham/ Worldwide Picture Library/Alamy Images; Figure 5.28: © David Paterson/ Alamy Images; Figure 5.29: © Trygve Bolstad/Panos Pictures.

Tables

Table 5.4: *Statistical Review of World Energy,* 2003, © BP p.l.c; Table 5.8: ChinaOnLine® (2000). Available at www.chinaonline.com/refer/ministry-profiles/threegorgesdam.asp.

Chapter 6

Figures

Figure 6.1: Media Library, EDF/Gerard Halary; Figures 6.5, 6.7–6.9, 6.14 and 6.15: *Tidal Power from the Severn Estuary, Volume 1*, Energy Paper 46, 1981, Department of Energy. Crown copyright material is reproduced under Class Licence Number C01W0000065 with the permission of the Controller of HMSO and the Queen's Printer for Scotland; Figure 6.6: Twidell, J. W. and Weir (1986) A. J. *Renewable Energy Sources,* E. & F. N. Spon. Reproduced courtesy of Routledge; Figure 6.10: *Report of the Legislative Assembly of Western Australia,* presented by Hon. Ian Thompson, November, 1991, with permission of the Department of Housing and Works, Australia; Figure 6.11: *Tidal Power from the Severn Estuary, Volume 1*, Energy Paper 46, 1981 Department of Energy. Crown copyright material is reproduced under Class Licence Number C01W0000065 with the permission of the Controller of HMSO and the Queen's Printer for Scotland; Figures 6.12 and 6.13: Reproduced courtesy of Mr L. Carson and Power Magazine, March 1978; Figure 6.16: © ETSU Department of Trade and Industry; Figures 6.17–6.19 and 6.21: *The Severn Barrage Project: General Report*, Energy Paper 57, 1989, Department of Energy. Crown copyright material is reproduced under Class Licence Number C01W0000065 with the permission of the Controller of HMSO and the Queen's Printer for Scotland; Figure 6.20: Laughton, M. A. *Renewable Energy Sources, Report 22,* E. & F. N. Spon. Reproduced courtesy of Routledge; Figure 6.23: Baker, A. C. (1991) *Tidal Power,* Peter Peregrinus Limited, with permission from IEE Publishing; Figure 6.24: Tidal Electric,

Inc., www.tidalelectric.com; Figures 6.25–6.27: Courtesy of Marine Current Turbines Limited, www.marineturbine.com; Figure 6.29: Based on an image supplied by the Engineering Business. The Patented Stingray tidal stream is being developed by the Engineering Business, who will provide project information on their website, updated at regular intervals. The first Stingray generator was installed in Shetland summer 2002, with 75% DTI funding. The Stingray demonstrator is about 20 m high and weighs about 40 tonnes with a nominal 150 kW output in a 4 knot current, www.engb.com; Figure 6.30: Courtesy of The Engineering Business, www.engb.com; Figure 6.34: Based on a map from Blue Energy; Figure 6.35: Slater, S. H. *Proposal for a Large, Vertical-Axis Tidal Stream Generator with Ring-Cam Hydraulics*, reproduced with permission.

Tables

Tables 6.1 and 6.2: *The Severn Barrage Project: General Report*, Energy Paper 57, 1989, Department of Energy. Crown copyright material is reproduced under Class Licence Number C01W0000065 with the permission of the Controller of HMSO and the Queen's Printer for Scotland; Table 6.3: From Fourth Report Session 1991–92, House of Commons Select Committee on Energy. Crown copyright material is reproduced under Class Licence Number C01W0000065 with the permission of the Controller of HMSO and the Queen's Printer for Scotland; Table 6.4: Baker, A. C. (1986) *ICE Symposium on Tidal Power*, Thomas Telford Ltd., London; Table 6.5: Adapted from World Energy Council website (2003).

Text

Box 6.8: Based on *The Severn Barrage Project: General Report*, Energy Paper 57, 1989, Department of Energy. Crown copyright material is reproduced under Class Licence Number C01W0000065 with the permission of the Controller of HMSO and the Queen's Printer for Scotland.

Chapter 7

Figures:

Figure 7.1: Marlec Engineering Ltd; Figure 7.2: Burroughs, W. J., Crowder, B., Robertson, E., Vallier-Talbot, E. and Whitaker, R. (1996) *Weather – The Ultimate Guide to the Elements*, HarperCollins; Figure 7.4: Farndon, J., (1992), *How the Earth Works*, Dorling Kindersley Ltd; Figure 7.8: Needham, J. with the collaboration of Ling, W. (1965) *Science and Civilisation in China* Vol. 4, Part II, Cambridge University Press; Figures 7.10, 7.16 and 7.39: © Derek Taylor/Altechnica; Figure 7.11: © Science Photo Library; Figures 7.12, 7.15, 7.17, 7.18 and 7.44: Courtesy of Bob Everett; Figure 7.13: Howden Wind Turbines; Figure 7.14: Courtesy of Stewart Boyle; Figure 7.33: *Time for Action: Wind Energy in Europe* (1991) European Wind Energy Association; Figure 7.35: With permission from Skyscraper Page.com; Figure 7.38: © North Energy Associates, courtesy of Gazelle Wind Turbines Ltd.; Figure 7.41: Bonus ES/Elkraft; Figure 7.42: © Adam Schmedes/Lokefilm, www.lokefilm.dk; Figure 7.43: AMEC Border Wind.

Tables

Table 7.2: © 2002 American Wind Energy Association; Tables 7.3, 7.4: *Wind Force 12*, © European Wind Energy Association and Greenpeace.

Chapter 8

Figures

Figures 8.1(b) and 8.22(a): Courtesy of Les Duckers; Figures 8.4 and 8.23(b): © Martin Bond/Science Photo Library; Figure 8.14: Falnes, J. and Løvseth, J. (1991), 'Ocean wave energy', first published in *Energy Policy*, vol. 19, no. 8, October 1991, pp. 768–75 and reproduced here with the permission of the authors and Butterworth-Heinemann, Oxford, UK; Figure 8.15: Courtesy of Jamie Taylor, Edinburgh University Wave Power Group; Figure 8.24: Courtesy of Ocean Power Delivery Ltd; Figure 8.25: Courtesy of AquaEnergy Group Ltd; Figure 8.32: Energetech Australia Pty Limited.

Chapter 9

Figures

Figures 9.2 and 9.3: © Alinari Archives; Figure 9.7: Reprinted from *Earth Science Reviews,* Vol. 19, No. 1, Henley and Ellis, 'Conceptual model of a typical geothermal system'. © 1983, with permission from Elsevier Science; Figure 9.13: Geoff Brown; Figure 9.14: Courtesy of EEIG Heat Mining; Figure 9.15: Courtesy of the Geothermal Education Office, California; Figure 9.16: IPR/31-12 British Geological Survey © NERC. All rights reserved.

Chapter 10

Figures

Figure 10.1: Courtesy of Peter Ruiter; Figure 10.2: Adapted from *Euro Gas Annual Report 2001,* courtesy of Euro Gas; Figure 10.3: Courtesy of Transco; Figure 10.5: *New and Renewable Energy Prospects for the 21st Century: Supporting Analysis,* March 1999, produced by ETSU for the Department of Trade and Industry; Figure 10.6: Courtesy of Løgstør Rør; Figures 10.9 and 10.13: Courtesy of Bob Everett; Figure 10.10: © Electricity Association; Figure 10.14: Courtesy of Regenesys Technologies; Figure 10.15: *Renewable Sources of Electricity in the SWEB Area: Future Prospects,* SWEB and the Department of Trade and Industry; Figure 10.17: Bodlund, B. *et al.* (1989) 'The challenge of choices: technology option for the Swedish electricity sector', in Johansson, T. B., Bodlund, B. and Williams, R. H. (eds) (1989) *Electricity – Efficient end use and new technologies and their planning implications,* Lund University Press; Figure 10.18: Courtesy of Ballard Power Systems; Figures 10.25 and 10.26: *Exploring the Future Energy Needs, Choices and Possibilities: Scenarios to 2050,* Global Business Environment, © Shell International 2001.

Table

Table 10.7: Exploring the Future Energy Needs, Choices and Possibilities: Scenarios to 2050, Global Business Environment, © Shell International 2001.

Every effort has been made to trace all copyright owners, but if any have been inadvertently overlooked, the publishers will be pleased to make the necessary arrangements at the first opportunity.

Index

A

B

C

D

R

S